BRIEF CONTENTS

CONTENTS

PREFACE

The Second Edition of this trigonometry book is a standard right triangle approach to trigonometry. Each section is written so that it can be discussed in a 45 to 50 minute class session. The text covers all the material usually taught in trigonometry. The emphasis of the book is on understanding the definitions and principles of trigonometry and their application to problem solving. However, when memorization is necessary, I say so.

Identities are introduced early in the book in Chapter 1. They are reviewed often and are then covered in more detail in Chapter 5. Also, exact values of the trigonometric functions are emphasized throughout the book. There are numerous calculator notes placed throughout the text that indicate how the examples and problems could be solved using a calculator.

TEXTBOOK FEATURES

Every section has been written so that it can be discussed in a single class session. A clean layout and conversational style is used by the authors to make it easy for students to read. Important information, such as definitions and properties, are also highlighted so that it can easily be located and referenced by students.

In addition, the following features provide both instructors and students a vast array of resources that can be used to enrich the learning environment and promote student success.

WHAT, WHY, WHERE (NEW TO THIS EDITION)

Each chapter begins with a *What, Why, Where* feature that enhances a student's understanding of why the concepts are essential and how to use them by creating engagement. The "What" part tells you the idea, so you get the basics. The "Why" part explains why it matters in the real world, making it easier to see its relevance. Lastly, the "Where" part shows you where you can use these ideas, clearly showing how to apply them. This feature imparts knowledge and empowers students to connect theoretical understanding with practical implementation, fostering a more comprehensive and applicable grasp of quantitative concepts.

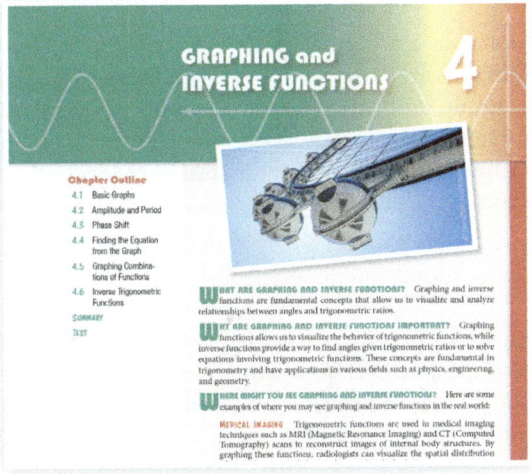

CONNECTING PRINT AND DIGITAL: QR CODES

We want students to get the most out of their course materials, which is why we think of the textbook as a "toolbox" for students. QR codes are integrated throughout the textbook to connect the printed version to the digital assets easily. As students read their printed book, support material is quickly and conveniently available via one scan of the accompanying QR code. Students don't need to hunt, search, or scroll to find an example video. We find that this direct link to additional support is the appropriate level of technology for all students taking this course; it integrates technology without being overwhelming.

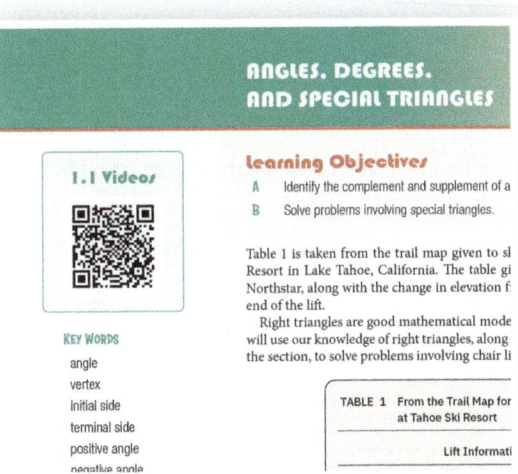

LEARNING OBJECTIVES & KEY WORDS (NEW TO THIS EDITION)

We have provided a list of the section's important concepts and vocabulary at the beginning of each section. The Learning Objectives are notated with a green letter, which is reiterated beside each learning objective header in the section. The Key Words are listed in the margin at the beginning of each section and repeated in bold throughout the text.

SCENARIOS (NEW TO THIS EDITION)

Scenarios are featured in each of our chapters and showcase real-world applications of the topics we will cover. Each scenario takes the trigonometry concepts from the chapter and applies them to unique and engaging situations. By incorporating these scenarios, you can better understand how trigonometry is used in various fields, such as engineering, navigation, architecture, robotics, and medical imaging. Our goal is to make the material more relevant and stimulating, helping you see the practical value of what you are learning and motivating your interest in these mathematical principles.

Scenario: Special Triangles and Nautical Distance

A ship leaves the Port of Los Angeles and travels due west for 100 nautical miles. Then, it changes course and travels due north for another 150 nautical miles. Using GPS, the ship's navigator wants to determine the direct distance from the ship's current position to the Port of Los Angeles. How can trigonometric functions help in calculating this distance?

1. Visualize the ship's journey on a coordinate plane. Starting from the Port of Los Angeles (considered as the origin), the ship first travels 100 nautical miles west (positive x-direction) and then 150 nautical miles north (positive y-direction).

2. Identify the ship's current position. The ship's current position forms a right triangle with the x-axis and y-axis. The direct distance from the ship's current position to the Port of Los Angeles represents the hypotenuse of this right triangle.

EXAMPLE PROBLEMS

Example problems, with work, explanations, and solutions are provided for every learning objective. Each example problem is also worked on video by multiple student instructors. Access to these videos is provided by scanning the QR code at the start of the section or when viewing the eBook.

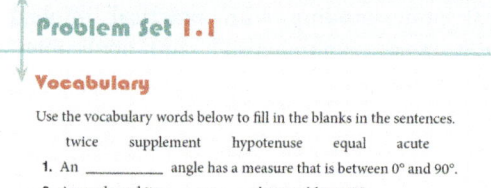

EXAMPLE 1   Give the complement and the supplement of each angle.

a. 40°   b. 110°   c. θ

Solution

a. The complement of 40° is 50° since 40° + 50° = 90°.
   The supplement of 40° is 140° since 40° + 140° = 180°.

b. The complement of 110° is −20° since 110° + (−20°) = 90°.
   The supplement of 110° is 70° since 110° + 70° = 180°.

c. The complement of θ is (90° − θ) since θ + (90° − θ) = 90°.
   The supplement of θ is (180° − θ) since θ + (180° − θ) = 180°.

VOCABULARY

We have provided fill-in-the-blank exercises at the beginning of each problem set, which give students the opportunity to practice and review important vocabulary from each section.

Problem Set 1.1

Vocabulary

Use the vocabulary words below to fill in the blanks in the sentences.

twice   supplement   hypotenuse   equal   acute

1. An _____ angle has a measure that is between 0° and 90°.

2. An angle and its _____ always add to 180°.

3. The longest side of a right triangle is called the _____ .

4. The longest side of a 30°-60°-90° triangle is _____ the shortest side.

5. The two legs in a 45°-45°-90° triangle are always _____ .

END-OF-CHAPTER SUMMARIES AND TESTS

Problems at the end of each chapter provide students with a comprehensive review of each chapter's concepts.

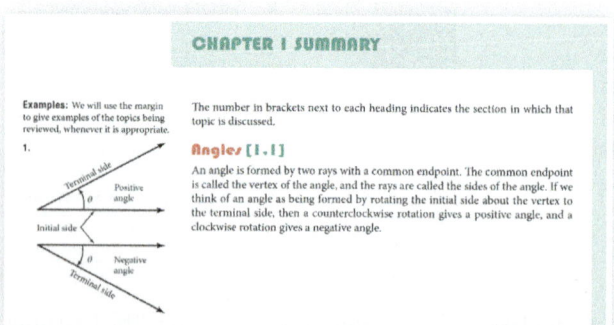

CHAPTER 1 SUMMARY

Examples: We will use the margin to give examples of the topics being reviewed, whenever it is appropriate.

1.

The number in brackets next to each heading indicates the section in which that topic is discussed.

Angles [1.1]

An angle is formed by two rays with a common endpoint. The common endpoint is called the vertex of the angle, and the rays are called the sides of the angle. If we think of an angle as being formed by rotating the initial side about the vertex to the terminal side, then a counterclockwise rotation gives a positive angle, and a clockwise rotation gives a negative angle.

CHAPTER 1 TEST

1. Find the other two sides of a 30°–60°–90° triangle if the shortest side is 5. [1.1]

2. Find the other two sides of a 45°–45°–90° triangle if the longest side is 12. [1.1]

3. Find the six trigonometric functions of θ if the point (4, −3) is on the terminal side of θ. [1.2]

4. In which quadrant must θ lie if sin θ < 0 and tan θ > 0? [1.2]

5. Find the six trigonometric functions of θ if the point $(-\sqrt{3}, 1)$ lies on the terminal side of θ, and θ terminates in QII. [1.2]

Find the exact value for each of the following: [1.3]

TRIGONOMETRY 2E

CHARLES P. McKEAGUE

*xyz*textbooks

Trigonometry 2E
Charles P. McKeague

Publisher: XYZ Textbooks

Design & Composition:
  Katherine Heistand Shields

Content Consultant:
  Ross Rueger

Proofreaders:
  Katherine Diefenderfer
  Rachael Hillman
  Amy Jacobs
  Julieta Carabantes
  Michael Carr, Mott Community College
  Brooke Jacobs
  Jana Mooney
  Kathryn ZagRodny

Printed in the United States of America

Student Edition   ISBN-13: 978-1-63098-428-1   /   ISBN-10: 1-63098-428-0

For product information and technology assistance, contact us at
XYZ Textbooks, 1-877-745-3499

For permission to use material from this text or product,
e-mail: **info@mathtv.com**

XYZ Textbooks
1339 Marsh Street
San Luis Obispo, CA 93401
USA

For your course and learning solutions, visit **www.xyztextbooks.com**

SPOTLIGHT ON SUCCESS

The "Spotlight on Success" feature offers students a variety of strategies from many different sources, and highlights both the challenges and triumphs common to students taking mathematics courses. Students and instructors share their journey and suggestions for success in a math course. These stories highlight the featured individuals' unique and relatable approaches to learning. Many of the spotlights feature the same peer tutors that students will see on the MathTV videos that accompany the text.

Spotlight on Success

Julieta, Student Instructor

Success is no accident. It is hard work, perseverance, learning, studying, sacrifice, and most of all, love of what you are doing or learning to do. —Pelé

Success really is no accident, nor is it something that happens overnight. Sure you may be sitting there wondering why you don't understand a certain lesson or topic, but you are not alone. There are many others who are sitting in your exact position. Throughout my first year in college (and more specifically in Calculus I) I learned that it is normal for any

ADDITIONAL TEXTBOOK FEATURES

HELPFUL NOTES   In many sections, you will find boxes labeled "Note" that correspond to the section's current discussion. These notes contain important reminders, fun facts, and helpful advice for working through problems.

COLORED BOXES   Throughout each section, we have highlighted important definitions, properties, rules, how to's, and strategies in colored boxes for easy reference.

FACTS FROM GEOMETRY   Throughout the book, we highlight specific geometric topics and show students how they apply to the concepts they are learning in the course.

USING TECHNOLOGY   Throughout the book, we provide students with step-by-step instructions for how to work some problems on a scientific or graphing calculator.

SUPPLEMENTS TO THE TEXT

Trigonometry is accompanied by a number of useful supplements. Please contact your sales representative or see **xyztextbooks.com/instructors_resources** for more info. All supplements, such as solutions, answers, and powerpoints slides are on the Instructor's Resource Site.

MATHTV.COM   Every example in every XYZ Textbook is worked on video by multiple student instructors. Students benefit from seeing multiple approaches and gain confidence in learning from their peers. These videos can be used to supplement class time or can be used for alternative instruction such as a flipped classroom approach.

EBOOK   Through the website, instructors can access the eBook for this course as well as other texts. Using the My Bookshelf feature, you can view the text as well as the resources available to both you and the student.

COMPLETE SOLUTIONS MANUAL   Available online, this manual contains complete solutions to all the exercises in the problem sets.

POWERPOINT PRESENTATIONS   These presentations can be opened in PowerPoint, Apple Keynote or Google Slides and have been provided as a template for classroom lectures. Instructors can add examples that they want to use in class or use these to supplement their presentation of the material.

SUPPLEMENTS FOR THE STUDENT

MathTV.com   Students have access to math instruction 24 hours a day, seven days a week. Within a few clicks or a scan of the QR code in the textbook, students can access multiple videos that correspond to each example in the text. Many of the video examples are also presented by Spanish-speaking instructors.

XYZ eBook   Students who purchase the textbook have one year of free access to the eBook. Students can also choose to purchase the eBook instead of the printed text. The eBook is tightly integrated with MathTV.com so that students can read the book and watch videos for each example.

QR Codes   QR codes quickly and easily connect the textbook to digital resources. By scanning the QR code located in each section, students will be taken directly to the accompanying MathTV videos for the examples in that section.

Student Solutions Manual   The Student Solutions Manual contains complete solutions to all the odd-numbered exercises in the text. It is available for purchase separately.

Prerequisites, Corequisites, or Review   If students need additional review of topics, please have them visit our website where they can access our free corequisite materials.

If you would like additional review topics added to your online homework course, please let us know.

SCAN FOR
XYZ CATALOG

A Note to Instructors and Students

We at XYZ Textbooks would like to extend our deepest gratitude to all the educators and students who have chosen our books for their academic journeys. Your trust and support mean the world to us.

Our team is committed to the success of all students. We develop materials aimed to excite and motivate learners, by delivering those light-bulb moments that spark an inner drive and passion for learning. We believe education should be engaging and enlightening, and we strive to create resources which achieve just that.

As a small, family-owned publisher, our roots are deeply intertwined with the educational community. Founded by a community college math teacher, our mission has always been to provide high-quality, accessible educational materials. We are passionate about what we do and hope our dedication shines through in every book and ebook we produce.

To the Instructors   who have integrated our materials into their teaching, we thank you for your invaluable contribution to the development of XYZ Textbooks. Your hard work and dedication inspire us to innovate and improve our resources.

To the Students who use our books, we are honored to be a part of your learning journey. Your achievements and progress drive us to keep pushing the boundaries of what educational materials can be.

We also want to share a personal touch that makes our work even more special. The names used in our examples are often those of people we know —family, friends, and colleagues who have inspired us. This small detail reflects our deep connection to the community and our desire to make learning personal and relatable.

Thank you for choosing XYZ Textbooks. We are excited to continue supporting your educational endeavors and look forward to many more years of collaboration and success.

If there is anything we can do better, please let us know.

In gratitude,

The XYZ Textbooks Team

WE WANT YOU TO SUCCEED.

Welcome to the XYZ Textbooks/MathTV community! We are dedicated to your success in this course. As you will see as you progress through this book, and access the other tools we have for you, we are different than other publishers. Here's how:

OUR AUTHORS ARE EXPERIENCED TEACHERS   We believe the best textbooks are written by the experienced instructors. These individuals can draw on their background teaching in the classroom and online learning environments and incorporate that expertise into our textbooks. Our objective is to provide the best instruction students can get in written form, produced by award-winning, experienced instructors.

INNOVATIVE PRODUCTS   The foundation of our products is the textbook, which is also a source of tools that you will need to do well in your math course. These include eBooks, videos, worksheets, and a variety of ways to access these resources, from QR codes built into our books, to our MathTV Mobile site.

PEER TUTORS   Learning mathematics can be a challenge. Sometimes your class time and textbook are not enough. We understand that, which is why we created MathTV, providing you with a set of instructional videos by students just like you, who have found a way to master the same material you are studying. Since these videos are available on the internet, you can access them whenever and wherever you may be studying. You'll also get to see how your peers solve each problem, sometimes offering a different view from how an instructor solves the problem, and other times, solving the problem in the same way your instructor does, giving you confidence that that is the way the problem should be solved.

FAIR PRICES   We're small, independently owned, and independently run. Why does that matter to you? Because we do not have the overhead and expenses that the larger publishers have. Yes, we want to be a profitable business, but we believe that we can keep our prices reasonable and still give you everything you need to be successful. Also, we want you to use this book, and the best way to make sure that happens is to make it affordable.

UNLIMITED ACCESS   When you purchase one of our products, we give you access to all of our products. Why? Because everything you need to know about math is not contained in one book. Suppose you need to review a topic from a math course you completed previously? No problem. Suppose you want to see an alternate approach? It's all yours. As a member of our XYZ Textbooks/MathTV community, you have access to everything we produce, including all our eBooks.

WE KNOW YOU CAN DO IT.

WE BELIEVE IN YOU.   We have seen students with all varieties of backgrounds and levels in mathematics do well in the courses we supply books and materials for. In fact, we have never run across a student who could not be successful in algebra. And that carries over to you: We believe in you; we believe you can be successful in whatever math class you are taking. Our job is to supply you with the tools you can use to attain success; you supply the drive and ambition.

WE KNOW COLLEGE CAN BE DIFFICULT.   It is not always the material you are studying that makes college difficult. We know that many of you are working, some part time, some full time. We know many of you have families to support, or look after. We understand that your time can be limited. We take all this to heart when we create the materials we think you will need. For example, we make our videos available on your smartphone, tablet, and on the internet. That way, no matter where you are, you will have access to help when you get stuck on a problem.

WE BELIEVE IN WHAT WE DO.   We are confident that you will see the value in the text and resources we have created. That's why the first chapter in every one of our eBooks is free, and so are all the resources that come with it. We want you to try us out for free. See what you think. We wouldn't do that if we didn't believe in what we do here.

HOW TO BE SUCCESSFUL IN MATHEMATICS

Mathematics is a challenging subject. Often, it includes abstract concepts that require critical thinking. Concepts also build on one another, requiring that you maintain skills and retain knowledge that you learned in the past. As a result, learning a topic in mathematics isn't always accomplished the first time through the material. If you don't understand a topic the first time you see it, that's perfectly normal. Our advice, though, is to stick with it! Understanding mathematics takes time. You may find that you need to read over new material a number of times before you can begin to work problems. The process of understanding requires reading the book, studying the examples, working problems, and getting your questions answered.

Here are some additional suggestions that will help you succeed in mathematics.

1. IF YOU ARE IN A LECTURE CLASS, BE SURE TO ATTEND ALL CLASS SESSIONS ON TIME.   You simply will not know exactly what went on in class unless you were there. Missing class and then expecting to find out what went on from someone else is not a good strategy. Make the time to be there—and to be attentive.

2. READ THE BOOK.   It is best to read the section that will be covered in class beforehand. It's OK if you don't fully understand everything you read! Reading in advance at least gives you a sense of what will be discussed, which puts you in a good position when you get to class.

3. WORK PROBLEMS EVERY DAY AND CHECK YOUR ANSWERS.   One secret to success in mathematics is working problems. The more problems you work, the better you will perform. It's really that simple. The answers to the odd-numbered problems in each section are given in the back of the book. When you have finished an assignment, be sure to compare your answers with those in the book. If you have made a mistake, find out what it is, and try to correct it.

4. DO IT ON YOUR OWN.   Having someone else show you how to work a problem is not the same as working the problem yourself. It is absolutely OK to get help when you are stuck. As a matter of fact, it is a good idea. Just be sure you do the work yourself. After all, when it's test time, it's all you! Get confident in every problem type, and you will do well.

5. REVIEW EVERY DAY.   After you have finished the problems your instructor has assigned, take another 15 minutes and review a section you have already completed. This simple trick works wonders. Studies have shown, the more you review, the longer you will retain the material you have learned. Since math topics build upon one another, this will help you throughout the term.

6. DON'T EXPECT TO UNDERSTAND EVERY NEW TOPIC THE FIRST TIME YOU SEE IT.   Sometimes it will come easy and sometimes it won't. Don't beat yourself up over it—this is perfectly normal. Expecting to understand each new topic the first time you see it can lead to disappointment and frustration. The process of understanding takes time and practice. It requires that you read the book, work problems, and get your questions answered.

7. SPEND AS MUCH TIME AS IT TAKES FOR YOU TO MASTER THE MATERIAL.   What's the exact amount of time you need to spend on mathematics to master it? There's no way to know except to do it. You will find out as you go what is or isn't enough time for you. Some sections may take less time, and some may take more. If you end up spending two or more hours on each section, OK. Then that's how much time it takes; trying to get by with less will not work.

8. KNOW WHAT RESOURCES ARE AVAILABLE TO YOU.   We have provided a multitude of resources within this text to help you succeed in this course. There are probably also resources at your college to help you as well. Does your instructor have time set aside to help students? Does your campus have a tutoring center with tutors? We want to make sure you have all the resources you need to be successful in your math course.

9. RELAX.   It's probably not as difficult as you think. You might get stuck at points. That's OK, everyone does. Take a break if you need to. Seek some outside help. Watch a MathTV video of the problem. There is a solution, and you will find it—even if it takes a while.

THE SIX TRIGONOMETRIC FUNCTIONS

1

WHAT IS A TRIGONOMETRIC FUNCTION?   The six trigonometric functions relate the angles of a right triangle to the ratios of its sides. These functions are fundamental in trigonometry and describe relationships between the angles and sides of right triangles. They help in solving problems related to angles and sides in right triangles.

WHY ARE TRIGONOMETRIC FUNCTIONS IMPORTANT?   Learning about trigonometric functions is essential for college students as it forms the mathematical foundation for various disciplines such as calculus, physics, and engineering, while also finding applications in fields like computer graphics, architecture, and navigation.

WHERE MIGHT YOU SEE TRIGONOMETRIC FUNCTIONS?   The following examples illustrate the diverse applications of trigonometric functions in various fields, highlighting their importance in solving practical problems and advancing technological innovations.

NAVIGATION AND GPS   Trigonometric functions are essential in navigation systems and GPS technology. Sine and cosine functions are used to determine the latitude and longitude of a location, while tangent functions help calculate the angle of direction between two points.

MECHANICAL ENGINEERING   Trigonometric functions play a crucial role in mechanical engineering, especially in designing machinery and mechanisms. Engineers use trigonometry to analyze forces, motion, and mechanical linkages in machines like engines, gears, and pulleys.

Scenario: Special Triangles and Nautical Distance

© istockphoto.com bify9

A ship leaves the Port of Hilo, Hawaii and travels due east for 100 nautical miles. Then, it changes course and travels due north for another 150 nautical miles. Using GPS, the ship's navigator wants to determine the direct distance from the ship's current position to the Port of Hilo. How can trigonometric functions help in calculating this distance?

1. **Visualize the ship's journey on a coordinate plane.** Starting from the Port of Hilo (considered as the origin), the ship first travels 100 nautical miles east (positive x-direction) and then 150 nautical miles north (positive y-direction).

2. **Identify the ship's current position.** The ship's current position forms a right triangle with the x-axis and y-axis. The direct distance from the ship's current position to the Port of Hilo represents the hypotenuse of this right triangle.

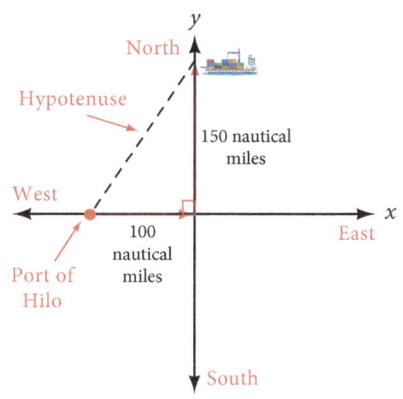

3. **Calculate the length of the hypotenuse.** Use the Pythagorean theorem to find the length of the hypotenuse.

$$\text{Direct distance}^2 = (\text{Distance traveled east})^2 + (\text{Distance traveled north})^2$$

$$= (100 \text{ nautical miles})^2 + (150 \text{ nautical miles})^2$$

$$= (10{,}000 \text{ nautical miles})^2 + (22{,}500 \text{ nautical miles})^2$$

$$= 32{,}500 \text{ nautical miles}^2$$

$$\text{Direct distance} = \sqrt{32{,}500} \text{ nautical miles}$$

$$\approx 180.28 \text{ nautical miles}$$

Solution   Using trigonometric functions, the ship's navigator can calculate that the distance from the ship's current position to the Port of Hilo is approximately 180.28 nautical miles.

1.1 Videos

Note The symbol θ is called **theta** and is a variable used to represent an angle. Later, you will see the variables α (called **alpha**) and β (called **beta**) also used to represent different angles.

Learning Objectives

A   Identify the complement and supplement of a given angle.

B   Solve problems involving special triangles.

Table 1 is taken from the trail map given to skiers at the Northstar at Tahoe Ski Resort in Lake Tahoe, California. The table gives the length of each chair lift at Northstar, along with the change in elevation from the beginning of the lift to the end of the lift.

Right triangles are good mathematical models for chair lifts. In this section, we will use our knowledge of right triangles, along with the new material developed in the section, to solve problems involving chair lifts and a variety of other examples.

TABLE 1	From the Trail Map for the Northstar at Tahoe Ski Resort	
Lift Information		
Lift	Vertical Rise (feet)	Length (feet)
Big Springs Gondola	480	4,100
Bear Paw Double	120	790
Vista Express	535	3,110
Lookout Double	960	4,330
Comstock Express Quad	1,250	5,900
Rendezvous Triple	650	2,900

A Angles in General

An angle is formed when two rays have a common endpoint. The common endpoint is called the **vertex** of the angle, and the rays are called the **sides** of the angle (Figure 1).

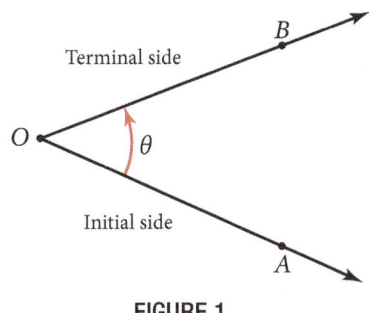

FIGURE 1

We can think of θ as having been formed by rotating OA about the vertex to side OB. In this case, we call side OA the **initial side** of θ and side OB the **terminal side** of θ.

When the rotation from the initial side to the terminal side takes place in a counterclockwise direction, the angle formed is considered a **positive angle**. If the rotation is in a clockwise direction, the angle formed is a **negative angle** (Figure 2).

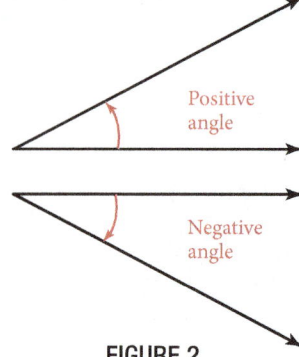

FIGURE 2

Degree Measure One degree of angle measure, written 1°, is $\frac{1}{360}$ of a complete rotation of a ray around its endpoint; there are 360° in one full rotation (Figure 3). (The number 360 was decided upon by early civilizations because it was believed that Earth was at the center of the universe and the sun would rotate once around Earth every 360 days.)

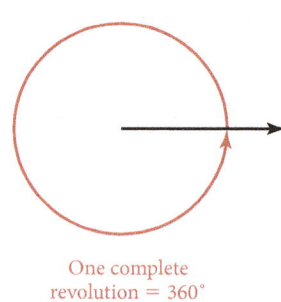

One complete revolution = 360°

FIGURE 3

Similarly, 180° is half of a complete rotation, and 90° is a quarter of a full rotation. Angles that measure 90° are called *right angles*, and angles that measure 180° are called *straight angles*. If an angle measures between 0° and 90° it is called an *acute angle*, and an angle that measures between 90° and 180° is an *obtuse angle*. Figure 4 illustrates further.

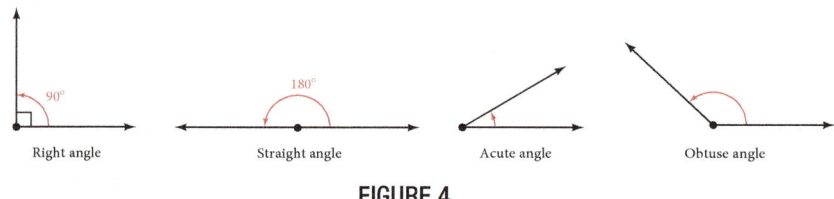

FIGURE 4

If two angles add up to 90°, we call them *complementary angles*, and each is called the *complement* of the other. If two angles have a sum of 180°, we call them *supplementary angles*, and each is called the *supplement* of the other. Figure 5 illustrates the relationship between angles that are complementary and angles that are supplementary.

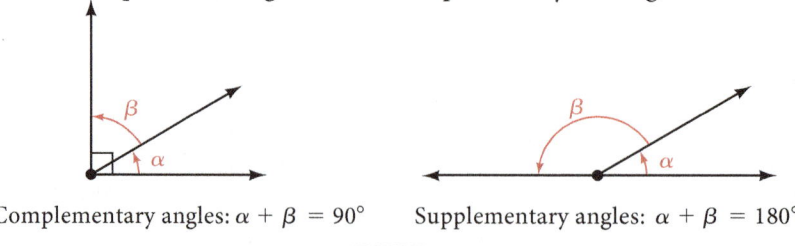

Complementary angles: $\alpha + \beta = 90°$ Supplementary angles: $\alpha + \beta = 180°$

FIGURE 5

EXAMPLE 1   Give the complement and the supplement of each angle.

  a. 40°   **b.** 110°   **c.** θ

Solution

a. The complement of 40° is 50° since 40° + 50° = 90°.
    The supplement of 40° is 140° since 40° + 140° = 180°.

b. The complement of 110° is −20° since 110° + (−20°) = 90°.
    The supplement of 110° is 70° since 110° + 70° = 180°.

c. The complement of θ is (90° − θ) since θ + (90° − θ) = 90°.
    The supplement of θ is (180° − θ) since θ + (180° − θ) = 180°.

B Special Triangles

A *right triangle* (Figure 6) is a triangle with an interior angle of 90°. The side opposite the right angle is called the *hypotenuse.* The other two sides are called *legs.* The *Pythagorean theorem* states a special relationship between the hypotenuse and legs of a right triangle.

Pythagorean Theorem
In any right triangle, the square of the longest side (hypotenuse) is equal to the sum of the squares of the other two sides (legs).

$$c^2 = a^2 + b^2$$

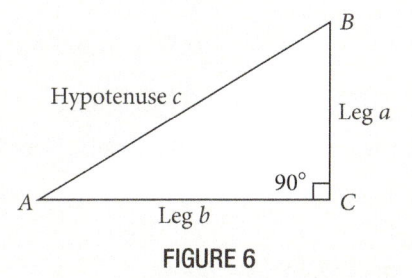

FIGURE 6

EXAMPLE 2   The hypotenuse of a right triangle is 5 inches, and the lengths of the two legs (the other two sides) are given by two consecutive integers. Find the lengths of the two legs.

Solution If we let x = the length of the shortest side, then the other side must be $x + 1$. A diagram of the triangle is shown in Figure 7.

  The Pythagorean theorem tells us that the square of the longest side, 5^2, is equal to the sum of the squares of the two shorter sides, $x^2 + (x + 1)^2$. Here is the equation:

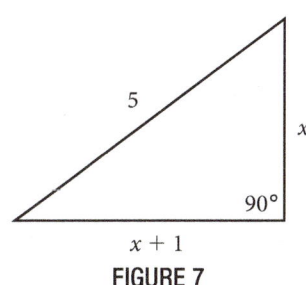

$5^2 = x^2 + (x + 1)^2$	Pythagorean theorem
$25 = x^2 + x^2 + 2x + 1$	Expand 5^2 and $(x + 1)^2$
$25 = 2x^2 + 2x + 1$	Simplify the right side
$0 = 2x^2 + 2x - 24$	Add −25 to each side
$0 = 2(x^2 + x - 12)$	Begin factoring
$0 = 2(x + 4)(x - 3)$	Factor completely
$x + 4 = 0$ or $x - 3 = 0$	Set variable factors to 0
$x = -4$ $x = 3$	

FIGURE 7

Since a triangle cannot have a side with a negative number for its length, we cannot use −4; therefore, the shortest side must be 3 inches. The next side is $x + 1 = 3 + 1 = 4$ inches. Since the hypotenuse is 5, we can check our solutions with the Pythagorean theorem as shown in Figure 8.

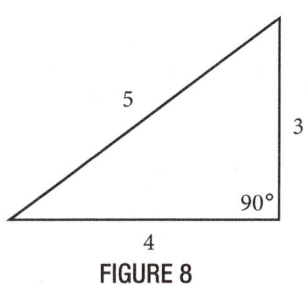

FIGURE 8

EXAMPLE 3 Suppose a chair lift at a ski resort has a vertical rise of 1,170 feet and a length of 5,750 feet. To the nearest foot, find the horizontal distance covered by a person riding this lift.

Solution Figure 9 is a model of the chair lift. A rider gets on the lift at point A and exits at point B. The length of the lift is AB.

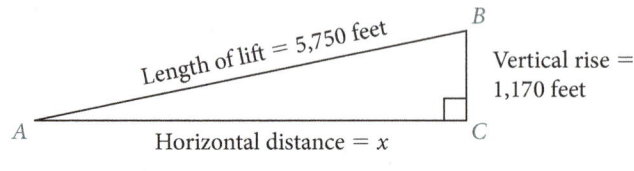

FIGURE 9

To find the horizontal distance covered by a person riding the chair lift, we use the Pythagorean theorem.

$5{,}750^2 = x^2 + 1{,}170^2$	Pythagorean theorem
$33{,}062{,}500 = x^2 + 1{,}368{,}900$	Simplify squares.
$x^2 = 33{,}062{,}500 - 1{,}368{,}900$	Solve for x^2.
$x^2 = 31{,}693{,}600$	Simplify the right side.
$x = \sqrt{31{,}693{,}600}$	Square Root Property for Equations
$= 5{,}630$ feet	To the nearest foot

A rider getting on the lift at point A and riding to point B will cover a horizontal distance of approximately 5,630 feet.

The 30°-60°-90° Triangle

In any right triangle which the two acute angles are 30° and 60°, the longest side (the hypotenuse) is always twice the shortest side(the side opposite the 30° angle), and the side of medium length (the side opposite the 60° angle) is always $\sqrt{3}$ times the shortest side (Figure 10).

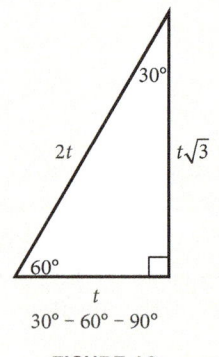

FIGURE 10

Note that the shortest side, t, is opposite the smallest angle, 30°. The longest side, $2t$, is opposite the largest angle, 90°. To verify the relationship between the sides in this triangle we draw an **equilateral triangle** (one in which all three sides are equal) and label half the base with t (Figure 11).

The altitude h (shown in red) bisects the base. We have two 30°– 60°– 90° triangles. The longest side in each is $2t$. We find that h is $t\sqrt{3}$ by applying the Pythagorean Theorem:

$$t^2 + h^2 = (2t)^2$$
$$h = \sqrt{4t^2 - t^2}$$
$$= \sqrt{3t^2}$$
$$= t\sqrt{3}$$

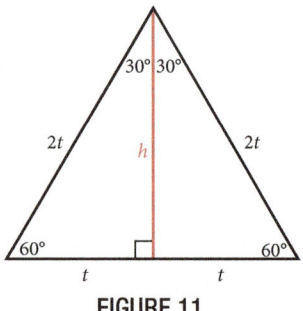

FIGURE 11

EXAMPLE 4   If the shortest side of a 30°– 60°– 90° triangle is 5, find the other two sides.

Solution The longest side is 10 (twice the shortest side), and the side opposite the 60° angle is $5\sqrt{3}$ (Figure 12).

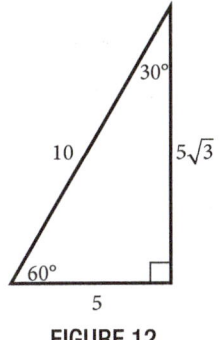

FIGURE 12

EXAMPLE 5   A ladder is leaning against a wall. The top of the ladder is 4 feet above the ground, and the bottom of the ladder makes an angle of 60° with the ground. How long is the ladder, and how far from the wall is the bottom of the ladder?

Solution The triangle formed by the ladder, the wall, and the ground is a 30°– 60°– 90° triangle. If we let x represent the distance from the bottom of the ladder to the wall, then the length of the ladder can be represented by $2x$ (Figure 13). The distance from the top of the ladder to the ground is $x\sqrt{3}$, since it is opposite the 60° angle. Therefore,

$$x\sqrt{3} = 4$$

$$x = \frac{4}{\sqrt{3}}$$

$$= \frac{4\sqrt{3}}{3}$$ *Rationalize the denominator by multiplying the numerator and denominator by $\sqrt{3}$*

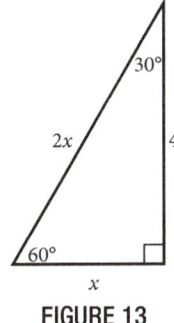

FIGURE 13

The distance from the bottom of the ladder to the wall, x, is $\frac{4\sqrt{3}}{3}$ feet, so the length of the ladder, $2x$, must be $\frac{8\sqrt{3}}{3}$ feet. Note that these lengths are given in exact values. If we want a decimal approximation for them, we can replace $\sqrt{3}$ with 1.732 to obtain

$$\frac{4\sqrt{3}}{3} \approx \frac{4(1.732)}{3} \approx 2.309 \text{ feet}$$

$$\frac{8\sqrt{3}}{3} \approx \frac{8(1.732)}{3} \approx 4.619 \text{ feet}$$

The other special right triangle is the 45°– 45°– 90° triangle.

The 45°-45°-90° Triangle

If the two acute angles in a right triangle are both 45°, then the two shorter sides (the legs) are equal, and the longest side (the hypotenuse) is $\sqrt{2}$ times as long as the shorter sides. That is, if the shorter sides are of length t, then the longest side has length $t\sqrt{2}$ (Figure 14).

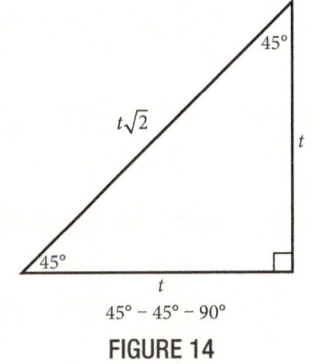

45° – 45° – 90°

FIGURE 14

To verify this relationship, we simply note that if the two acute angles are equal, then the sides opposite them are also equal. We apply the Pythagorean Theorem to them to find the length of the hypotenuse:

$$\text{Hypotenuse} = \sqrt{t^2 + t^2}$$
$$= \sqrt{2t^2} = t\sqrt{2}$$

EXAMPLE 6 A 10-foot rope connects the top of a tent pole to the ground. If the rope makes an angle of 45° with the ground, find the length of the tent pole.

Solution Assuming that the tent pole forms an angle of 90° with the ground, the triangle formed by rope, tent pole, and the ground is a 45°– 45°– 90° triangle. (See Figure 15.)

If we let x represent the length of the tent pole, then the length of the rope, in terms of x, is $x\sqrt{2}$. It is also given as 10 feet. Therefore,

$$x\sqrt{2} = 10$$
$$x = \frac{10}{\sqrt{2}} = 5\sqrt{2}$$

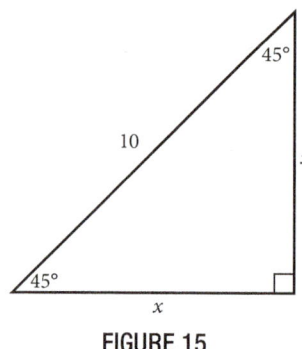

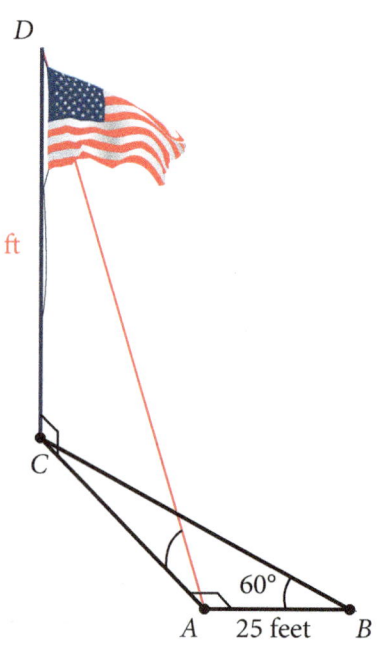

FIGURE 15

The length of the tent pole is $5\sqrt{2}$ feet. Again, $5\sqrt{2}$ is the exact value of the length of the tent pole. To find a decimal approximation, we replace $\sqrt{2}$ with 1.414 to obtain

$$5\sqrt{2} \approx 5(1.414) = 7.07 \text{ feet}$$

Our next example has two right triangles that share a common side. It has a three-dimensional look to it and is drawn in perspective.

EXAMPLE 7 Figure 16 shows a 75-foot flagpole. If angle B is 60° and AB is 25 feet, find the distance from A to D, the top of the flagpole. (Round to the nearest foot.)

Solution Because the bottom triangle is a 30°-60°-90° triangle, and side AB is 25 feet, then the longer side, AC, must be $25\sqrt{3}$ feet. Using the Pythagorean theorem in triangle ACD, we have

$$x^2 = (25\sqrt{3})^2 + 75^2$$
$$= 1{,}875 + 5{,}625$$
$$= 7{,}500$$

Taking the square root, we have

$$x = \sqrt{7{,}500} \approx 87 \text{ feet}$$

The distance to the top of the flagpole is approximately 87 feet.

FIGURE 16

Vocabulary

Use the vocabulary words below to fill in the blanks in the sentences.

twice   supplement   hypotenuse   equal   acute

1. An _____ angle has a measure that is between 0° and 90°.

2. An angle and its _____ always add to 180°.

3. The longest side of a right triangle is called the _____ .

4. The longest side of a 30°-60°-90° triangle is _____ the shortest side.

5. The two legs in a 45°-45°-90° triangle are always _____ .

Indicate which of the angles below are acute angles and which are obtuse angles. Then give the complement and the supplement of each angle.

1. 10°   **2.** 50°   **3.** 45°   **4.** 90°

5. 120°   **6.** 160°   **7.** x **8.** y

Problems 9 through 14 refer to Figure 17. (*Remember:* The sum of the three angles in any triangle is always 180°.)

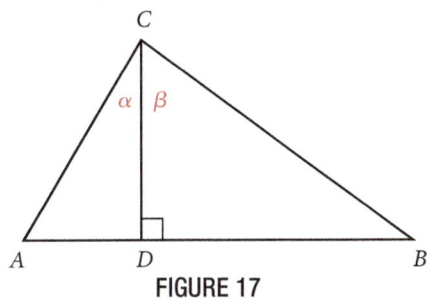

FIGURE 17

9. Find α if $A = 30°$.   **10.** Find B if $\beta = 45°$.

11. Find α if $A = \alpha$.   **12.** Find α if $A = 2\alpha$.

13. Find A if $B = 30°$ and $\alpha + \beta = 100°$.   **14.** Find B if $\alpha + \beta = 80°$ and $A = 80°$.

Problems 15 and 16 refer to Figure 18.

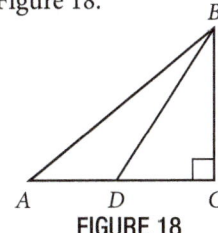

FIGURE 18

15. Find AB if $BC = 4$, $BD = 5$, and $AD = 2$.

16. Find BD if $BC = 5$, $AB = 13$, and $AD = 2$.

17. What acute positive angle does the line $y = x$ make with the positive x-axis?

18. What obtuse positive angle does the line $y = -x$ make with the positive x-axis?

19. Through how many degrees does the hour hand of a clock move in 4 hours?

20. Through how many degrees does the minute hand of a clock move in 40 minutes?

9

21. It takes the earth 24 hours to make one complete revolution on its axis. Through how many degrees does the earth turn in 12 hours?

22. Through how many degrees does the earth turn in 6 hours?

Each of the following problems refers to a right triangle ABC with $C = 90°$.

23. If $a = 4$ and $b = 3$, find c.

24. If $a = 1$ and $b = 2$, find c.

25. If $a = 7$ and $c = 25$, find b.

26. If $a = 2$ and $c = 6$, find b.

27. If $b = 12$ and $c = 13$, find a

28. If $b = 6$ and $c = 10$, find a.

Solve for x in each of the following right triangles:

29.

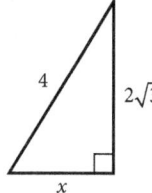

4   $2\sqrt{3}$

x

30.

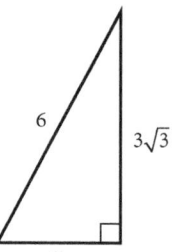

6   $3\sqrt{3}$

x

31.

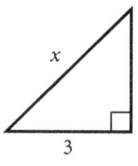

x 3

3

32.

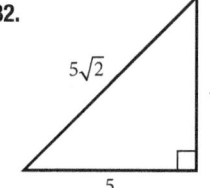

$5\sqrt{2}$ x

5

33.

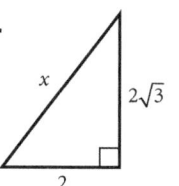

x $2\sqrt{3}$

2

34.

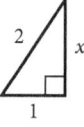

2   x

1

35.

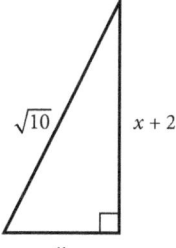

$\sqrt{10}$ $x + 2$

x

36.

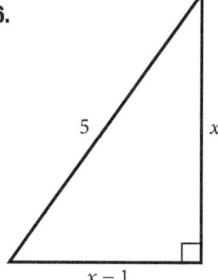

5   x

$x - 1$

37. Construction The roof of a house is to extend up 13.5 feet above the ceiling, which is 36 feet across. If the edge of the roof is to extend 2 feet beyond the side of the house and 1.5 feet below the ceiling, find the length of one side of the roof.

38. Surveying A surveyor is attempting to find the distance across a pond. From a point on one side of the pond he walks 25 yards to the end of the pond and then makes a 90° turn and walks another 60 yards before coming to a point directly across the pond from the point at which he started. What is the distance across the pond?

Find the remaining sides of a 30°–60°–90° triangle if:

39. The shortest side is 1.

40. The shortest side is 3.

41. The longest side is 8.

42. The longest side is 5.

43. The longest side is $\frac{2}{3}$.

44. The longest side is 24.

45. The medium side is $3\sqrt{3}$.

46. The medium side is $2\sqrt{3}$.

47. The medium side is 6.

48. The medium side is 4.

Find the remaining sides of a 45°–45°–90° triangle if:

49. The shorter sides are each 1.   **50.** The shorter sides are each 5.

51. The shorter sides are each $\frac{4}{5}$.   **52.** The shorter sides are each $\frac{1}{2}$.

53. The longest side is $8\sqrt{2}$.   **54.** The longest side is $5\sqrt{2}$.

55. The longest side is 4.   **56.** The longest side is 12.

Distance   Figure 19 shows a 75-foot flagpole. Use the information in the diagram to work Problems 57 and 58.

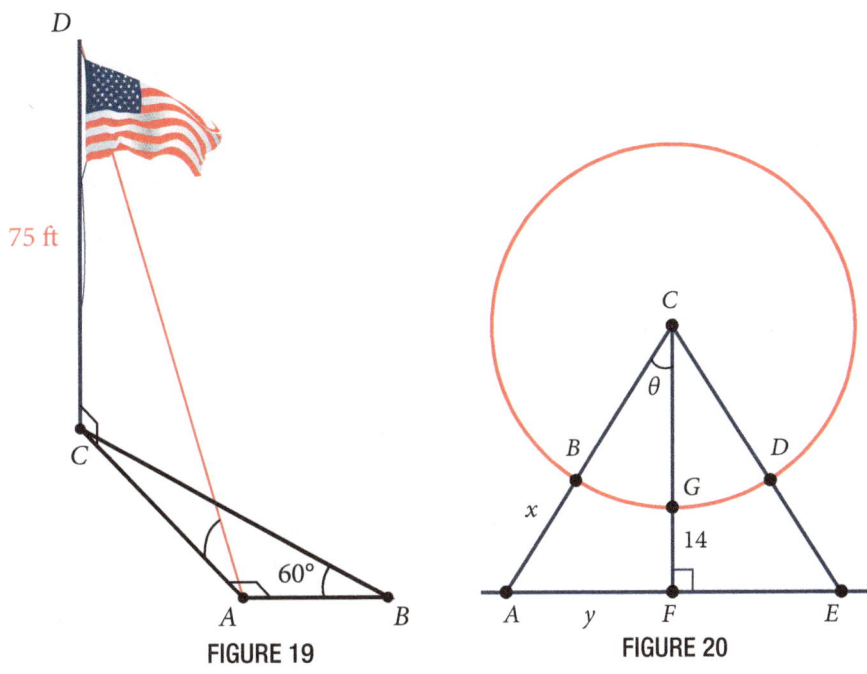

FIGURE 19   FIGURE 20

57. a. If angle $ABC = 60°$ and $AB = 35$ feet, find AC.

   b. Using part **a**, find AD.

58. a. If angle $CAD = 60°$, find AC.

   b. Using part **a**, and angle $ABC = 60°$, find BC.

Ferris Wheel   As we progress through the book, we will be looking at a number of problems that involve Ferris wheels. Figure 20 is a simple model for a Ferris wheel. The bottom of the wheel is 14 feet above the ground. Use the information in the diagram to work Problems 59 and 60. Round your answers to the nearest tenth of a foot.

59. a. Find y if the radius of the circle is 100 feet, and $\theta = 30°$.

   b. Use the information from Part a to find x.

60. a. Find y if the radius of the circle is 100 feet, and $\theta = 45°$.

   b. Use the information from Part a to find x.

61. Distance a Bullet Travels   A bullet is fired into the air at an angle of 45°. How far does it travel before it is 1,000 ft above the ground? (Assume that the bullet travels in a straight line, neglect the forces of gravity, and give your answer to the nearest foot.)

62. Time a Bullet Travels   If the bullet in Problem 61 is traveling at 2,828 ft/sec, how long does it take for the bullet to reach a height of 1,000 ft?

63. The Spiral of Roots   The introduction to this chapter shows the Spiral of Roots. The following three figures (Figures 19, 20, and 21) show the first three stages in the construction of the Spiral of Roots. Using graph paper and a ruler, construct the Spiral of Roots, labeling each diagonal as you draw it, to the point where you can see a line segment with a length of $\sqrt{10}$.

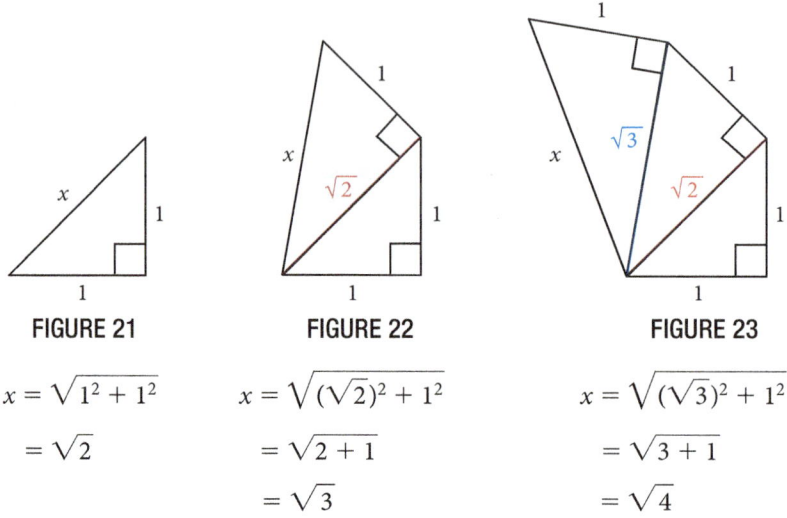

FIGURE 21	FIGURE 22	FIGURE 23

$$x = \sqrt{1^2 + 1^2} \qquad x = \sqrt{(\sqrt{2})^2 + 1^2} \qquad x = \sqrt{(\sqrt{3})^2 + 1^2}$$

$$= \sqrt{2} \qquad\qquad = \sqrt{2 + 1} \qquad\qquad = \sqrt{3 + 1}$$

$$\qquad\qquad\qquad = \sqrt{3} \qquad\qquad = \sqrt{4}$$

64. The Golden Ratio   Rectangle *ACEF* (Figure 24) is a golden rectangle. It is constructed from square *ACDB* by holding line segment *OB* fixed at point *O* and then letting point *B* drop down until *OB* aligns with *CD*. The ratio of the length to the width in the golden rectangle is called the *golden ratio*. Find the lengths below to arrive at the golden ratio.

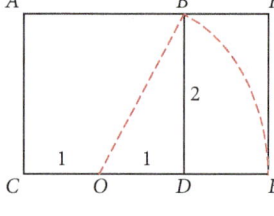

a. Find the length of *OB*.

b. Find the length of *OE*.

c. Find the length of *CE*.

d. Find the ratio $\dfrac{CE}{EF}$.

FIGURE 24

Getting Ready for the Next Section

These are problems that you must be able to work in order to understand the material in the next section. The problems below are exactly the type of problems you will see in the explanations and examples in the next section.

65. Work each problem according to the instructions given.

    a. Solve: $2x + 5 = 10$ **b.** Find x when y is 0: $2x + 5y = 10$

    c. Find y when x is 0: $2x + 5y = 10$ **d.** Graph: $2x + 5y = 10$

    e. Solve for y: $2x + 5y = 10$

66. Work each problem according to the instructions given.

    a. Solve: $x - 2 = 6$ **b.** Find x when y is 0: $x - 2y = 6$

    c. Find y when x is 0: $x - 2y = 6$ **d.** Graph: $x - 2y = 6$

    e. Solve for y: $x - 2y = 6$

1.2 Videos

KEY WORDS

ordered pair
x-coordinate
y-coordinate
rectangular coordinate
    system
axes
origin
x-axis
y-axis
quadrant
analytic geometry
x-intercept
y-intercept
standard position
quadrantal angle
coterminal angles

Learning Objectives

A   Graph ordered pairs and equations.

B   Find the equation of a graph.

C   Sketch the graph of a circle given an equation.

D   Solve problems using the distance formula.

E   Solve problems involving angles in standard position, terminal sides, and coterminal angles.

A Ordered Pairs

Paired data play an important role in equations that contain two variables. Working with these equations is easier if we standardize in trigonometry the terminology and notation associated with paired data. So here is a definition that will do just that.

> A pair of numbers enclosed in parentheses and separated by a comma, such as $(-2, 1)$, is called an *ordered pair* of numbers. The first number in the pair is called the *x-coordinate* of the ordered pair; the second number is called the *y-coordinate*. For the ordered pair $(-2, 1)$, the *x*-coordinate is -2 and the *y*-coordinate is 1.

Rectangular Coordinate System

A *rectangular coordinate system* is made by drawing two real number lines at right angles to each other. The two number lines, called *axes*, cross each other at 0. This point is called the *origin*. Positive directions are to the right and up. Negative directions are to the left and down. The rectangular coordinate system is shown in Figure 1.

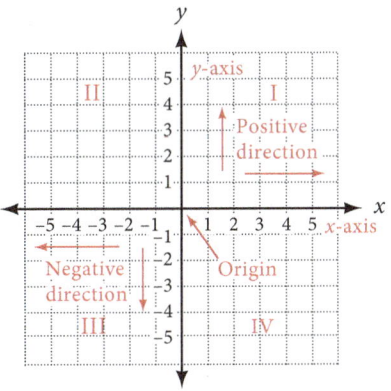

FIGURE 1

The horizontal number line is called the *x-axis*, and the vertical number line is called the *y-axis*. The two number lines divide the coordinate system into four *quadrants*, which we number I through IV in a counterclockwise direction. Points on the axes are not considered as being in any quadrant.

Note A rectangular coordinate system allows us to connect algebra and geometry by associating geometric shapes (the curves shown in the diagrams) with algebraic equations. The French philosopher and mathematician René Descartes (1596–1650) is usually credited with the invention of the rectangular coordinate system, which is often referred to as the Cartesian coordinate system in his honor. As a philosopher, Descartes is responsible for the statement, "I think, therefore, I am." Until Descartes invented his coordinate system in 1637, algebra and geometry were treated as separate subjects.

Graphing Ordered Pairs

To graph the ordered pair (a, b) on a rectangular coordinate system, we start at the origin and move a units right or left (right if a is positive, left if a is negative). Then we move b units up or down (up if b is positive, down if b is negative). The point where we end up is the graph of the ordered pair (a, b).

EXAMPLE 1   Graph the ordered pairs $(1, -3)$, $\left(\frac{1}{2}, 2\right)$, $(3, 0)$, $(0, -2)$, $(-1, 0)$, and $(0, 5)$.

Solution

From Figure 2, we see that any point on the x-axis has a y-coordinate of 0 (it has no vertical displacement), and any point on the y-axis has an x-coordinate of 0 (no horizontal displacement).

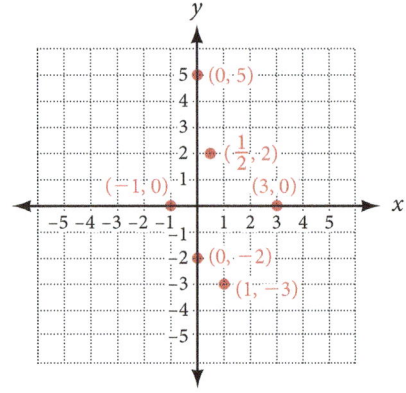

FIGURE 2

Graphing Equations

We can plot a single point from an ordered pair, but to draw a line or a curve, we need more points. To graph an equation in two variables, we draw a line or smooth curve through all the points whose coordinates satisfy the equation.

EXAMPLE 2   Graph the equation $y = -\frac{1}{3}x$.

Solution The graph of this equation will be a straight line. We need to find three ordered pairs that satisfy the equation. To do so, we can let x equal any numbers we choose and find corresponding values of y.

However, because every value of x we substitute into the equation is going to be multiplied by $-\frac{1}{3}$, let's use numbers for x that are divisible by 3, like -3, 0, and 3. That way, when we multiply them by $-\frac{1}{3}$, the result will be an integer.

Let $x = -3$; $y = -\frac{1}{3}(-3) = 1$

The ordered pair $(-3, 1)$ is one solution.

Let $x = 0$; $y = -\frac{1}{3}(0) = 0$

The ordered pair $(0, 0)$ is a second solution.

Let $x = 3$; $y = -\frac{1}{3}(3) = -1$

The ordered pair $(3, -1)$ is a third solution.

In table form

x	y
-3	1
0	0
3	-1

Plotting the ordered pairs $(-3, 1)$, $(0, 0)$, and $(3, -1)$ and drawing a straight line through their graphs, we have the graph of the equation $y = -\frac{1}{3}x$ as shown in Figure 3.

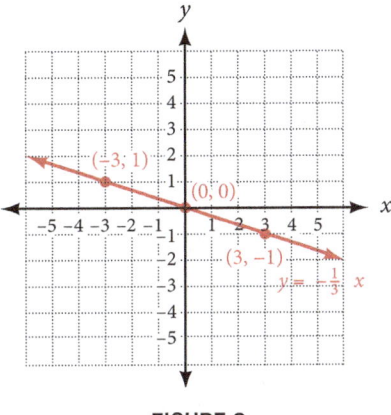

FIGURE 3

> *Note* It takes only two points to determine a straight line. We have included a third point for "insurance." If all three points do not line up in a straight line, we have made a mistake.

Example 3 illustrates again the connection between algebra and geometry that we mentioned earlier in this section. Descartes's rectangular coordinate system allows us to associate the equation $y = -\frac{1}{3}x$ (an algebraic concept) with a specific straight line (a geometric concept). The study of the relationship between equations in algebra and their associated geometric figures is called *analytic geometry*.

Intercepts

Two important points on the graph of a straight line, if they exist, are the points where the graph crosses the axes.

> **Intercepts**
> An *x-intercept* of the graph of an equation is the x-coordinate of a point where the graph crosses the x-axis. The *y-intercept* is defined similarly.

Because any point on the x-axis has a y-coordinate of 0, we can find the x-intercept by letting $y = 0$ and solving the equation for x. We find the y-intercept by letting $x = 0$ and solving for y.

EXAMPLE 3 Find the x- and y-intercepts for $2x + 3y = 6$; then graph the equation.
Solution To find the y-intercept, we let $x = 0$.

$$\begin{aligned} \text{When} \qquad x &= 0 \\ \text{we have} \qquad 2(0) + 3y &= 6 \\ 3y &= 6 \\ y &= 2 \end{aligned}$$

The y-intercept is 2 so the graph crosses the y-axis at the point $(0, 2)$. To find the x-intercept, we let $y = 0$.

$$\begin{aligned} \text{When} \qquad y &= 0 \\ \text{we have} \qquad 2x + 3(0) &= 6 \\ 2x &= 6 \\ x &= 3 \end{aligned}$$

The x-intercept is 3, so the graph crosses the x-axis at the point (3, 0). We use these results to graph the solution set for $2x + 3y = 6$. The graph is shown in Figure 4.

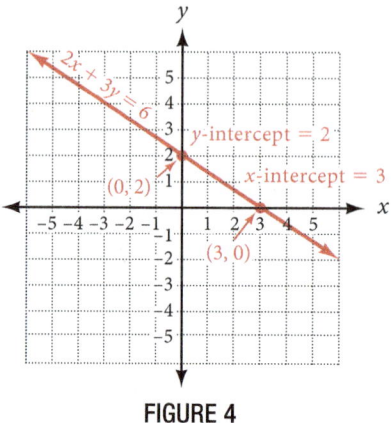

FIGURE 4

> **Note** Graphing straight lines by finding the intercepts works best when the coefficients of x and y are factors of the constant term.

Graphing Parabolas

EXAMPLE 4   Graph the equation $y = x^2$.

Solution We input values of x then calculate using the equation to output values of y. The result is a set of ordered pairs that we plot and then connect with a smooth curve. The table below shows our calculations.

Input x	Calculate Using the Equation	Output y	Form Ordered Pairs
-3	$y = (-3)^2 =$	9	$(-3, 9)$
-2	$y = (-2)^2 =$	4	$(-2, 4)$
-1	$y = (-1)^2 =$	1	$(-1, 1)$
0	$y = (0)^2 =$	0	$(0, 0)$
1	$y = (1)^2 =$	1	$(1, 1)$
2	$y = (2)^2 =$	4	$(2, 4)$
3	$y = (3)^2 =$	9	$(3, 9)$

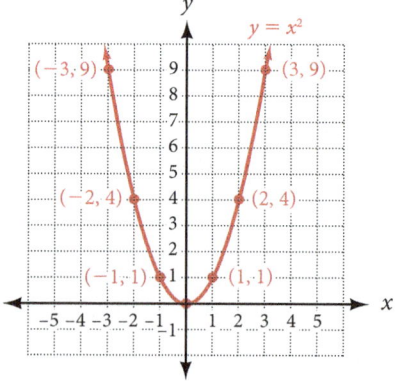

FIGURE 5

EXAMPLE 5 Graph $y = x^2 - 4$.

Solution We make a table as we did in the previous example. We have included the graph at $y = x^2$ for reference. As you will see, for a given value of x the value of y on the red graph is 4 units below the corresponding value of y on the blue graph.

Input x	Calculate Using the Equation	Output y	Form Ordered Pairs
-3	$y = (-3)^2 - 4 =$	5	$(-3, 5)$
-2	$y = (-2)^2 - 4 =$	0	$(-2, 0)$
-1	$y = (-1)^2 - 4 =$	-3	$(-1, -3)$
0	$y = (0)^2 - 4 =$	-4	$(0, -4)$
1	$y = (1)^2 - 4 =$	-3	$(1, -3)$
2	$y = (2)^2 - 4 =$	0	$(2, 0)$
3	$y = (3)^2 - 4 =$	5	$(3, 5)$

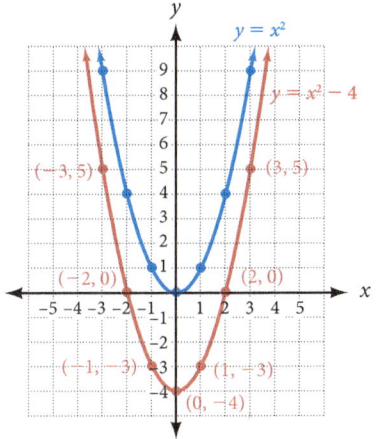

FIGURE 6

As you can see, the graph of $y = x^2 - 4$ is the graph of $y = x^2$ translated down 4 units. We generalize this as follows.

If K is a positive number, then:	
The Graph of	Is the Graph of $y = x^2$ translated
$y = x^2 + K$	K units up
$y = x^2 - K$	K units down

To generalize even further, your previous work in algebra gave you the following information on parabolas in general.

Graphing Parabolas
The graph of
$$y = a(x - h)^2 + k, a \neq 0$$
will be a parabola with a vertex at (h, k). The vertex will be the highest point on the graph when $a < 0$, and the lowest point on the graph when $a > 0$.

B Finding the Equation from the Graph

EXAMPLE 6 At the 1997 Washington County Fair in Oregon, David Smith, Jr., The Bullet, was shot from a cannon. As a human cannonball, he reached a height of 70 feet before landing in a net 160 feet from the cannon. Sketch the graph of his path, and then find the equation of the graph.

Solution We assume that the path taken by the human cannonball is a parabola. If the origin of the coordinate system is at the opening of the cannon, then the net that catches him will be at 160 on the x-axis. Figure 7 shows the graph.

Because the curve is a parabola, we know the equation will have the form

$$y = a(x - h)^2 + k$$

Because the vertex of the parabola is at (80, 70), we can fill in two of the three constants in our equation, giving us

$$y = a(x - 80)^2 + 70$$

To find a, we note that the landing point will be (160, 0). Substituting the coordinates of this point into the equation, we solve for a:

$$0 = a(160 - 80)^2 + 70$$

$$0 = a(80)^2 + 70$$

$$0 = 6{,}400a + 70$$

$$a = -\frac{70}{6{,}400} = -\frac{7}{640}$$

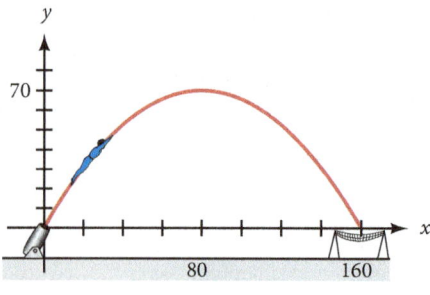

FIGURE 7

The equation that describes the path of the human cannonball is

$$y = -\frac{7}{640}(x - 80)^2 + 70 \quad \text{for} \quad 0 \le x \le 160$$

Using Technology

Graph the equation found in Example 7 on a graphing calculator using the window shown here. (We will use this graph later in the book to find the angle between the cannon and the horizontal.)

Window: X from 0 to 180, increment 20
Y from 0 to 80, increment 10

On the TI-83, an increment of 20 for X means Xscl = 20.

C Graphing Circles

The equation of any circle with its center at the origin and radius r will be

$$x^2 + y^2 = r^2$$

EXAMPLE 7   Sketch the graph of $x^2 + y^2 = 9$.

Solution Because the equation has the form above, it must have its center at (0, 0) and a radius of 3. The graph is shown in Figure 8.

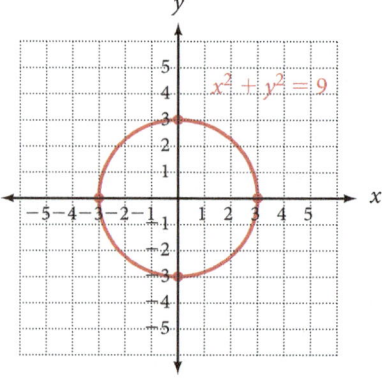

FIGURE 8

The unit circle is the circle with its center at the origin and a radius of 1. The equation of the unit circle is $x^2 + y^2 = r^2$ (Figure 9).

From your previous math classes, you know you can generalize the relationship between the equation of a circle and its graph, as follows.

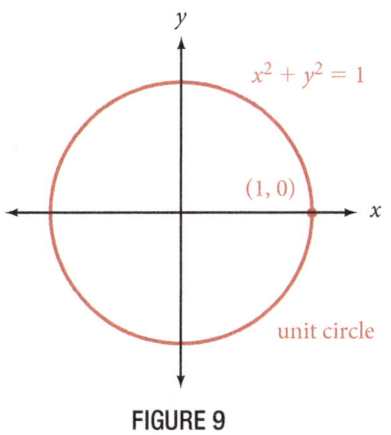

FIGURE 9

Graphing Circles
The graph of the equation $(x - a)^2 + (y - b)^2 = r^2$ will be a circle with a center at (a, b) and a radius of r.

EXAMPLE 8   Find the center and radius, and sketch the graph of the circle whose equation is

$$(x - 1)^2 + (y + 3)^2 = 4$$

Solution Writing the equation in the form

$$(x - a)^2 + (y - b)^2 = r^2$$

we have

$$(x - 1)^2 + [y - (-3)]^2 = 2^2$$

The center is at $(1, -3)$, and the radius is 2. The graph is shown in Figure 10.

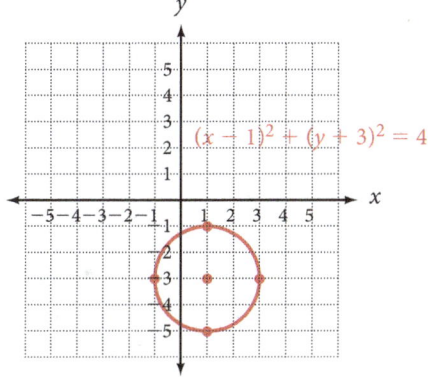

FIGURE 10

As we progress through the book, we will be looking at a number of problems that involve Ferris wheels. The next example will get us started. Notice in the example how we use our rectangular coordinate system to help us work the problem.

EXAMPLE 9   Suppose a Ferris wheel has a radius of 50 feet, with the bottom of the wheel 10 feet above the ground. Let t represent the time (in minutes). If the wheel completes one revolution in 2 minutes, find the distance a rider is above the ground when the ride starts ($t = 0$), after 30 seconds ($t = \frac{1}{2}$), at one minute ($t = 1$), 30 seconds later ($t = \frac{3}{2}$), and at 2 minutes ($t = 2$).

Solution We can simplify our work by taking away all the unnecessary items form the diagram to obtain a simplified model.

Ferris Wheel Diagram

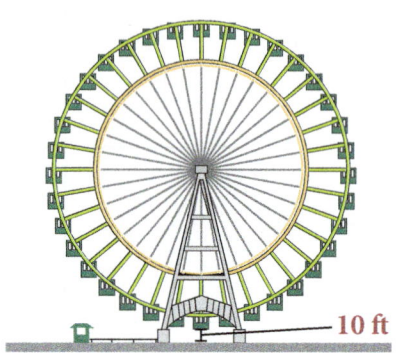

Simplified Mathematical Model

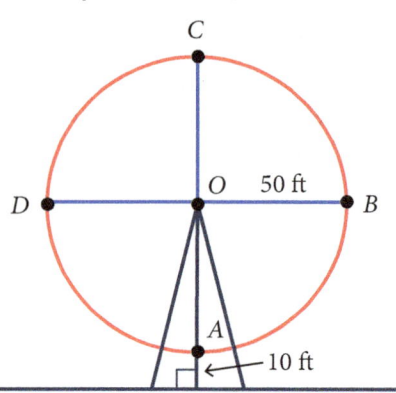

We can superimpose a coordinate system on our simplified model and place the origin wherever we choose. Let's put the origin at the center of the circle.

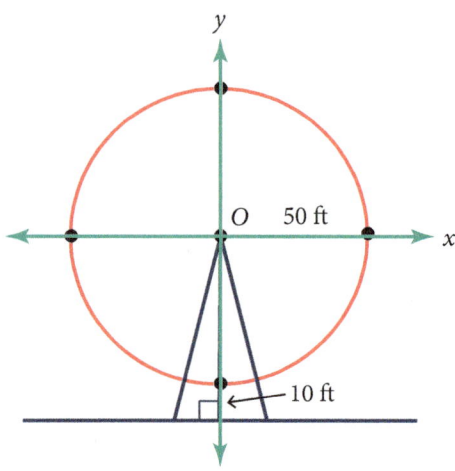

Point on the Wheel	Time in Minutes	Height in Feet
A	0	10
B	1/2	60
C	1	110
D	3/2	60
A	2	10

Now let's make a table showing the height a rider is above the ground when $t = 0$, $t = \frac{1}{2}$, $t = 1$, $t = \frac{3}{2}$, and $t = 2$. At $t = 0$, the rider is at the bottom of the wheel, 10 feet above the ground. After 30 seconds, the rider is at point B where his height above the ground is the radius of the circle plus the initial height, or $50 + 10 = 60$ feet. Likewise, at Point C the rider is $50 + 50 + 10 = 110$ feet above the ground. Continuing in this manner we have the numbers in the table below.

As you will see as we progress through the book, the graph of the distance the rider is above the ground as a function of time turns out to be a trigonometric function.

Note that the equation of the circle in our model with the superimposed coordinate system is $x^2 + y^2 = 50^2$. If we had superimposed our coordinate system so the origin was on the ground, then the equation would be $x^2 + (y - 60)^2 = 50^2$. The following diagram illustrates.

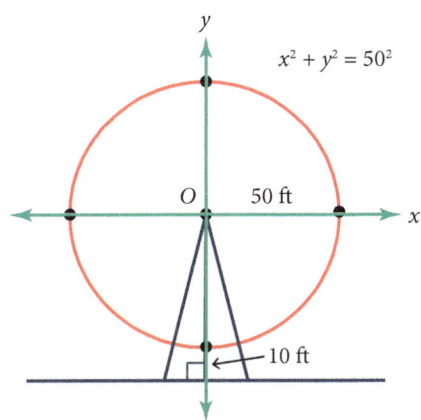

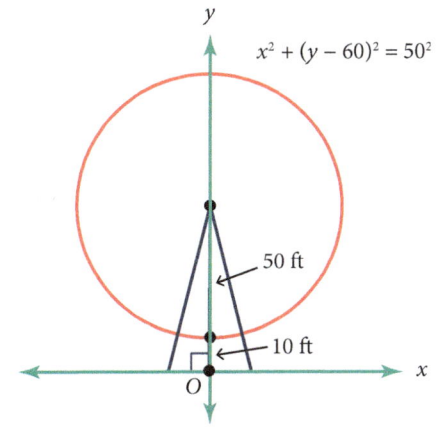

D The Distance Formula

Suppose (x_1, y_1) and (x_2, y_2) are any two points in the first quadrant. (Actually, we could choose the two points to be anywhere on the coordinate plane. It is just more convenient to have them in the first quadrant.) We can name the points P_1 and P_2, respectively, and draw the diagram shown in Figure 11. Notice the coordinates of point Q. The x-coordinate is x_2 because Q is directly below point P_2. The y-coordinate of Q is y_1 because Q is directly across from point P_1. It is evident from the diagram that the length of $\overline{P_2Q}$ is $y_2 - y_1$ and the length of $\overline{P_1Q}$ is $x_2 - x_1$.

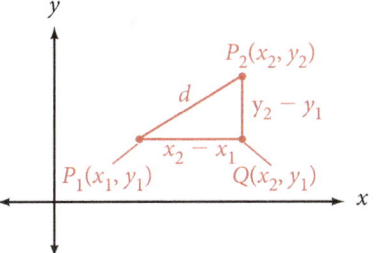

FIGURE 11

Using the Pythagorean theorem, we have

$$(\overline{P_1P_2})^2 = (\overline{P_1Q})^2 + (\overline{P_2Q})^2$$

or

$$d^2 = (x_2 - x_1)^2 + (y_2 - y_1)^2$$

Taking the square root of both sides, we have

$$d = \sqrt{(x_2 - x_1)^2 + (y_2 - y_1)^2}$$

We know this is the positive square root, because d is the distance from P_1 to P_2 and must therefore be positive. This formula is called the *distance formula*.

EXAMPLE 10 Find the distance between $(3, 5)$ and $(2, -1)$.

Solution If we let $(3, 5)$ be (x_1, y_1) and $(2, -1)$ be (x_2, y_2) and apply the distance formula, we have

$$d = \sqrt{(2 - 3)^2 + (-1 - 5)^2}$$

$$= \sqrt{(-1)^2 + (-6)^2}$$

$$= \sqrt{1 + 36}$$

$$= \sqrt{37}$$

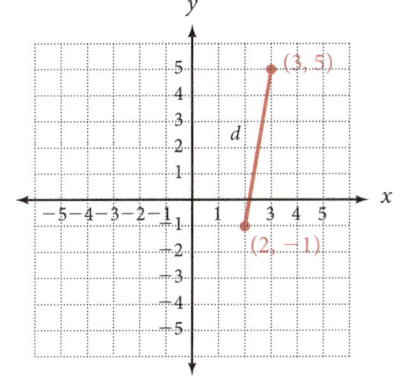

FIGURE 12

EXAMPLE 11   Find the distance from the origin to the point (x, y).

Solution The coordinates of the origin are $(0, 0)$. Applying the distance formula, we have

$$r = \sqrt{(x - 0)^2 + (y - 0)^2}$$

$$r = \sqrt{x^2 + y^2}$$

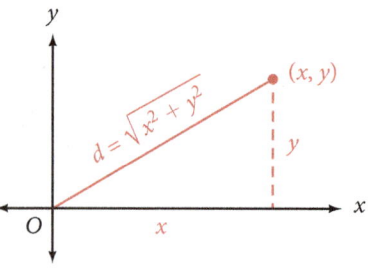

FIGURE 13

E Angles in Standard Position

Angles in Standard Position
An angle is said to be in *standard position* if its initial side is along the positive x-axis and its vertex is at the origin.

EXAMPLE 12   Draw an angle of 45° in standard position and find a point on the terminal side.

Solution If we draw 45° in standard position we see that the terminal side is along the line $y = x$ in quadrant I (Figure 14).

Since the terminal side of 45° lies along the line $y = x$ in the first quadrant, any point on the terminal side will have positive coordinates that satisfy the equation $y = x$. Here are some of the points that do just that.

$$(1, 1) \quad (2, 2) \quad (3, 3) \quad \left(\sqrt{2}, \sqrt{2}\right) \quad \left(\frac{1}{2}, \frac{1}{2}\right) \quad \left(\frac{7}{8}, \frac{7}{8}\right)$$

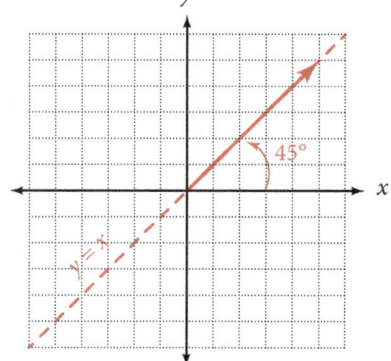

FIGURE 14

Vocabulary
If an angle, say θ, is in standard position and the terminal side of θ lies in quadrant I, then we say θ lies in the quadrant I and abbreviate it like this

$$\theta \in \text{QI}$$

Likewise, $\theta \in \text{QII}$ means θ is in standard position with its terminal side in quadrant II.

If the terminal side of an angle in standard position lies along one of the axes, then that angle is called a *quadrantal angle*. For example, an angle of 90° drawn in standard position would be a quadrantal angle, since the terminal side would lie along the positive *y*-axis. Likewise, 270° in standard position is a quadrantal angle because the terminal side would like along the negative *y*-axis (Figure 15).

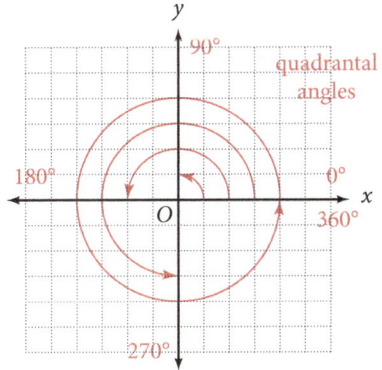

FIGURE 15

Two angles in standard position with the same terminal side are called *coterminal angles*. Figure 15 shows that 60° an −300° are coterminal angles when they are in standard position.

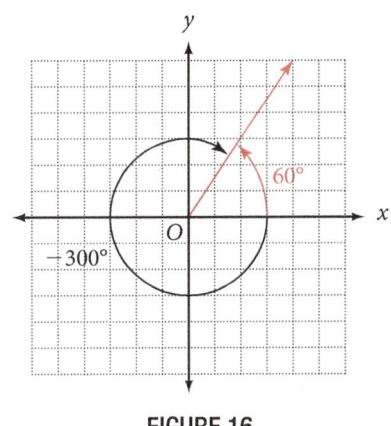

FIGURE 16

EXAMPLE 13 Draw −90° in standard position, name some of the points on the terminal side, and find an angle between 0° and 360° that is coterminal with −90°.

Solution Figure 16 shows −90° in standard position. Since the terminal side is along the negative *y*-axis, the points $(0, -1)$, $(0, -2)$, $\left(0, -\frac{5}{2}\right)$, and, in general, $(0, b)$ where b is a negative number, are all on the terminal side. The angle between 0 and 360 that is coterminal with −90° is 270°.

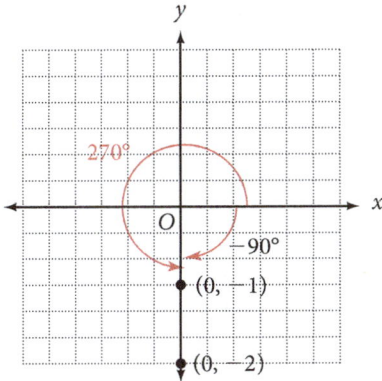

FIGURE 17

Problem Set 1.2

Vocabulary

Use the vocabulary words below to fill in the blanks in the sentences.

standard 2 vertex *x*-axis radius

1. In a rectangular coordinate system, the _____ is the horizontal axis.

2. The *y*-intercept for $2x + 3y = 6$ is _____ .

3. The general form for the equation of a parabola is $y = (x - h)^2 + k$. The ordered pair (h, k) gives the coordinate of the _____ of the parabola.

4. The general form for the equation of a circle centered at the origin is $x^2 + y^2 = r^2$. In this case, *r* represents the _____ of the circle.

5. An angle that has its vertex at the origin of a rectangular coordinate system, and its initial side along the positive *x*-axis is said to be in _____ position.

Graph each of the following ordered pairs on a rectangular coordinate system.

1. **a.** $(-1, 2)$ **b.** $(-1, -2)$ **c.** $(5, 0)$ **d.** $(0, 2)$ **e.** $(-5, -5)$ **f.** $\left(\frac{1}{2}, 2\right)$

2. **a.** $(1, 2)$ **b.** $(1, -2)$ **c.** $(0, -3)$ **d.** $(4, 0)$ **e.** $(-4, -1)$ **f.** $\left(3, \frac{1}{4}\right)$

Give the coordinates of each point.

3.

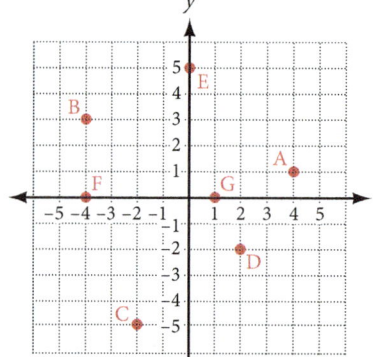

4.

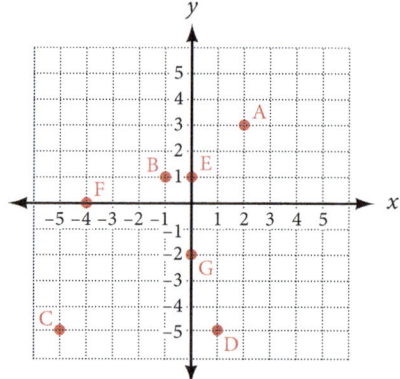

5. Which of the following tables could be produced from the equation $y = 2x - 6$?

a.

x	y
0	6
1	4
2	2
3	0

b.

x	y
0	−6
1	−4
2	−2
3	0

c.

x	y
0	−6
1	−5
2	−4
3	−3

6. Which of the following tables could be produced from the equation $3x - 5y = 15$?

a.

x	y
0	5
−3	0
10	3

b.

x	y
0	−3
5	0
10	3

c.

x	y
0	−3
−5	0
10	−3

7. The graph shown here is the graph of which of the following equations?

   a. $y = \dfrac{3}{2}x - 3$

   b. $y = \dfrac{2}{3}x - 2$

   c. $y = -\dfrac{2}{3}x + 2$

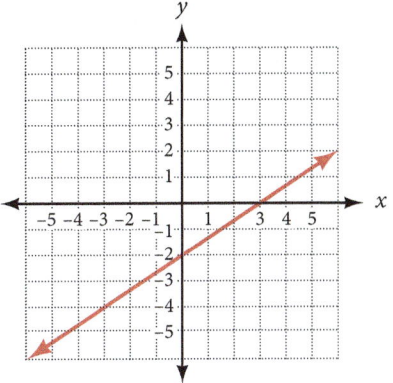

8. The graph shown here is the graph of which of the following equations?

   a. $3x - 2y = 8$

   b. $2x - 3y = 8$

   c. $2x + 3y = 8$

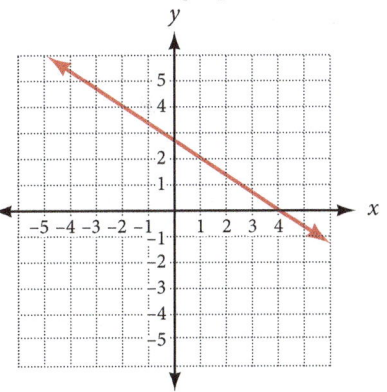

Graph each of the following. Use one coordinate system for each problem.

9. a. $y = 2x$ **b.** $y = 2x + 3$ **c.** $y = 2x - 5$

10. a. $y = \dfrac{1}{3}x$ **b.** $y = \dfrac{1}{3}x + 1$ **c.** $y = \dfrac{1}{3}x - 3$

11. a. $y = \dfrac{1}{2}x^2$ **b.** $y = \dfrac{1}{2}x^2 - 2$ **c.** $y = \dfrac{1}{2}x^2 + 2$

12. a. $y = 2x^2$ **b.** $y = 2x^2 - 8$ **c.** $y = 2x^2 + 1$

Graph each parabola. Label the vertex and any intercepts that exist.

13. $y = 2(x - 1)^2 + 3$ **14.** $y = 2(x + 1)^2 - 3$

Give the center and radius, and sketch the graph of each of the following circles.

15. $x^2 + y^2 = 4$ **16.** $x^2 + y^2 = 16$

17. $(x - 1)^2 + (y - 3)^2 = 25$ **18.** $(x - 4)^2 + (y - 1)^2 = 36$

19. $(x + 2)^2 + (y - 4)^2 = 8$ **20.** $(x - 3)^2 + (y + 1)^2 = 12$

21. Find the equations of circles A, B, and C in the following diagram. The three points are the centers of the three circles.

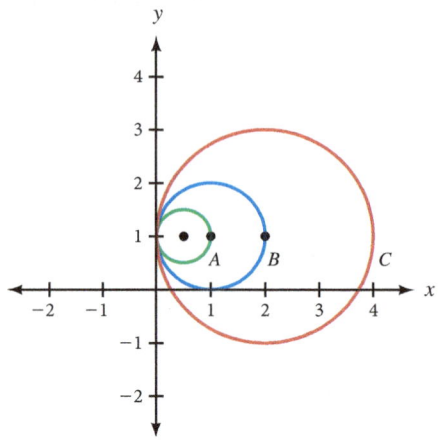

22. Each of the following circles passes through the origin. The centers are along the line $y = x$ at a 45° angle with the x-axis. Find the equation of each circle.

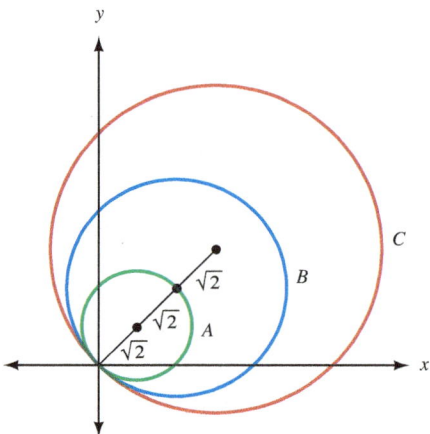

Find the distance between the following points.

23. $(3, 7)$ and $(6, 3)$ **24.** $(4, 7)$ and $(8, 1)$

25. $(0, 9)$ and $(5, 0)$ **26.** $(-3, 0)$ and $(0, 4)$

27. $(3, -5)$ and $(-2, 1)$ **28.** $(-8, 9)$ and $(-3, -2)$

29. $(-1, -2)$ and $(-10, 5)$ **30.** $(-3, -8)$ and $(-1, 6)$

31. Find x so the distance between $(x, 2)$ and $(1, 5)$ is $\sqrt{13}$.

32. Find x so the distance between $(-2, 3)$ and $(x, 1)$ is 3.

33. Find x so the distance between $(x, 5)$ and $(3, 9)$ is 5.

34. Find y so the distance between $(-4, y)$ and $(2, 1)$ is 8.

35. **Flight Path** An airplane is approaching Los Angeles International Airport at an altitude of 2,640 feet. If the plane is 1.2 miles from the runway (this is the horizontal distance to the runway), use the Pythagorean theorem to find the distance from the plane to the runway. (5,280 feet equals 1 mile.)

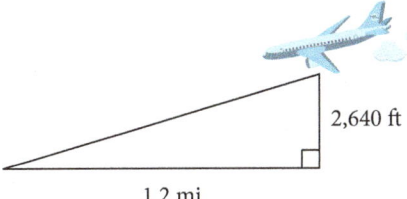

2,640 ft

1.2 mi

36. **Softball Diamond** In softball, the distance from home plate to first base is 60 feet, as is the distance from first base to second base. If the lines joining home plate to first base and first base to second base form a right angle, how far does a catcher standing on home plate have to throw the ball so that it reaches the shortstop standing on second base?

second

60 ft

third   90°   first

60 ft

home

37. **Softball Diamond** If a coordinate system is superimposed on the softball diamond in Problem 30 with the x-axis along the line from home plate to first base and the y-axis on the line from home plate to third base, what would be the coordinates of home plate, first base, second base, and third base?

38. **Softball Diamond** If a coordinate system is superimposed on the softball diamond in Problem 36 with the origin on home plate and the positive x-axis along the line joining home plate to second base, what would be the coordinates of first base and third base?

39. **Ferris Wheel** Figure 18 is a model of the Ferris wheel known as the Reisenrad. The diameter of the wheel is 240 feet, and one complete revolution takes 16 minutes. The bottom of the wheel is 12 feet above the ground. Refer to the diagram below to complete the table to give the height a rider is above the ground at the given times.

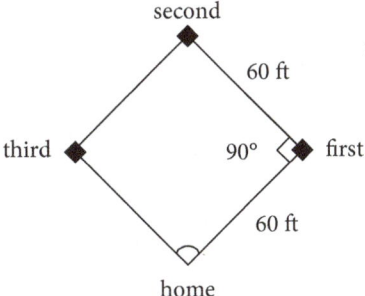

12 ft

Point on the Wheel	Time in Minutes	Height in Feet
A	0	
B	4	
C	8	
D	12	
A	16	

40. Ferris Wheel The diameter of a Ferris wheel is 160 feet, it rotates at 1/2 a revolution every minute, and the bottom of the wheel is 9 feet above the ground. Fill in the table to show the height a rider is above the ground at the given times. Assume the passenger starts the ride at the bottom of the wheel.

Time in Minutes	Height in Feet
0	
1/2	
1	
3/2	
2	

Draw each of the following angles in standard position, find a point on the terminal side, and name one other angle that is coterminal with it.

41. 135°   **42.** 45°   **43.** 225°   **44.** 315°

45. 90°   **46.** 360°   **47.** −45°   **48.** −90°

49. Draw an angle of 30° in standard position. Then find a if the point $(a, 1)$ is on the terminal side of 30°.

50. Draw 60° in standard position. Then find b if the point $(2, b)$ is on the terminal side of 60°.

51. Draw an angle in standard position whose terminal side contains the point $(3, -2)$. Find the distance from the origin to this point.

52. Draw an angle in standard position whose terminal side contains the point $(2, -3)$. Find the distance from the origin to this point.

53. Plot the points $(0, 0)$, $(5, 0)$, and $(5, 12)$ and show that, when connected, they are the vertices of a right triangle.

54. Plot the points $(0, 2)$, $(-3, 2)$, and $(-3, -2)$ and show that they form the vertices of a right triangle.

Getting Ready for the Next Section

These are problems that you must be able to work in order to understand the material in the next section. The problems below are exactly the type of problems you will see in the explanations and examples in the next section.

Simplify each expression.

55. $\dfrac{-1}{1}$ **56.** $\dfrac{0}{1}$ **57.** $\dfrac{-1}{0}$ **58.** $\dfrac{0}{-1}$ **59.** $\dfrac{1}{-1}$ **60.** $\dfrac{1}{0}$

Rationalize the denominator.

61. $\dfrac{3}{\sqrt{13}}$ **62.** $\dfrac{1}{\sqrt{2}}$

63. Find the distance from the origin to $(1, 1)$.

64. Find the distance from the origin to $(-2, 3)$.

Solve for x.

65. $x^2 + 25 = 169$ **66.** $x^2 + 16 = 25$

Give the reciprocal of each number, or expression. Assume all variables represent positive numbers.

67. 2   **68.** $\dfrac{1}{2}$ **69.** $\dfrac{4}{3}$ **70.** $\dfrac{3}{4}$ **71.** $\dfrac{a}{b}$ **72.** $\dfrac{y}{r}$

1.3 Videos

KEY WORDS

sine

cosine

tangent

cotangent

secant

cosecant

reference triangle

Learning Objectives

A Find the six trigonometric functions for a given angle.

B Find the algebraic sign of a given trigonometric function.

Now we can define the trigonometric functions. For us, the definition stated below is very important. What should you do with it? Memorize it. Remember, in mathematics, definitions are simply accepted. That is, unlike theorems, there is no proof associated with a definition. We simply accept them exactly as they are written, memorize them, and then use them. In Chapter 2, we will use an alternative definition (Definition II) to define the trigonometric functions. When you are finished with this section, be sure you have memorized this definition. It is the most valuable thing you can do for yourself at this point in your study of trigonometry.

A Six Trigonometric Functions

Definition I

If θ is an angle in standard position, and the point (x, y) is any point on the terminal side of θ other than the origin, then the six trigonometric functions of angle θ are defined as follows:

Function	Abbreviation	Definition
sine of θ	$= \sin \theta$	$= \dfrac{y}{r}$
cosine of θ	$= \cos \theta$	$= \dfrac{x}{r}$
tangent of θ	$= \tan \theta$	$= \dfrac{y}{x}$
cotangent of θ	$= \cot \theta$	$= \dfrac{x}{y}$
secant of θ	$= \sec \theta$	$= \dfrac{r}{x}$
cosecant of θ	$= \csc \theta$	$= \dfrac{r}{y}$

Where $x^2 + y^2 = r^2$, or $r = \sqrt{x^2 + y^2}$; that is, r is the distance from the origin to (x, y).

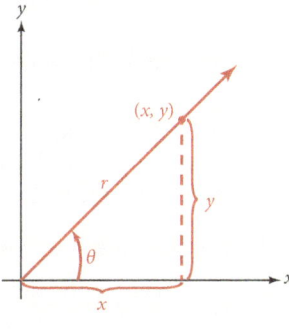

FIGURE 1

As you can see in Definition I, the six trigonometric functions are simply names given to the six possible ratios that can be made from the numbers x, y, and r as shown in Figure 1.

EXAMPLE 1 Find the six trigonometric functions for θ if θ is in standard position and the point $(-2, 3)$ is on the terminal side of θ.

Solution Figure 2 shows θ, the point $(-2, 3)$, and the distance r from the origin to $(-2, 3)$.

Applying the definition for the six trigonometric functions with $x = -2, y = 3$, and $r = \sqrt{13}$, we have

$$\sin \theta = \frac{y}{r} = \frac{3}{\sqrt{13}}$$

$$\cos \theta = \frac{x}{r} = -\frac{2}{\sqrt{13}}$$

$$\tan \theta = \frac{y}{x} = -\frac{3}{2}$$

$$\csc \theta = \frac{r}{y} = \frac{\sqrt{13}}{3}$$

$$\sec \theta = \frac{r}{x} = -\frac{\sqrt{13}}{2}$$

$$\cot \theta = \frac{x}{y} = -\frac{2}{3}$$

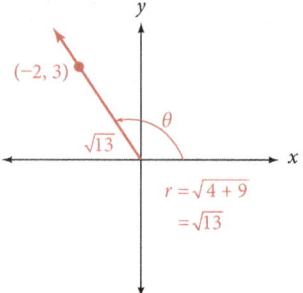

FIGURE 2

Note In algebra, when we encounter expressions like $\frac{3}{\sqrt{13}}$ that contain a radical in the denominator, we usually rationalize the denominator. In this case we could multiply the numerator and denominator by $\sqrt{13}$:

$$\frac{3}{\sqrt{13}} = \frac{3}{\sqrt{13}} \cdot \frac{\sqrt{13}}{\sqrt{13}} = \frac{3\sqrt{13}}{13}$$

In trigonometry, it is sometimes convenient to use $\frac{3\sqrt{13}}{13}$, but other times it is easier to use $\frac{3}{\sqrt{13}}$. For now, lets agree not to rationalize any denominators unless we are told to do so.

From Definition I (or from Example 1), you may have noticed that the sine and the cosecant functions are reciprocals. That is,

$$\csc \theta = \frac{1}{\sin \theta} \text{ because } \frac{1}{\sin \theta} = \frac{1}{y/r} = \frac{r}{y} = \csc \theta$$

We can also write this in another form as

$$\sin \theta = \frac{1}{\csc \theta} \text{ because } \frac{1}{\csc \theta} = \frac{1}{r/y} = \frac{y}{r} = \sin \theta$$

From this discussion and from the definition of $\cos \theta$, $\sec \theta$, $\tan \theta$. and $\cot \theta$, it is apparent that $\sec \theta$ is the reciprocal of $\cos \theta$, and $\cot \theta$ is the reciprocal of $\tan \theta$.

EXAMPLE 2 Find sine, cosine, and tangent of 45°.

Solution To apply the definition of sine, cosine, and tangent, we need a point on the terminal side of an angle with a measure of 45° when it is drawn in standard position. A convenient point is (1, 1). The distance from the origin to (1, 1) is $\sqrt{2}$ (see Figure 3).

$$\sin 45° = \frac{1}{\sqrt{2}}$$

$$\cos 45° = \frac{1}{\sqrt{2}}$$

$$\tan 45° = \frac{1}{1} = 1$$

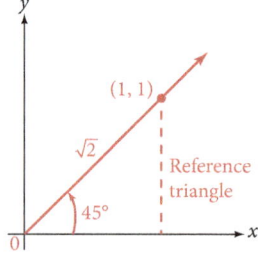

FIGURE 3

Note The triangle formed by drawing the dashed line from the point on the terminal side to the x-axis in Figure 3 is called the **reference triangle**.

EXAMPLE 3   Find the six trigonometric ratios of 270°.

Solution From Figure 4, we see that the terminal side of 270° is along the negative y-axis. Since we are free to choose any point on the terminal side of 270°, we choose the point $(0, -1)$ because it is easy to work with. The distance from the origin down to $(0, -1)$ is $r = 1$.

$$\sin 270° = \frac{y}{r} = \frac{-1}{1} = -1$$

$$\cos 270° = \frac{x}{r} = \frac{0}{1} = 0$$

$$\tan 270° = \frac{y}{x} = \frac{-1}{0} = \text{Undefined}$$

$$\cot 270° = \frac{x}{y} = \frac{0}{-1} = 0$$

$$\sec 270° = \frac{r}{x} = \frac{1}{0} = \text{Undefined}$$

$$\csc 270° = \frac{r}{y} = \frac{1}{-1} = -1$$

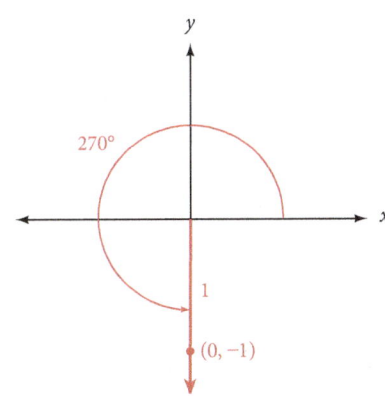

FIGURE 4

Note that tangent and secant are undefined since division by 0 is undefined.

B Algebraic Signs of Trigonometric Functions

The algebraic sign, $+$ or $-$, of each of the six trigonometric functions will depend on the quadrant in which θ terminates. Since trigonometric functions are defined in terms of x, y, and r, where (x, y) is a point on the terminal side of θ, and we know r is always positive, we can look to the algebraic signs, $+$ or $-$, of x and y to determine the signs of the trigonometric functions. In quadrant I, x and y are both positive, so all six trigonometric functions will be positive there also. On the other hand, in quadrant II, only $\sin \theta$ and $\csc \theta$ are positive since they are defined in terms of y and y is positive in quadrant II, while x is negative there. Table 2 shows the signs of all of the trigonometric functions in each of the four quadrants.

Table 2

	For θ in Quadrant			
	QI	QII	QIII	QIV
$\sin \theta = \dfrac{y}{r}$ and $\csc \theta = \dfrac{r}{y}$	$+$	$+$	$-$	$-$
$\cos \theta = \dfrac{x}{r}$ and $\sec \theta = \dfrac{r}{x}$	$+$	$-$	$-$	$+$
$\tan \theta = \dfrac{y}{x}$ and $\cot \theta = \dfrac{x}{y}$	$+$	$-$	$+$	$-$

EXAMPLE 4   If $\sin \theta = -\frac{5}{13}$, and θ terminates in quadrant III, find $\cos \theta$ and $\tan \theta$.

Solution Since $\sin \theta = -\frac{5}{13}$, we know the ratio of y to r, or $\frac{y}{r}$, is $-\frac{5}{13}$. We can let y be -5 and r be 13 and use these values of y and r to find x.

> *Note* We are not saying that if $\frac{y}{r} = -\frac{5}{13}$, then y *must* be -5 and r *must* be 13. We know from algebra that there are many pairs of numbers whose ratio is $-\frac{5}{13}$, not just -5 and 13. For example, y could be -10 and r could be 26, with the ratio $\frac{y}{r} = \frac{-10}{26} = -\frac{5}{13}$. Our definition for sine and cosine, however, indicates we can choose *any* point on the terminal side of θ to find $\sin \theta$ and $\cos \theta$.

To find x we use the fact that $x^2 + y^2 = r^2$:

$$x^2 + y^2 = r^2$$
$$x^2 + (-5)^2 = 13^2$$
$$x^2 + 25 = 169$$
$$x^2 = 144$$
$$x = \pm 12$$

Is x the number -12 or $+12$? Since θ terminates in quadrant III, we know any point on its terminal side will have a negative x-coordinate; therefore, $x = -12$. Using $x = -12$, $y = -5$, and $r = 13$ in our original definition, we have

$$\cos \theta = \frac{x}{r} = \frac{-12}{13} = -\frac{12}{13} \qquad \text{and} \qquad \tan \theta = \frac{y}{x} = \frac{-5}{-12} = \frac{5}{12}$$

Figure 5 is a diagram of the situation.

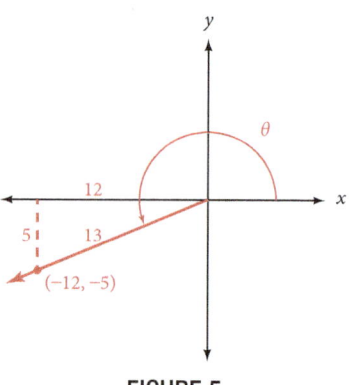

FIGURE 5

EXAMPLE 5   Suppose a Ferris wheel has a radius of 50 feet, with the bottom of the wheel 10 feet above the ground. If the wheel completes one revolution in 2 minutes, find the distance a rider is above the ground when the rider is at point A on the wheel.

Solution As we did in the previous section, we can simplify our work by taking away all the unnecessary items from the diagram, then we superimpose a coordinate system so that its origin is at the center of the circle to obtain a simplified model.

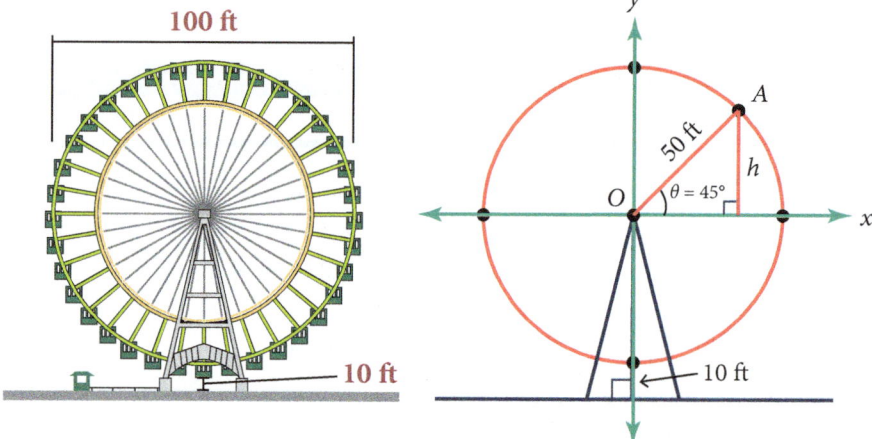

Because angle θ is 45°, and the radius of the wheel is 50 feet, we can find h.

$$\sin 45° = \frac{h}{50}$$
$$h = 50 \sin 45°$$
$$= 50 \cdot \frac{1}{\sqrt{2}}$$
$$\approx 50(0.7071)$$
$$\approx 35.4 \text{ feet}$$

To find the rider's height above the ground, we add 60 to this number.

$$\text{Rider's height} = 60 + 35.4 = 95.4 \text{ feet}$$

Problem Set 1.3

Vocabulary

Use the vocabulary words below to fill in the blanks in the sentences.

rationalized   reciprocals   standard position   Quadrant I

1. Our definition for $\sin \theta$ requires that θ is an angle in _____ .

2. For any angle θ, $\tan \theta$ and $\cot \theta$ will always be _____ .

3. The algebraic sign of all six trigonometric functions will be positive in _____ .

4. When we rewrite $\frac{3}{\sqrt{13}}$ as $\frac{3\sqrt{13}}{13}$, we have _____ the denominator.

Find all six trigonometric functions of θ if the given point is on the terminal side of θ.

1. $(3, 4)$ 2. $(-3, -4)$ 3. $(-5, 12)$ 4. $(-12, 5)$

5. $(-1, -3)$ 6. $(1, -3)$ 7. (a, b) 8. (m, n)

Draw each angle below in standard position, and then locate a convenient point on the terminal side of the angle. Use the coordinates of the point to find sine, cosine, and tangent of the angle.

9. $180°$ 10. $90°$ 11. $135°$ 12. $225°$

13. $-45°$ 14. $-90°$ 15. $0°$ 16. $-135°$

Indicate the quadrants in which the terminal side of θ must lie in order that:

17. $\cos \theta$ is positive   18. $\cos \theta$ is negative

19. $\sin \theta$ is negative   20. $\sin \theta$ is positive

21. $\sin \theta$ is negative and $\tan \theta$ is positive

22. $\sin \theta$ is positive and $\cos \theta$ is negative

23. In which quadrants must the terminal side of θ lie if $\sin \theta$ and $\tan \theta$ are to have the same sign?

24. In which quadrants must the terminal side of θ lie if $\cos \theta$ and $\cot \theta$ are to have the same sign?

Find the remaining trigonometric functions of θ if:

25. $\sin \theta = \dfrac{12}{13}$ and θ terminates in QI

26. $\sin \theta = \dfrac{12}{13}$ and θ terminates in QII

27. $\cos \theta = -\dfrac{1}{2}$ and θ terminates in QII

33

28. $\cos \theta = -\dfrac{\sqrt{3}}{2}$ and θ terminates in QIII

29. $\cos \theta = -\dfrac{3}{5}$ and θ is not in QII

30. $\tan \theta = \dfrac{3}{4}$ and θ is not in QIII

31. $\tan \theta = -\dfrac{4}{3}$ and θ terminates in QIV **32.** $\tan \theta = \dfrac{12}{5}$ and θ terminates in QIII

33. $\sec \theta = \dfrac{5}{4}$ and θ terminates in QIV **34.** $\sec \theta = -\dfrac{3}{2}$ and θ terminates in QIII

35. $\cot \theta = -\dfrac{1}{2}$ and θ terminates in QII **36.** $\cot \theta = \dfrac{3}{2}$ and θ terminates in QIII

37. $\csc \theta = 3$ and θ terminates in QII **38.** $\csc \theta = -\dfrac{4}{3}$ and θ terminates in QIV

39. $\tan \theta = \dfrac{5}{12}$ and $\sin \theta < 0$ **40.** $\cot \theta = -\dfrac{3}{4}$ and $\cos \theta > 0$

41. Find $\sin \theta$ and $\cos \theta$ if the terminal side of θ lies along the line $y = 2x$ in quadrant I.

42. Find $\sin \theta$ and $\cos \theta$ if the terminal side of θ lies along the line $y = 2x$ in quadrant III.

43. Draw $45°$ and $-45°$ in standard position and then show that $\cos(-45°) = \cos 45°$.

44. Draw $45°$ and $-45°$ in standard position and then show that $\sin(-45°) = \sin -45°$.

Find the exact value of the indicated trigonometric function, using the given information.

45. $\tan \theta$ if $\cos \theta = -\dfrac{1}{3}$; terminal side of θ in Quadrant III

46. $\cot \theta$ if $\sin \theta = \dfrac{2}{5}$; terminal side of θ in Quadrant II

47. $\sec \theta$ if $\sin \theta = -0.8$; terminal side of θ in Quadrant IV

48. $\csc \theta$ if $\cos \theta = -0.6$; terminal side of θ in Quadrant II

49. Suppose a Ferris wheel has a radius of 50 feet, with the bottom of the wheel 10 feet above the ground. If the wheel completes one revolution in 2 minutes, complete the table to find the distance a rider is above the ground when the rider is at each of the points given in the table. Round to the nearest tenth of a foot.

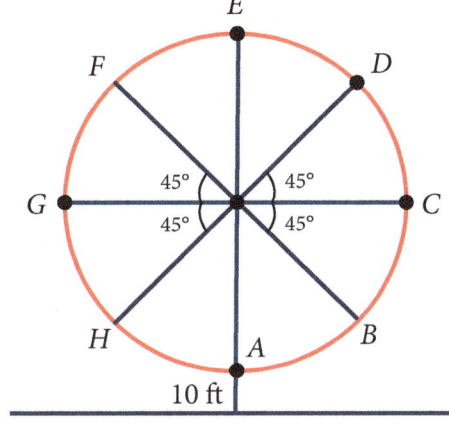

Point on the Wheel	Time in Minutes	Height in Feet
A	0	10
B	0.25	
C	0.5	60
D	0.75	95.4
E	1	
F	1.25	
G	1.5	
H	1.75	
A	2	

50. The diameter of a Ferris wheel is 160 feet, it rotates at 1/2 a revolution every minute, and the bottom of the wheel is 9 feet above the ground. Fill in the table to show the height a rider is above the ground at the given times. Assume the passenger starts the ride at the bottom of the wheel.

Point on the Wheel	Time in Minutes	Height in Feet
A	0	9
B	0.25	
C	0.5	89
D	0.75	145.6
E	1	
F	1.25	
G	1.5	
H	1.75	
A	2	

Getting Ready for the Next Section

These are problems that you must be able to work in order to understand the material in the next section. The problems below are exactly the type of problems you will see in the explanations and examples in the next section.

Simplify.

51. $\dfrac{1}{\frac{1}{2}}$ **52.** $\dfrac{1}{\frac{5}{3}}$ **53.** $\dfrac{1}{\frac{3}{5}}$ **54.** $\dfrac{1}{\frac{4}{3}}$

55. $\dfrac{1}{\frac{x}{r}}$ **56.** $\dfrac{1}{\frac{r}{x}}$ **57.** $\dfrac{1}{\frac{y}{x}}$ **58.** $\dfrac{1}{\frac{x}{y}}$

59. $\dfrac{-\frac{3}{5}}{\frac{4}{5}}$ **60.** $\dfrac{\frac{4}{5}}{-\frac{3}{5}}$ **61.** $\dfrac{-\frac{\sqrt{3}}{2}}{\frac{1}{2}}$ **62.** $\dfrac{-\frac{2}{\sqrt{3}}}{-\frac{1}{\sqrt{3}}}$

63. $\left(\dfrac{3}{5}\right)^2$ **64.** $-\left(\dfrac{1}{2}\right)^3$ **65.** $\sqrt{\dfrac{16}{25}}$ **66.** $\sqrt{\dfrac{4}{9}}$

67. $\sqrt{1-\left(\dfrac{3}{5}\right)^2}$ **68.** $\sqrt{1-\left(\dfrac{1}{2}\right)^2}$

69. $\sqrt{1-\left(\dfrac{\sqrt{3}}{2}\right)^2}$ **70.** $\sqrt{1-\left(-\dfrac{4}{5}\right)^2}$

71. $-\sqrt{1-\left(-\dfrac{3}{5}\right)^2}$ **72.** $-\sqrt{1-\left(\dfrac{12}{13}\right)^2}$

Learning Objectives

A   Find the value of a trigonometric function, given the value of its reciprocal.

B   Find the ratio identity of a given trigonometric function.

C   Solve problems involving Pythagorean identities.

In this section, we will turn our attention to identities. In algebra, statements such as $2x = x + x$, $x^3 = x \cdot x \cdot x$, and $\frac{x}{4x} = \frac{1}{4}$ are called **identities**. They are identities because they are true for all replacements of the variable for which they are defined.

The eight basic **trigonometric identities** are listed in Table 1. As we will see, they are all derived from the definition of the trigonometric functions. Since many of the trigonometric identities have more than one form, we list the basic identity first and then give the most common equivalent forms.

Table 1

	Basic Identities	Common Equivalent Forms
Reciprocal	$\csc \theta = \dfrac{1}{\sin \theta}$	$\sin \theta = \dfrac{1}{\csc \theta}$
	$\sec \theta = \dfrac{1}{\cos \theta}$	$\cos \theta = \dfrac{1}{\sec \theta}$
	$\cot \theta = \dfrac{1}{\tan \theta}$	$\tan \theta = \dfrac{1}{\cot \theta}$
Ratio	$\tan \theta = \dfrac{\sin \theta}{\cos \theta}$	
	$\cot \theta = \dfrac{\cos \theta}{\sin \theta}$	
Pythagorean	$\cos^2\theta + \sin^2\theta = 1$	$\sin^2 \theta = 1 - \cos^2\theta$
	$1 + \tan^2\theta = \sec^2\theta$	$\sin \theta = \pm \sqrt{1 - \cos^2\theta}$
	$1 + \cot^2\theta = \csc^2\theta$	$\cos^2 \theta = 1 - \sin^2\theta$
		$\cos \theta = \pm \sqrt{1 - \sin^2\theta}$

A Reciprocal Identities

Note that, in Table 1, the eight basic identities are grouped in categories. For example, since $\csc \theta = \frac{1}{\sin \theta}$, cosecant and sine must be reciprocals. It is for this reason that we call the identities in this categories **reciprocal identities**.

As we mentioned above, the eight basic identities are all derived from the definition of the six trigonometric functions. To derive the first reciprocal identity, we use the definition of $\sin \theta$ to write

$$\frac{1}{\sin \theta} = \frac{1}{y/r} = \frac{r}{y} = \csc \theta$$

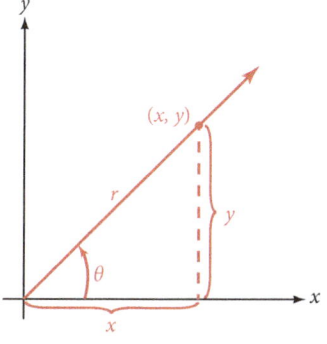

Note that we can write this same relationship between $\sin\theta$ and $\csc\theta$ as

$$\sin\theta = \frac{1}{\csc\theta}$$

because

$$\frac{1}{\csc\theta} = \frac{1}{r/y} = \frac{y}{r} = \sin\theta$$

The first identity we wrote, $\csc\theta = \frac{1}{\sin\theta}$, is the basic identity. The second one, $\sin\theta = \frac{1}{\csc\theta}$, is an equivalent form of the first one.

The other reciprocal identities and their common equivalent forms are derived in a similar manner. Examples 1–6 show how we use the reciprocal identities to find the value of one trigonometric function, given the value of its reciprocal.

EXAMPLES

1. If $\sin\theta = \frac{3}{5}$, then $\csc\theta = \frac{5}{3}$, because $\csc\theta = \dfrac{1}{\sin\theta} = \dfrac{1}{\frac{3}{5}} = \dfrac{5}{3}$.

2. If $\cos\theta = -\dfrac{\sqrt{3}}{2}$, then $\sec\theta = -\dfrac{2}{\sqrt{3}}$. (*Remember:* Reciprocals always have the same algebraic sign.)

3. If $\tan\theta = 2$, then $\cot\theta = \dfrac{1}{2}$.

4. If $\csc\theta = a$, then $\sin\theta = \dfrac{1}{a}$.

5. If $\sec\theta = 1$, then $\cos\theta = 1$.

6. If $\cot\theta = -1$, then $\tan\theta = -1$.

B Ratio Identities

Unlike the reciprocal identities, the ratio identities do not have any common equivalent forms. Here is how we derive the ratio identity for $\tan\theta$:

$$\frac{\sin\theta}{\cos\theta} = \frac{\frac{y}{r}}{\frac{x}{r}} = \frac{y}{x} = \tan\theta$$

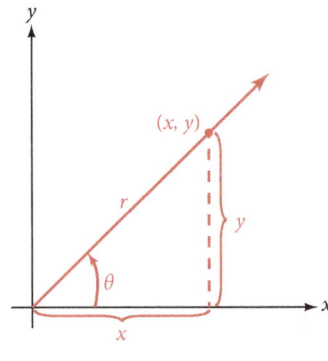

EXAMPLE 7   If $\sin\theta = -\dfrac{3}{5}$ and $\cos\theta = \dfrac{4}{5}$, find $\tan\theta$ and $\cot\theta$.

Solution   Using the ratio identities we have

$$\tan\theta = \frac{\sin\theta}{\cos\theta} = \frac{-\frac{3}{5}}{\frac{4}{5}} = -\frac{3}{4} \qquad \cot\theta = \frac{\cos\theta}{\sin\theta} = \frac{\frac{4}{5}}{-\frac{3}{5}} = -\frac{4}{3}$$

Note that, once we found $\tan\theta$, we could have used a reciprocal identity to find $\cot\theta$:

$$\cot\theta = \frac{1}{\tan\theta} = \frac{1}{-\frac{3}{4}} = -\frac{4}{3}$$

Notation

The notation is $\sin^2\theta$ is a shorthand notation for $(\sin\theta)^2$. It indicates we are to square the number that is the sine of θ.

EXAMPLES

8. If $\sin \theta = \dfrac{3}{5}$, then $\sin^2 \theta = \left(\dfrac{3}{5}\right)^2 = \dfrac{9}{25}$.

9. If $\cos \theta = -\dfrac{1}{2}$, then $\cos^3 \theta = \left(-\dfrac{1}{2}\right)^3 = -\dfrac{1}{8}$.

C Pythagorean Identities

The identity $\cos^2 \theta + \sin^2 \theta = 1$ is called the Pythagorean identity because it is derived from the Pythagorean Theorem. Recall from the definition of $\sin \theta$ and $\cos \theta$ that if (x, y) is a point on the terminal side of θ and r is the distance to (x, y) from the origin, the relationship between x, y, and r is $x^2 + y^2 = r^2$. This relationship comes from the Pythagorean Theorem. Here is how we use it to derive the first Pythagorean identity.

$$x^2 + y^2 = r^2$$

$$\frac{x^2}{r^2} + \frac{y^2}{r^2} = 1 \qquad \textit{Divide each side by } r^2.$$

$$\left(\frac{x}{r}\right)^2 + \left(\frac{y}{r}\right)^2 = 1 \qquad \textit{Property of exponents.}$$

$$(\cos \theta)^2 + (\sin \theta)^2 = 1 \qquad \textit{Definition of } \sin \theta \textit{ and } \cos \theta.$$

$$\cos^2 \theta + \sin^2 \theta = 1 \qquad \textit{Notation}$$

There are four very useful equivalent forms of the first Pythagorean identity. Two of the forms occur when we solve $\cos^2 \theta + \sin^2 \theta = 1$ for $\cos \theta$, while the other two forms are the result of solving for $\sin \theta$. Solving $\cos^2 \theta + \sin^2 \theta = 1$ for $\cos \theta$, we have

$$\cos^2 \theta + \sin^2 \theta = 1$$

$$\cos^2 \theta = 1 - \sin^2 \theta \qquad \textit{Add } -\sin^2 \theta \textit{ to each side.}$$

$$\cos \theta = \pm\sqrt{1 - \sin^2 \theta} \qquad \textit{Take the square root of each side.}$$

Similarly, solving for $\sin \theta$ gives us

$$\sin^2 \theta = 1 - \cos^2 \theta \qquad \text{and} \qquad \sin \theta = \pm\sqrt{1 - \cos^2 \theta}$$

EXAMPLE 10   If $\sin \theta = \dfrac{3}{5}$ and θ terminates in quadrant II, find $\cos \theta$.

Solution We can obtain $\cos \theta$ from $\sin \theta$ by using the identity

$$\cos \theta = \pm\sqrt{1 - \sin^2 \theta}$$

If $\sin \theta = \dfrac{3}{5}$, the identity becomes

$$\cos \theta = \pm\sqrt{1 - \left(\frac{3}{5}\right)^2} \qquad \textit{Substitute } \tfrac{3}{5} \textit{ for } \sin \theta.$$

$$= \pm\sqrt{1 - \frac{9}{25}} \qquad \textit{Square } \tfrac{3}{5} \textit{ to get } \tfrac{9}{25}$$

$$= \pm\sqrt{\frac{16}{25}} \qquad \textit{Subtract.}$$

$$= \pm\frac{4}{5} \qquad \textit{Take the square root of the numerator and denominator separately.}$$

Now we know that $\cos \theta$ is either $+\frac{4}{5}$ or $-\frac{4}{5}$. Looking back to the original statement of the problem, however, we see that θ terminates in quadrant II; therefore, $\cos \theta$ must be negative.

$$\cos \theta = -\frac{4}{5}$$

EXAMPLE 11   If $\cos \theta = \frac{1}{2}$ and θ terminates in quadrant IV, find the remaining trigonometric ratios for θ.

Solution The first, and easiest, ratio to find is $\sec \theta$, because it is the reciprocal of $\cos \theta$.

$$\sec \theta = \frac{1}{\cos \theta} = \frac{1}{\frac{1}{2}} = 2$$

Next, we find $\sin \theta$. Since θ terminates in QIV, $\sin \theta$ will be negative. Using one of the equivalent forms of the Pythagorean identity, we have

$$\sin \theta = -\sqrt{1 - \cos^2 \theta} \qquad \text{Negative sign because } \theta \text{ is in QIV.}$$

$$= -\sqrt{1 - \left(\frac{1}{2}\right)^2} \qquad \text{Substitute } \frac{1}{2} \text{ for } \cos \theta.$$

$$= -\sqrt{1 - \frac{1}{4}} \qquad \text{Square } \frac{1}{2} \text{ to get } \frac{1}{4}$$

$$= -\sqrt{\frac{3}{4}} \qquad \text{Subtract.}$$

$$= -\frac{\sqrt{3}}{2} \qquad \text{Take the square root of the numerator and denominator separately.}$$

Now that we have $\sin \theta$ and $\cos \theta$, we can find $\tan \theta$ by using a ratio identity.

$$\tan \theta = \frac{\sin \theta}{\cos \theta} = \frac{-\frac{\sqrt{3}}{2}}{\frac{1}{2}} = -\sqrt{3}$$

Next, $\cot \theta$ and $\csc \theta$ are the reciprocals of $\tan \theta$ and $\sin \theta$, respectively. Therefore,

$$\cot \theta = \frac{1}{\tan \theta} = -\frac{1}{\sqrt{3}} \qquad \csc \theta = \frac{1}{\sin \theta} = -\frac{2}{\sqrt{3}}$$

Here are all six ratios together:

$$\sin \theta = -\frac{\sqrt{3}}{2} \qquad\qquad \csc \theta = -\frac{2}{\sqrt{3}}$$

$$\cos \theta = \frac{1}{2} \qquad\qquad \sec \theta = 2$$

$$\tan \theta = -\sqrt{3} \qquad\qquad \cot \theta = -\frac{1}{\sqrt{3}}$$

As a final note, we should mention that the six basic identities we have derived here, along with their equivalent forms, are very important in the study of trigonometry. It is essential that you memorize them. It may be a good idea to practice writing them from memory until you can write each of the six, and their equivalent forms, perfectly. As time goes by, we will increase our list of identities, so you will want to keep up with them as we go along.

Problem Set 1.4

Vocabulary

Use the vocabulary words below to fill in the blanks in the sentences.

ratio   reciprocal   Pythagorean   identity

1. The statement $\tan \theta = \frac{\sin \theta}{\cos \theta}$ is called an _____ because it is true for all replacements of the variable for which the expressions are defined.

2. The statement $\sin^2 \theta + \cos^2 \theta = 1$ is a _____ identity.

3. The statement $\sin \theta = \frac{1}{\csc \theta}$ is a _____ identity.

4. The statement $\cot \theta = \frac{\cos \theta}{\sin \theta}$ is a _____ identity.

Use the reciprocal identities in the following problems.

1. If $\sin \theta = \frac{4}{5}$, find $\csc \theta$.

2. If $\cos \theta = \frac{\sqrt{3}}{2}$, find $\sec \theta$.

3. If $\sec \theta = -2$, find $\cos \theta$.

4. If $\csc \theta = -\frac{13}{12}$, find $\sin \theta$.

5. If $\tan \theta = a(a \neq 0)$, find $\cot \theta$.

6. If $\cot \theta = -b(b \neq 0)$, find $\tan \theta$.

Use a ratio identity to find $\tan \theta$ if:

7. $\sin \theta = \frac{3}{5}$ and $\cos \theta = -\frac{4}{5}$

8. $\sin \theta = \frac{2}{\sqrt{5}}$ and $\cos \theta = \frac{1}{\sqrt{5}}$

Use a ratio identity to find $\cot \theta$ if:

9. $\sin \theta = -\frac{5}{13}$ and $\cos \theta = -\frac{12}{13}$

10. $\sin \theta = \frac{2}{\sqrt{13}}$ and $\cos \theta = \frac{3}{\sqrt{13}}$

Use the equivalent forms of the Pythagorean identity on Problems 11–20.

11. Find $\sin \theta$ if $\cos \theta = \frac{3}{5}$ and θ terminates in QI.

12. Find $\sin \theta$ if $\cos \theta = \frac{5}{13}$ and θ terminates in QI.

13. Find $\cos \theta$ if $\sin \theta = \frac{1}{3}$ and θ terminates in QII.

14. Find $\cos \theta$ if $\sin \theta = \frac{\sqrt{3}}{2}$ and θ terminates in QII.

15. If $\sin \theta = -\frac{4}{5}$ and θ terminates in QIII, find $\cos \theta$.

16. If $\sin \theta = -\dfrac{4}{5}$ and terminates in QIV, find $\cos \theta$.

17. If $\cos \theta = \dfrac{\sqrt{3}}{2}$ and θ terminates in QI, find $\sin \theta$.

18. If $\cos \theta = -\dfrac{1}{2}$ and θ terminates in QII, find $\sin \theta$.

19. If $\sin \theta = \dfrac{1}{\sqrt{5}}$ and $\theta \in$ QII, find $\cos \theta$.

20. If $\cos \theta = -\dfrac{1}{\sqrt{10}}$ and $\theta \in$ QIII, find $\sin \theta$.

Find the remaining trigonometric ratios of θ if:

21. $\cos \theta = \dfrac{12}{13}$ and θ terminates in QI

22. $\cos \theta = \dfrac{2}{\sqrt{13}}$ and θ terminates in QIV

23. $\cos \theta = -\dfrac{1}{3}$ and $\theta \in$ QIII

24. $\cos \theta = -\dfrac{3}{\sqrt{13}}$ and $\theta \in$ QII

25. $\sin \theta = \dfrac{5}{13}$ and θ terminates in QI

26. $\sin \theta = \dfrac{3}{\sqrt{10}}$ and θ terminates in QII

27. $\sin \theta = -\dfrac{1}{2}$ and $\theta \in$ QIV

28. $\sin \theta = -\dfrac{3}{4}$ and $\theta \in$ QIII

29. $\tan \theta = \dfrac{3}{4}$ and θ terminates in QI

30. $\tan \theta = 2$ and θ terminates in QIII

31. $\tan \theta = -3$ and $\theta \in$ QII

32. $\tan \theta = -\dfrac{2}{3}$ and $\theta \in$ QIV

33. $\sec \theta = 2$ and θ terminates in QI

34. $\sec \theta = 5$ and θ terminates in QIV

35. $\sec \theta = -3$ and $\theta \in$ QIII

36. $\sec \theta = -4$ and $\theta \in$ QII

37. $\csc \theta = \dfrac{5}{4}$ and θ terminates in QII

38. $\csc \theta = \dfrac{5}{3}$ and θ terminates in QI

39. $\csc \theta = -\dfrac{13}{5}$ and $\theta \in$ QIII   **40.** $\csc \theta = -\dfrac{4}{3}$ and $\theta \in$ QIV

41. $\cot \theta = \dfrac{1}{2}$ and θ terminates in QIII   **42.** $\cot \theta = \dfrac{2}{3}$ and θ terminates in QI

43. $\cot \theta = -2$ and $\theta \in$ QII   **44.** $\cot \theta = -\dfrac{1}{4}$ and $\theta \in$ QIV

Find the required trigonometric ratio of θ using the given information.

45. $\tan \theta$; $\sin \theta = \dfrac{2}{3}$ and $\theta \in$ QI   **46.** $\tan \theta$; $\cos \theta = \dfrac{3}{4}$ and $\theta \in$ QI

47. $\cot \theta$; $\sin \theta = -\dfrac{4}{5}$ and $\theta \in$ QIII   **48.** $\cot \theta$; $\cos \theta = -\dfrac{3}{5}$ and $\theta \in$ QII

49. $\sec \theta$; $\tan \theta = \dfrac{5}{2}$ and $\theta \in$ QI   **50.** $\sec \theta$; $\tan \theta = -\dfrac{5}{3}$ and $\theta \in$ QII

51. $\csc \theta$; $\cot \theta = -\dfrac{3}{4}$ and $\theta \in$ QIV   **52.** $\csc \theta$; $\cot \theta = -3$ and $\theta \in$ QII

53. $\cos \theta$; $\csc \theta = \dfrac{5}{4}$ and $\theta \in$ QII   **54.** $\cos \theta$; $\csc \theta = -2$ and $\theta \in$ QIV

55. $\sin \theta$; $\sec \theta = 3$ and $\theta \in$ QIV   **56.** $\sin \theta$; $\sec \theta = -\dfrac{3}{2}$ and $\theta \in$ QII

57. The line $y = 3x$ passes through the points $(0, 0)$ and $(1, 3)$. Find its slope.

58. Suppose the angle formed by the line $y = 3x$ and the positive x-axis is θ. Find the tangent of θ.

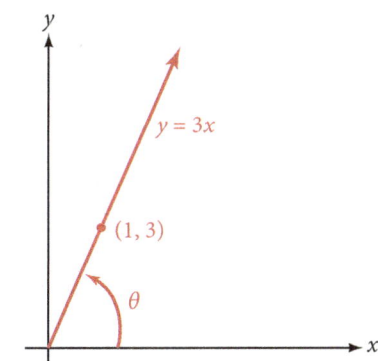

59. Find the slope of the line $y = mx$. (It passes through the origin and the point $(1, m)$.)

60. Find $\tan \theta$ if θ is the angle formed by the line $y = mx$ and the positive x-axis.

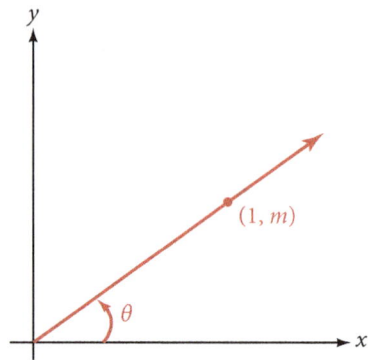

Getting Ready for the Next Section

These are problems that you must be able to work in order to understand the material in the next section. The problems below are exactly the type of problems you will see in the explanations and examples in the next section.

61. Solve $\sin^2 \theta + \cos^2 \theta = 1$ for $\sin \theta$.   **62.** Solve $1 + \tan^2 \theta = \sec^2 \theta$ for $\tan \theta$.

63. Add $\dfrac{1}{2} + \dfrac{1}{3}$ **64.** Add $\dfrac{1}{x} + \dfrac{1}{2}$

65. Multiply $(x + 2)(x - 5)$ **66.** Multiply $(x - 2)(x + 5)$

67. Factor $x^2 - 4$ **68.** Factor $4 - x^2$

Learning Objectives

A   Write a trigonometric function in the terms of another trigonometric function.

B   Simplify algebraic expressions involving trigonometric functions.

C   Prove trigonometric identities using the six basic identities and algebra.

The topics we will cover in this section are an extension of the work we did with identities in Section 1.4. The first topic involves writing any of the six trigonometric functions in terms of any of the others. Let's look at an example.

A   Writing Trigonometric Functions in Terms of Another

EXAMPLE 1   Write $\tan \theta$ in terms of $\sin \theta$.

Solution   When we say we want $\tan \theta$ written in terms of $\sin \theta$, we mean that we want to write an expression that is equivalent to $\tan \theta$ but involves no trigonometric function other than $\sin \theta$. Let's begin by using a ratio identity to write $\tan \theta$ in terms of $\sin \theta$ and $\cos \theta$:

$$\tan \theta = \frac{\sin \theta}{\cos \theta}$$

Now we need to replace $\cos \theta$ with an expression involving only $\sin \theta$. Since $\cos = \pm\sqrt{1 - \sin^2\theta}$, we have

$$\tan \theta = \frac{\sin \theta}{\cos \theta}$$

$$= \frac{\sin \theta}{\pm\sqrt{1 - \sin^2\theta}}$$

$$= \pm\frac{\sin \theta}{\sqrt{1 - \sin^2\theta}}$$

This last expression is equivalent to $\tan \theta$ and is written in terms of $\sin \theta$ only. (In a problem like this it is okay to include numbers and algebraic symbols with $\sin \theta$—just no other trigonometric functions.)

Here is another example. This one involves simplification of the product of two trigonometric functions.

EXAMPLE 2   Write $\sec \theta \tan \theta$ in terms of $\sin \theta$ and $\cos \theta$, then simplify.

Solution   Since $\sec \theta = \frac{1}{\cos \theta}$ and $\tan \theta = \frac{\sin \theta}{\cos \theta}$, we have

$$\sec \theta \tan \theta = \frac{1}{\cos \theta} \cdot \frac{\sin \theta}{\cos \theta}$$

$$= \frac{\sin \theta}{\cos^2 \theta}$$

> **Note**   The notation $\sec \theta \tan \theta$ means $\sec \theta \cdot \tan \theta$.

If Example 2 asked to write $\sec \theta \tan \theta$ in terms of $\sin \theta$ solely, we would use the identity $\cos^2 \theta = 1 - \sin^2 \theta$ to write $\sec \theta \tan \theta = \frac{\sin \theta}{1 - \sin^2 \theta}$.

The next examples show how we manipulate trigonometric expressions using algebraic techniques.

B Algebra with Expressions Involving Trigonometric Functions

EXAMPLE 3 Add $\dfrac{1}{\sin \theta} + \dfrac{1}{\cos \theta}$.

Solution We can add these two expressions in the same way we would add $\frac{1}{3}$ and $\frac{1}{4}$, by first finding the least common denominator, and then writing each expression again with the LCD for its denominator.

$$\frac{1}{\sin \theta} + \frac{1}{\cos \theta} = \frac{1}{\sin \theta} \cdot \frac{\cos \theta}{\cos \theta} + \frac{1}{\cos \theta} \cdot \frac{\sin \theta}{\sin \theta} \qquad \text{The LCD is } \sin \theta \cos \theta.$$

$$= \frac{\cos \theta}{\sin \theta \cos \theta} + \frac{\sin \theta}{\cos \theta \sin \theta}$$

$$= \frac{\cos \theta + \sin \theta}{\sin \theta \cos \theta}$$

EXAMPLE 4 Multiply $(\sin \theta + 2)(\sin \theta - 5)$.

Solution We multiply these two expressions in the same way we would multiply $(x + 2)(x - 5)$.

$$\overset{\text{F} \qquad \text{O} \qquad \text{I} \qquad \text{L}}{(\sin \theta + 2)(\sin \theta - 5) = \sin \theta \sin \theta - 5 \sin \theta + 2 \sin \theta - 10}$$

$$= \sin^2 \theta - 3 \sin \theta - 10$$

As you can see in Example 4, we treat $\sin \theta$ in that example the same way we treat x in the expression $(x + 2)(x - 5)$. The next two examples illustrate this concept further.

EXAMPLE 5 Factor $9 - x^2$.

Solution The expression has the form of the difference of two squares, $a^2 - b^2$, which factors as

$$a^2 - b^2 = (a + b)(a - b)$$

Therefore, our expression factors in the same way.

$$9 - x^2 = (3 + x)(3 - x)$$

EXAMPLE 6 Factor $9 - \sin^2 \theta$.

Solution Our expression has the form of the difference of two squares because it can be written as

$$3^2 - (\sin \theta)^2$$

Therefore, it factors like this:

$$9 - \sin^2 \theta = 3^2 - (\sin \theta)^2$$
$$= (3 + \sin \theta)(3 - \sin \theta) \quad \text{Factor}$$

EXAMPLE 7   Show that the following statement is true by transforming the left side into the right side.
$$\cos \theta \tan \theta = \sin \theta$$

Solution We begin by writing everything on the left side in terms of $\sin \theta$ and $\cos \theta$.

$$\cos \theta \tan \theta = \cos \theta \cdot \frac{\sin \theta}{\cos \theta}$$
$$= \frac{\cos \theta \sin \theta}{\cos \theta}$$
$$= \sin \theta \qquad \text{\textcolor{gray}{Divide out the $\cos \theta$ common to the numerator and denominator.}}$$

Since we have succeeded in transforming the left side into the right side, we have shown that the statement $\cos \theta \tan \theta = \sin \theta$ is an identity.

> **Note** You may be wondering at this point, what good all of this is. As we progress through the book, you will see that identities play a very important part in trigonometry (and also in classes for which trigonometry is a prerequisite). This is an introduction to proving identities. In Chapter 5, we will introduce a more formal method of proving identities. For now, we are simply practicing changing trigonometric expressions into other, equivalent, trigonometric expressions.

EXAMPLE 8   Prove the identity
$$(\sin \theta + \cos \theta)^2 = 1 + 2 \sin \theta \cos \theta$$

Solution Let's agree to prove the identities in this section, and the problem set that follows, by transforming the left side into the right side. In this case, we begin by expanding $(\sin + \cos)^2$. (Remember from algebra, $(a + b)^2 = a^2 + 2ab + b^2$.)

$$(\sin \theta + \cos \theta)^2 = \sin^2\theta + 2 \sin \theta \cos \theta + \cos^2\theta$$

Now we can rearrange the terms on the right side to get $\sin^2\theta$ and $\cos^2\theta$ together.

$$= (\sin^2\theta + \cos^2\theta) + 2 \sin \theta \cos \theta$$
$$= 1 + 2 \sin \theta \cos \theta$$

As a final note, we should mention that the ability to prove identities in trigonometry is not always obtained immediately. It usually requires a lot of practice. The more you work at it, the better you will become at it. In the meantime, if you are having trouble, check first to see that you have memorized the six basic identities—reciprocal, ratio, Pythagorean, and their equivalent forms, as given in Section 1.4.

EXAMPLE 9   Simplify the expression $\sqrt{x^2 + 4}$ as much as possible after substituting $2 \tan \theta$ for x. Assume $0° < \theta < 90°$.

Solution Our goal is to write the expression $\sqrt{x^2 + 4}$ without a square root by first making the substitution $x = 2 \tan \theta$.

If $\qquad\qquad\qquad x = 2 \tan \theta$

then the expression $\qquad \sqrt{x^2 + 4}$

becomes $\qquad\qquad \sqrt{(2 \tan \theta)^2 + 4} = \sqrt{4 \tan^2 \theta + 4}$
$$= \sqrt{4(\tan^2 \theta + 1)}$$
$$= \sqrt{4 \sec^2 \theta}$$
$$= 2 |\sec \theta|$$

Because of the restriction $0° < \theta < 90°$, we know that $\sec \theta$ is positive. Without the restriction, we would need absolute value symbols and the answer would be $2 |\sec \theta|$. Remember in algebra, $\sqrt{a^2} = a$ only when a is positive or 0. If it is possible that a is negative, then $\sqrt{a^2} = |a|$.

Problem Set 1.5

Write each of the following in terms of $\sin \theta$ and $\cos \theta$, and then simplify if possible. Leave all answers in terms of sine and cosine.

1. $\csc \theta \cot \theta$

2. $\sec \theta \cot \theta$

3. $\csc \theta \tan \theta$

4. $\sec \theta \tan \theta \csc \theta$

5. $\dfrac{\sec \theta}{\csc \theta}$

6. $\dfrac{\csc \theta}{\sec \theta}$

7. $\dfrac{\sin \theta}{\csc \theta}$

8. $\dfrac{\cos \theta}{\sec \theta}$

9. $\tan \theta + \sec \theta$

10. $\cot \theta - \csc \theta$

11. $\sin \theta \cot \theta + \cos \theta$

12. $\cos \theta \tan \theta + \sin \theta$

Add and subtract as indicated. Then simplify your answers if possible. Leave all answers in terms of $\sin \theta$ and/or $\cos \theta$.

13. $\dfrac{\sin \theta}{\cos \theta} + \dfrac{1}{\sin \theta}$

14. $\dfrac{\cos \theta}{\sin \theta} + \dfrac{\sin \theta}{\cos \theta}$

15. $\dfrac{1}{\sin \theta} - \dfrac{1}{\cos \theta}$

16. $\dfrac{1}{\cos \theta} - \dfrac{1}{\sin \theta}$

17. $\sin \theta + \dfrac{1}{\cos \theta}$

18. $\cos \theta + \dfrac{1}{\sin \theta}$

19. $\dfrac{1}{\sin \theta} - \sin \theta$

20. $\dfrac{1}{\cos \theta} - \cos \theta$

Multiply.

21. $(\sin \theta + 4)(\sin \theta + 3)$

22. $(\cos \theta + 2)(\cos \theta - 5)$

23. $(2 \cos \theta + 3)(4 \cos \theta - 5)$

24. $(3 \sin \theta - 2)(5 \sin \theta - 4)$

25. $(1 - \sin \theta)(1 + \sin \theta)$

26. $(1 - \cos \theta)(1 + \cos \theta)$

27. $(1 - \tan \theta)(1 + \tan \theta)$

28. $(1 - \cot \theta)(1 + \cot \theta)$

29. $(\sin \theta - \cos \theta)^2$

30. $(\cos \theta + \sin \theta)^2$

31. $(\sin \theta - 4)^2$

32. $(\cos \theta - 2)^2$

Factor.

33. $x^2 - 25$

34. $25 - x^2$

35. $1 - x^2$

36. $x^2 - 1$

37. $\sin^2 \theta - 25$

38. $25 - \sin^2 \theta$

39. $1 - \cos^2 \theta$

40. $\cos^2 \theta - 1$

41. $\sin^2 \theta - \cos^2 \theta$

42. $\cos^2 \theta - \sin^2 \theta$

Show that each of the following statements is an identity by transforming the left side of each one into the right side:

43. $\cos\theta\tan\theta = \sin\theta$

44. $\sin\theta\cot\theta = \cos\theta$

45. $\sin\theta\sec\theta\cot\theta = 1$

46. $\cos\theta\csc\theta\tan\theta = 1$

47. $\dfrac{\csc\theta}{\cot\theta} = \sec\theta$

48. $\dfrac{\sec\theta}{\tan\theta} = \csc\theta$

49. $\dfrac{\sec\theta}{\csc\theta} = \tan\theta$

50. $\dfrac{\csc\theta}{\sec\theta} = \cot\theta$

51. $\sin\theta\tan\theta + \cos\theta = \sec\theta$

52. $\cos\theta\cot\theta + \sin\theta = \csc\theta$

53. $\tan\theta + \cot\theta = \sec\theta\csc\theta$

54. $\tan^2\theta + 1 = \sec^2\theta$

55. $\csc\theta - \sin\theta = \dfrac{\cos^2\theta}{\sin\theta}$

56. $\sec\theta - \cos\theta = \dfrac{\sin^2\theta}{\cos\theta}$

57. $1 - \sin^2\theta = \cos^2\theta$

58. $1 - \cos^2\theta = \sin^2\theta$

59. $(1 - \cos\theta)(1 + \cos\theta) = \sin^2\theta$

60. $(1 + \sin\theta)(1 - \sin\theta) = \cos^2\theta$

61. $(\sin\theta - \cos\theta)^2 - 1 = -2\sin\theta\cos\theta$

62. $(\cos\theta + \sin\theta)^2 - 1 = 2\sin\theta\cos\theta$

63. $\sin\theta(\sec\theta + \csc\theta) = \tan\theta + 1$

64. $\sec\theta(\sin\theta + \cos\theta) = \tan\theta + 1$

In calculus, it is often desired to replace radical expressions with trigonometric expressions. The next group of exercises illustrates this. Simplify the given expression as much as possible by making the given substitution. Assume $0 < \theta < 90°$.

65. $\sqrt{x^2 + 9}$; $x = 3\tan\theta$

66. $\sqrt{x^2 + 16}$; $x = 4\tan\theta$

67. $\sqrt{36 - x^2}$; $x = 6\sin\theta$

68. $\sqrt{100 - x^2}$; $x = 10\sin\theta$

69. $\sqrt{x^2 - 25}$; $x = 5\sec\theta$

70. $\sqrt{x^2 - 49}$; $x = 7\sec\theta$

71. $\sqrt{4x^2 + 25}$; $x = \dfrac{5}{2}\tan\theta$

72. $\sqrt{9x^2 + 1}$; $x = \dfrac{1}{3}\tan\theta$

73. $\sqrt{16 - 9x^2}$; $x = \dfrac{4}{3}\sin\theta$

74. $\sqrt{10 - x^2}$; $x = \sqrt{10}\sin\theta$

75. $\sqrt{4x^2 - 5}$; $x = \dfrac{\sqrt{5}}{2}\sec\theta$

76. $\sqrt{25x^2 - 16}$; $x = \dfrac{4}{5}\sec\theta$

Spotlight on Success

Julieta, Student Instructor

Success is no accident. It is hard work, perseverance, learning, studying, sacrifice, and most of all, love of what you are doing or learning to do. —Pelé

Success really is no accident, nor is it something that happens overnight. Sure you may be sitting there wondering why you don't understand a certain lesson or topic, but you are not alone. There are many others who are sitting in your exact position. Throughout my first year in college (and more specifically in Calculus I) I learned that it is normal for any student to feel stumped every now and then. The students who do well are the ones who keep working, even when they are confused.

Pelé wasn't just born with all that legendary talent. It took dedication and hard work as well. Don't ever feel bad because there's something you don't understand—it's not worth it. Stick with it 100% and just keep working problems; I'm sure you'll be successful with whatever you set your mind to achieve in this course.

Examples: We will use the margin to give examples of the topics being reviewed, whenever it is appropriate.

The number in brackets next to each heading indicates the section in which that topic is discussed.

Angles [1.1]

1.

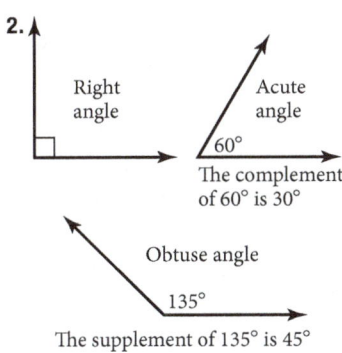

An angle is formed by two rays with a common endpoint. The common endpoint is called the vertex of the angle, and the rays are called the sides of the angle. If we think of an angle as being formed by rotating the initial side about the vertex to the terminal side, then a counterclockwise rotation gives a positive angle, and a clockwise rotation gives a negative angle.

Degree Measure [1.1]

2.

There are 360° in a full rotation. This means 1° is $\frac{1}{360}$ of a full rotation. An angle that measures 90° is a right angle. An angle with measure between 0° and 90° is an acute angle, while an angle with measure between 90° and 180° is an obtuse angle. If the sum of two angles is 90°, then the two angles are called complementary angles. If the sum of two angles is 180°, the angles are called supplementary angles.

Distance Formula [1.1]

The distance r between the points (x_1, y_1) and (x_2, y_2) is given by the formula

$$r = \sqrt{(x_1 - x_2)^2 + (y_2 - y_1)^2}$$

Pythagorean Theorem [1.1]

In any right triangle, the square of the length of the longest side (the hypotenuse) is equal to the sum of the squares of the lengths of the other two sides (legs).

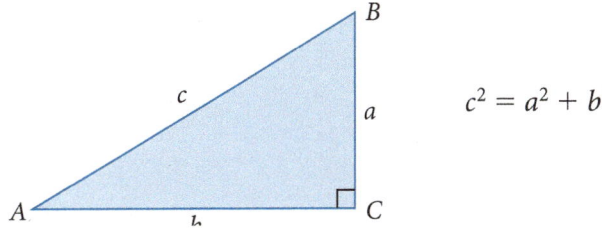

$$c^2 = a^2 + b^2$$

Standard Position for Angles [1.2]

3. 135° in standard position is:

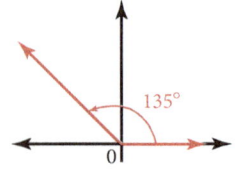

An angle is said to be in standard position if its vertex is at the origin and its initial side is along the positive x-axis.

Trigonometric Functions [1.3]

4. If $(-3, 4)$ is on the terminal side of θ, then

$$r = \sqrt{9 + 16} = 5$$

and

$\sin \theta = \dfrac{4}{5}$ $\csc \theta = \dfrac{5}{4}$

$\cos \theta = -\dfrac{3}{5}$ $\sec \theta = -\dfrac{5}{3}$

$\tan \theta = -\dfrac{4}{3}$ $\cot \theta = -\dfrac{3}{4}$

If θ is an angle in standard position and (x, y) is any point on the terminal side of θ (other than the origin), then

$\sin \theta = \dfrac{y}{r}$ $\csc \theta = \dfrac{r}{y}$

$\cos \theta = \dfrac{x}{r}$ $\sec \theta = \dfrac{r}{x}$

$\tan \theta = \dfrac{y}{x}$ $\cot \theta = \dfrac{x}{y}$

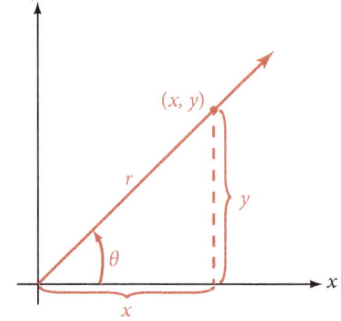

where $x^2 + y^2 = r^2$, or $r = \sqrt{x^2 + y^2}$. That is, r is the distance from the origin to (x, y).

Signs of the Trigonometric Functions [1.3]

5. If $\sin \theta > 0$ and $\cos \theta > 0$, then θ must terminate in QI.
If $\sin \theta > 0$ and $\cos \theta < 0$, then θ must terminate in QII.
If $\sin \theta < 0$ and $\cos \theta < 0$, then θ must terminate in QIII.
If $\sin \theta < 0$ and $\cos \theta > 0$, then θ must terminate in QIV.

The algebraic signs, $+$ or $-$, of the six trigonometric functions depend on the quadrant in which θ terminates.

	QI	QII	QIII	QIV
$\sin \theta$ and $\csc \theta$	$+$	$+$	$-$	$-$
$\cos \theta$ and $\sec \theta$	$+$	$-$	$-$	$+$
$\tan \theta$ and $\cot \theta$	$+$	$-$	$+$	$-$

Basic Identities [1.4]

6. If $\sin \theta = \frac{1}{2}$ and θ terminates in QI, then

$$\cos \theta = \sqrt{1 - \sin^2 \theta}$$

$$= \sqrt{1 - \frac{1}{4}}$$

$$= \frac{\sqrt{3}}{2}$$

$$\tan \theta = \frac{\sin \theta}{\cos \theta}$$

$$= \frac{\frac{1}{2}}{\frac{\sqrt{3}}{2}} = \frac{1}{\sqrt{3}}$$

$$\cot \theta = \frac{1}{\tan \theta} = \sqrt{3}$$

$$\sec \theta = \frac{1}{\cos \theta} = \frac{2}{\sqrt{3}}$$

$$\csc \theta = \frac{1}{\sin \theta} = 2$$

	Basic Identities	Common Equivalent Forms
Reciprocal	$\csc \theta = \dfrac{1}{\sin \theta}$	$\sin \theta = \dfrac{1}{\csc \theta}$
	$\sec \theta = \dfrac{1}{\cos \theta}$	$\cos \theta = \dfrac{1}{\sec \theta}$
	$\cot \theta = \dfrac{1}{\tan \theta}$	$\tan \theta = \dfrac{1}{\cot \theta}$
Ratio	$\tan \theta = \dfrac{\sin \theta}{\cos \theta}$	
	$\cot \theta = \dfrac{\cos \theta}{\sin \theta}$	
Pythagorean	$\cos^2\theta + \sin^2\theta = 1$	$\sin^2\theta = 1 - \cos^2\theta$
	$1 + \tan^2\theta = \sec^2\theta$	$\sin \theta = \pm\sqrt{1 - \cos^2 - \theta}$
	$1 + \cot^2\theta = \csc^2\theta$	$\cos^2\theta = 1 - \sin^2\theta$
		$\cos \theta = \pm\sqrt{1 - \sin^2\theta}$

1. Find the other two sides of a 30°–60°–90° triangle if the shortest side is 5. [1.1]

2. Find the other two sides of a 45°–45°–90° triangle if the longest side is 12. [1.1]

3. Find the six trigonometric functions of θ if the point $(4, -3)$ is on the terminal side of θ. [1.2]

4. In which quadrant must θ lie if $\sin \theta < 0$ and $\tan \theta > 0$? [1.2]

5. Find the six trigonometric functions of θ if the point $(-\sqrt{3}, 1)$ lies on the terminal side of θ, and θ terminates in QII. [1.2]

Find the exact value for each of the following: [1.3]

6. $\cos 240°$

7. $\sin 210°$

Solve for x in each of the following right triangles:

8.

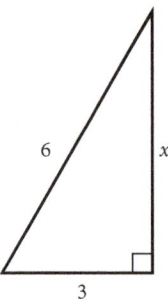

9.

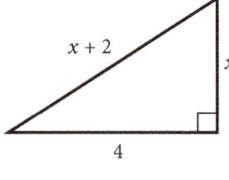

Use the information in the diagram below to work Problem 10 and 11. If rounding is necessary, round to the nearest tenth.

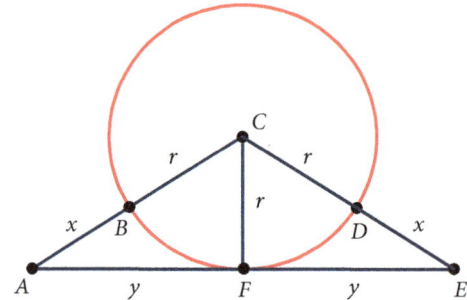

10. Find r if $x = 5$ and $y = 10$.

11. Find r if $x = 10$ and $y = 30$.

Graph each of the following lines:

12. $2x + 3y = 6$

13. $y = 3x - 1$

14. Find the distance between the points $(4, -2)$ and $(-1, 10)$.

15. Find the distance from the origin to the point (a, b).

16. Find x so that the distance between $(-2, 3)$ and $(x, 1)$ is $\sqrt{13}$.

Find $\sin \theta$, $\cos \theta$, and $\tan \theta$ for each of the following values of θ:

17. $90°$

18. $-45°$

In which quadrant will θ lie if

19. $\sin \theta < 0$ and $\cos \theta > 0$

20. $\csc \theta > 0$ and $\cos \theta < 0$

Find all six trigonometric functions for θ if the given point lies on the terminal side of θ in standard position:

21. $(-6, 8)$

22. $(-3, -1)$

Find the remaining trigonometric functions of θ if

23. $\sin \theta = \dfrac{1}{2}$ and terminates θ in QII

24. $\tan \theta = \dfrac{12}{5}$ and terminates θ in QIII

25. Find $\sin \theta$ and $\cos \theta$ if the terminal side of θ lies along the line $y = -2x$ in quadrant IV.

26. If $\sin \theta = -\dfrac{3}{4}$, find $\csc \theta$.

If $\sin \theta = \dfrac{1}{3}$ with θ in QI, find each of the following:

27. $\cos \theta$ and $\sin \theta$

28. $\tan \theta$ and $\cot \theta$

29. $1 + \cot^2 \theta$ and $\csc^2 \theta$

30. $1 + \tan^2 \theta$ and $\sec^2 \theta$

31. $1 + 27 \sin^3 \theta$

32. $2 \csc^3 \theta - 1$

33. Express $\sec \theta \tan \theta$ in terms of $\cos \theta$.   **34.** Multiply $(\sin \theta + 3)(\sin \theta - 7)$.

35. Expand and simplify $(\cos \theta - \sin \theta)^2$.

36. Subtract $\dfrac{1}{\sin \theta} - \sin \theta$

Show that each of the following statements is an identity by transforming the left side of each one into the right side:

37. $\sin \theta \cot \theta \sec \theta = 1$

38. $\dfrac{\cot \theta}{\csc \theta} = \cos \theta$

39. $\cot \theta + \tan \theta = \csc \theta \sec \theta$

40. $\cos \theta(\csc \theta + \sec \theta) = \cot \theta + 1$

41. $(1 - \sin \theta)(1 + \sin \theta) = \cos^2\theta$

42. $\sin \theta(\csc \theta + \cot \theta) = 1 + \cos \theta$

43. Make the substitution $x = 7 \tan \theta$ to simplify the expression $\sqrt{x^2 + 49}$. Assume $0 < \theta < 90°$.

44. Make the substitution $x = 5 \sin \theta$ to simplify the expression $\sqrt{25 - x^2}$. Assume $0 < \theta < 90°$.

45. Make the substitution $x = \frac{1}{2} \sec \theta$ to simplify the expression $\sqrt{4x^2 - 1}$. Assume $0 < \theta < 90°$.

46. Make the substitution $x = \frac{3}{8} \tan \theta$ to simplify the expression $\sqrt{64x^2 + 9}$. Assume $0 < \theta < 90°$.

47. Magic Rings A magician is holding two rings that seem to lie in the same plane and intersect in two points. Each ring is 10 inches in diameter.

    a. Find the equation of each ring if a coordinate system is placed with its origin at the center of the first ring and the x-axis contains the center of the second ring.

    b. Find the equation of each ring if a coordinate system is placed with its origin at the center of the second ring and the x-axis contains the center of the first ring.

RIGHT TRIANGLE TRIGONOMETRY

2

WHAT IS RIGHT TRIANGLE TRIGONOMETRY? Right triangle trigonometry is a branch of mathematics that deals specifically with the relationships between the angles and sides of a right triangle. In a right triangle, one of the angles measures 90 degrees (a right angle), and the other two angles are acute (less than 90 degrees).

WHY IS RIGHT TRIANGLE TRIGONOMETRY IMPORTANT? Right triangle trigonometry is foundational in various fields such as engineering, physics, architecture, and navigation, where it's used to solve problems involving angles and distances.

WHERE MIGHT YOU SEE RIGHT TRIANGLE TRIGONOMETRY? Here are some examples of where you may see right triangle trigonometry in the real world:

STAIRCASE DESIGN When designing a staircase, architects use trigonometry to determine the angle of inclination for each step. This ensures that the staircase is safe and comfortable to use, and complies with existing building codes.

TILT ANGLE OF SOLAR PANELS Solar panels are often installed at an angle to maximize sunlight exposure. Trigonometry can be used to calculate the optimal tilt angle for solar panels based on the location's latitude and the time of year.

Scenario: The Right Triangle and Staircase Design

A contractor needs to design a staircase for a 3-story building with a total height of 35 feet. Each step is rise 6 inches and run 11 inches. Determine the number of steps required and the angle of inclination for each step.

Solution

1. Calculate the Number of Steps. Divide the total height of the building by the rise of each step. Convert all measurements to inches for consistency.

$$\text{Number of steps} = \frac{\text{Total height}}{\text{Rise of each step step}} = \frac{35 \text{ feet} \cdot \dfrac{12 \text{ inches}}{\text{foot}}}{6 \dfrac{\text{inches}}{\text{step}}} = 70 \text{ steps}$$

2. Calculate the angle of inclination (θ). Use the tangent function, which is the ratio of the rise to the run.

$$\text{Angle of inclination } (\theta) = \arctan\left(\frac{\text{rise}}{\text{run}}\right) = \arctan\left(\frac{6 \text{ inches}}{11 \text{ inches}}\right)$$

Using a calculator, we find

$$\theta \approx 28.61°$$

Now, we can proceed with the solution: The contractor needs to design 70 steps for the staircase, and each step will have an angle of inclination of approximately 28.61°.

2.1 Videos

KEY WORDS

right triangle trigonometry

cofunction

table of exact values

Learning Objectives

A   Find the six trigonometric functions of an angle given the sides of a right triangle.

B   Find trigonometric functions using the cofunction theorem.

C   Simplify trigonometric expressions using a table of exact values.

In Chapter 1, we gave a definition for the six trigonometric functions for any angle in standard position. In this section we will give a second definition for the six trigonometric functions in terms of the sides and angles in a right triangle.

A Right Triangle Trigonometry

Definition II

If triangle ABC is a right triangle with $C = 90°$, then the six trigonometric functions for A are defined as follows:

$$\sin A = \frac{\text{side opposite } A}{\text{hypotenuse}} = \frac{a}{c}$$

$$\cos A = \frac{\text{side adjacent } A}{\text{hypotenuse}} = \frac{b}{c}$$

$$\tan A = \frac{\text{side opposite } A}{\text{side adjacent } A} = \frac{a}{b}$$

$$\cot A = \frac{\text{side adjacent } A}{\text{side opposite } A} = \frac{b}{a}$$

$$\sec A = \frac{\text{hypotenuse}}{\text{side adjacent } A} = \frac{c}{b}$$

$$\csc A = \frac{\text{hypotenuse}}{\text{side opposite } A} = \frac{c}{a}$$

Side opposite A and adjacent to B

Side opposite B and adjacent to A

FIGURE 1

EXAMPLE 1   Triangle ABC is a right triangle with $C = 90°$. If $a = 6$ and $c = 10$, find the six trigonometric functions of A.

Solution We begin by making a diagram of ABC (Figure 2) and then use the given information and the Pythagorean Theorem to solve for b:

$$b = \sqrt{c^2 - a^2}$$
$$= \sqrt{100 - 36}$$
$$= \sqrt{64}$$
$$= 8$$

Now we write the six trigonometric functions of A using $a = 6$, $b = 8$, and $c = 10$:

$$\sin A = \frac{a}{c} = \frac{6}{10} = \frac{3}{5}$$

$$\cos A = \frac{b}{c} = \frac{8}{10} = \frac{4}{5}$$

$$\tan A = \frac{a}{b} = \frac{6}{8} = \frac{3}{4}$$

$$\csc A = \frac{c}{a} = \frac{5}{3}$$

$$\sec A = \frac{c}{b} = \frac{5}{4}$$

$$\cot A = \frac{b}{a} = \frac{4}{3}$$

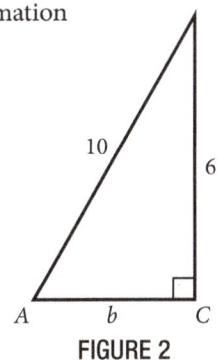

FIGURE 2

Now that we have done an example using our new definition, let's see how our new definition compares to Definition I from Section 1.3. We can place right triangle ABC on a rectangular coordinate system so that A is in standard position (Figure 3). We then note that a point on the terminal side of A is (b, a).

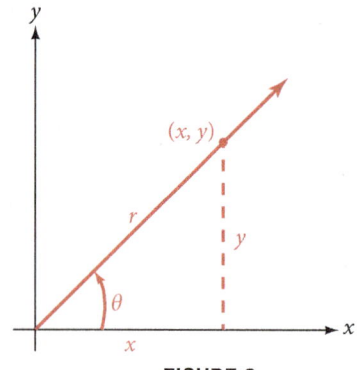

FIGURE 3a

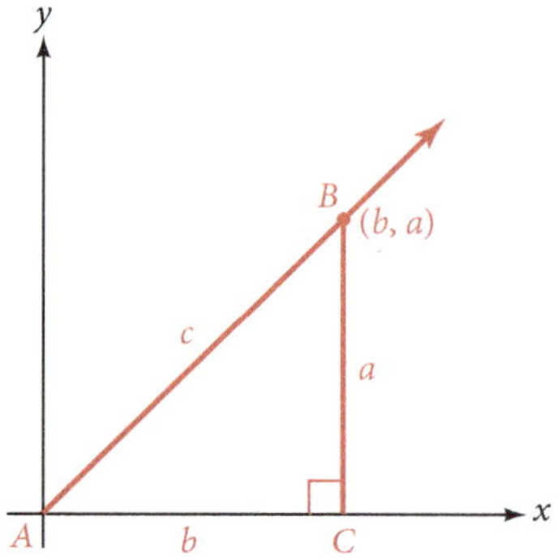

FIGURE 3b

From Definition I, we have:

$$\sin A = \frac{y}{r}$$

$$\cos A = \frac{x}{r}$$

$$\tan A = \frac{y}{x}$$

From Definition II, we have:

$$\sin A = \frac{a}{c}$$

$$\cos A = \frac{b}{c}$$

$$\tan A = \frac{a}{b}$$

The two definitions agree as long as A is an acute angle. If A is not an acute angle, then Definition II does not apply, since in right triangle ABC, A must be an acute angle.

Here is another definition that we will need before we can take Definition II any further.

Definition
Sine and *co*sine are **cofunctions**, as are tangent and *co*tangent, and secant and *co*secant. We say sine is the cofunction of cosine, and cosine is the cofunction of sine.

B The Cofunction Theorem

Now let's see what happens when we apply Definition II to B in right triangle ABC.

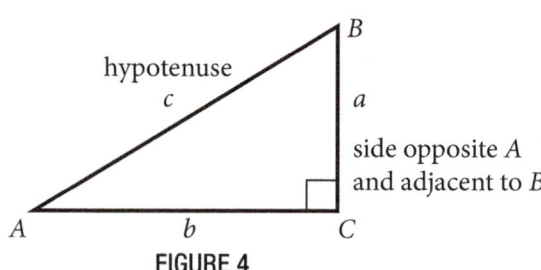

hypotenuse

side opposite A
and adjacent to B

FIGURE 4

$$\sin B = \frac{\text{side opposite } B}{\text{hypotenuse}} = \frac{b}{c} = \cos A$$

$$\tan B = \frac{\text{side opposite } B}{\text{side adjacent } B} = \frac{b}{a} = \cot A$$

$$\sec B = \frac{\text{hypotenuse}}{\text{side adjacent } B} = \frac{c}{a} = \csc A$$

$$\cos B = \frac{\text{side adjacent } B}{\text{hypotenuse}} = \frac{a}{c} = \sin A$$

$$\cot B = \frac{\text{side adjacent } B}{\text{side opposite } B} = \frac{a}{b} = \tan A$$

$$\csc B = \frac{\text{hypotenuse}}{\text{side opposite } B} = \frac{c}{b} = \sec A$$

We can summarize the information above with the following theorem:

> ### Cofunction Theorem
> A trigonometric function of an angle is always equal to the cofunction of the complement of the angle. That is, if x is an acute angle, then, since x and $90° - x$ are complementary angles, it must be true that
>
> $$\sin x = \cos (90° - x)$$
> $$\cos x = \sin (90° - x)$$
> $$\tan x = \cot (90° - x)$$
> $$\cot x = \tan (90° - x)$$
> $$\sec x = \csc (90° - x)$$
> $$\csc x = \sec (90° - x)$$

EXAMPLE 2   Fill in the blanks so that each expression becomes a true statement.

a. $\sin \underline{\hspace{1cm}} = \cos 30°$ **b.** $\tan y = \cot \underline{\hspace{1cm}}$ **c.** $\sec 75° = \csc \underline{\hspace{1cm}}$

Solution   Using the theorem on cofunctions of complementary angles, we fill in the blanks as follows:

a. $\sin \underline{\;\;60°\;\;} = \cos 30°$ *Since sine and cosine are cofunctions and $60° + 30° = 90°$*

b. $\tan y = \cot \underline{(90° - y)}$ *Since tangent and cotangent are cofunctions and $y + (90 - y) = 90°$*

c. $\sec 75° = \csc \underline{\;\;15°\;\;}$ *Since secant and cosecant are cofunctions and $75° + 15° = 90°$*

For our next application of Definition II, we need to recall the two special triangles we introduced in Chapter 1. They are the 30°-60°-90° triangle, and the 45°-45°-90° triangle.

We can use these two special triangles to find the trigonometric functions of 30°, 45°, and 60°. For example,

$$\sin 60° = \frac{\text{side opposite } 60°}{\text{hypotenuse}} = \frac{t\sqrt{3}}{2t} = \frac{\sqrt{3}}{2}$$

Dividing out the common factor t

$$\tan 30° = \frac{\text{side opposite } 30°}{\text{side adjacent } 30°} = \frac{t}{t\sqrt{3}} = \frac{1}{\sqrt{3}}$$

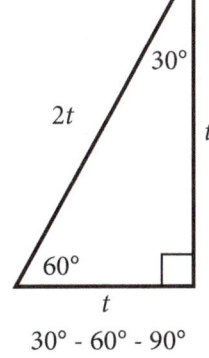

30° - 60° - 90°

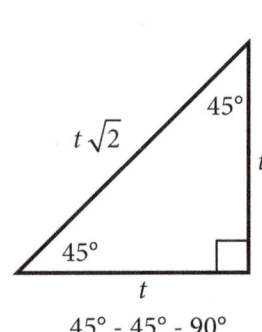

45° - 45° - 90°

FIGURE 5

We could go on in this manner—using Definition II and the two special triangles—to find the six trigonometric ratios for 30°, 45°, and 60°. Instead, let us vary things a little and use the information just obtained for sin 60° and tan 30° and theorem on cofunctions of complementary angles to find cos 30° and cot 60°.

$$\cos 30° = \sin 60° = \frac{\sqrt{3}}{2}$$

Cofunction theorem

$$\cot 60° = \tan 30° = \frac{1}{\sqrt{3}}$$

To vary things even more, we can use some reciprocal identities to find csc 60° and cot 30°.

$$\csc 60° = \frac{1}{\sin 60°} = \frac{2}{\sqrt{3}} \text{ or } \frac{2\sqrt{3}}{3}$$

If denominator is rationalized

$$\cot 30° = \frac{1}{\tan 30°} = \sqrt{3}$$

The idea behind using the different methods listed above is not to make things confusing. We have a number of tools at hand, and it does not hurt to show the different ways they can be used.

C Table of Exact Values

If we were to continue finding sine, cosine, and tangent for these special angles, we would obtain the results summarized in Table 1.

Table 1 Exact Values			
0°	**30°**	**45°**	**60°**
$\sin \theta$	$\frac{1}{2}$	$\frac{1}{\sqrt{2}}$ or $\frac{\sqrt{2}}{2}$	$\frac{\sqrt{3}}{2}$
$\cos \theta$	$\frac{\sqrt{3}}{2}$	$\frac{1}{\sqrt{2}}$ or $\frac{\sqrt{2}}{2}$	$\frac{1}{2}$
$\tan \theta$	$\frac{1}{\sqrt{3}}$ or $\frac{\sqrt{3}}{3}$	1	$\sqrt{3}$

Table 1 is called a ***table of exact values*** to distinguish it from a table of approximate values. Later in this chapter we will work with tables of approximate values.

Calculator Note Calculators vary by brand and model, so it is important that you become familiar with using the one you own. First, be sure your calculator is in degree (not radian) mode. Most calculators will have DEG on the screen to indicate degree mode. If yours has RAD on the screen, you will need to change to degree mode. Second, to calculate sin 30°, experiment to determine if the sequence is

$$\boxed{\sin}\ 30 \quad \text{or} \quad 30\ \boxed{\sin}$$

The calculator should display 0.5 or $\frac{1}{2}$.

The calculator will then display 0.5. The values the calculator gives for the trigonometric functions of an angle are approximations, except for a few cases such as sin 30° = 0.5. If you use a calculator to find sin 60°, the calculator will display 0.866025403, which is a nine digit decimal approximation of $\frac{\sqrt{3}}{2}$. You can check this by using your calculator to find an approximation to $\frac{\sqrt{3}}{2}$ to see that it agrees with the calculator value of sin 60°. In the meantime, remember that if you are asked to find the exact value of a trignometric function, you must use the values given in the table of exact values.

EXAMPLE 3   Use the exact values from the table to show that the following are true.

a. $\cos^2 30° + \sin^2 30° = 1$ **b.** $\cos^2 45° + \sin^2 45° = 1$

Solution

a. $\cos^2 30° + \sin^2 30° = \left(\dfrac{\sqrt{3}}{2}\right)^2 + \left(\dfrac{1}{2}\right)^2 = \dfrac{3}{4} + \dfrac{1}{4} = 1$

b. $\cos^2 45° + \sin^2 45° = \left(\dfrac{1}{\sqrt{2}}\right)^2 + \left(\dfrac{1}{\sqrt{2}}\right)^2 = \dfrac{1}{2} + \dfrac{1}{2} = 1$

EXAMPLE 4   Let $x = 30°$ and $y = 45°$ in each of the expressions that follow, and then simplify each expression as much as possible.

a. $2 \sin x$ **b.** $\sin 2y$ **c.** $4 \sin(3x - 90°)$

Solution

a. $2 \sin x = 2 \sin 30° = 2\left(\dfrac{1}{2}\right) = 1$

b. $\sin 2y = \sin 2(45°) = \sin 90° = 1$

c. $4 \sin(3x - 90°) = 4 \sin [3(30°) - 90°] = 4 \sin 0° = 4(0) = 0$

Problem Set 2.1

Vocabulary

Use the vocabulary words below to fill in the blanks in the sentences.

adjacent   opposite   cofunctions   hypotenuse   complement

1. Our usual way of labeling the sides and angles in triangle ABC is to have angle C as the right angle, and side c as the _____ .

2. Sine and cosine are _____ , as are tangent and cotangent, and secant and cosecant.

3. A trigonometric function of an angle is always equal to the cofunction of the _____ of the angle.

4. The sine of an acute angle in a right triangle is always the side _____ the angle divided by the hypotenuse.

5. The tangent of an acute angle in a right triangle is always the side opposite the angle divided by the side _____ to the angle.

Problems 1–6 refer to right triangle ABC with $C = 90°$. In each case, use the given information to find the six trigonometric functions of A.

1. $b = 3, c = 5$ 2. $b = 5, c = 13$

3. $a = 2, b = 1$ 4. $a = 3, b = 2$

5. $a = 2, c = 4$ 6. $a = 3, c = 6$

In each triangle below, find $\sin A$, $\cos A$, $\sin B$, and $\cos B$.

7.
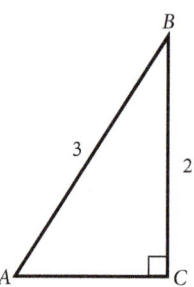

8.
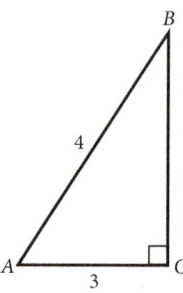

9.
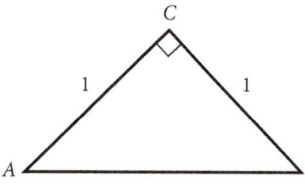

10.
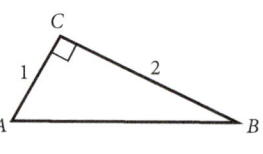

11.
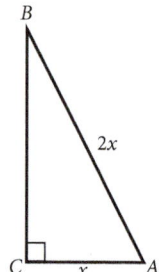

12.
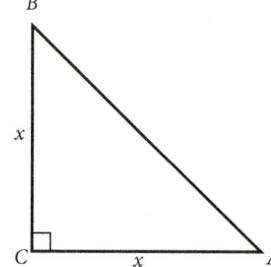

Use the Cofunction Theorem to fill in the blanks so each expression becomes a true statement.

13. $\sin 10° = \cos$ _____

14. $\cos 40° = \sin$ _____

15. $\tan 8° = \cot$ _____

16. $\cot 12° = \tan$ _____

17. $\sec 73° = \csc$ _____

18. $\csc 63° = \sec$ _____

19. $\sin x = \cos$ _____

20. $\sin y = \cos$ _____

21. $\tan (90° - x) = \cot$ _____

22. $\tan (90° - y) = \cot$ _____

Simplify each expression by first substituting values from the table of exact value and then simplifying the resulting expression.

23. $4 \sin 30°$

24. $5 \sin^2 30°$

25. $(2 \cos 30°)^2$

26. $\sin^3 30°$

27. $\sin 30° \cos 45°$

28. $\sin 30° + \cos 45°$

29. $(\sin 60° + \cos 60°)^2$

30. $\sin^2 60° + \cos^2 60°$

31. $\sin^2 45° - 2 \sin 45° \cos 45° + \cos^2 45°$ **32.** $(\sin 45° - \cos 45°)^2$

33. $(\tan 45° + \tan 60°)^2$

34. $\tan^2 45° + \tan^2 60°$

For each expression that follows, replace x with 30°, y with 45°, and z with 60°, and then simplify as much as possible.

35. $2 \sin x$

36. $4 \cos y$

37. $4 \cos(z - 30°)$

38. $-2 \sin(y + 45°)$

39. $-3 \sin 2x$

40. $3 \sin 2y$

41. $2 \cos (3x - 45°)$

42. $2 \sin (90° - z)$

Find exact values for each of the following:

43. $\sec 30°$

44. $\csc 30°$

45. $\csc 60°$

46. $\sec 60°$

47. $\cot 45°$

48. $\cot 30°$

49. $\sec 45°$

50. $\csc 45°$

Use a calculator to find an approximation to each of the following expressions. Round to four decimal places.

51. sin 60°   **52.** cos 30°

53. $\dfrac{\sqrt{3}}{2}$ **54.** $\dfrac{\sqrt{2}}{2}$

55. sin 45°   **56.** cos 45°

57. $\dfrac{1}{\sqrt{2}}$ **58.** $\dfrac{1}{\sqrt{3}}$

59. tan 45°   **60.** tan 30°

Getting Ready for the Next Section

Perform the indicated operations.

61. $48 + 72$ **62.** $49 + 26$ **63.** $89 - 24$

64. $60 - 14$ **65.** $0.25(60)$ **66.** $0.50(60)$

67. $45/60$ **68.** $15/60$

Use your calculator to find the following. Round to the nearest hundredth.

69. $1 \div 1.08$ **70.** $1 \div 0.98$ **71.** $1/1.25$

72. $1/0.75$ **73.** $1/1.5$ **74.** $1/0.5$

Learning Objectives

A Solve problems involving degree measure.

B Convert decimal degrees to degrees and minutes.

C Find decimal approximations for trigonometric functions using a calculator.

We previously defined 1 degree (1°) to be $\frac{1}{360}$ of a full rotation. A degree itself can be broken down further. If we divide 1° into 60 equal parts, each one of the parts is called 1 minute, denoted 1′. One minute is 1/60 of a degree; in other words, there are 60 minutes in every degree. The next smallest unit of angle measure is a second. One second, 1″, is 1/60 of a minute. There are 60 seconds in every minute.

$$1° = 60' \quad \text{or} \quad 1' = \left(\frac{1}{60}\right)^{°}$$

$$1' = 60'' \quad \text{or} \quad 1'' = \left(\frac{1}{60}\right)^{'}$$

A Degree Measure

Table 1 shows how to read angles written in degree measure.

Table 1	
The Expression	**Is Read**
52° 10′	52 degrees, 10 minutes
5° 27′ 30″	5 degrees, 27 minutes, 30 seconds
13° 24′ 75″	13 degrees, 24 minutes, 75 seconds

EXAMPLE 1 Add 48° 49′ and 72° 26′.

Solution We can add in columns with degrees in the first column and minutes in the second column.

$$\begin{array}{r} 48° \ 49' \\ + \ 72° \ 26' \\ \hline 120° \ 75' \end{array}$$

Since 60 minutes is equal to 1 degree, we can carry 1 degree from the minutes column to the degree column.

$$120° \ 75' = 121° \ 15'$$

EXAMPLE 2 Subtract 24° 14′ from 90°.

Solution In order to subtract 24° 14′ from 90° we will have to "borrow" 1° and write that 1° as 60′.

$$\begin{array}{rcl} 90° & = & 89° \ 60' \qquad \textit{(Still 90°)} \\ -24° \ 14' & & -24° \ 14' \\ \hline & & 65° \ 46' \end{array}$$

B Decimal Degrees

An alternative to using minutes and seconds to break down degrees into smaller units is *decimal degrees*. For example, 30.5°, 101.75°, and 62.831° are the measures of angles written in decimal degrees.

To convert from decimal degrees to degrees and minutes, we simply multiply the fractional part (the part to the right of the decimal point) of the angle by 60 to convert it to minutes.

EXAMPLE 3   Change 27.25° to degrees and minutes.

Solution   Multiplying 0.25 by 60 we have the number of minutes equivalent to 0.25°.

$$27.25° = 27° + 0.25°$$

$$= 27° + 0.25(60')$$

$$= 27° + 15'$$

$$= 27° \, 15'$$

Of course in actual practice, we would not show all of these steps. They are shown here simply to indicate why we multiply only the decimal part of the decimal degree by 60 to change to degrees and minutes.

> *Calculator Note*   Some scientific calculators have a key that automatically converts angles given in decimal degrees to degrees and minutes. Consult the manual that came with your calculator to see if yours has this key.

EXAMPLE 4   Change 10° 45' to decimal degrees.

Solution   We have to reverse the process we used in Example 3. To change 45' to a decimal we must divide by 60.

$$10° \, 45' = 10° + 45'$$

$$= 10° + 45\left(\frac{1}{60}\right)°$$

$$= 10° + \frac{45°}{60°}$$

$$= 10° + 0.75°$$

$$= 10.75°$$

> *Calculator Note*   On a calculator, the result given in Example 4 is accomplished as follows:
>
> 45 ÷ 60 + 10 =

The process of converting back and forth between decimal degrees and degrees and minutes can become more complicated when we use decimal numbers with more digits or when we convert to degrees, minutes, and seconds. In this book, most of the angles written in decimal degrees will be written to the the nearest tenth or, at most, the nearest hundredth. The angles written in degrees, minutes, and seconds will rarely go beyond the minutes column.

Table 2 lists the most common conversions between decimal degrees and minutes.

Table 2	
Decimal Degree	**Minutes**
0.1°	6′
0.2°	12′
0.3°	18′
0.4°	24′
0.5°	30′
0.6°	36′
0.7°	42′
0.8°	48′
0.9°	54′
1.0°	60′

C Decimal Approximations of Trigonometric Functions

Until now, we have been able to determine trigonometric functions only for angles for which we could find a point on the terminal side or angles that were part of special triangles. We can find decimal approximations for trigonometric functions of any acute angle by using a calculator with keys for sine, cosine, and tangent.

EXAMPLE 5   Use a calculator to find cos 37.8°.

Solution   First, be sure your calculator is in degree mode. Then, press the indicated keys.

Your calculator will display a number that rounds to 0.7902. The number 0.7902 is just an approximation of cos 37.8°, which is actually an irrational number, as are the trigonometric functions of most angles.

> *Note* Notice the "(and others)" text after the Graphing Calculator title above. It is there because there are many other calculators that use the same logic as a graphing calculator, but are not graphing calculators. You must know which of the two logics your calculator uses and work problems on your calculator accordingly.

> *Note* We will give answers accurate to four places past the decimal point. You can set your calculator to four-place fixed-point mode, and it will show you the same result without you having to round your answers mentally.

EXAMPLE 6   Find sin 58.75°.

Solution This time we use the $\boxed{\text{sin}}$ key.

Scientific Calculator	Graphing Calculator
58.75 $\boxed{\text{sin}}$	$\boxed{\text{sin}}$ 58.75 $\boxed{\text{ENTER}}$

Rounding to four digits past the decimal point, we have

$$\sin 58.75° \approx 0.8549$$

EXAMPLE 7   Find sec 78°.

Solution Standard calculators rarely have a secant function, so we use the reciprocal of the cosine.

Since sec 78° = 1/cos 78°, the calculator sequence is:

Scientific Calculator	Graphing Calculator
78 $\boxed{\text{cos}}$ $\boxed{\text{1/x}}$	1 $\boxed{\div}$ $\boxed{\text{cos}}$ 78 $\boxed{\text{ENTER}}$
or	or
1 $\boxed{\div}$ 78 $\boxed{\text{cos}}$ $\boxed{=}$	$\boxed{\text{cos}}$ 78 $\boxed{\text{ENTER}}$ $\boxed{\text{x}^{-1}}$ $\boxed{\text{ENTER}}$

Rounding to four digits past the decimal point, we have

$$\sec 78° \approx 4.8097$$

Next we want to use a calculator to find an angle, given the value of one of the trigonometric functions of the angle. This process is somewhat like going in the opposite direction from that shown in the examples above.

EXAMPLE 8   Find the acute angle θ for which tan θ = 3.152. Round your answer to the nearest tenth of a degree.

Solution We are looking for the angle whose tangent is 3.152. To find this angle on a calculator, we must use the $\tan^{-1}$ or arctan function. This is usually accomplished by pressing the key labeled $\boxed{\text{inv}}$ or $\boxed{\text{2nd}}$, and then the $\boxed{\text{tan}}$ key. (Check your manual to see how your calculator does this. In the index, look under inverse trigonometric functions.) In this book we use the following sequences. (Your calculator may require a different sequence.)

Scientific Calculator	Graphing Calculator
3.152 $\boxed{\tan^{-1}}$	$\boxed{\tan^{-1}}$ 3.152 $\boxed{\text{ENTER}}$

To the nearest tenth of a degree the answer is 72.4°. That is, if tan θ = 3.152, then $\theta \approx 72.4°$.

EXAMPLE 9   Find the acute angle A for which sin A = 0.3733. Round your answer to the nearest tenth of a degree.

Solution The sequences are:

Scientific Calculator	Graphing Calculator
0.3733 $\boxed{\sin^{-1}}$	$\boxed{\sin^{-1}}$ 0.3733 $\boxed{\text{ENTER}}$

The result is $A \approx 21.9°$.

EXAMPLE 10   To the nearest hundredth of a degree, find the acute angle B for which sec $B = 1.0768$.

Solution   Since we do not have a secant key on the calculator, we must use a reciprocal to first see how we can convert this problem into a problem involving cos B (as in Example 3).

If sec $B = 1.0768$, then

$$\frac{1}{\sec B} = \frac{1}{1.0768}$$ *Take the reciprocal of each side.*

$$\cos B = \frac{1}{1.0768}$$ *Since the cosine is the reciprocal of the secant.*

From the last line we see that the keys to press are:

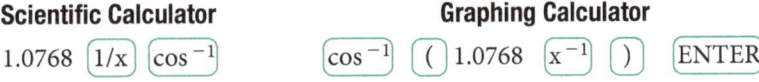

To the nearest hundredth of a degree our answer is $B \approx 21.77°$.

EXAMPLE 11   Find the acute angle C for which cot $C = 0.0975$. Round to the nearest degree.

Solution   First, we rewrite the problem in terms of tan C.

If cot $C = 0.0975$, then

$$\frac{1}{\cot C} = \frac{1}{0.0975}$$ *Take the reciprocal of each side.*

$$\tan C = \frac{1}{0.0975}$$ *Since the tangent is the reciprocal of the cotangent*

From this last line we see that the keys to press are:

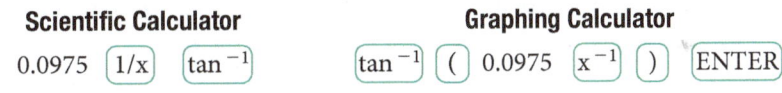

To the nearest degree our answer is $C \approx 84°$.

Problem Set 2.2

Vocabulary

Use the vocabulary words below to fill in the blanks in the sentences.

minute   rotation   degrees   decimal   multiply

1. One degree (1°) is defined to be 1/360 of a full _____ .

2. One _____ (1′) is defined to be 1/60 of a degree.

3. 52°30′ is written in _____ and minutes.

4. 52.5° is written in _____ degrees.

5. To convert 52.5° to degrees and minutes, we _____ 0.5 by 60 to find the number of minutes.

Add and subtract as indicated.

1. $(37° 45′) + (26° 24′)$ 2. $(41° 20′) + (32° 16′)$

3. $(51° 55′) + (37° 45′)$ 4. $(63° 38′) + (24° 52′)$

5. $(61° 33′) + (45° 16′)$ 6. $(77° 21′) + (23° 16′)$

7. $90° − (34° 12′)$ 8. $90° − (62° 25′)$

9. $180° − (120° 17′)$ 10. $180° − (112° 19′)$

11. $(76° 24′) − (22° 34′)$ 12. $(89° 38′) − (28° 58′)$

13. $(70° 40′) − (30° 50′)$ 14. $(80° 50′) − (50° 56′)$

Convert each of the following to degrees and minutes.

15. 35.4°   16. 63.2°

17. 16.25°   18. 18.75°

19. 92.55°   20. 34.45°

21. 19.9°   22. 18.8°

23. 28.35°   24. 76.85°

Change each of the following to decimal degrees.

25. 45° 12′   26. 74° 18′

27. 62° 36′   28. 21° 48′

29. 17° 20′   30. 29° 40′

31. 48° 27′   32. 78° 21′

Use a calculator to find each of the following. Round all answers to four places past the decimal point.

33. sin 27.2°

34. cos 82.9°

35. cos 18°

36. sin 42°

37. tan 87.32°

38. tan 81.43°

39. cot 31°

40. cot 24°

41. sec 48.2°

42. sec 71.8°

43. csc 14.15°

44. csc 12.21°

45. cos 24° 30′

46. sin 35° 10′

47. tan 42° 10′

48. cot 19° 40′

49. sin 56° 40′

50. cos 66° 40′

51. cos 70° 20′

52. sin 80° 50′

53. cot 88° 50′

54. tan 50° 10′

55. sec 51° 12′

56. sec 10° 45′

57. csc 32° 48′

58. csc 67° 24′

59. cot 55° 12′

60. cot 76° 15′

61. csc 10° 25′

62. sec 5° 25′

Find θ if θ is between 0° and 90°. Round your answer to the nearest tenth of a degree.

63. $\cos \theta = 0.9770$

64. $\sin \theta = 0.3971$

65. $\tan \theta = 0.6873$

66. $\cos \theta = 0.5490$

67. $\sin \theta = 0.9813$

68. $\tan \theta = 0.6273$

69. $\sin \theta = 0.7038$

70. $\cos \theta = 0.9153$

71. $\cos \theta = 0.4112$

72. $\sin \theta = 0.9954$

73. $\tan \theta = 1.1953$

74. $\tan \theta = 1.7391$

75. $\csc \theta = 3.9451$

76. $\sec \theta = 2.1609$

77. $\cot \theta = 5.5764$

78. $\cot \theta = 4.6252$

79. $\sec \theta = 1.0129$

80. $\csc \theta = 7.0683$

Use your calculator to simplify each expression. Round your answers to the nearest hundredths.

81. $\cos^2 30°$

82. $\sin^2 30°$

83. $\tan^4 60°$

84. $\tan^3 60°$

Simplify without using a calculator.

85. $\sin^2 45° + \cos^2 45°$

86. $(\sin 45° + \cos 45°)^2$

87. $\sin^2 60° + \cos^2 60°$

88. $(\sin 60° + \cos 60°)^2$

Getting Ready for the Next Section

Simplify. Round to two decimal places if necessary.

89. $12(0.7660)$

90. $\sqrt{84.71}$

91. $\dfrac{2.73}{3.41}$

92. $\dfrac{17.04}{3.87}$

93. $\dfrac{21}{0.8693}$

94. $\dfrac{21}{0.6561}$

95. $\dfrac{20}{\tan 60°}$

96. $\dfrac{20}{\tan 50°}$

97. $\dfrac{2.73}{\sin 38.7°}$

98. $\dfrac{21}{\sin 41°}$

Find θ to the nearest degree, if

99. $\tan \theta = 2.73/3.41$

100. $\tan \theta = 2$

Learning Objectives

A   Identify significant digits for a given number.

B   Solve right triangles using Definition II for trigonometric functions.

In this section we will use Definition II for trigonometric functions of an acute angle, along with our tables (or calculators), to find the missing parts to some right triangles. Before we begin, however, we need to talk about significant digits.

A Significant Digits

> **Definition**
>
> The number of *significant digits* (or figures) in a number is found by counting the number of digits from left to right beginning with the first nonzero digit on the left.
>
> If the number in question does not contain a decimal point, then trailing zeros are not significant.

EXAMPLE 1   Give the number of significant digits in the side given in each of the following triangles.

a. 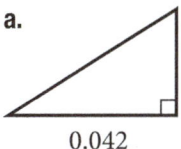   b. 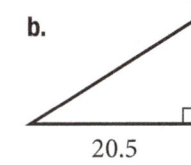   c. 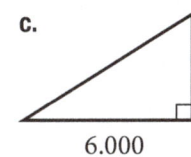   d.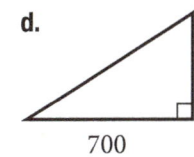

0.042   20.5   6.000   700

FIGURE 1

Solution

a. 0.042 has two significant digits.

b. 20.5 has three significant digits.

c. 6.000 has four significant digits.

d. 700 has one significant digit.

Note that, in Part d, if the number was written with a decimal point as 700., it would have three significant digits.

The relationship between the accuracy of the sides of a triangle and the accuracy of the angles in the same triangle is shown in the following table.

Accuracy of Sides	Accuracy of Angles
Two significant digits	Nearest degree
Three significant digits	Nearest 10 minutes or tenth of a degree
Four significant digits	Nearest minute or hundredth of a degree

EXAMPLE 2   For each triangle, round the number in parentheses so that its accuracy corresponds with the accuracy of the other numbers in the triangle.

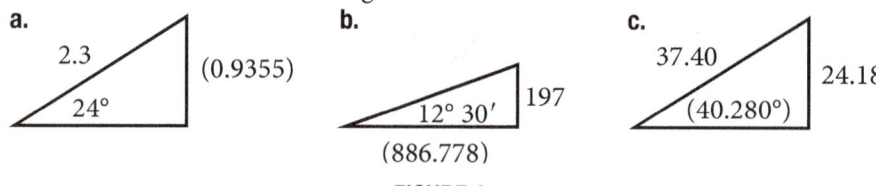

a.   **b.**   **c.**

FIGURE 2

Solution

a. Since 2.3 is accurate to two significant digits, we must round 0.9355 to that accuracy also.

$0.9355 = 0.94$ to two significant digits   *Remember: We do not count leading zeros as significant.*

b. Since 197 has three significant digits, we must round 886.778 to that accuracy also.
$886.778 = 887$ to three significant digits

c. Since both of the given sides, 37.40 and 24.18, are accurate to four significant digits, we must round 40.280° to the nearest hundredth of a degree.

$40.280° = 40.28°$ to the nearest hundredth of a degree

B Solving Right Triangles

We are now ready to use Definition II, along with our calculators, to solve some right triangles. We solve a right triangle by using the information given about it to find all the missing sides and angles. In Example 3, we are given the values of one side and one of the acute angles and asked to find the remaining two sides and the other acute angle. In the examples and the Problem Set that follow, let's assume C is the right angle in all our right triangles.

EXAMPLE 3   In right triangle ABC, $A = 40°$ and $c = 12$ centimeters. Find a, b, and B.

Solution We begin by making a diagram of the situation in Figure 3.
To find B, we use the fact that the sum of the two acute angles in any right triangle is 90°:

$$B = 90° - A$$
$$= 90° - 40° = 50°$$

To find a, we can use the definition for $\sin A$:

$$\sin A = \frac{a}{c}$$

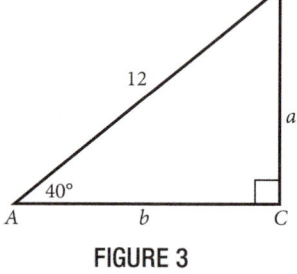

FIGURE 3

Multiplying both sides of this equation by c and then substituting in our given values for A and c, we have

$$a = c \sin A$$
$$= 12 \sin 40°   \qquad \sin 40° \approx 0.6428$$
$$\approx 7.7   \qquad \text{Rounded to two-digit accuracy.}$$

There is more than one way to find b:

Using $\cos A = \dfrac{b}{c}$ we have	**Using the Pythagorean Theorem, we have**
$b = c \cos A$	$c^2 = a^2 + b^2 \qquad b = \sqrt{c^2 - a^2}$
$\quad = 12 \cos 40°$	$\qquad = \sqrt{12^2 - (7.7)^2}$
$\quad \approx 9.2$	$\qquad = \sqrt{84.71}$
	$\qquad \approx 9.2$

So we have $a \approx 7.7$ centimeters and $b \approx 9.2$ centimeters. Note that using the cosine ratio to find b is the best approach because it uses the exact numbers given in the problem. The Pythagorean theorem method uses a rounded value of a, which is not as accurate as the cosine method.

In Example 4, we are given two sides and asked to find the remaining parts of a right triangle.

EXAMPLE 4   In right triangle ABC, $a = 2.73$ and $b = 3.41$. Find the remaining side and angles.

Solution   Figure 4 is a diagram of the triangle.
We can find A by using the definition for $\tan A$:

$$\tan A = \frac{a}{b}$$

$$= \frac{2.73}{3.41}$$

$$\approx 0.8006$$

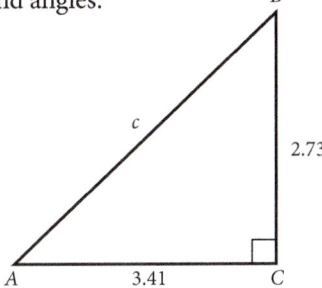

FIGURE 4

Now, to find A, we use a calculator to look for the angle whose tangent is 0.8006:

$$A = \tan^{-1}(0.8006) \approx 38.7°$$

Next, we find B:

$$B = 90° - A$$

$$\approx 90° - 38.7° = 51.3°$$

We can find c using the Pythagorean Theorem or one of the trigonometric functions. Let's use a trigonometric function.

$$\text{If} \qquad \sin A = \frac{a}{c}$$

$$\text{then} \qquad c = \frac{a}{\sin A}$$

$$\approx \frac{2.73}{\sin 38.7°}$$

$$\approx 4.37$$

EXAMPLE 5   The center of the circle in Figure 5 is at C. If the radius is 21 inches, and angle A is 41°, find y, then find x.

Solution   First we find y using the tangent:

$$\tan 41° = \frac{21}{y}$$

$$y \tan 41° = 21$$

$$y = \frac{21}{\tan 41°}$$

$$\approx 24 \text{ inches} \qquad \text{To two significant digits}$$

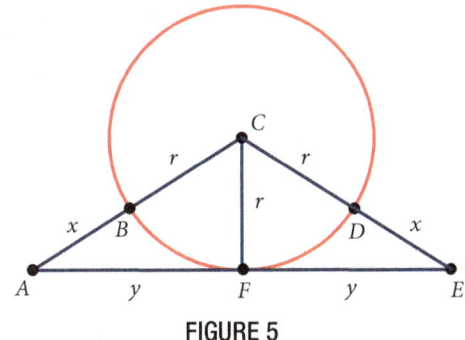

FIGURE 5

We have some choices for finding x. Because we have y, we could use the Pythagorean theorem. Instead, let's use the sine ratio.

$$\sin 41° = \frac{21}{x + 21}$$

$(x + 21) \sin 41° = 21$ Multiplying each side by x + 21

$$x + 21 = \frac{21}{\sin 41°}$$ Dividing each side by sin 41°

$$x = \frac{21}{\sin 41°} - 21$$ Subtract 21 from each side

$$\approx 32 - 21$$
$$\approx 11 \text{ inches}$$

EXAMPLE 6   A man is walking away from a 20-foot flagpole. When he reaches point B, the angle to the top of the pole is 60°, and when he reaches point C the angle to the top of the flagpole is 50°. Find his distance from the flagpole at point B and at point C. (Round to the nearest tenth.)

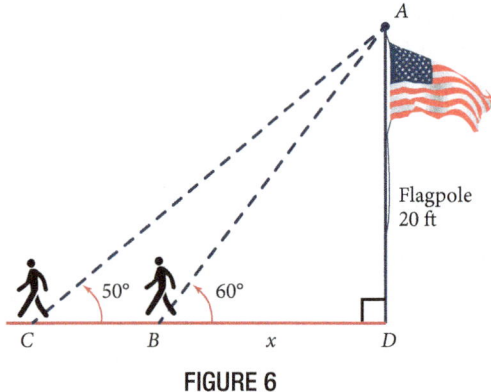

Flagpole
20 ft

FIGURE 6

Solution   Let's find x first, his distance from the flagpole when he reaches point B.

$$\tan 60° = \frac{20}{x}$$

$x \tan 60° = 20$ Multiply each side by x

$$x = \frac{20}{\tan 60°}$$ Divide each side by tan 60°

$$\approx 11.5 \text{ feet}$$ Divide

Next, we find his distance from the flagpole at point C.

$$\tan 50° = \frac{20}{CD}$$

$CD \tan 50° = 20$ Multiply each side by CD

$$CD = \frac{20}{\tan 50°}$$ Divide each side by tan 50°

$$\approx \frac{20}{1.1918}$$ Evaluate tan 50°

$$\approx 16.8 \text{ feet}$$ Divide

Problem Set 2.3

Vocabulary

Use the vocabulary words below to fill in the blanks in the sentences. Vocabulary words may be used more than once.

significant   minutes   tenth   two

1. If the sides in a triangle are accurate to _____ significant digits, then the angles must be accurate to the nearest degree.

2. If the sides in a triangle are accurate to three significant digits, then the angles must be accurate to the nearest ten _____ or _____ of a degree.

3. To find the number of _____ digits in a number, we begin by counting the number of digits from left to right, beginning with the first nonzero digit on the left.

4. The first number below has _____ significant digits, while the second number has four _____ digits.

    9,400   9,400.

Give the number of significant digits in each of the following numbers:

1. 42.3   **2.** 2.05

3. 0.76   **4.** 0.84

5. 3.5723   **6.** 4.189

7. 2.400   **8.** 4010

9. 7900   **10.** 0.0076

If each of the following angles were an angle in a triangle, how many significant digits would a side in the triangle contain?

11. 24°   **12.** 4.3°

13. 76.2°   **14.** 23.45°

15. 36° 24′   **16.** 23° 10′

17. 41.22°   **18.** 45.7°

Problems 19–26 refer to right triangle ABC with C = 90°. Write all answers that are angles to the nearest tenth of a degree.

19. If $A = 40°$ and $c = 16$ cm, find a.   **20.** If $A = 20°$ and $c = 42$ cm, find a.

21. If $B = 20°$ and $b = 12$ ft, find a.   **22.** If $B = 20°$ and $a = 35$ ft, find b.

23. If $a = 5$ mi. and $b = 10$ mi., find A.   **24.** If $a = 6$ mi. and $b = 10$ mi., find B.

25. If $a = 12$ m and $c = 20$ m, find B.   **26.** If $a = 15$ m and $c = 25$ m, find A.

Problems 27–34 refer to right triangle *ABC* with C = 90°. In each case, solve for all the missing parts using the given information.

27. $A = 20°$, $c = 24$ in.   **28.** $A = 40°$, $c = 36$ in.

29. $B = 76°$, $c = 5.8$ mi.   **30.** $B = 21°$, $c = 4.2$ mi.

31. $a = 35$ m, $b = 97$ m   **32.** $a = 78$ m, $b = 83$ m

33. $b = 150$ cm, $c = 200$ cm   **34.** $b = 320$ cm, $c = 650$ cm

35. $B = 26° \, 30'$, $b = 324$ mm   **36.** $B = 53° \, 30'$, $b = 725$ mm

37. $B = 23.45°$, $a = 5.432$ mi.   **38.** $B = 44.44°$, $a = 5.555$ mi.

39. $a = 2.75$ cm, $c = 4.05$ cm   **40.** $a = 62.3$ cm, $c = 73.6$ cm

41. $b = 12.21$ in., $c = 25.52$ in.   **42.** $b = 377.3$ in., $c = 588.5$ in.

In problems 29–32, use the information given in the diagrams to find angle A.

43.

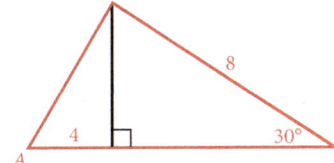

44.

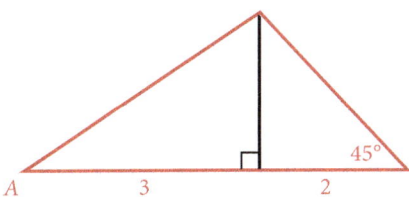

45.

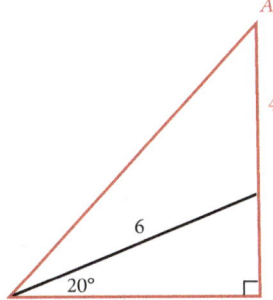

46.

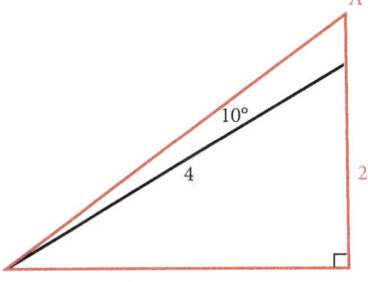

Given the following figure

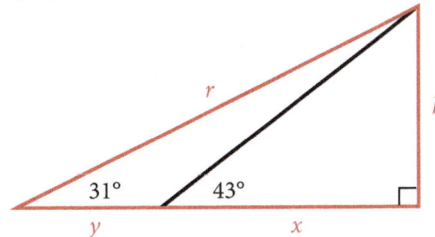

47. If $h = 12$, find x and y.

48. If $x = 20$, find h and y.

49. If $y = 10$, find x and h.

50. If $y = 15$, find r.

Problems 51 through 58 refer to the diagram below.

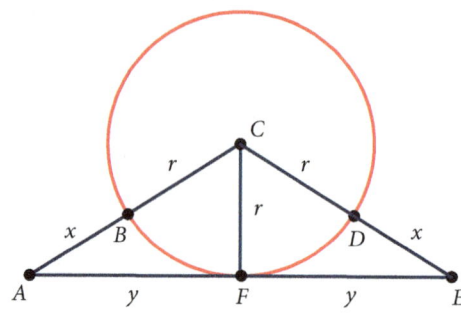

51. If $A = 31°$, find Angle *ECF*.

52. If Angle *ACF* is 51°, find *E*.

53. If $AE = 20$ and $x = 5$, find *r*.

54. If $AC = 24$ and $y = 18$, find *r*.

55. If $A = 35°$ and $r = 12$, find *y*, then find *x*.

56. If $FCE = 26°$ and $r = 20$, find *y*, then find *x*.

57. If $A = 45°$ and $x = 15$, find *r*.

58. If $E = 65°$ and $x = 22$, find *r*.

Getting Ready for the Next Section

Find the following. Round to the nearest tenth.

59. 24 sin 52°

60. 24 cos 52°

61. 213 sin 52.6°

62. 35 tan 61.7°

Find θ. Round to the nearest tenth.

63. $\tan \theta = \dfrac{75}{43}$

64. $\tan \theta = \dfrac{120}{600}$

65. $\sin \theta = \dfrac{24}{25}$

66. $\cos \theta = \dfrac{24}{25}$

KEY WORDS

angle of elevation

angle of depression

bearing of a line

Learning Objectives

A Solve application problems involving right triangles.

B Solve problems involving angles of elevation and depression.

C Solve problems involving the bearing of a line.

We are now ready to put our knowledge of solving right triangles to work to solve some application problems.

A Applications with Right Triangles

EXAMPLE 1 The two equal sides of an isosceles triangle are each 24 centimeters. If each of the two equal angles measure 52°, find the length of the base and the length of the altitude.

Solution An isosceles triangle is any triangle with two equal sides. The angles opposite the two equal sides are called the base angles, and they are always equal. Figure 1 is a picture of our isosceles triangle.

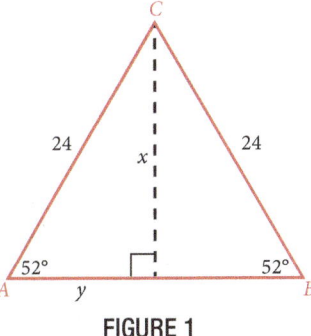

FIGURE 1

We have labeled the altitude x. We can solve for x using a sine ratio.

$$\text{If} \qquad \sin 52° = \frac{x}{24}$$

$$\text{then} \qquad x = 24 \sin 52°$$

$$\approx 19 \text{ centimeters}$$

We have labeled half the base with y. To solve for y, we can use a cosine ratio.

$$\text{If} \qquad \cos 52° = \frac{y}{24}$$

$$\text{then} \qquad y = 24 \cos 52°$$

$$\approx 15 \text{ centimeters}$$

The base is $2y \approx 2(15) = 30$ centimeters

B Angles of Elevation and Depression

For our next applications, we need the following definition.

Definition
An angle measured from the horizontal up is called an **angle of elevation** (or inclination). An angle measured from the horizontal down is called an **angle of depression** (see Figure 2).

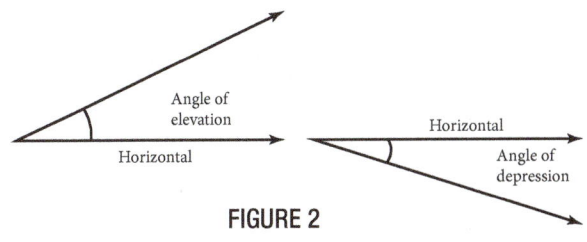

FIGURE 2

These angles of elevation and depression are always considered positive angles.

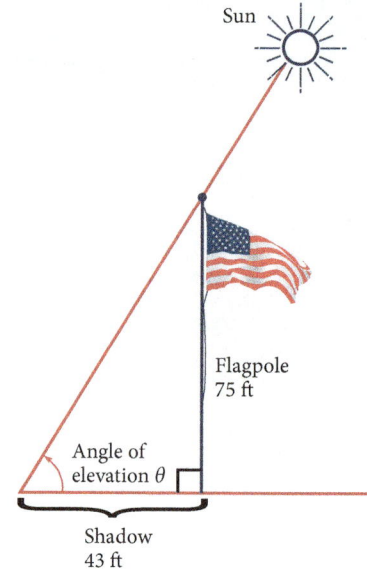

EXAMPLE 2 If a flagpole 75 feet tall casts a shadow 43 feet long, what is the angle of elevation of the sun from the tip of the shadow?

Solution We begin by making a diagram of the situation (Figure 3).

If we let θ = the angle of elevation of the sun, then

$$\tan \theta = \frac{75}{43}$$

$$\tan \theta \approx 1.7442$$

which means $\theta = \tan^{-1}(1.7442) \approx 60°$ *To the nearest degree*

FIGURE 3

EXAMPLE 3 A man climbs 213 meters up the side of a pyramid and finds the angle of depression to his starting point is 52.6°. How high off the ground is he?

Solution Again, we begin by making a diagram of the situation (Figure 4). If x is his height above the ground, we can solve for x using a sine ratio.

If $$\sin 52.6° = \frac{x}{213}$$

then $$x = 213 \sin 52.6°$$

$$= 169 \text{ meters}$$

The man is 169 meters above the ground.

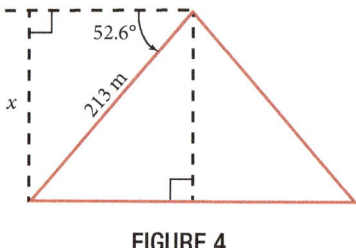

FIGURE 4

EXAMPLE 4 Figure 5 is a diagram that shows how Diane estimates the height h of a flagpole. She can't measure the distance between herself and the flagpole directly because there is a fence in the way. So she stands at point A facing the pole and finds the angle of elevation from point A to the top of the pole to be 61.7°. Then she turns 90° and walks 25.0 feet to point B, where she measures the angle between her path and the base of the pole. She finds that angle is 54.5°. Use this information to find the height of the pole.

Solution First we find x in right triangle ABC using a tangent ratio:

$$\tan 54.5° = \frac{x}{25.0}$$

$$x = 25.0(\tan 54.5°)$$

$$\approx 35.0487 \text{ feet}$$

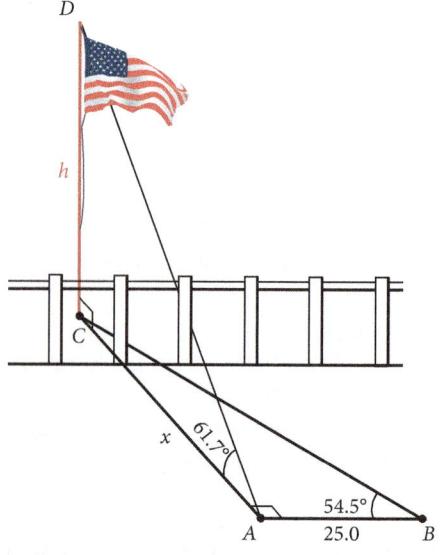

FIGURE 5

Without rounding, we use this value for x to find h in right triangle ACD, using another tangent ratio:

$$\tan 61.7° = \frac{h}{35.0487}$$

$$h \approx 35.0487(\tan 61.7°)$$

$$\approx 65.1 \text{ feet}$$

Note that if it weren't for the fence, she could have measured x directly and used just one triangle to find the height of the flagpole.

EXAMPLE 5 Figure 6 shows the topographic map we mentioned in the introduction to this chapter. Suppose Stacey and Amy are climbing Bishop's Peak.
Stacey is at position S, and Amy is at position A. Find the angle of elevation from Amy to Stacey.

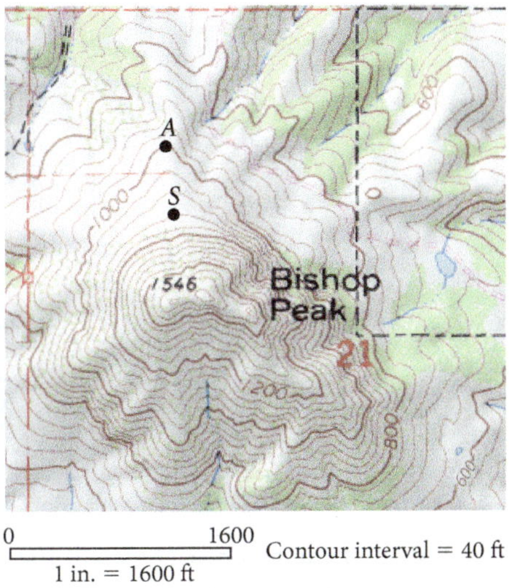

Solution To solve this problem, we have to use two pieces of information from the legend on the map. First, we need to find the horizontal distance between the two people. The legend indicates that 1 inch on the map corresponds to an actual horizontal distance of 1,600 feet. If we measure the distance from Amy to Stacy with a ruler, we find it is $\frac{3}{8}$ inch. Multiplying this by 1,600, we have

$$\frac{3}{8} \cdot 1,600 = 600 \text{ feet}$$

This is the actual horizontal distance from Amy to Stacey.

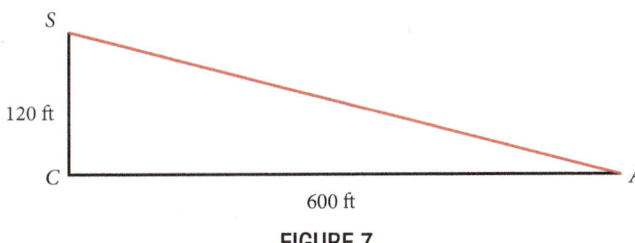

FIGURE 6

Next, we need the vertical distance between them—Stacey is above Amy. We find this by counting the number of contour intervals between them. There are three. From the legend on the map we know that the elevation changes by 40 feet between any two contour lines. Therefore, Stacey is 120 feet above Amy. Figure 7 is a triangle that models the information we have so far:

FIGURE 7

The angle of elevation from Amy to Stacey is angle A. To find A, we use the tangent ratio:

$$\tan A = \frac{120}{600} = 0.2$$

$$A = \tan^{-1}(0.2) \approx 11.3° \qquad \textit{To the nearest tenth of a degree}$$

Amy must look up at an angle of 11.3° from straight ahead to see Stacey.

EXAMPLE 6   The London Eye, pictured below, is a Ferris wheel that opened in London in the year 2000. It has 32 capsules, each holding 25 passengers. It is suspended from two struts that are anchored to the ground at an angle of 65°. If the diameter of the wheel is 120 meters, and the bottom of the wheel is 15 meters above the ground, find the distance *AB*.

Solution   Figure 8 is a simplified model of the London Eye, labeled with the information in the problem.

Ferris Wheel Diagram

Simplified Mathematical Model

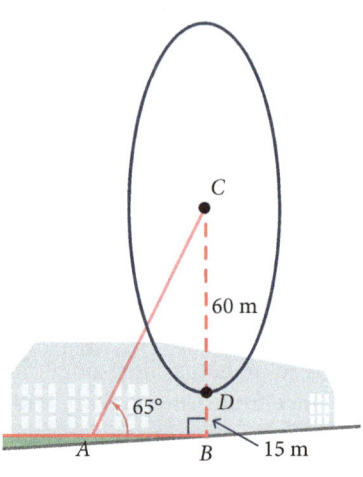

FIGURE 8

The radius of the wheel, *CD*, is half the diameter, or 60 meters. The distance *BC* is the radius plus 15 meters, or 75 meters. Using our tangent function, we have

$$\tan 65° = \frac{75}{AB}$$

$$AB \tan 65° = 75$$

$$AB = \frac{75}{\tan 65°}$$

$$\approx 35 \text{ meters} \quad \textit{To two significant digits}$$

EXAMPLE 7   A woman standing on the street looks up to the top of a building and finds the angle of elevation is 38°. She then walks one block farther away (440 feet) and finds the angle of elevation to the top of the building is now 28°. How far away from the building is she when she makes her second observation? (See Figure 9.)

Solution   There are two right triangles in the diagram. The vertical leg of each triangle is *h*. The horizontal legs are *x* and *x* + 440. If we use the tangent function, we can write two equations that involve both *x* and *h*, which is a system of equations. Here is how we proceed.

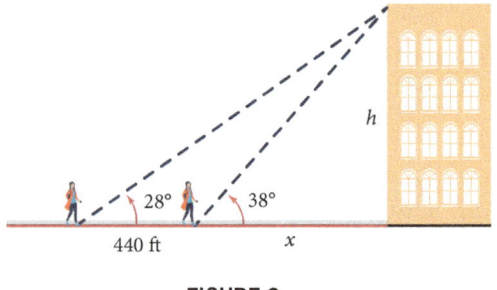

FIGURE 9

Two equations involving both *x* and *h*
$$\begin{cases} \tan 38° = \dfrac{h}{x} & \Rightarrow & h = x \tan 38° \\[2mm] \tan 28° = \dfrac{h}{x + 440} & \Rightarrow & h = (x + 440) \tan 28° \end{cases}$$
Solve each equation for *h*

Setting the two expressions for h equal to each other gives us an equation that involves only x.

Since $$h = h$$

we have $$x \tan 38° = (x + 440) \tan 28°$$

$$x \tan 38° = x \tan 28° + 440 \tan 28°$$ *Distributive property*

$$x \tan 38° - x \tan 28° = 440 \tan 28°$$ *Subtract x tan 38 from each side*

$$x (\tan 38° - \tan 28°) = 440 \tan 28°$$ *Factor x from each term on the left side*

$$x = \frac{440 \tan 28°}{\tan 38° - \tan 28°}$$ *Divide each side by the coefficient of x*

$$= 937 \text{ feet}$$ *Three significant digits*

The person is $937 + 440 = 1,377$ feet away.

C Bearing of a Line

Our next applications are concerned with what is called the **bearing of a line**. It is used in navigation and surveying.

> **Definition**
> The **bearing of a line** *l* is the acute angle formed by the north–south line and the line *l*. The notation used to designate the bearing of a line begins with N or S (for north or south), followed by the number of degrees in the angle, and ends with E or W (for east or west). Some examples are shown in Figure 9.

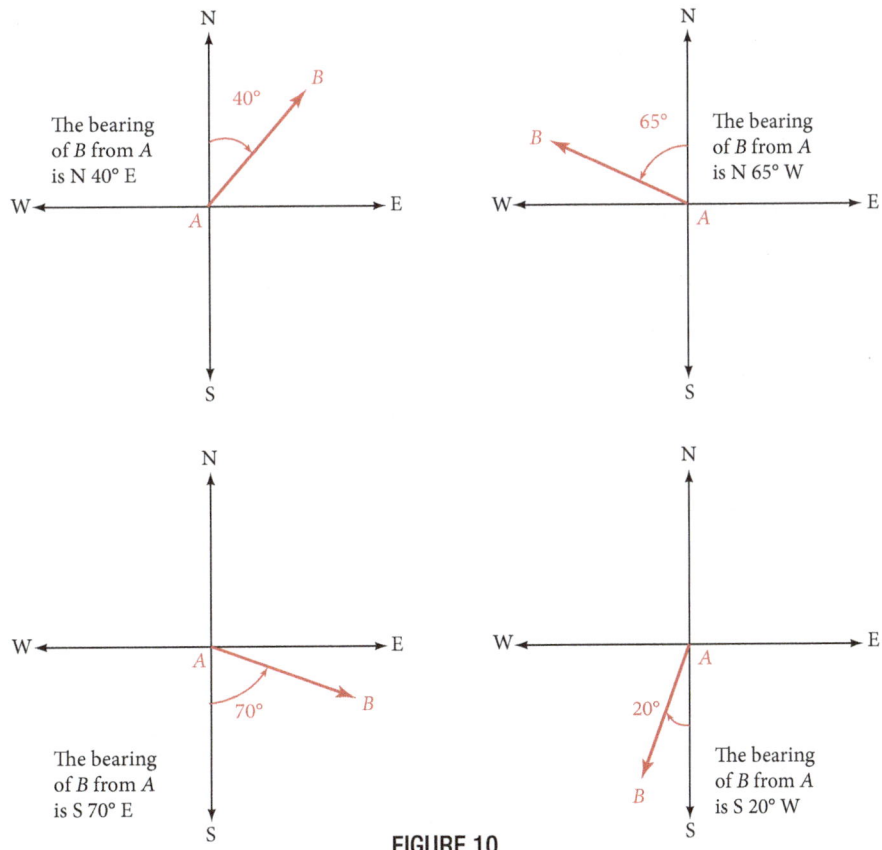

FIGURE 10

EXAMPLE 8   A boat travels on a course of bearing N 52.5° E for a distance of 238 miles. How many miles north and how many miles east has the boat traveled?

Solution In Figure 11 we show a diagram of the situation, where we put our N–S–E–W system at the boat's starting point.

Solving for x with a sine ratio and y with a cosine ratio and rounding our answers to three significant digits, we have:

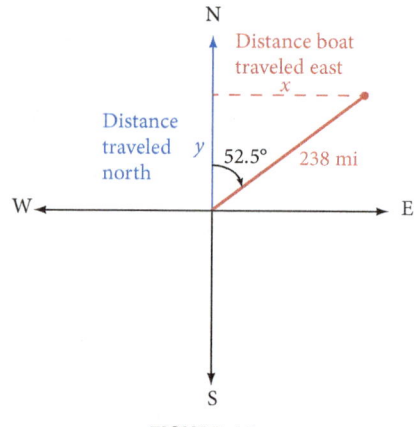

If	$\sin 52.5° = \dfrac{x}{238}$	If	$\cos 52.5° = \dfrac{y}{238}$
then	$x \approx 238(\sin 52.5°)$	then	$x \approx 238(\cos 52.5°)$
	≈ 189 miles		≈ 145 miles

Traveling 238 miles on a line that is N 52.5° E will get you to the same place as traveling 145 miles north and then 189 miles east.

FIGURE 11

EXAMPLE 9   San Luis Obispo, California, is 12 miles due north of Grover Beach. If Arroyo Grande is 4.6 miles due east of Grover Beach, what is the bearing of San Luis Obispo from Arroyo Grande?

Solution Since we are looking for the bearing of San Luis Obispo *from* Arroyo Grande, we will put our N-S-E-W system on Arroyo Grande.

We solve for θ using the tangent ratio.

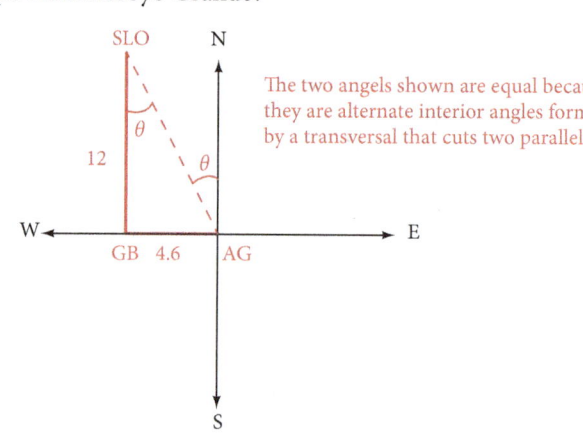

$$\tan \theta = \frac{4.6}{12}$$

$$\theta = \tan^{-1}\left(\frac{4.6}{12}\right)$$

which means $\theta = 21°$ to the nearest degree

The bearing of San Luis Obispo from Arroyo Grande is N 21° W.

The two angles shown are equal because they are alternate interior angles formed by a transversal that cuts two parallel lines.

FIGURE 12

Problem Set 2.4

Vocabulary

Use the vocabulary words below to fill in the blanks in the sentences.

elevation   depression   bearing

1. An angle measured from the horizontal up is called an angle of _____ .

2. An angle measured from the horizontal down is called an angle of _____ .

3. The _____ of a line *l* is the acute angle formed by the north-south line and the line *l*.

Solve each of the following problems. In each case, be sure to make a diagram of the situation with all the given information labeled.

1. **Isosceles Triangle** The two equal sides of an isosceles triangle are each 42 centimeters. If the base measures 30 centimeters, find the height and the measure of the two equal angles. Round your answer to the nearest tenth.

2. **Equilateral Triangle** An equilateral triangle (one with all sides the same length) has an altitude of 4.3 inches. Find the length of the sides.

3. **Length of an Escalator** How long should an escalator be if it is to make an angle of 33° with the floor and carry people a vertical distance of 21 feet between floors? Round to the nearest tenth.

4. **Height of a Hill** A road up a hill makes an angle of 5° with the horizontal. If the road from the bottom of the hill to the top of the hill is 2.5 miles long, how high is the hill?

5. **Circus Tent** A rope from the top of a circus tent pole is 72.5 feet long and is anchored to the ground 43.2 feet from the bottom of the pole. What angle does the rope make with the pole? (Give your answer to the nearest tenth of a degree.)

6. **Leaning Ladder** A ladder is leaning against the top of a wall that is 7.0 feet high. If the bottom of the ladder is 4.5 feet from the wall, what is the angle between the ladder and the wall? (Give your answer to the nearest tenth of a degree.)

7. **Angle of Elevation** If a flagpole that is 73 feet tall casts a shadow 51 feet long, what is the angle of elevation of the sun? (Give your answer to the nearest tenth of a degree.)

8. **Angle of Elevation** If the angle of elevation of the sun is 63.4° when a building casts a shadow of 37.5 feet, what is the height of the building?

9. **Height of a Door** From a point on the floor, the angle of elevation to the top of a door is 47°, while the angle of elevation to the ceiling above the door is 59°. If the ceiling is 10 feet above the floor, what is the vertical dimension of the door? (See Figure 13.)

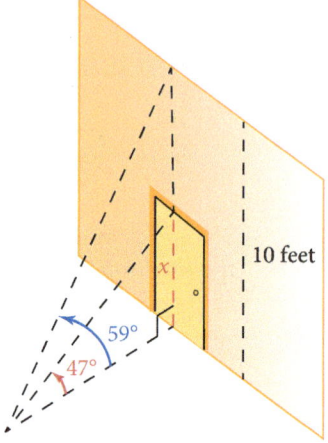

FIGURE 13

10. **Height of a Building** A man standing on the roof of a building 60.0 feet high looks down to the building next door. He finds the angle of depression to the roof of that building from the roof of his building to be 34.5°, while the angle of depression from the roof of his building to the bottom of the building next door is 63.2°. How tall is the building next door?

11. **Height of a Tree** An ecologist wishes to find the height of a redwood tree that is on the other side of a creek, as shown in Figure 14. From point A he finds the angle of elevation to the top of the tree is 10.7°. He then walks 24.8 feet at a right angle from point A and finds angle B is 86.6°. What is the height of the tree?

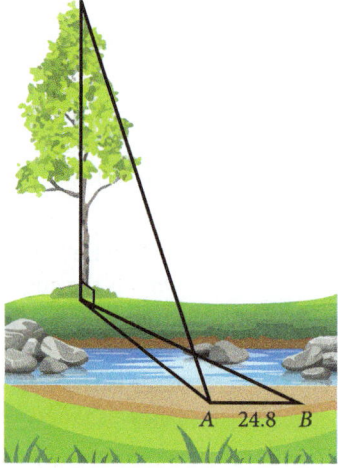

FIGURE 14

12. **Distance to Accident** A helicopter makes a forced landing at sea. The last radio signal received at station C gives the bearing of the helicopter from C as N 57.5° E at an altitude of 426 feet. An observer at C sighted the helicopter and gives $\angle DCB$ as 12.3°. How far will a rescue boat at A have to travel to reach any survivors at B? (See Figure 15.)

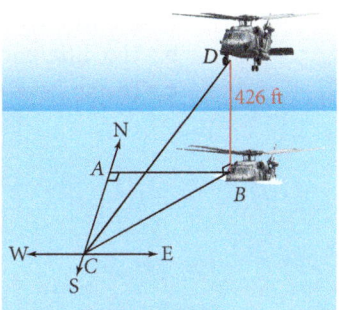

FIGURE 15

13. **Dimensions of a Mirror** A person standing 150 centimeters from a mirror notices that the angle of depression from his eyes to the bottom of the mirror is 12°, while the angle of elevation to the top of the mirror is 11°. Find the vertical dimension of the mirror. (See Figure 16.)

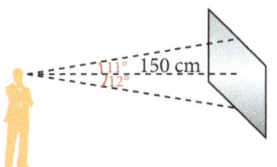

FIGURE 16

14. **Height of Sand Pile** A person standing on top of a sand pile that is 15 feet high wishes to estimate the width of the pile. He visually locates two rocks on the ground below, at the base of the sand pile. The rocks are on opposite sides of the sand pile, and he and the two rocks are in line with one another. If the angles of depression from the top of the sand pile to each of the rocks are 27° and 19°, how far apart are the rocks?

15. **Angle of Depression** From a plane 3,000 feet in the air, the angle of depression to the beginning of the runway is 3° 50′. What is the horizontal distance between the plane and runway? (That is, if the plane flew at the same altitude, how far would it be to the runway?) Give your answer to the nearest tenth of a mile. (5,280 feet = 1 mile)

16. **Angle of Depression** From the edge of a platform 25.1 meters high, a diver notices the angles of depression to each end of the pool are 14.7° and 78.9°. What is the length of the pool?

Topographic Map   Figure 17 shows the topographic map we used in Example 8 of this section. Recall that Stacey is at position *S* and Amy is at position *A*. In Figure 19, Travis, a third hiker, is at position *T*.

17. If the distance between *A* and *T* on the map in Figure 19 is 0.5 inch, find each of the following:

    a. The horizontal distance between Amy and Travis

    b. The difference in elevation between Amy and Travis

    c. The angle of elevation from Travis to Amy

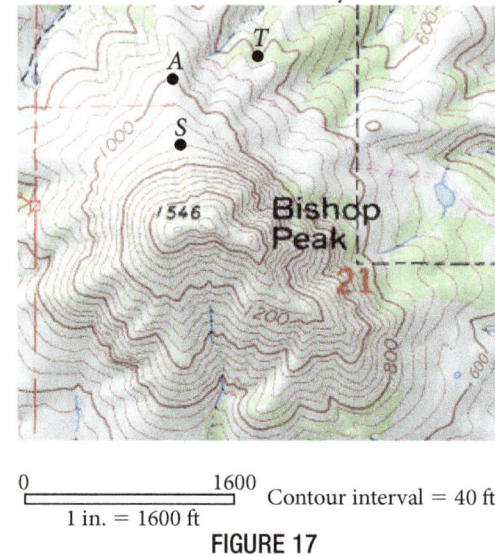

0   1600   Contour interval = 40 ft

1 in. = 1600 ft

FIGURE 17

18. If the distance between *S* and *T* on the map in Figure 17 is $\frac{5}{8}$ inch, find each of the following:

    a. The horizontal distance between Stacey and Travis

    b. The difference in elevation between Stacey and Travis

    c. The angle of elevation from Travis to Stacey

19. Lompoc, California, is 18 miles due south of Nipomo. Buellton, California, is due east of Lompoc and S 65° E from Nipomo. How far is Lompoc from Buellton?

20. A tree on one side of a river is due west of a rock on the other side of the river. From a stake 21 yards north of the rock, the bearing of the tree is S 18.2° W. How far is it from the rock to the tree?

21. Bearing A boat leaves the harbor entrance and travels 25 miles in the direction N 42° E. The captain then turns the boat 90° and travels another 18 miles in the direction S 48° E. At that time, how far is the boat from the harbor entrance, and what is the bearing of the boat from the harbor entrance? (See Figure 18.)

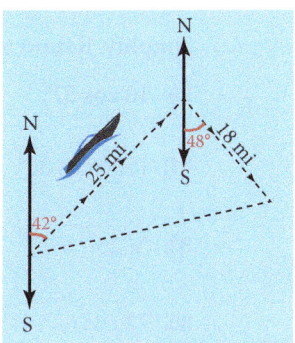

FIGURE 18

22. Bearing A man wandering in the desert walks for 2.3 miles in the direction S 31° W. He then turns 90° and walks 3.5 miles in the direction N 59° W. At that time, how far is he from his starting point, and what is his bearing from his starting point?

23. Bearing A boat travels on a course of bearing N 37° W for 79.5 miles. How many miles north and how many miles west has the boat traveled?

24. Bearing A boat travels on a course of bearing S 63° E for 100 miles. How many miles south and how many miles east has the boat traveled?

25. Ferris Wheel Referring to the model of the London Eye Ferris wheel in Figure 8, find the distance from A to C. Round to the nearest foot.

26. Ferris Wheel The photo and model below are of the Ferris wheel known as the Reisenrad, that is located in Vienna. The diameter of the wheel is 197 feet, and the bottom of the wheel is 12 feet above the ground. Find angle A if AB is 40 feet. Round to the nearest degree.

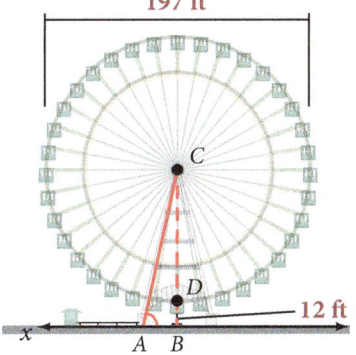

Getting Ready for the Next Section

Find θ. Round to nearest degree.

27. $\tan \theta = 5$

28. $\tan \theta = 0.6$

Simplify. Round to two significant figures.

29. $10 \cos 52°$

30. $10 \sin 52°$

31. $1500 \cos 30°$

32. $1500 \sin 30°$

33. $\sqrt{25^2 + 15^2}$

34. $\sqrt{5^2 + 12^2}$

35. $72 \cos 63° + 37 \cos 35°$

36. $72 \sin 63° + 37 \sin 35°$

37. $\dfrac{10}{\sin 79°}$

38. $\dfrac{20}{\cos 59°}$

2.5 Videos

Key Words

vector quantities

scalars

resultant vector

exact values

Learning Objectives

A   Solve problems involving vectors.

B   Find the horizontal and vertical components of a vector.

C   Solve problems involving navigation.

Many of the quantities that describe the world around us have both magnitude and direction, while others have only magnitude. Quantities that have magnitude and direction are called *vector quantities*, while quantities with magnitude only are called *scalars*. Some examples of vector quantities are force, velocity, and acceleration. For example, a car traveling 50 miles per hour due south has a different velocity then another car traveling due north at 50 miles per hour, while a third car traveling at 25 miles per hour due north has a velocity that is different from either of the first two.

One way to represent vector quantities is with arrows. The direction of the arrow represents the direction of the vector quantity and the length of the arrow corresponds to the magnitude. For example, the velocities of the three cars we mentioned above could be represented like this

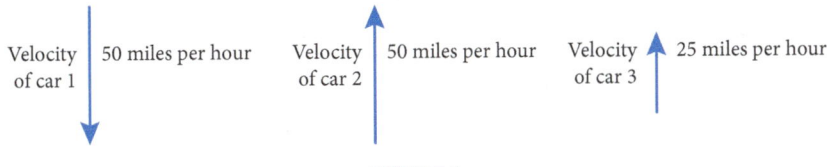

Velocity of car 1   50 miles per hour   Velocity of car 2   50 miles per hour   Velocity of car 3   25 miles per hour

FIGURE 1

Notation   To distinguish between vectors and scalars we will write the letters used to represent vectors with bold face type, like $\mathbf{U}$ or $\mathbf{V}$. (When you write them on paper, put an arrow above them like this: $\vec{U}$ or $\vec{V}$.) The magnitude of a vector is represented with absolute value symbols. For example, the magnitude of $\mathbf{V}$ is written $|\mathbf{V}|$. Table 1 illustrates further.

Table 1	
Notation	**The Quantity Is**
$\mathbf{V}$	a vector
$\vec{V}$	a vector
$\overrightarrow{AB}$	a vector
x	a scalar
$|\mathbf{V}|$	the magnitude of vector V, a scalar

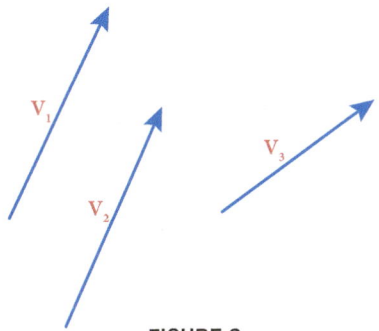

FIGURE 2

A   Equality for Vectors

The position of a vector in space is unimportant. Two vectors are equivalent if they have the same magnitude and direction. In Figure 2, $\mathbf{V}_1 = \mathbf{V}_2 \neq \mathbf{V}_3$. The vectors $\mathbf{V}_1$ and $\mathbf{V}_2$ are equivalent because they have the same magnitude and the same direction.

Addition of Vectors

The sum of the vectors **U** and **V**, written **U** + **V** and sometimes called the *resultant vector*, is the vector that extends from the tail of **U** to the tip of **V** when the tail of **V** coincides with the tip of **U**. Figure 3 illustrates.

The vector sum **U** + **V** can also be represented by the diagonal of the parallelogram, the adjacent sides of which are formed by putting the tails of **U** and **V** together (Figure 4).

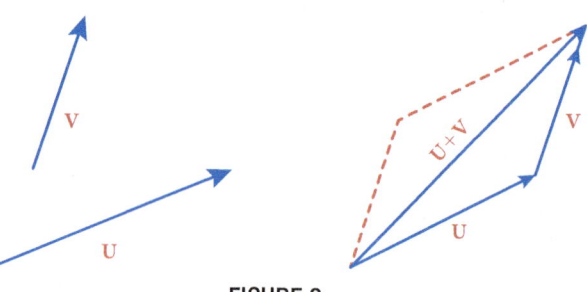

FIGURE 3

Example 1 illustrates how vectors can be used to solve motion problems.

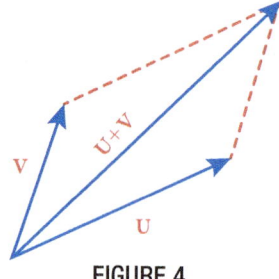

FIGURE 4

EXAMPLE 1   A boat is crossing a river that runs due north. The boat is pointed due east and is moving through the water at 10.0 miles per hour. If the current of the river is a constant 2.0 miles per hour, find the true course of the boat through the water to three significant digits.

Solution Problems like this are a little difficult to read the first time they are encountered. Even though the boat is "headed" due east, the current is pushing it a little toward the north, so it is actually on a course that will take it east and a little north. By representing the heading of the boat and the current of the water with vectors, we can find the true course of the boat.

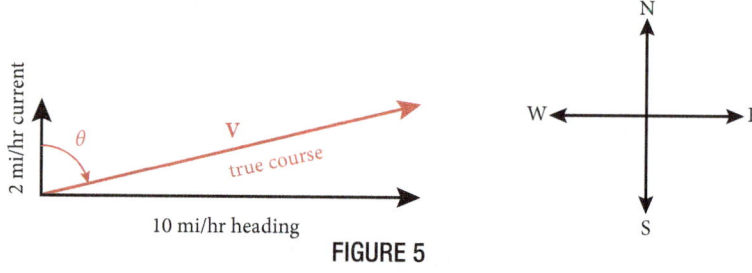

FIGURE 5

We find θ using a tangent ratio:

$$\tan \theta = \frac{10}{2} = 5 \quad \text{so} \quad \theta = 78.7° \text{ to the nearest tenth}$$

If we let **V** represent the true course of the boat, then we can find the magnitude of **V** using the Pythagorean theorem or a trigonometric ratio. Using the sine ratio, we have

$$\sin \theta = \frac{10}{|\mathbf{V}|}$$

$$|\mathbf{V}| = \frac{10}{\sin \theta}$$

$$= \frac{10}{\sin 78.7°}$$

$$= 10.2$$

The true course of the boat is 10.2 miles per hour at N 78.7° E. That is, the vector **V**, which represents the motion of the boat with respect to the banks of the river, has magnitude of 10.2 miles per hour and a direction N 78.7° E.

B Horizontal and Vertical Components

Many times it is convenient to write vectors in terms of their horizontal and vertical components. To do so, we first superimpose a coordinate system on the vector in question so that the tail of the vector is at the origin. Figure 6 shows a vector with magnitude 10 making an angle of 52° with the horizontal.

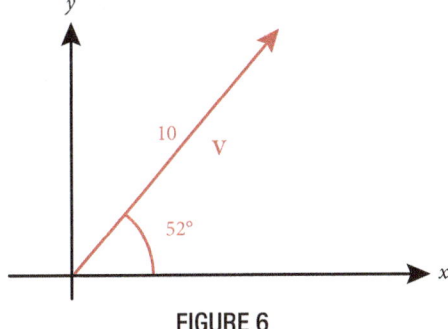

FIGURE 6

The horizontal and vertical components of vector **V** in Figure 6 are the horizontal and vertical vectors whose sum is **V**. They are shown in Figure 7.

Note that in Figure 7, we labeled the horizontal component of **V** as $\mathbf{V}_x$ and the vertical component as $\mathbf{V}_y$. We can find the magnitudes of the components by using sine and cosine ratios.

$|\mathbf{V}_x| = |\mathbf{V}| \cos 52° = 10(\cos 52°) = 6.2$ to the nearest tenth

$|\mathbf{V}_y| = |\mathbf{V}| \sin 52° = 10(\sin 52°) = 7.9$ to the nearest tenth

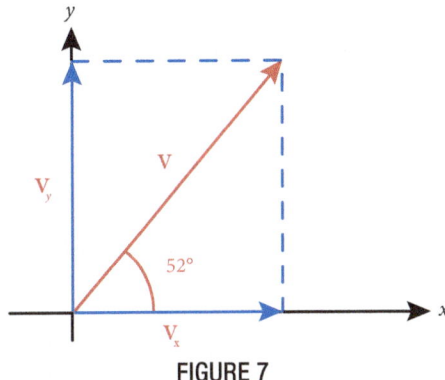

FIGURE 7

EXAMPLE 2   A bullet is fired into the air with an initial velocity of 1500 feet per second at an angle of 30° from the horizontal. Find the horizontal and vertical components of the velocity vector.

Solution Figure 8 is a diagram of the situation. The magnitudes of $\mathbf{V}_x$ and $\mathbf{V}_y$ from Figure 8 to two significant digits are as follows:

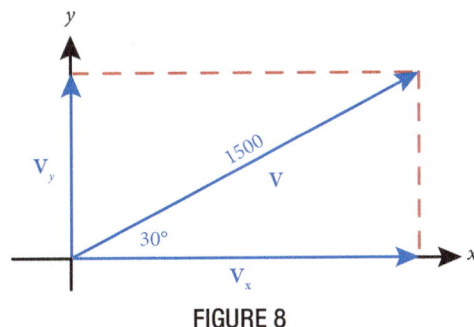

FIGURE 8

$|\mathbf{V}_x| = 1500 \cos 30° \approx 1300$ feet per second

$|\mathbf{V}_y| = 1500 \sin 30° = 750$ feet per second

The bullet has a horizontal velocity of 1300 feet per second and a vertical velocity of 750 feet per second.

The magnitude of a vector can be written in terms of the magnitude of its horizontal and vertical components.

By the Pythagorean theorem we have,

$$|\mathbf{V}| = \sqrt{|\mathbf{V}_x|^2 + |\mathbf{V}_y|^2}$$

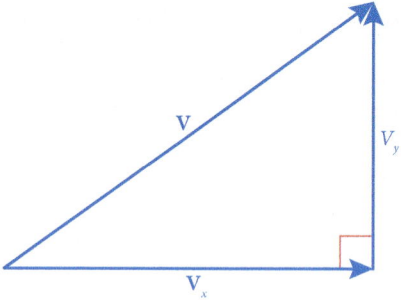

FIGURE 9

EXAMPLE 3 At the 1997 Washington County Fair in Oregon, David Smith, Jr., The Bullet, was shot from a cannon. Suppose his initial velocity as he exits the cannon can be represented by a vector with magnitude 54 miles/hour, at an angle of 45 with the horizontal, find the magnitude of the horizontal and vertical components of his initial velocity.

Solution The diagram in Figure 10 shows all three velocities.

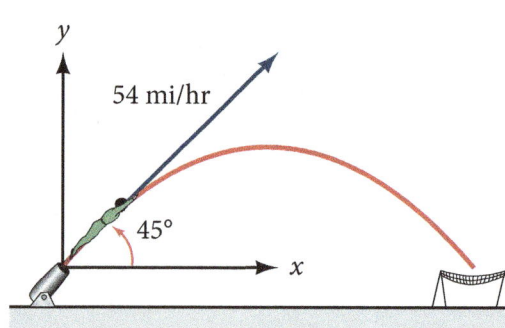

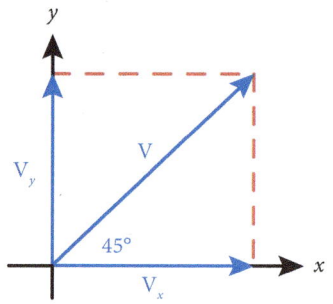

FIGURE 10

Proceeding as we did in Example 2, we have

$$|\mathbf{V}_x| = 54 \cos 45° \approx 38 \text{ feet per second}$$

$$|\mathbf{V}_y| = 54 \sin 45° \approx 38 \text{ feet per second}$$

As you would expect, when the angle of the cannon to the horizontal is 45°, the magnitude of the horizontal and vertical components of the initial velocity will be equal.

EXAMPLE 4   An arrow is shot into the air so that its horizontal velocity is 25 feet per second and its vertical velocity is 15 feet per second. Find the velocity of the arrow.

Solution Figure 10 shows the velocity vector and its components along with the angle of elevation of the velocity vector.
The magnitude of the velocity is given by

$$|V| = \sqrt{25^2 + 15^2}$$

$$\approx 29 \text{ feet per second to the nearest whole number}$$

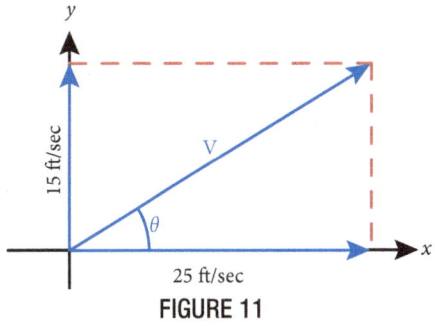

FIGURE 11

We can find the angle of elevation using a tangent ratio

$$\tan \theta = \frac{|V_y|}{|V_x|}$$

$$= \frac{15}{25}$$

$$\tan \theta = 0.6$$

so $\theta = \tan^{-1}(0.6) \approx 31°$ to the nearest degree

The arrow was shot into the air at 29 feet per second at an angle of elevation of 31°.

C Navigation

Another way to specify the bearing of a moving object is by giving the angle between the north-south line and the vector representing the velocity of the object. Figure 12 shows two vectors that represent the velocities of two ships, one traveling at 14 miles per hour with bearing 120° and another traveling at 12 miles per hour with a bearing of 240°. Note that this method of specifying bearing is different than the method we used previous to this.

With this method, every course is given with respect to the north-south line, and the bearing angle is always measured clockwise from due north. This new method of giving the direction of a moving object is a little simpler. Under our previous method, we would specify the direction of the ship traveling at 14 miles per hour as S 60° E.

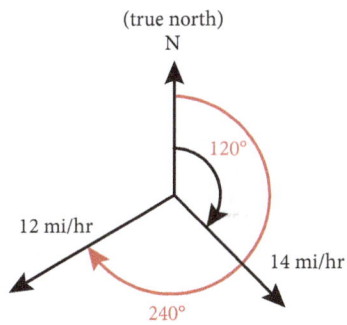

FIGURE 12

EXAMPLE 5   A boat travels for 72 miles on a course with bearing 27° and then changes its course to travel 37 miles on a course with bearing 55°. How far north and how far east has the boat traveled on this 109 mile trip?

Solution We can solve this problem by representing each part of the trip with a vector and then writing each vector in terms of its horizontal and vertical components. Figure 13 shows the vectors that represent the two parts of the trip, along with their horizontal and vertical components.

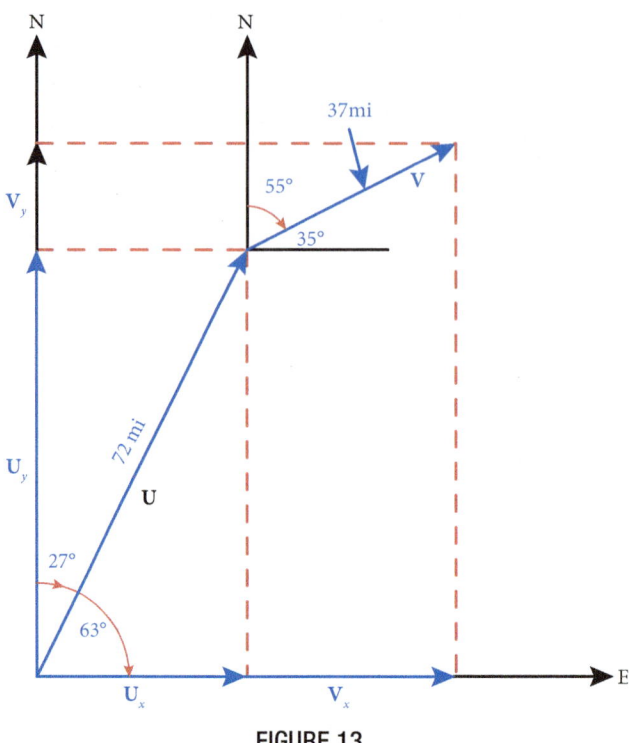

FIGURE 13

As Figure 13 indicates, the total distance traveled east is given by the sum of the horizontal components, while the total distance traveled north is given by the sum of the vertical components.

$$\text{Total distance traveled east} = |U_x| + |V_x|$$
$$= 72 \cos 63° + 37 \cos 35°$$
$$= 63.0 \text{ miles to the nearest tenth}$$

$$\text{Total distance traveled north} = |U_y| + |V_y|$$
$$= 72 \sin 63° + 37 \sin 35°$$
$$= 85.4 \text{ miles to the nearest tenth}$$

Problem Set 2.5

Vocabulary

Use the vocabulary words below to fill in the blanks in the sentences.

velocity   components   horizontal   scalars   magnitude

1. Quantities that have _____ and direction are called vector quantities, while quantities with magnitude only are called _____ .

2. Many times it is convenient to write vectors in terms of their _____ and vertical _____ .

3. Another way to specify the bearing of a moving object is by giving the angle between the north-south line and the vector representing the _____ of the object.

Draw vectors representing the following velocities:

1. 30 miles per hour due north   2. 30 miles per hour due south

3. 30 miles per hour due east   4. 30 miles per hour due west

5. 50 centimeters per second N 30° W   6. 50 centimeters per second N 30° E

7. 20 feet per minute S 60° E   8. 20 feet per minute S 60° W

Draw vectors representing the course of a ship that travels

9. 75 miles on a course with bearing 30°

10. 75 miles on a course with bearing 330°

11. 25 miles on a course with bearing 135°

12. 25 miles on a course with bearing 225°

13. **Heading** A boat is crossing a river that runs due north. The heading of the boat (the direction the boat is pointed) is due east and it is moving through the water at 12.0 miles per hour. If the current of the river is a constant 3 miles per hour, find the true course of the boat to three significant digits.

14. **Heading** A boat is crossing a river that runs due east. The heading of the boat is due south and its speed is 11.0 feet per second. If the current of the river is 2 feet per second, find the true course of the boat to three significant digits.

15. **Heading** An airplane is headed N 30.0° E and is traveling at 200 miles per hour through the air. The air currents are moving at a constant 30 miles per hour in the direction N 60.0° W. Find the true course of the plane. (That is, find its speed with respect to the ground in miles per hour and its direction with respect to the ground.) Use three significant digits.

16. **Heading** An airplane is headed S 50.0° E and is moving through the air at 140 miles per hour. The air currents are moving at a constant 24 miles per hour in the direction S 40.0° W. Find the true course of the plane. Use three significant digits.

17. **Heading** A ship headed due west is moving through the water at a constant 12 miles per hour. However, the true course of the ship is in the direction N 78° W. If the current of the water is running due north at a constant rate of speed, find the speed of the current.

18. **Heading** A plane headed due east is moving through the air at a constant 180 miles per hour. Its true course, however, is in the direction N 65.0° E. If the wind currents are moving due north at a constant rate, find the speed of these currents.

Each problem below refers to a vector V with magnitude $|V|$ that forms an angle θ with the positive x-axis. In each case, give the magnitude of the horizontal and vertical components of V.

19. $|V| = 40, \theta = 30°$ 20. $|V| = 40, \theta = 60°$

21. $|V| = 13.8, \theta = 24.2°$ 22. $|V| = 17.6, \theta = 67.2°$

23. $|V| = 420, \theta = 36° \, 10'$ 24. $|V| = 380, \theta = 16° \, 40'$

25. $|V| = 64, \theta = 150°$ 26. $|V| = 48, \theta = 120°$

For each problem below, the magnitude of the horizontal and vertical components of vector V are given. In each case find the magnitude of V.

27. $|V_x| = 30, |V_y| = 40$

28. $|V_x| = 8, |V_y| = 6$

29. $|V_x| = 35.0, |V_y| = 26.0$

30. $|V_x| = 45.0, |V_y| = 15.0$

31. $|V_x| = 4.5, |V_y| = 3.8$

32. $|V_x| = 2.2, |V_y| = 5.8$

33. Velocity A bullet is fired into the air with an initial velocity of 1,200 feet per second at an angle of 45° from the horizontal. Find the magnitude of the horizontal and vertical components of the velocity vector.

34. Velocity A bullet is fired into the air with an initial velocity of 1,800 feet per second at an angle of 60° from the horizontal. Find the horizontal and vertical components of the velocity.

35. Velocity Use the results of Problem 33 to find the horizontal distance traveled by the bullet in 3 seconds. (Neglect the resistance of air on the bullet.)

36. Velocity Use the results of Problem 34 to find the horizontal distance traveled by the bullet in 2 seconds.

37. Velocity An arrow is shot into the air so that its horizontal velocity is 35.0 feet per second and its vertical velocity is 15.0 feet per second. Find the velocity of the arrow.

38. Velocity The horizontal and vertical components of the velocity of an arrow shot into the air are 15.0 feet per second and 25.0 feet per second, respectively. Find the velocity of the arrow.

39. Bearing A ship travels for 135 kilometers on a course with bearing 138°. How far east and how far south has it traveled?

40. Bearing A plane flies for 3 hours at 230 kilometers per hour on a course with bearing 215°. How far west and how far south does it travel in the 3 hours?

41. Bearing A plane travels for 175 miles on a course with bearing 18° and then changes its course to 49° and travels another 120 miles. Find the total distance traveled north and the total distance traveled east.

42. Bearing A ship travels on a course with bearing 168° for 68 miles and then changes its course to 120° and travels for another 112 miles. Find the total distance south and the total distance east that the ship traveled.

Maintaining Your Skills

The problems that follow review material we covered in Chapter 1.

43. Draw 135° in standard position, locate a convenient point on the terminal side, and then find sin 135°, cos 135°, and tan 135°.

44. Draw −270° in standard position, locate a convenient point on the terminal side, and then find sine, cosine, and tangent of −270°.

45. Find $\sin \theta$ and $\cos \theta$ if the terminal side of θ lies along the line $y = 2x$ in quadrant I.

46. Find $\sin \theta$ and $\cos \theta$ if the terminal side of θ lies along the line $y = -x$ in quadrant II.

47. Find x if the point $(x, -8)$ is on the terminal side of θ and $\sin \theta = -4/5$.

48. Find y if the point $(-6, y)$ is on the terminal side of θ and $\cos \theta = -3/5$.

Examples

1.

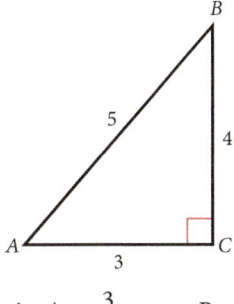

$$\sin A = \frac{3}{5} = \cos B$$

$$\cos A = \frac{4}{5} = \sin B$$

$$\tan A = \frac{3}{4} = \cot B$$

$$\cot A = \frac{4}{3} = \tan B$$

$$\sec A = \frac{5}{4} = \csc B$$

$$\csc A = \frac{5}{3} = \sec B$$

Definition II: Trigonometric Functions [2.1]

If triangle ABC is a right triangle with $C = 90°$, then the six trigonometric functions for A are

$$\sin A = \frac{\text{Side opposite } A}{\text{Hypotenuse}} = \frac{a}{c}$$

$$\cos A = \frac{\text{Side adjacent } A}{\text{Hypotenuse}} = \frac{b}{c}$$

$$\tan A = \frac{\text{Side opposite } A}{\text{Side adjacent } A} = \frac{a}{b}$$

$$\cot A = \frac{\text{Side adjacent } A}{\text{Side opposite } A} = \frac{b}{a}$$

$$\sec A = \frac{\text{Hypotenuse}}{\text{Side adjacent } A} = \frac{c}{b}$$

$$\csc A = \frac{\text{Hypotenuse}}{\text{Side opposite } A} = \frac{c}{a}$$

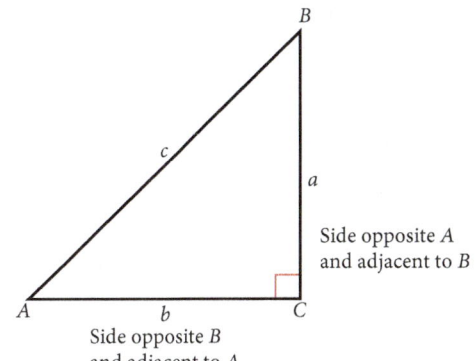

Side opposite A and adjacent to B

Side opposite B and adjacent to A

2. $\sin 3° = \cos 87°$

$\cos 10° = \sin 80°$

$\tan 15° = \cot 75°$

$\cot A = \tan(90° - A)$

$\sec 30° = \csc 60°$

$\csc 45° = \sec 45°$

Cofunction Theorem [2.1]

A trigonometric function of an angle is always equal to the cofunction of its complement. In symbols, since the complement of x is $90° - x$, we have

$$\sin x = \cos(90° - x)$$

$$\cos x = \sin(90° - x)$$

$$\tan x = \cot(90° - x)$$

3. The values given in the table are called *exact values* because they are not decimal approximations as you would find in Tables II or III.

Trigonometric Functions of Special Angles [2.1]

θ	30°	45°	60°
$\sin \theta$	$\frac{1}{2}$	$\frac{1}{\sqrt{2}}$ or $\frac{\sqrt{2}}{2}$	$\frac{\sqrt{3}}{2}$
$\cos \theta$	$\frac{\sqrt{3}}{2}$	$\frac{1}{\sqrt{2}}$ or $\frac{\sqrt{2}}{2}$	$\frac{1}{2}$
$\tan \theta$	$\frac{1}{\sqrt{3}}$ or $\frac{\sqrt{3}}{3}$	1	$\sqrt{3}$

4. $47° \, 30'$
     $+ \, 23° \, 50'$
     $\overline{70° \, 80'} = 71° \, 20'$

5. $74.3° = 74° + 0.3°$
        $= 70° + 0.3(60')$
        $= 70° + 18'$
        $= 70° \, 18'$

   $42° \, 48' = 42° + \dfrac{48°}{60}$
          $= 42° + 0.8°$
          $= 42.8°$

6. These angles and sides
correspond in accuracy.

$a = 24$	$A = 39°$
$a = 5.8$	$A = 45°$
$a = 62.3$	$A = 31.3°$
$a = 0.498$	$A = 42.9°$
$a = 2.77$	$A = 37° \, 10'$
$a = 49.87$	$A = 43° \, 18'$
$a = 6.932$	$A = 24.81°$

7.

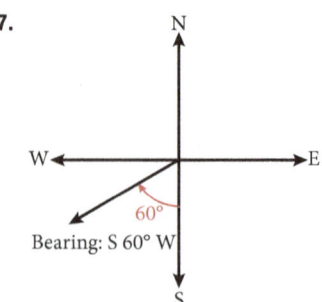

Bearing: S 60° W

Degrees, Minutes, and Seconds [2.2]

There are 360° (degrees) in one complete rotation, 60′ (minutes) in one degree, and 60″ (seconds) in one minute. This is equivalent to saying 1 minute is 1/60 of a degree, and 1 second is 1/60 of a minute.

Converting to and from Decimal Degrees [2.2]

To convert from decimal degrees to degrees and minutes, multiply the fractional part of the angle (that which follows the decimal point) by 60 to get minutes.

   To convert from degrees and minutes to decimal degrees, divide minutes by 60 to get the fractional part of the angle.

Significant Digits [2.3]

The number of *significant digits* (or figures) in a number that has digits to the right of the decimal point is found by counting the number of digits from left to right, beginning with the first nonzero digit on the left and disregarding the decimal point. If the number is an integer with no digits to the right of the decimal point, we use the same process but disregard all ending zeros unless we have other information.

   The relationship between the accuracy of the sides in a triangle and the accuracy of the angles in the same triangle is given below.

Accuracy of Sides	Accuracy of Angles
Two significant digits	Nearest degree
Three significant digits	Nearest 10 minutes or tenth of a degree
Four significant digits	Nearest minute or hundredth of a degree

Bearing [2.4]

We can specify the *bearing of a line* by giving the acute angle formed by the north–south line and the given line. The notation used to designate the bearing in this way begins with N or S, followed by the number of degrees in the angle, and ends with E or W, as in S 60° W.

8. If car 1 is traveling at 50 miles per hour due south, car 2 at 50 miles per hour due north, and car 3 at 25 mites per hour due north, then the velocities of the three cars can be represented with vectors.

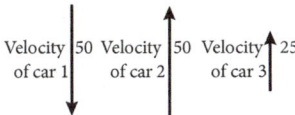

Velocity │50 Velocity │50 Velocity ↑25
of car 1 │   of car 2 │   of car 3 │

Vectors [2.5]

Quantities that have both magnitude and direction are called *vector quantities*, while quantities that have only magnitude are called *scalar quantities*. We represent vectors graphically by using arrows. The length of the arrow corresponds to the magnitude of the vector, and the direction of the arrow corresponds to the direction of the vector. In symbols, we denote the magnitude of vector **V** with |**V**|.

9.

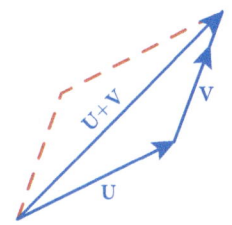

Addition of Vectors [2.5]

The sum of the vectors **U** and **V**, written **U** + **V**, is the vector that extends from the tail of **U** to the tip of **V** when the tail of **V** coincides with the tip of **U**.

10.

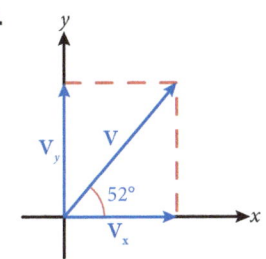

Horizontal and Vertical Components of a Vector [2.5]

The horizontal and vertical components of vector **V** are the horizontal and vertical vectors whose sum is **V**. The horizontal component is denoted by V_x, and the vertical component is denoted by V_y.

Find sin A, cos A, tan A, and sin B, cos B, and tan B in right triangle ABC, with $C = 90°$, if

1. $a = 1$ and $b = 2$
2. $b = 3$ and $c = 6$
3. $a = 3$ and $c = 5$
4. $a = 5$ and $b = 12$

Fill in the blanks to make each statement true.

5. $\sin 14° = \cos$ _____
6. $\sec$ _____ $= \csc 73°$

Simplify each expression as much as possible. Do not use tables.

7. $\sin^2 45° + \cos^2 30°$
8. $\tan 45° + \cot 45°$
9. $\sin^2 60° - \cos^2 30°$
10. $\dfrac{1}{\sec 30°}$
11. Add $48° \, 18'$ and $24° \, 52'$.
12. Subtract $15° \, 32'$ from $25° \, 15'$.

Convert to degrees and minutes.

13. $73.2°$
14. $16.45°$

Convert to decimal degrees.

15. $2° \, 48'$
16. $79° \, 30'$

Use a calculator to find the following:

17. $\sin 24° \, 20'$
18. $\cos 37.8°$
19. $\tan 63° \, 50'$
20. $\cot 71° \, 20'$

Use a calculator to find θ if

21. $\tan \theta = 0.0816$
22. $\sec \theta = 1.923$
23. $\sin \theta = 0.9465$
24. $\cos \theta = 0.9730$

Give the number of significant digits in each number.

25. 49.35
26. 0.0028

The following problems refer to right triangle ABC with $C = 90°$. In each case, find all the missing parts. (Use Table II for problems 27 through 28.)

27. $a = 68.0$ and $b = 104$
28. $a = 24.3$ and $c = 48.1$
29. $b = 305$ and $B = 24.9°$
30. $c = 0.462$ and $A = 35° \, 30'$

31. Geometry If the altitude of an isosceles triangle is 25 centimeters and each of the two equal angles measures 17°, how long are the two equal sides?

32. Angle of Elevation If the angle of elevation of the sun is 75° 30′, how tall is a post that casts a shadow 1.5 feet long?

33. Angle of Depression From a 7 foot lifeguard tower located on the long side of a pool, a lifeguard notices the angles of depression to the ends of the pool are 31° and 4°. How long is the pool? Give your answer to the nearest foot.

34. Navigation A boat travels on a course of bearing S 48° 50′ W for 128 miles. How far south and how far west has the boat traveled?

35. Vectors If vector V has magnitude 5.0 and makes an angle of 30° with the positive x-axis, find the magnitude of the horizontal and vertical components of V.

36. Vectors Vector V has a horizontal component with magnitude 1 and a vertical component with magnitude 3. What is the angle formed by V and the positive x-axis?

37. Velocity A bullet is fired into the air with an initial velocity of 800 feet per second at an angle of 60° from the horizontal. Find the magnitude of the horizontal and vertical components of the velocity vector.

38. Navigation A ship travels for 120 miles on a course with bearing 120°. How far east and how far south has the ship traveled?

Spotlight on Success

Edwin, Student Instructor

You never fail until you stop trying.
—*Albert Einstein*

Coming to the United States at the age of 10 and not knowing how to speak English was a very difficult hurdle to overcome. However, with hard work and dedication I was able to rise above those obstacles. When I came to the U.S. our school did not have a strong English development program as it was known at that time, English as a Second Language (ESL). The approach back then was "sink or swim." When my self-esteem was low, my mom and my three older sisters were always there for me and they would always encourage me to do well. My mom was a single parent, and her number one priority was that we would receive a good education. My mother's perseverance is what has made me the person I am today. At a young age I was able to see that she had overcome more than what my situation was, and I would always tell myself, "if Mom can do it, I could also do it." Not only did she not have an education, but she also saved us from a civil war that was happening in my home country of El Salvador.

When things in school got hard, I would always reflect on all the hard work, sacrifice and effort of mother. I would just tell myself that I should not have any excuses and that I needed to keep going. If my mother, who worked as a housekeeper, could send all four of her kids to college doesn't motivate you, I don't know what does. It definitely motivated me. The day everything began to change for me was when I was in eighth grade. I was sitting in my biology class not paying attention to the teacher because I was really focusing on a piece of paper on the wall. It said, "You never fail until you stop trying." I read it over and over, trying to digest what the quote meant. With my limited English I was doing my best to translate what it meant in my native language. It finally clicked! I was able to figure out what those seven words meant. I memorized the quote and began to apply it to my academics and to real-life situations. I began to really focus in my studies. I wanted to do well in school, and most important I wanted to improve my English. To this day I always reflect to that quote when I feel I can't do something.

I was able to finish junior high successfully. Going to high school was a lot easier and I ended up with very good grades and eventually I was accepted to an excellent college. I was never the smartest student on campus, but I always did well because I never quit. I earned my college degree and now I teach at a dual immersion elementary school. I have that same quote in my classroom and I constantly remind my students to never stop trying.

RADIAN MEASURE 3

WHAT IS RADIAN MEASURE? Radian measure is a unit used to measure angles in trigonometry and calculus. It is based on the concept of the radius of a circle. One radian is defined as the angle subtended at the center of a circle by an arc whose length is equal to the radius of the circle. If you were to wrap the radius of a circle around the circumference, it would span an angle of one radian at the center. This angle is approximately 57.3 degrees.

WHY IS RADIAN MEASURE IMPORTANT? Radian measure is preferred in many mathematical contexts because it simplifies calculations involving trigonometric functions, derivatives, and integrals. It is considered a "natural" unit for measuring angles in mathematics and physics.

WHERE MIGHT YOU SEE RADIAN MEASURE? Here are some examples of where you may see radian measure in the real world:

GEODESY AND GEOPHYSICS In geodesy and geophysics, radian measure is used to model the curvature of the Earth's surface and to calculate distances, areas, and volumes on the Earth's surface. Geographic coordinates, such as latitude and longitude, are often expressed in radians for precise positioning.

COMPUTER GRAPHICS AND ANIMATION In computer graphics and animation, radian measure is used to represent rotations and transformations of objects in three-dimensional space. Graphics software and game engines use radians to specify the orientation and movement of virtual objects, cameras, and light sources.

Scenario: Horizon Distance on a Planet

Abby stands at the top of Mount Everest, the highest peak on Earth, with an altitude of approximately 8,848 meters above sea level. Abby wants to calculate the distance to the horizon from her position. Assuming the Earth is a perfect sphere with a radius of approximately 3,959 miles, determine the distance to the horizon from Abby's position in miles, neglecting atmospheric refraction.

1. Find the Angle to the Horizon

Using the formula:

$$\theta = \arcsin\left(\frac{R}{h + R}\right)$$

where h is the height above the earth's surface, we'll calculate the angle θ to the horizon.

2. Substitute Values

The radius of the Earth R is approximately 3,959 miles. Abby's altitude is approximately 5.50 miles (converted from 8,848 meters to miles). Therefore, Abby's distance from the center of the Earth is

$$\text{Abby's distance} = 5.50 \text{ miles} + 3,959 \text{ miles} = 3964.50 \text{ miles}$$

3. Calculate the Distance to the Horizon

We'll use the formula $d = R \cdot \tan(\theta)$ to find the distance to the horizon.

4. Solve for Distance to the Horizon

First, we calculate θ using

$$\theta = \arcsin\left(\frac{R}{h + R}\right) = \arcsin\left(\frac{3959}{3964.50}\right)$$

$$\theta \approx 1.5181 \text{ radians}$$

Then we find d:

$$d = \frac{3959}{\tan(1.5181)} \approx 208.82 \text{ miles}$$

Solution   The distance to the horizon from the summit of Mount Everest for Abby is approximately 208.82 miles, taking into account Abby's altitude and the curvature of the Earth.

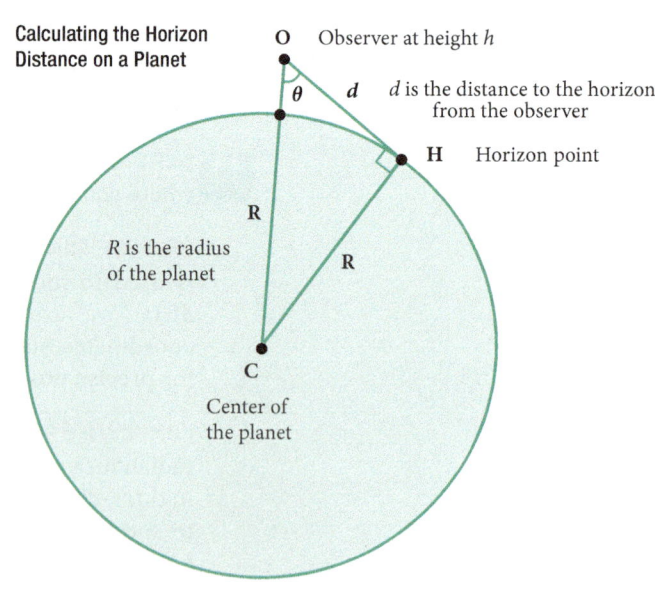

Calculating the Horizon Distance on a Planet

O   Observer at height h

θ d d is the distance to the horizon from the observer

H   Horizon point

R is the radius of the planet

R

R

C

Center of the planet

Learning Objectives

A   Name the reference angle for a given angle.

B   Find the coterminal angle of a given angle.

C   Approximate angles using trigonometric functions.

We are going to begin this section with a look at how we can find approximate values for trigonometric functions of any angle, not just those between 0° and 90°.

> **Definition**
> The **reference angle** (sometimes called **related angle**) for any angle θ in standard position is the positive acute angle between the terminal side of θ and the x-axis. In this book, we will denote the reference angle for θ by $\hat{\theta}$.

Note that, for this definition, $\hat{\theta}$ is always positive and always between 0° and 90°. That is, a reference angle is always an acute angle.

A Reference Angle Theorem

EXAMPLE 1   Name the reference angle for each of the following angles:

a. 30°   **b.** 135°   **c.** 240°   **d.** 330°

Solution   We draw each angle in standard position. The reference angle is the positive acute angle formed by the terminal side of the angle in question and the x-axis.

(a)

(b)

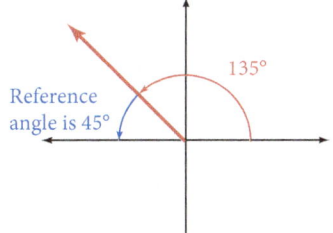

(c)

(d)

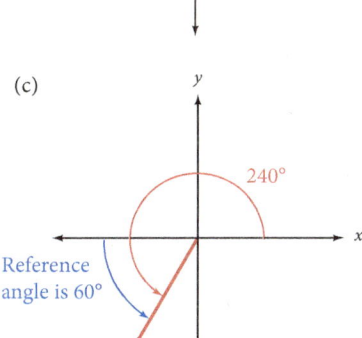

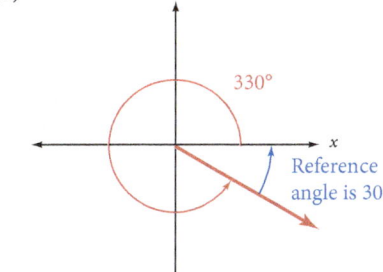

We can generalize the results of Example 8 as follows: If θ is a positive angle between 0° and 360°, then:

$$\text{If } \quad \theta \in \text{QI}, \qquad \text{then } \hat{\theta} = \theta$$

$$\text{If } \quad \theta \in \text{QII}, \qquad \text{then } \hat{\theta} = 180° - \theta$$

$$\text{If } \quad \theta \in \text{QIII}, \qquad \text{then } \hat{\theta} = \theta - 180°$$

$$\text{If } \quad \theta \in \text{QIV}, \qquad \text{then } \hat{\theta} = 360° - \theta$$

We can use our information on reference angles and the signs of the trigonometric functions to write the following theorem.

> **Reference Angle Theorem**
> A trigonometric function of an angle and its reference angle differ at most in sign.

We will not give a detailed proof of this theorem but, rather, justify it by example. Let's look at the sines of all the angles between 0° and 360° that have a reference angle of 30°. These angles are 30°, 150°, 210°, and 330°.

From Figure 1 we have the following:

$$\sin 150° = \sin 30° = \frac{1}{2} \qquad \rightarrow$$

$$\qquad\qquad\qquad\qquad\qquad \textit{They differ in sign only.}$$

$$\sin 210° = \sin 330° = -\frac{1}{2} \qquad \rightarrow$$

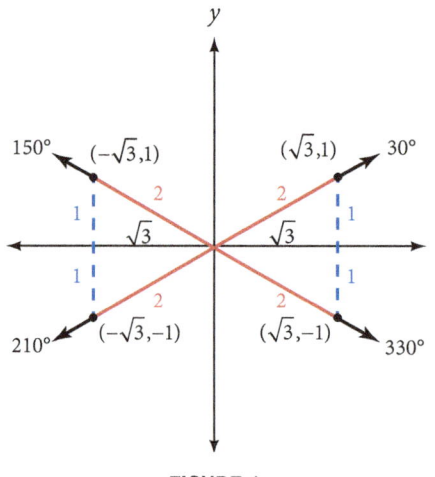

FIGURE 1

As you can see, any angle with a reference angle of 30° will have a sine of $\frac{1}{2}$ or $-\frac{1}{2}$. The sign, $+$ or $-$, will depend on the quadrant in which the angle terminates. Using this discussion as justification, we write the following steps used to find trigonometric functions of angles between 0° and 360°.

Step 1: Find $\hat{\theta}$, the reference angle.

Step 2: Determine the sign of the trigonometric function based on the quadrant in which θ terminates.

Step 3: Write the original trigonometric function of θ in terms of the same trigonometric function of $\hat{\theta}$.

Step 4: Find the trigonometric function of $\hat{\theta}$.

EXAMPLE 2 Find the exact value of sin 240°.

Solution For this first example, we will list the steps given on the preceding page as we use them. Figure 2 is a diagram of the situation.

Step 1: We find $\hat{\theta}$ by subtracting 180° from θ.

$$240° - 180° = 60°$$

Step 2: Since θ terminates in quadrant III, and the sine function is negative in quadrant III, our answer will be negative. That is, in this case, $\sin \theta = -\sin \hat{\theta}$.

Step 3: Using the results of Steps 1 and 2, we write

$$\sin 240° = -\sin 60°$$

Step 4: We finish by finding sin 60°:

$$\sin 240° = -\sin 60° = -\frac{\sqrt{3}}{2}$$

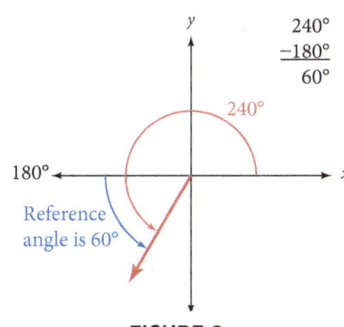

$$\begin{array}{r} 240° \\ -180° \\ \hline 60° \end{array}$$

FIGURE 2

EXAMPLE 3 Find the exact value of tan 315°.

Solution The reference angle is 360° − 315° = 45° (see Figure 3). Since 315° terminates in quadrant IV, its tangent will be negative.

$\tan 315° = -\tan 45°$ *Because tangent is now negative in QIV*
$\qquad\quad = -1$

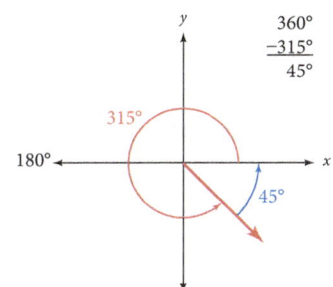

$$\begin{array}{r} 360° \\ -315° \\ \hline 45° \end{array}$$

FIGURE 3

EXAMPLE 4 Find the exact value of csc 300°.

Solution The reference angle is 360° − 300° = 60° (see Figure 4). To find the exact value of csc 60°, we use the fact that cosecant and sine are reciprocals.

$\csc 300° = -\csc 60°$ *Because cosecant is negative in QIV*

$$= -\frac{1}{\sin 60°}$$

$$= -\frac{1}{\sqrt{3}/2}$$

$$= -\frac{2}{\sqrt{3}}$$

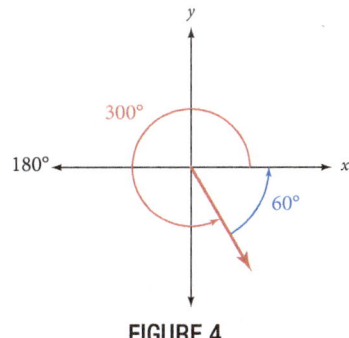

FIGURE 4

B Coterminal Angles

Definition
Two angles with the same terminal side are called **coterminal angles**.

Coterminal angles always differ from each other by multiples of 360°. For example, 10° and 370° are coterminal, as are −45° and 315° (Figure 5).

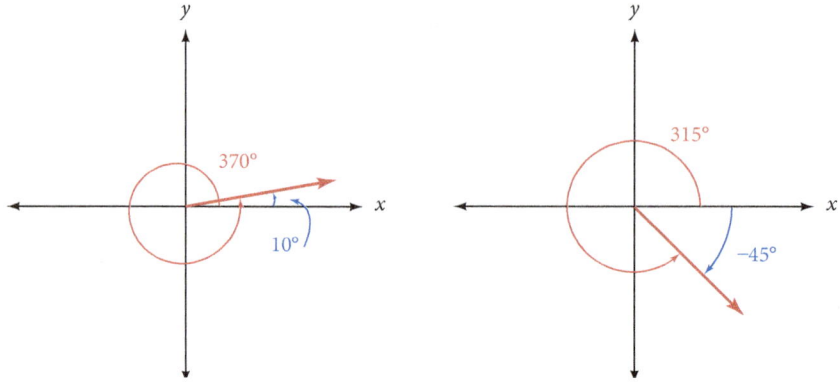

FIGURE 5

The trigonometric functions of an angle and any angle coterminal with it are always equal. For sine and cosine, we can write this in symbols as follows:

For any integer *k*,

$$\sin(\theta + 360°k) = \sin \theta \quad \text{and} \quad \cos(\theta + 360°k) = \cos \theta$$

To find values of trigonometric functions for an angle larger than 360° or smaller than 0°, we simply find an angle between 0° and 360° that is coterminal with it, and then use the steps outlined in Examples 2–4.

EXAMPLE 5 Find the exact value of cos 495°.

Solution By subtracting 360° from 495°, we obtain 135°, which is coterminal with 495°. The reference angle for 135° is 45° (see Figure 6). Since 495° terminates in quadrant II, its cosine is negative.

$$\cos 495° = \cos 135° \qquad \textcolor{green}{\textit{495° and 135° are coterminal.}}$$

$$= -\cos 45° \qquad \textcolor{green}{\textit{In QII cos } \theta = -\cos \hat{\theta}}$$

$$= -\frac{1}{\sqrt{2}} \qquad \textcolor{green}{\textit{Exact value}}$$

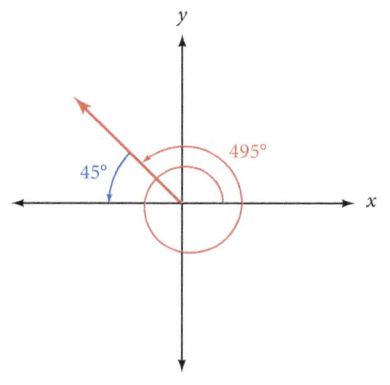

FIGURE 6

C Approximations

To find trigonometric functions of angles that do not lend themselves to exact values, we use a calculator. To find an approximation for sin θ, cos θ, or tan θ, we press the appropriate keys on the calculator. Check to see that you can obtain the following values for sine, cosine, and tangent of 250° and −160° in your calculator. (These answers are rounded to the nearest ten thousandth.)

$$\sin 250° \approx -0.9397 \qquad \sin(-160°) \approx -0.3420$$
$$\cos 250° \approx -0.3420 \qquad \cos(-160°) \approx -0.9397$$
$$\tan 250° \approx 2.7475 \qquad \tan(-160°) \approx 0.3640$$

To find csc 250°, sec 250°, and cot 250°, we must use the reciprocals of sin 250°, cos 250°, and tan 250°.

	Scientific Calculator	**Graphing Calculator**
$\csc 250° = \dfrac{1}{\sin 250°} \approx -1.0642$	250 [sin] [1/x]	1 [÷] [sin] 250 [ENTER]
$\sec 250° = \dfrac{1}{\cos 250°} \approx -2.9238$	250 [cos] [1/x]	1 [÷] [cos] 250 [ENTER]
$\cot 250° = \dfrac{1}{\tan 250°} \approx 0.3640$	250 [tan] [1/x]	1 [÷] [tan] 250 [ENTER]

Next we use a calculator to find an approximation for θ, given one of the trigonometric functions of θ and the quadrant in which θ terminates.

EXAMPLE 6 Find θ if sin θ = −0.5592 and θ terminates in QIII with 0° < θ < 360°.

Solution In this example, we must use a calculator in the reverse direction from the way we used it in the discussion above. Using the [sin⁻¹] key, we find that 34° is the angle whose sine is 0.5592. That is,

$$\hat{\theta} = \sin^{-1}(0.5592) \approx 34°$$

This is our reference angle, $\hat{\theta}$. The angle in quadrant III whose reference angle is 34° is θ = 180° + 34° = 214° (see Figure 7).

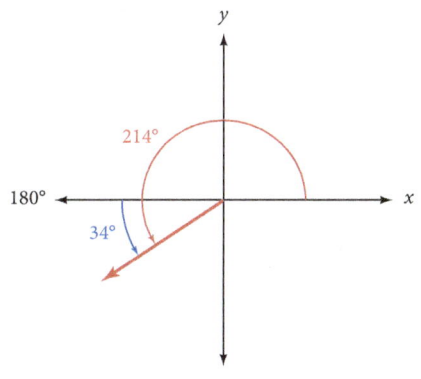

FIGURE 7

Calculator Note If you were to try Example 6 on your calculator by using −0.5592 and the [sin⁻¹] key, you would not obtain 214° for your answer. Instead you would get approximately −34° for the answer, which is wrong. To see why this happens, you will have to wait until we cover inverse trigonometric functions. In the meantime, to use a calculator on this kind of problem, use it to find the reference angle and then proceed as we did in Example 6.

If $\sin \theta = -0.5592$ and θ terminates in QIII, then

$$\theta = 180° + 34°$$
$$= 214°$$

If we wanted to list *all* the angles that terminate in quadrant III and have a sine of -0.5592, we would write

$$\theta = 214° + 360°k$$

where $k =$ any integer. This gives us all angles coterminal with $214°$.

EXAMPLE 7   Find θ to the nearest tenth of a degree if $\tan \theta = -0.8541$ and θ terminates in QIV with $0° < \theta < 360°$.

Solution Using 0.8541 and the $\boxed{\tan^{-1}}$ key gives the reference angle as $\hat{\theta} = \tan^{-1}(0.8541) = 40.5°$. The angle in QIV with a reference angle of $40.5°$ is

$$\theta = 360° - 40.5° = 319.5°$$

Again, if we wanted to list all angles in the quadrant IV with a tangent of -0.8541, we would write

$$\theta = 319.5° + 360°k \qquad\qquad k = \text{any integer}$$

to include not only $319.5°$ but all angles coterminal with it.

EXAMPLE 8   Find θ if $\sin \theta = -\frac{1}{2}$ with θ in QIII and $0° < \theta < 360°$.

Solution From the table of exact values, we find that the angle whose sine is $\frac{1}{2}$ is $30°$. This is the reference angle. The angle in quadrant III with a reference angle of $30°$ is $180° + 30° = 210°$.

EXAMPLE 9   Find θ if $\cot \theta = -1.1106$ and θ terminates in QIV, with $0° < \theta < 360°$. Round to the nearest tenth of a degree.

Solution To find the reference angle on a calculator we ignore the negative sign in -1.1106 and use the fact that $\cot \theta$ is the reciprocal of $\tan \theta$.

$$\text{If } \cot \theta = 1.1106, \text{ then } \tan \theta = \frac{1}{1.1106}$$

Therefore, we enter 1.1106 and press the reciprocal key, $\boxed{1/x}$. Then we press the $\boxed{\tan^{-1}}$ key to find our reference angle.

$$\boxed{\tan^{-1}}\ \boxed{(}\ \boxed{1/x}\ 1.1106\ \boxed{)}$$

To the nearest tenth of a degree, the reference angle is $\hat{\theta} = 42.0°$. Since we want θ to terminate in QIV, we subtract $42.0°$ from $360°$ to get $\theta = 318.0°$.

Again, we can check our result on a calculator by entering $318°$, finding its tangent, and then finding the reciprocal of the result.

$$\boxed{1/x}\ \boxed{\tan}\ \boxed{(}\ 318.0\ \boxed{)}\ \text{gives} -1.1106$$

EXAMPLE 10   Find θ if $\csc \theta = 1.2361$ and θ terminates in QII, with $0° < \theta < 360°$. Round to the nearest degree.

Solution To find the reference angle on a calculator we must use the fact that $\csc \theta$ is the reciprocal of $\sin \theta$. That is,

$$\text{If } \csc \theta = 1.2361, \text{ then } \sin \theta = \frac{1}{1.2361}$$

Therefore, we enter 1.2361 and press the reciprocal key, $\boxed{1/x}$. Then we press the $\boxed{\sin^{-1}}$ key to find our reference angle.

$$\boxed{\sin^{-1}}\ \boxed{(}\ \boxed{1/x}\ 1.2361\ \boxed{)}$$

To the nearest degree, the reference angle is $\hat{\theta} = 54°$. Since we want θ to terminate in QIII, we subtract $54°$ from $180°$ to get

$$180° - 54° = 126°$$

We can check our result on a calculator by entering $126°$, finding its sine, and then finding the reciprocal of the result.

$$\boxed{1/x}\ \boxed{\sin}\ \boxed{(}\ 126\ \boxed{)}\ \text{gives } 1.2361$$

Problem Set 3.1

Vocabulary

Use the vocabulary words below to fill in the blanks in the sentences. Vocabulary words may be used more than once.

sign   cofunctions   reference   acute   coterminal

1. The _____ angle is the positive _____ angle between the terminal side of an angle in standard position and the *x*-axis.

2. The _____ angle for 135° is 45°.

3. A trigonometric function of an angle and its _____ angle differ at most in _____ .

4. Two angles in standard position with the same terminal side are called _____ angles.

Draw each of the following angles in standard position and then name the reference angle:

1. 210°   2. 150°

3. 143.4°   4. 253.8°

5. 311.7°   6. 93.2°

7. 195° 10′   8. 171° 40′

9. 331° 20′   10. 252° 50′

11. −300°   12. −330°

13. −120°   14. −150°

Use a calculator to find the following. If you use a calculator, use it only to find the trigonometric functions of the reference angle. Remember, we are learning the relationships that exist between an angle, its reference angle, and the trigonometric functions of both. Following the steps in the examples in this section with help us understand these relationships.

15. cos 347°   16. cos 238°

17. cos 101.8°   18. sin 166.7°

19. tan 143.4°   **20.** tan 253.8°

21. sec 311.7°   **22.** cos 93.2°

23. cot 390°   **24.** cot 420°

25. cos 575.4°   **26.** sin 590.9°

Use a calculator to find the following. (Again, find the reference angle first, even if you are using a calculator. Also, remember, if you use a calculator, you must first convert to decimal degrees before you use the trigonometric function keys.)

27. sin 210°   **28.** cos 150°

29. cos (−315°)   **30.** sin (−225°)

31. tan 195° 10′   **32.** tan 171° 40′

33. sec 314° 40′   **34.** csc 670° 20′

35. csc 410° 10′   **36.** sec 380° 50′

37. sin (−120°)   **38.** cos (−150°)

39. cot (−300° 20′)   **40.** cot (−330° 30′)

Use a calculator to find θ, to the nearest tenth of a degree, if $0° < \theta < 360°$ and

41. $\sin \theta = -0.3090$ with θ in QIII   **42.** $\sin \theta = -0.3090$ with θ in QIV

43. $\cos \theta = -0.7660$ with θ in QII   **44.** $\cos \theta = -0.7660$ with θ in QIII

45. $\tan \theta = 0.5890$ with θ in QIII   **46.** $\tan \theta = 0.5890$ with θ in QI

47. $\cos \theta = 0.2644$ with θ in QI   **48.** $\cos \theta = 0.2644$ with θ in QIV

49. $\sin \theta = 0.9652$ with θ in QII   **50.** $\sin \theta = 0.9652$ with θ in QI

51. $\sec \theta = 1.5263$ with θ in QIV   **52.** $\sec \theta = 1.5263$ with θ in QI

53. $\sec \theta = -2.0387$ with θ in QII   **54.** $\sec \theta = -2.0387$ with θ in QIII

55. $\csc \theta = -2.6587$ with θ in QIII   **56.** $\csc \theta = -2.6587$ with θ in QIV

57. $\cot \theta = -1.3862$ with θ in QII   **58.** $\cot \theta = -1.3862$ with θ in QIV

59. $\cot \theta = -0.6452$ with θ in QIV   **60.** $\cot \theta = -0.6452$ with θ in QII

Find exact values for each of the following:

61. $\sin 120°$ **62.** $\sin 210°$

63. $\tan 135°$ **64.** $\tan 315°$

65. $\cos 240°$ **66.** $\cos 150°$

67. $\csc 330°$ **68.** $\sec 330°$

69. $\sec 300°$ **70.** $\csc 300°$

71. $\sin 390°$ **72.** $\cos 420°$

73. $\cot 480°$ **74.** $\cot 510°$

Find θ, if $0° < \theta < 360°$ and

75. $\sin \theta = -\dfrac{\sqrt{3}}{2}$ and θ in QIII   **76.** $\sin \theta = -\dfrac{1}{\sqrt{2}}$ with θ in QIII

77. $\cos \theta = -\dfrac{1}{\sqrt{2}}$ with θ in QII   **78.** $\cos \theta = -\dfrac{\sqrt{3}}{2}$ and θ in QIII

79. $\sin \theta = -\dfrac{\sqrt{3}}{2}$ and θ in QIV   **80.** $\sin \theta = \dfrac{1}{\sqrt{2}}$ with θ in QII

81. $\tan \theta = \sqrt{3}$ with θ in QIII   **82.** $\tan \theta = \dfrac{1}{\sqrt{3}}$ with θ in QIII

83. Use a calculator to find θ if sec $\theta = 2.3931$ and θ is an acute angle. (Give your answer to the nearest tenth of a degree.)

84. Use a calculator to find θ if csc $\theta = 1.1164$ and θ is an acute angle. (Give your answer to the nearest tenth of a degree.)

85. We know that tan 90° is undefined. If we try to find tan 90° on a calculator, we get an error message. To begin to get an idea of why this happens, we can use a calculator to find the tangent of angles close to 90°. Use a calculator to find tan θ, if θ takes on values of 85°, 87°, 89°, 89.9°, and 89.99°.

86. Follow the instructions given in Problem 75 but use the secant function instead of the tangent function. (This will require that you use the cos key and the $1/x$ key on your calculator.)

Getting Ready for the Next Section

Let $\pi = 3.14$ and find the following. Round to the three significant digits, if rounding is necessary.

87. $\dfrac{180}{\pi}$ **88.** $\dfrac{\pi}{180}$

89. $\dfrac{\pi}{4}$ **90.** $\dfrac{\pi}{3}$

91. $\dfrac{\pi}{3} \cdot \dfrac{180}{\pi}$ **92.** $\dfrac{3\pi}{4} \cdot \dfrac{180}{\pi}$

Simplify. Leave your answers in terms of π.

93. $45 \cdot \dfrac{\pi}{180}$ **94.** $30 \cdot \dfrac{\pi}{180}$

95. $270 \cdot \dfrac{\pi}{180}$ **96.** $360 \cdot \dfrac{\pi}{180}$

97. $90 \cdot \dfrac{\pi}{180}$ **98.** $180 \cdot \dfrac{\pi}{180}$

3.2 Videos

KEY WORDS

central angle θ

radian

Learning Objectives

A Measure a given angle in radians.

B Compare angles in radians and degrees.

C Convert from degrees to radians.

D Convert from radians to degrees.

We begin this section with the definition for the radian measure of an angle. As you will see, specifying the measure of an angle with radian measure gives us a way to associate the measure of an angle with real numbers, rather than degrees.

A Radian Measure

To understand the definition for radian measure, we have to recall from geometry that a **central angle** is an angle with its vertex at the center of a circle.

Definition

In a circle, a central angle that cuts off an arc equal in length to the radius of the circle has a measure of 1 radian. That is, in a circle of radius r, a central angle that measures 1 radian will cut off an arc of length r (Figure 1).

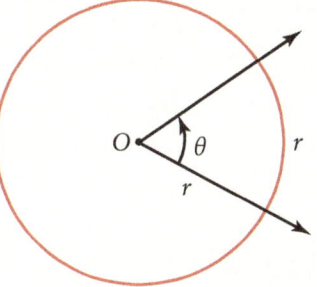

Angle θ has a measure of 1 radian

The vertex of θ is at the center of the circle; the arc cut off by θ is equal in length to the radius

FIGURE 1

Note It is common practice to omit the word *radian* when using radian measure. If no units are showing, an angle is understood to be measured in radians; with degree measure, the degree symbol ° must be written.

$$\theta = 2 \text{ means the measure of } \theta \text{ is 2 radians}$$

$$\theta = 2° \text{ means the measure of } \theta \text{ is 2 degrees}$$

To find the radian of *any* central angle, we must find how many radii there are in the arc it cuts off. To do so, we divide the arc length by radius. If the radius is 2 centimeters and the arc cut off by central angle θ is 6 centimeters, then the radian measures of θ is $\frac{6}{2} = 3$ radians.

Here is the formal definition:

Definition
If a central angle θ in a circle of radius r cuts off an arc of length s, then the **measure of θ in radians** is given by $\frac{s}{r}$. (See Figure 2.)

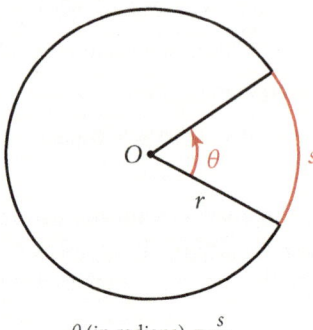

$$\theta \text{ (in radians)} = \frac{s}{r}$$

FIGURE 2

EXAMPLE 1   A central angle θ in a circle of radius 3 centimeters cuts off an arc of length 6 centimeters. What is the radian measure of θ?

Solution   We have $r = 3$ centimeters and $s = 6$ centimeters, therefore,

$$\theta \text{ (in radians)} = \frac{s}{r}$$

$$= \frac{6 \text{ centimeters}}{3 \text{ centimeters}}$$

$$= 2$$

We say the radian measure of θ is 2 or $\theta = 2$ radians.

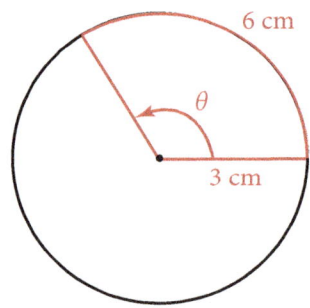

FIGURE 3

B Comparing Degrees and Radians

To see the relationship between degrees and radians, we can compare the number of degrees and the number of radians in one full rotation.

The angle formed by one full rotation about the center of a circle of radius r will cut off an arc equal to the circumference of the circle (Figure 4).

Since the circumference of a circle of radius r is $2\pi r$, we have

θ measures one full rotation. $\theta = \dfrac{2\pi r}{r} = 2\pi$ The measure of θ in radians is 2π.

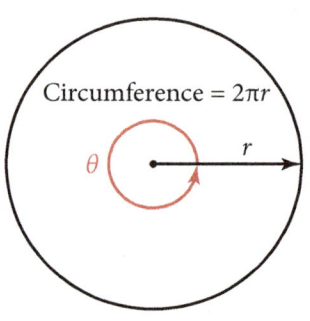

FIGURE 4

Since one full rotation in degrees is 360°, we have the following relationship between radians and degrees:

$$360° = 2\pi \text{ radians}$$

Dividing both sides by 2, we have

$$180° = \pi \text{ radians}$$

To obtain conversion factors that will allow us to change back and forth between degrees and radians, we divide both sides of this last equation alternately by 180 and by π:

Divide both sides by 180. ⟶ $180° = \pi$ radians ⟶ Divide both sides by π.

$$1° = \frac{\pi}{180} \text{ radians} \qquad \left(\frac{180}{\pi}\right)° = 1 \text{ radian}$$

To gain some insight into the relationship between degrees and radians, we can approximate π as 3.14 to obtain the approximate number of degrees in 1 radian:

$$1 \text{ radian} = 1\left(\frac{180}{\pi}\right)°$$

$$\approx 1\left(\frac{180}{3.14}\right)°$$

$$\approx 57.3°$$

We see that 1 radian is approximately 57°. A radian is much larger than a degree. Figure 5 illustrates the relationship between 20° and 20 radians.

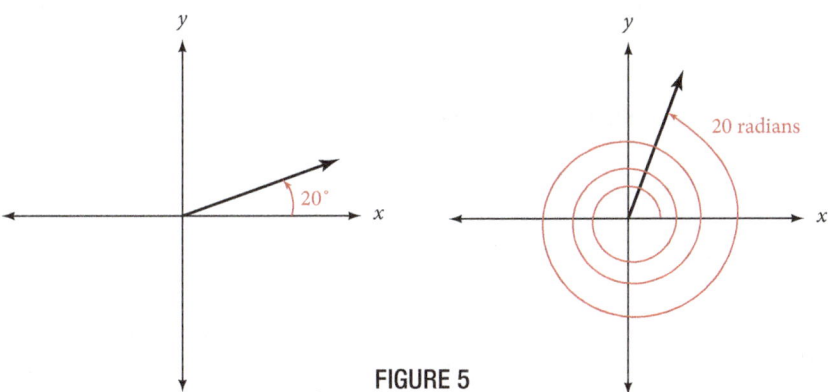

FIGURE 5

Here are some further conversions between degrees and radians.

C Converting from Degrees to Radians

EXAMPLE 2   Convert 45° to radians.

Solution Since $1° = \frac{\pi}{180}$ radians, and 45° is the same as 45(1°), we have $45° = 45\left(\frac{\pi}{180}\right)$ radians $= \frac{\pi}{4}$ radians as illustrated in Figure 6.

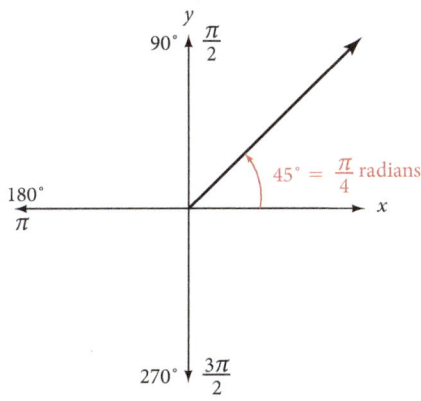

FIGURE 6

When we leave our answer in terms of π, as in $\frac{\pi}{4}$, we are writing an exact value. If we wanted a decimal approximation, we would substitute 3.14 for π.

$$\text{Exact value} \quad \frac{\pi}{4} \approx \frac{3.14}{4} = 0.785 \quad \text{Approximate value}$$

Note also that if we wanted the radian equivalent of 90°, we could simply multiply $\frac{\pi}{4}$ by 2, since 90° = 2 × 45°:

$$90° = 2 \times 45° = 2 \times \frac{\pi}{4} = \frac{\pi}{2}$$

EXAMPLE 3 Convert 450° to radians.

Solution Multiplying by $\frac{\pi}{180}$, we have $450° = 450\left(\frac{\pi}{180}\right) = \frac{5\pi}{2}$ radians. (See Figure 7.)

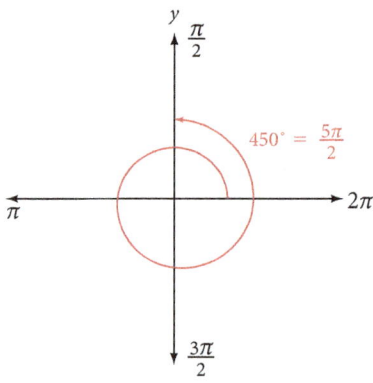

FIGURE 7

Again, $\frac{5\pi}{2}$ is the exact value. If we wanted a decimal approximation, we would substitute 3.14 for π.

$$\text{Exact value} \quad \frac{5\pi}{2} \approx \frac{5(3.14)}{2} = 7.85 \quad \text{Approximate value}$$

D Converting from Radians to Degrees

EXAMPLE 4 Convert $\frac{\pi}{6}$ to degrees.

Solution To convert from radians to degrees, we multiply by $\frac{180}{\pi}$.

$$\frac{\pi}{6} \text{ (radians)} = \frac{\pi}{6}\left(\frac{180}{\pi}\right)°$$
$$= 30°$$

Note that 60° is twice 30°, so $2\left(\frac{\pi}{6}\right) = \frac{\pi}{3}$ must be the radian equivalent of 60°. Figure 8 illustrates this.

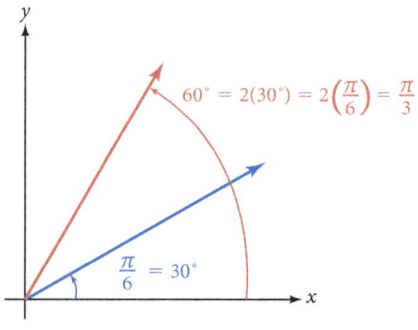

FIGURE 8

EXAMPLE 5   Convert $\frac{4\pi}{3}$ degrees.

Solution   Multiplying by $\frac{180}{\pi}$, we have $\frac{4\pi}{3}$ (radians) $= \frac{4\pi}{3}\left(\frac{180}{\pi}\right)^{\circ}$

$$= 240^{\circ}$$

See Figure 9.

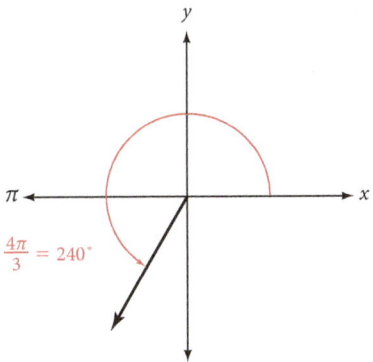

FIGURE 9

As is apparent from the preceding examples, changing from degrees to radians and from radians to degrees is simply a matter of multiplying by the appropriate conversion factors.

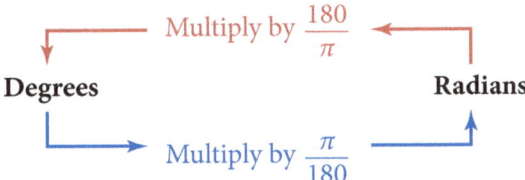

Figure 10 shows the most common angles written in both degrees and radians. Table 1 gives approximations to some of the exact radian measures, accurate to the nearest hundredth of a radian.

	Radians	
Degrees	**Exact Values**	**Approximations**
0°	0	0
30°	$\frac{\pi}{6}$	0.52
45°	$\frac{\pi}{4}$	0.79
60°	$\frac{\pi}{3}$	1.05
90°	$\frac{\pi}{2}$	1.57
180°	π	3.14
270°	$\frac{3\pi}{2}$	4.71
360°	2π	6.28

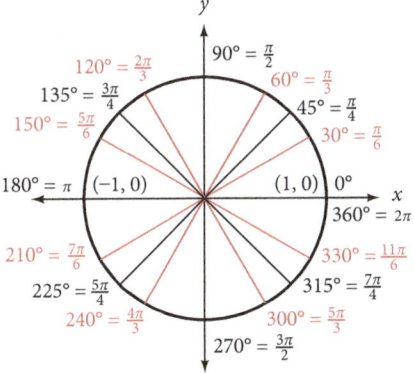

FIGURE 10

EXAMPLE 6 Find $\sin \frac{\pi}{6}$.

Solution Since $\frac{\pi}{6}$ and 30° are equivalent, so are their sines.

$$\sin \frac{\pi}{6} = \frac{1}{2}$$

Calculator Note To work this problem on a calculator, we must first put the calculator in radian mode. (Consult the manual that came with your calculator to see how to do this.) Here is the sequence to key in your calculator to work the problem given in Example 6.

$$\boxed{\text{Rad}} \; \boxed{\pi} \; \boxed{\div} \; 6 \; \boxed{=} \; \boxed{\sin}$$

EXAMPLE 7 Find $4 \sin \frac{7\pi}{6}$.

Solution Since $\frac{7\pi}{6}$ terminates in QIII, its sine will be negative. The reference angle is $\frac{7\pi}{6} - \pi = \frac{\pi}{6}$.

$$4 \sin \frac{7\pi}{6} = -4 \sin \frac{\pi}{6}$$

$$= -4\left(\frac{1}{2}\right)$$

$$= -2$$

EXAMPLE 8 Evaluate $4 \sin(2x + \pi)$ when $x = \frac{\pi}{6}$.

Solution Substituting $\frac{\pi}{6}$ for x and simplifying, we have

$$4 \sin\left(2 \cdot \frac{\pi}{6} + \pi\right) = 4 \sin\left(\frac{\pi}{3} + \pi\right)$$

$$= 4 \sin\left(\frac{4\pi}{3}\right)$$

$$= -4 \sin\left(\frac{\pi}{3}\right)$$

$$= -4\left(-\frac{\sqrt{3}}{2}\right)$$

$$= -2\sqrt{3}$$

Problem Set 3.2

Vocabulary

Use the vocabulary words below to fill in the blanks in the sentences. Vocabulary words may be used more than once.

center   radians   arc   degrees

1. We have two types of measures for angle: _____ and _____ .

2. A central angle in a circle has its vertex at the _____ of the circle.

3. If a central angle θ in a circle with radius r cuts off an _____ of length s, then the radian measure of θ is s/r.

4. One radian is approximately equal to 57.3 _____ .

5. 180° is equal to π _____ .

Find the radian measure of angle θ, if θ is a central angle in a circle of radius r, and θ cuts off an arc of length s.

1. $r = 3$ centimeters, $s = 9$ centimeters 2. $r = 6$ centimeters, $s = 3$ centimeters

3. $r = 10$ inches, $s = 5$ inches 4. $r = 5$ inches, $s = 10$ inches

5. $r = 4$ inches, $s = 12\pi$ inches 6. $r = 3$ inches, $s = 12$ inches

7. $r = \frac{1}{4}$ centimeters, $s = \frac{1}{2}$ centimeters 8. $r = \frac{1}{4}$ centimeters, $s = \frac{1}{8}$ centimeters

9. **Angle between Cities** Los Angeles and San Francisco are approximately 450 miles apart on the surface of the Earth. Assuming that the radius of the Earth is 4,000 miles, find the radian measure of the central angle with vertex at the center of the Earth that has Los Angeles on one side and San Francisco on the other side.

10. **Angle between Cities** Los Angeles and New York City are approximately 2,500 miles apart on the surface of the Earth. Assuming that the radius of the Earth is 4,000 miles, find the radian measure of the central angle with vertex at the center of the Earth that has Los Angeles on one side and New York City on the other side.

Convert each of the following from degree measure to radian measure. Write each answer as an exact value and as an approximation to the nearest hundredth.

11. 30°   12. 60°

13. 90°   14. 270°

15. 260°   16. 340°

17. −150°   18. −210°

19. 420°   20. 390°

21. −135°   22. −120°

For Problems 23–26, use 3.1416 for π or use the key marked π on your calculator.

23. Use a calculator to convert 120° 40′ to radians. Round your answer to the nearest hundredth. (First convert to decimal degrees, then multiply by the appropriate conversion factor to convert to radians.)

24. Use a calculator to convert 256° 20′ to radians to the nearest hundredth of a radian.

25. Use a calculator to convert 1′ (1 minute) to radians to three significant digits.

26. Use a calculator to convert 1° to radians to three significant digits.

27. Nautical Miles If a central angle with its vertex at the center of the earth has a measure of 1′, then the arc on the surface of the earth that is cut off by this angle has a measure of 1 nautical mile. Find the number of regular (statute) miles in 1 nautical mile to the nearest hundredth of a mile. (Use 4,000 miles for the radius of the earth.)

28. Nautical Miles If two ships are 20 nautical miles apart on the ocean, how many statute miles apart are they? (Use the results of Problem 27 to do the calculations.)

Convert each of the following from radian measure to degree measure:

29. $\dfrac{\pi}{3}$ **30.** $\dfrac{\pi}{4}$

31. $\dfrac{2\pi}{3}$ **32.** $\dfrac{3\pi}{4}$

33. $-\dfrac{7\pi}{6}$ **34.** $-\dfrac{5\pi}{6}$

35. $\dfrac{10\pi}{6}$ **36.** $\dfrac{7\pi}{3}$

37. 4π **38.** 3π

39. $\dfrac{\pi}{12}$ **40.** $\dfrac{5\pi}{12}$

41. $-\dfrac{7\pi}{18}$ **42.** $-\dfrac{11\pi}{18}$

Use a calculator to convert each of the following to degree measure to the nearest tenth of a degree:

43. 1.3   **44.** 2.4

45. 0.75   **46.** 0.25

47. 5   **48.** 6

Give the exact value of each of the following:

49. $\sin \dfrac{4\pi}{3}$

50. $\cos \dfrac{4\pi}{3}$

51. $\tan \dfrac{\pi}{6}$

52. $\cot \dfrac{\pi}{3}$

53. $\sec \dfrac{2\pi}{3}$

54. $\csc \dfrac{3\pi}{2}$

55. $\csc \dfrac{5\pi}{6}$

56. $\sec \dfrac{5\pi}{6}$

57. $4 \sin \left(-\dfrac{\pi}{4} \right)$

58. $4 \cos \left(-\dfrac{\pi}{4} \right)$

59. $-\sin \dfrac{\pi}{4}$

60. $-\cos \dfrac{\pi}{4}$

61. $2 \cos \dfrac{\pi}{6}$

62. $2 \sin \dfrac{\pi}{6}$

Evaluate each of the following expressions when x is $\dfrac{\pi}{6}$. In each case, use exact values.

63. $\sin 2x$

64. $\sin 3x$

65. $6 \cos 3x$

66. $6 \cos 2x$

67. $\sin \left(x + \dfrac{\pi}{2} \right)$

68. $\sin \left(x - \dfrac{\pi}{2} \right)$

69. $4 \cos \left(2x + \dfrac{\pi}{3} \right)$

70. $4 \cos \left(3x + \dfrac{\pi}{6} \right)$

For the following expressions, find the value of y that corresponds to each value of x, then write your results as ordered pairs (x, y).

71. $y = \sin x$ for $x = 0, \dfrac{\pi}{4}, \dfrac{\pi}{2}, \dfrac{3\pi}{4}, \pi$

72. $y = \cos x$ for $x = 0, \dfrac{\pi}{4}, \dfrac{\pi}{2}, \dfrac{3\pi}{4}, \pi$

73. $y = 2 \sin x$ for $x = 0, \dfrac{\pi}{2}, \pi, \dfrac{3\pi}{2}, 2\pi$

74. $y = \dfrac{1}{2} \cos x$ for $x = 0, \dfrac{\pi}{2}, \pi, \dfrac{3\pi}{2}, 2\pi$

75. $y = \sin 2x$ for $x = 0, \dfrac{\pi}{4}, \dfrac{\pi}{2}, \dfrac{3\pi}{4}, \pi$

76. $y = \cos 3x$ for $x = 0, \dfrac{\pi}{6}, \dfrac{\pi}{3}, \dfrac{\pi}{2}, \dfrac{2\pi}{3}$

77. $y = \sin\left(x - \dfrac{\pi}{2}\right)$ for $x = \dfrac{\pi}{2}, \pi, \dfrac{3\pi}{2}, 2\pi, \dfrac{5\pi}{2}$

78. $y = \cos\left(x - \dfrac{\pi}{6}\right)$ for $x = \dfrac{\pi}{6}, \dfrac{\pi}{3}, \dfrac{2\pi}{3}, \pi, \dfrac{7\pi}{6}$

79. $y = 3\sin\left(2x + \dfrac{\pi}{2}\right)$ for $x = -\dfrac{\pi}{4}, 0, \dfrac{\pi}{4}, \dfrac{\pi}{2}, \dfrac{3\pi}{4}$

80. $y = 5\cos\left(2x - \dfrac{\pi}{3}\right)$ for $x = \dfrac{\pi}{6}, \dfrac{\pi}{3}, \dfrac{2\pi}{3}, \pi, \dfrac{7\pi}{6}$

Getting Ready for the Next Section

Find all six trigonometric functions of θ, if the given point is on the terminal side of θ.

81. $(1, 0)$ **82.** $(0, 1)$ **83.** $\left(\dfrac{1}{2}, \dfrac{\sqrt{3}}{2}\right)$ **84.** $\left(\dfrac{1}{\sqrt{2}}, \dfrac{1}{\sqrt{2}}\right)$

85. Find θ, if $\sin\theta = \dfrac{1}{2}$ and θ terminates in QII.

86. Find θ, if $\cos\theta = -\dfrac{1}{\sqrt{2}}$ and θ terminates in QII.

87. Graph the unit circle $x^2 + y^2 = 1$.

88. Graph $x^2 + y^2 = 9$.

3.3 Videos

KEY WORDS

unit circle

even function

odd function

Learning Objectives

A Find the value of a given angle using the unit circle.

B Solve problems involving even and odd functions.

We will begin this section by using the unit circle and our table of exact values to make a diagram that summarizes what we know about exact values, degrees, and radians.

A The Unit Circle

The unit circle is the circle with its center at the origin and a radius of 1. The equation of the unit circle is $x^2 + y^2 = 1$ (Figure 1).

Suppose the terminal side of angle θ, in standard position, intersects the unit circle at point (x, y). Then (x, y) is a point on the terminal side of θ. The distance from the origin to (x, y) is 1, since the radius of the unit circle is 1. Therefore

$$\cos \theta = \frac{x}{r} = \frac{x}{1} = x$$

$$\sin \theta = \frac{y}{r} = \frac{y}{1} = y$$

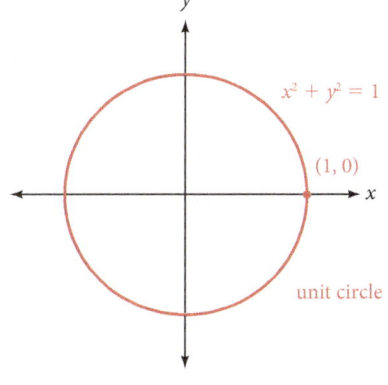

FIGURE 1

The coordinates of the point (x, y) are $\cos \theta$ and $\sin \theta$ (Figure 2). Since the radius of the unit circle is 1, the distance from the origin to the point (x, y) is 1. We can use our first definition for the trigonometric functions from Section 1.3 to write

$$\cos \theta = \frac{x}{r} = \frac{x}{1} = x$$

$$\sin \theta = \frac{y}{r} = \frac{y}{1} = y$$

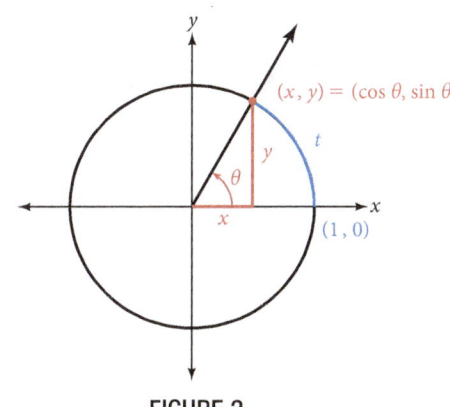

FIGURE 2

In other words, the ordered pair (x, y) shown in Figure 2 can be written $(\cos \theta, \sin \theta)$. We can arrive at this same result by applying our second definition for the trigonometric functions (from Section 2.1). Since θ is an acute angle in a right triangle, by Definition II we have

$$\cos \theta = \frac{\text{side adjacent } \theta}{\text{hypotenuse}} = \frac{x}{1} = x$$

$$\sin \theta = \frac{\text{side opposite } \theta}{\text{hypotenuse}} = \frac{y}{1} = y$$

As you can see, both of our first two definitions lead to the conclusion that the point (x, y) shown in Figure 2 can be written as $(\cos \theta, \sin \theta)$.

Next, consider the length of the arc from $(1, 0)$ to (x, y), which we have labeled t in Figure 2. If θ is measured in radians, then by definition

$$\theta = \frac{t}{r} = \frac{t}{1} = t$$

The length of the arc from $(1, 0)$ to (x, y) is exactly the same as the radian measure of angle θ. Therefore, we can write

$$\cos \theta = \cos t = x \qquad \text{and} \qquad \sin \theta = \sin t = y$$

These results give rise to a third definition for the trigonometric functions.

Definition III: Circular Functions

If (x, y) is any point on the unit circle, and t is the distance from $(1, 0)$ to (x, y) along the circumference of the unit circle, then

$$\cos t = x$$

$$\sin t = y$$

$$\tan t = \frac{y}{x} \quad (x \neq 0)$$

$$\cot t = \frac{x}{y} \quad (y \neq 0)$$

$$\csc t = \frac{1}{y} \quad (y \neq 0)$$

$$\sec t = \frac{1}{x} \quad (x \neq 0)$$

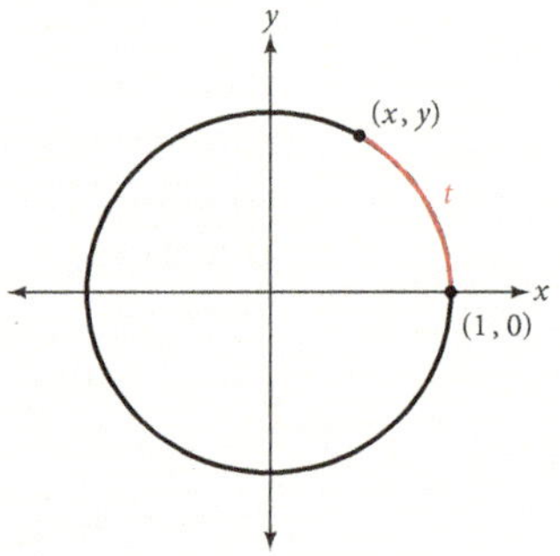

FIGURE 3

As we travel around the unit circle, starting at the point $(1, 0)$, the points we come across all have coordinates $(\cos \theta, \sin \theta)$, where t is the distance we have traveled. When we define the trigonometric functions this way, we call them circular functions because of their relationship to the unit circle.

Notice that the identities we worked with in Chapter 1 are very easy to see with this new definition. For example our ratio identity for tangent is

$$\tan t = \frac{y}{x} = \frac{\sin t}{\cos t}$$

and our reciprocal identity for secant is

$$\sec t = \frac{1}{x} = \frac{1}{\cos t}$$

Figure 4 shows an expanded version of the unit circle with multiples of 30° and 45° marked off in both degrees and radians. The cosine and sine of each angle are the x- and y-coordinates of the point where the terminal side of the angle intersects the unit circle.

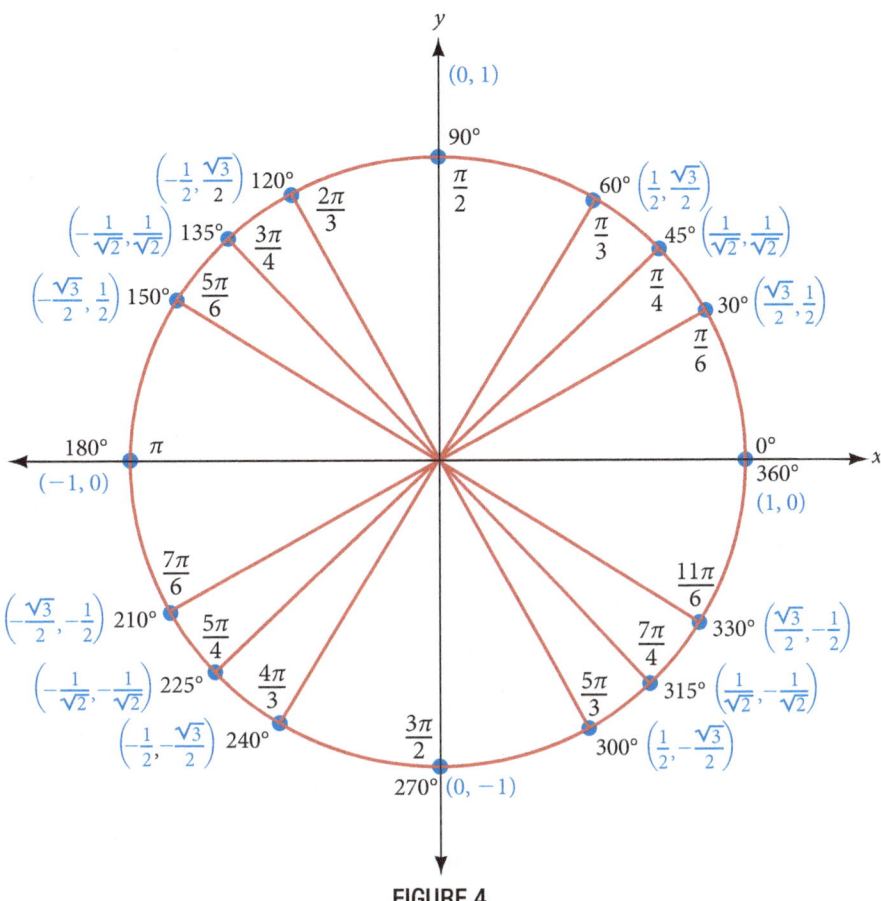

FIGURE 4

EXAMPLE 1 Use Figure 4 to find the six trigonometric functions of $\frac{5\pi}{6}$.

Solution We obtain cosine and sine directly from Figure 3. The other trigonometric functions of $\frac{5\pi}{6}$ are found by using the ratio and reciprocal identities.

$$\sin \frac{5\pi}{6} = \frac{1}{2}$$

$$\cos \frac{5\pi}{6} = -\frac{\sqrt{3}}{2}$$

$$\tan \frac{5\pi}{6} = \frac{\sin \frac{5\pi}{6}}{\cos \frac{5\pi}{6}} = \frac{1/2}{-\sqrt{3}/2} = -\frac{1}{\sqrt{3}}$$

$$\cot \frac{5\pi}{6} = \frac{\cos \frac{5\pi}{6}}{\sin \frac{5\pi}{6}} = \frac{-\frac{\sqrt{3}}{2}}{\frac{1}{2}} = -\sqrt{3}$$

$$\sec \frac{5\pi}{6} = \frac{1}{\cos \frac{5\pi}{6}} = \frac{1}{-\sqrt{3}/2} = -\frac{2}{\sqrt{3}}$$

$$\csc \frac{5\pi}{6} = \frac{1}{\sin \frac{5\pi}{6}} = \frac{1}{1/2} = 2$$

Figure 4 is very helpful in visualizing the relationships among the angles shown and the trigonometric functions of those angles. You may want to make a larger copy of this diagram yourself. In the process of doing so you will become more familiar with the relationship between degrees and radians and the exact values of the angles in the diagram.

EXAMPLE 2   Use the unit circle to find all values of θ between 0° and 360° for which $\cos \theta = \frac{1}{2}$.

Solution We look for all ordered pairs on the unit circle with an x-coordinate of $\frac{1}{2}$. The angles associated with these points are the angles for which $\cos \theta = \frac{1}{2}$. They are $\theta = 60°$ or $\frac{\pi}{3}$ and $\theta = 300°$ or $\frac{5\pi}{3}$.

B Even and Odd Functions

Recall from algebra the definitions of even and odd functions.

Definition

An *even function* is a function for which

$$f(-x) = f(x) \text{ for all } x \text{ in the domain of } f$$

An even function is a function for which replacing x with $-x$ leaves the equation that defines the function unchanged. If a function is even, then every time the point (x, y) is on the graph, so is the point $(-x, y)$. The function $f(x) = x^2 + 3$ is an even function since

$$f(-x) = (-x)^2 + 3 = x^2 + 3 = f(x)$$

Definition

An *odd function* is a function for which

$$f(-x) = -f(x) \text{ for all } x \text{ in the domain of } f$$

An odd function is a function for which replacing x with $-x$ changes the sign of the equation that defines the function. If a function is odd, then every time the point (x, y) is on the graph, so is the point $(-x, -y)$. The function $f(x) = x^3 - x$ is an odd function since

$$f(-x) = (-x)^3 - (-x) = -x^3 + x = -(x^3 - x) = -f(x)$$

From the unit circle it is apparent that sine is an odd function and cosine is an even function. To begin to see that this is true, we locate $\frac{\pi}{6}$ and $-\frac{\pi}{6}$ ($-\frac{\pi}{6}$ is coterminal with $\frac{11\pi}{6}$) on the unit circle and notice that

$$\cos\left(-\frac{\pi}{6}\right) = \frac{\sqrt{3}}{2} = \cos\frac{\pi}{6}$$

and

$$\sin\left(-\frac{\pi}{6}\right) = -\frac{1}{2} = -\sin\frac{\pi}{6}$$

We can generalize this result by drawing an angle θ and its opposite $-\theta$ in standard position, and then labeling the points where their terminal sides intersect the unit circle with (x, y) and $(x, -y)$, respectively. (Can you see from Figure 5 why we label these two points in this way? That is, does it make sense that if (x, y) is on the terminal side of θ, then $(x, -y)$ must be on the terminal side of $-\theta$?)

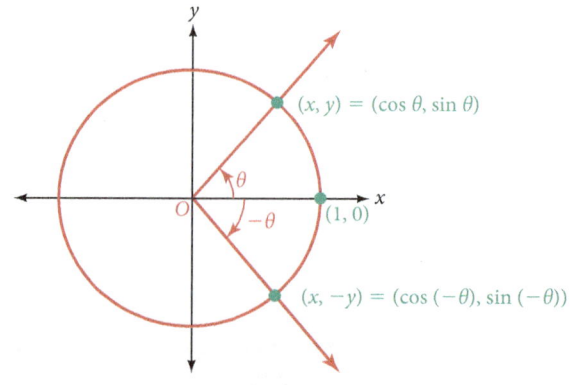

FIGURE 5

Since, on the unit circle, $\cos \theta = x$ and $\sin \theta = y$, we have

$$\cos(-\theta) = x = \cos \theta \qquad \textit{Indicating that cosine is an even function}$$

and

$$\sin(-\theta) = -y = -\sin \theta \qquad \textit{Indicating that sine is an odd function}$$

Now that we have established that sine is an odd function and cosine is an even function, we can use our ratio and reciprocal identities to find which of the other trigonometric functions are even or odd. Example 3 shows how this is done for the cosecant function.

EXAMPLE 3   Show that cosecant is an odd function.

Solution   We must prove that $\csc(-\theta) = -\csc \theta$. That is, we must turn $\csc(-\theta)$ into $-\csc \theta$. Here is how it goes

$$\csc(-\theta) = \frac{1}{\sin(-\theta)} \qquad \textit{Reciprocal identity}$$

$$= \frac{1}{-\sin \theta} \qquad \textit{Sine is a odd function}$$

$$= -\frac{1}{\sin \theta} \qquad \textit{Algebra}$$

$$= -\csc \theta \qquad \textit{Recriprocal identity}$$

EXAMPLE 4   Use the even and odd function relationships to find exact values for each of the following.

a. $\sin(-60°)$ 　　　　　 **b.** $\cos\left(-\dfrac{2\pi}{3}\right)$ 　　　　　 **c.** $\sec(-225°)$

Solution

a. $\sin(-60°) = -\sin 60° \qquad \textit{Sine is an odd function}$

$$= -\frac{\sqrt{3}}{2} \qquad \textit{Unit circle}$$

b. $\cos\left(-\dfrac{2\pi}{3}\right) = \cos\left(\dfrac{2\pi}{3}\right) \qquad \textit{Cosine is an even function}$

$$= -\frac{1}{2} \qquad \textit{Unit circle}$$

c. $\csc(-225°) = \dfrac{1}{\sin(-225°)} \qquad \textit{Reciprocal functions}$

$$= \frac{1}{-\sin 225°} \qquad \textit{Sine is odd function}$$

$$= \frac{1}{-(-1/\sqrt{2})} \qquad \textit{Unit circle}$$

$$= \sqrt{2}$$

An important relationship exists between the radian measure of a central angle in the unit circle and the length of its intercepted arc. Since the radian measure of a central angle is defined as $\theta = s/r$, and the radius of the unit circle is $r = 1$, we have

$$\theta = \frac{s}{r} = \frac{s}{1} = s$$

Thus, the radian measure of a central angle in the unit circle and the length of the arc it cuts off are equal. If the radius of the unit circle were 1 foot, then an angle of 2 radians would cut off an arc of length 2 feet. Likewise, if the radius of the unit circle were 1 centimeter, then an angle of 3 radians would cut off an arc of length 3 centimeters.

There is an interesting diagram we can draw using the ideas given in the preceding paragraph. Figure 6 shows a point $P(x, y)$ that is t units from the point $(1, 0)$ on the circumference of the unit circle. The central angle that cuts off the arc t is in standard position with the point $P(x, y)$ on the terminal side. Therefore, $\cos t = x$ and $\sin t = y$. (Can you see why QB is labeled tan t? It has to do with the fact that $\tan t$ is the ratio of $\sin t$ to $\cos t$ and the similar triangles POA and QOB.)

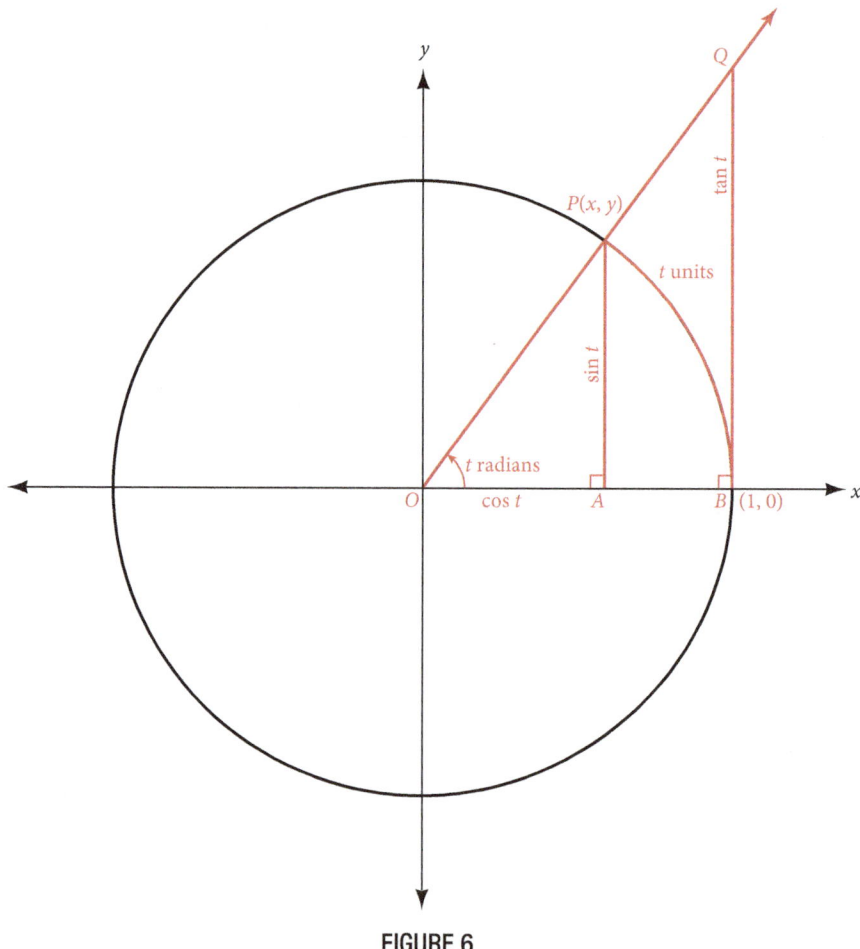

FIGURE 6

There are many concepts that can be visualized from Figure 6. One of the more important is the variations that occur in sin t, cos t, and tan t as P travels around the unit circle. To illustrate, imagine P traveling once around the unit circle starting at $(1, 0)$ and ending 2π units later at the same point. As P moves from $(1, 0)$ to $(0, 1)$, t increases from 0 to $\frac{\pi}{2}$. At the same time, sin t increases from 0 to 1, and cos t decreases from 1 down to 0. As we continue around the unit circle, sin t and cos t simply oscillate between -1 and 1, and further, everywhere sin t is 1 or -1, cos t is 0. We will look at these oscillations in more detail in Chapter 4. For now, it is enough to notice that the unit circle can be used to visualize the relationships that exist among angles measured in radians, the arcs associated with these angles, and the trigonometric functions of both.

Problem Set 3.3

Vocabulary

Use the vocabulary words below to fill in the blanks in the sentences.

cosine   radians   even   odd

1. If (x, y) is a point on the unit circle that is t units from the origin measured in a counterclockwise direction, then x is the _____ of t.

2. Because $\sin(-\theta) = -\sin\theta$, we say the sine function is an _____ function.

3. The cosine function is an _____ function.

4. There are 2π _____ in one complete rotation around the unit circle.

Use the unit circle to find the six trigonometric functions of each angle.

1. $150°$
2. $135°$

3. $\dfrac{11\pi}{6}$
4. $\dfrac{5\pi}{3}$

5. $180°$
6. $270°$

7. $\dfrac{3\pi}{4}$
8. $\dfrac{5\pi}{4}$

Use the unit circle and the fact that cosine is an even function to find each of the following:

9. $\cos(-60°)$
10. $\cos(-120°)$

11. $\cos\left(-\dfrac{5\pi}{6}\right)$
12. $\cos\left(-\dfrac{4\pi}{3}\right)$

Use the unit circle and the fact that sine is an odd function to find each of the following:

13. $\sin(-30°)$ **14.** $\sin(-90°)$

15. $\sin\left(-\dfrac{3\pi}{4}\right)$ **16.** $\sin\left(-\dfrac{7\pi}{4}\right)$

Use the unit circle to find all values of θ between 0 and 2π for which

17. $\sin\theta = \dfrac{1}{2}$ **18.** $\sin\theta = -\dfrac{1}{2}$

19. $\cos\theta = -\dfrac{\sqrt{3}}{2}$ **20.** $\cos\theta = 0$

21. $\tan\theta = -\sqrt{3}$ **22.** $\cot\theta = \sqrt{3}$

23. If angle θ is in standard position and intersects the unit circle at $\left(\dfrac{1}{\sqrt{5}}, -\dfrac{2}{\sqrt{5}}\right)$, find $\sin\theta$, $\cos\theta$, and $\tan\theta$.

24. If angle θ is in standard position and intersects the unit circle at $\left(-\dfrac{1}{\sqrt{10}}, -\dfrac{3}{\sqrt{10}}\right)$, find $\sin\theta$, $\cos\theta$, and $\tan\theta$.

25. If $\sin\theta = -\dfrac{1}{3}$, find $\sin(-\theta)$.   **26.** If $\cos\theta = -\dfrac{1}{3}$, find $\cos(-\theta)$.

If we start at the point $(1, 0)$ and travel once around the unit circle, we travel a distance of 2π units and arrive back where we started at the point $(1, 0)$. If we continue around the unit circle a second time, we will repeat all the values of x and y that occurred during our first trip around. Use this discussion to evaluate the following expressions:

27. $\sin\left(2\pi + \dfrac{\pi}{2}\right)$ **28.** $\cos\left(2\pi + \dfrac{\pi}{2}\right)$

29. $\sin\left(2\pi + \dfrac{\pi}{6}\right)$

30. $\cos\left(2\pi + \dfrac{\pi}{3}\right)$

31. $\sin\dfrac{5\pi}{2}$

32. $\cos\dfrac{5\pi}{2}$

33. $\sin\dfrac{13\pi}{6}$

34. $\cos\dfrac{7\pi}{3}$

Make a diagram of the unit circle with an angle θ in quadrant I and its complement $180° - \theta$ in quadrant II. Label the point on the terminal side of θ and the unit circle with (x, y) and the point on the terminal side of $180° - \theta$ and the unit circle with $(-x, y)$. Use this diagram to show that

35. $\sin(180° - \theta) = \sin\theta$

36. $\cos(180° - \theta) = -\cos\theta$

37. Show that tangent is an odd function.

38. Show that cotangent is an odd function.

Prove each identity.

39. $\sin(-\theta)\cot(-\theta) = \cos\theta$

40. $\cos(-\theta)\tan\theta = \sin\theta$

41. $\sin(-\theta)\sec(-\theta)\cot(-\theta) = 1$

42. $\cos(-\theta)\csc(-\theta)\tan(-\theta) = 1$

43. $\csc\theta + \sin(-\theta) = \dfrac{\cos^2\theta}{\sin\theta}$

44. $\sec\theta - \cos(-\theta) = \dfrac{\sin^2\theta}{\cos\theta}$

45. Redraw the diagram in Figure 6 from this section and label the line segment that corresponds to sec t.

46. Make a diagram similar to the diagram in Figure 6 from this section, but instead of labeling the point $(1, 0)$ with B, label the point $(0, 1)$ with B. Then place Q on the line OP and connect Q to B so that QB is perpendicular to the y-axis. Now, if $P(x, y)$ is t units from $(1, 0)$, label the line segments that correspond to sin t, cos t, cot t, and csc t.

Getting Ready for the Next Section

Simplify. Round answers to three significant digits.

47. $\dfrac{2.4(3.14)}{3}$

48. $1{,}800 \cdot \dfrac{3.14}{180}$

49. $\dfrac{230(180)}{(0.45)(3.14)}$

50. $\dfrac{1}{2}(2.1)^2(1.4)$

51. $\dfrac{1}{2}(30)^2(1.57)$

52. $\dfrac{\pi}{3} \cdot \dfrac{24}{\pi}$

Learning Objectives

A   Find the length of an arc.

B   Approximate chords with arcs.

C   Find the area of a sector.

3.4 Videos

KEY WORDS

arc length

chord

sector

In Section 3.2 we found that if a central angle θ, measured in radians, in a circle of radius r, cuts off an arc of length s, then the relationship between s, r, and θ can be written as

$$\theta = \frac{s}{r}$$

If we multiply both sides of this equation by r, we will obtain the equation that gives arc length s, in terms of r and θ.

$$s = r\theta \qquad (\theta \text{ in radians})$$

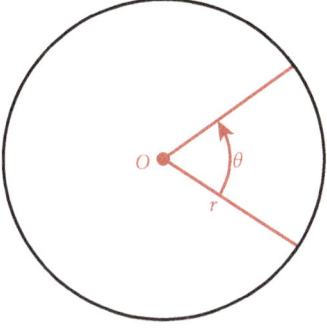

FIGURE 1

A Arc Length

EXAMPLE 1   Give the length of the arc cut off by a central angle of 2 radians in a circle of radius 4.3 inches.

Solution We have $\theta = 2$ and $r = 4.3$ inches. Applying the formula $s = r\theta$ gives us

$$s = r\theta$$
$$= 4.3(2)$$
$$= 8.6 \text{ inches}$$

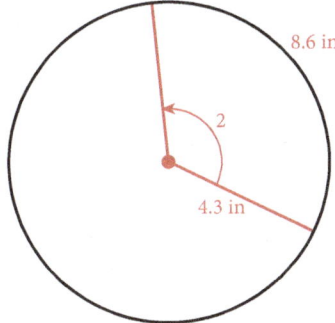

FIGURE 2

EXAMPLE 2   Find the length of the arc cut off by an angle of 45° in a circle of radius 2 centimeters.

Solution The formula $s = r\theta$ holds only when θ is measured in radians. If θ is measured in degrees, we must convert to radians by multiplying θ by $\frac{\pi}{180°}$.

$$s = r\theta$$
$$= 2(45°)\left(\frac{\pi}{180°}\right)$$
$$= \frac{\pi}{2} \text{ centimeters}$$

To find a decimal approximation for s, we substitute 3.14 for π

$$s \approx \frac{3.14}{2} \text{ centimeters}$$
$$= 1.57 \text{ centimeters}$$

EXAMPLE 3 The minute hand of a clock is 1.2 centimeters long. To two significant digits, how far does the tip of the minute hand move in 20 minutes?

Solution We have $r = 1.2$ centimeters. Since we are looking for s, we need to find θ. We can use a proportion to find θ. Since one complete rotation is 60 minutes and 2π radians, we say θ is to 2π as 20 minutes is to 60 minutes, or

If $\quad \dfrac{\theta}{2\pi} = \dfrac{20}{60}$

then $\quad \theta = \dfrac{2\pi}{3}$

Now we can find s.

$s = r\theta$

$\quad = 1.2\left(\dfrac{2\pi}{3}\right)$

$\quad = \dfrac{2.4\pi}{3}$

$\quad \approx \dfrac{2.4(3.14)}{3}$

$\quad = 2.5$ centimeters

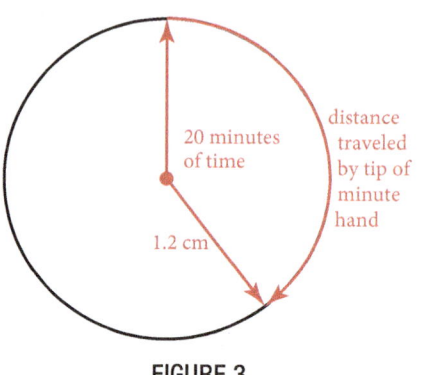

FIGURE 3

The tip of the minute hand will travel approximately 2.5 centimeters every 20 minutes.

EXAMPLE 4 Figure 4 is a model of the Ferris wheel known as the Reisenrad. The diameter of the wheel is 240 feet, and one complete revolution takes 16 minutes. The bottom of the wheel is 12 feet above the ground. Angle θ is the central angle formed as the rider travels from point P_0 to point P_1 on the wheel. Find the distance traveled by a rider if $\theta = 30°$, $\theta = 90°$, $\theta = 135°$.

Solution The formula for arc length, $s = r\theta$, requires θ to be given in radians. Since θ is given in degrees, we must multiply it by $\frac{\pi}{180°}$ to convert to radians. Also, since the diameter of the wheel is 240 feet, the radius is 120 feet.

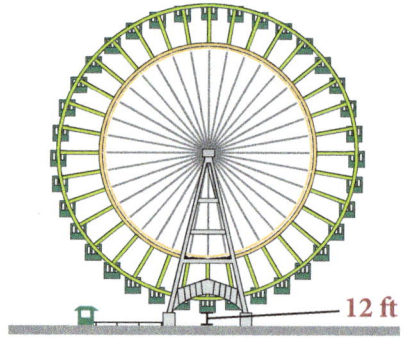

For $\theta = 30°$: $\quad s = r\theta$

$\qquad = 120(30°)\left(\dfrac{\pi}{180°}\right)$

$\qquad = 20\pi$

$\qquad = 62.8$ feet $\qquad$ *To the nearest tenth*

For $\theta = 90°$: $\quad s = r\theta$

$\qquad = 120(90°)\left(\dfrac{\pi}{180°}\right)$

$\qquad = 60\pi$

$\qquad = 188.5$ feet $\qquad$ *To the nearest tenth*

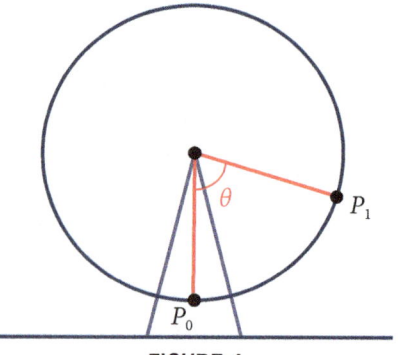

FIGURE 4

For $\theta = 135°$: $\quad s = r\theta$

$\qquad = 120(135°)\left(\dfrac{\pi}{180°}\right)$

$\qquad = 90\pi$

$\qquad = 282.7$ feet $\qquad$ *To the nearest tenth*

B Approximating

If we are working with relatively small central angles in circles with large radii, we can use the length of the intercepted arc to approximate the length of the associated chord. For example, Figure 4 shows a central angle of 1° in a circle of radius 1,800 feet, along with the arc and chord cut off by 1°. (Figure 5 is not drawn to scale.)

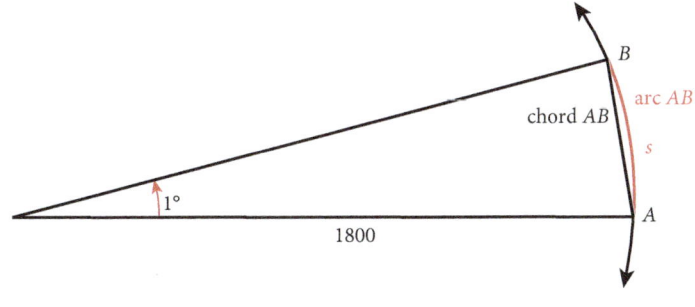

FIGURE 5

To find the length of arc AB we convert θ to radians by multiplying by $\frac{\pi}{180°}$. Then we apply the formula $s = r\theta$.

$$s = r\theta = 1{,}800(1°)\left(\frac{\pi}{180°}\right) = 10\pi \approx 31.4 \text{ feet}$$

If we were to carry out the calculation of arc AB to six significant digits we would have obtained $s = 31.4159$. The length of the chord AB is 31.4155 to six significant digits (found by using the law of sines which we will cover in Chapter 7). As you can see, the first five digits in each number are the same. It seems reasonable then to approximate the length of chord AB with the length of arc AB.

As our next example illustrates, we can also use the procedure outlined above in the reverse order to find the radius of a circle by approximating arc length with the length of the associated chord.

EXAMPLE 5 A person standing on the earth notices that a 747 Jumbo Jet flying overhead subtends an angle of 0.45°. If the length of the jet is 230 feet, find its altitude to the nearest thousand feet.

Solution Figure 6 is a diagram of the situation. Since we are working with a relatively small angle in a circle with a large radius, we use the length of the airplane (chord AB in Figure 6) as an approximation of the length of the arc AB.

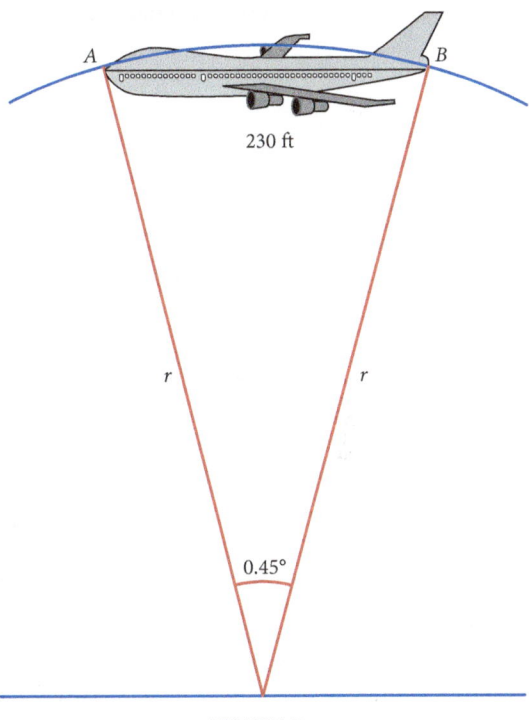

Since $s = r\theta$, $r = \dfrac{s}{\theta}$

so $r = \dfrac{230}{(0.45°)(\pi/180°)}$

$\approx \dfrac{230(180)}{(0.45)(3.14)}$

$= 29{,}000$ feet to the nearest thousand feet

FIGURE 6

C Area of a Sector

Next we want to derive the formula for the area of the sector formed by a central angle θ (Figure 7).

If we let A represent the area of the sector formed by central angle θ, we can find A by setting up a proportion as follows: We say the area A of the sector is to the area of the circle as θ is to one full rotation. That is,

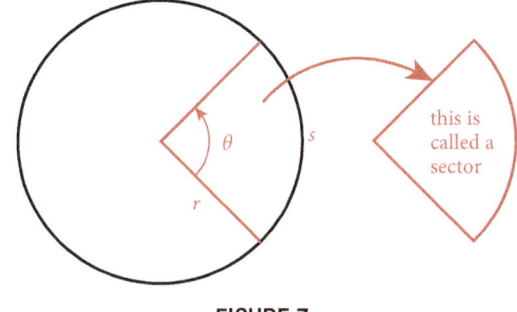

FIGURE 7

Area of sector $\longrightarrow$ $\dfrac{A}{\pi r^2} = \dfrac{\theta}{2\pi}$ $\longleftarrow$ Central angle θ
Area of circle $\longrightarrow$ $\phantom{\dfrac{A}{\pi r^2} = }$ $\longleftarrow$ One full rotation

We solve for A by multiplying both sides of the proportion by πr^2.

$$\pi r^2 \cdot \frac{A}{\pi r^2} = \frac{\theta}{2\pi} \cdot \pi r^2$$

$$A = \frac{1}{2} r^2 \theta$$

EXAMPLE 6   Find the area of the sector formed by a central angle of 1.4 radians in a circle of radius 2.1 meters.

Solution We have $r = 2.1$ meters and $\theta = 1.4$. Applying the formula for A gives us

$$A = \frac{1}{2} r^2 \theta = \frac{1}{2} (2.1)^2 (1.4) = 3.087 \text{ meters}^2$$

Remember Area is measured in square units. When $r = 2.1$ meters, $r^2 = (2.1 \text{ meters})^2 = 4.41 \text{ meters}^2$.

EXAMPLE 7   If the sector formed by a central angle of 15° has an area of $\frac{\pi}{3}$ centimeters^2, find the radius of the circle.

Solution We first convert 15° to radians.

$$\theta = 15° \left(\frac{\pi}{180°} \right) = \frac{\pi}{12}$$

Then we substitute $\theta = \frac{\pi}{12}$ and $A = \frac{\pi}{3}$ into the formula for A, and then solve for r.

$$A = \frac{1}{2} r^2 \theta$$

$$\frac{\pi}{3} = \frac{1}{2} r^2 \frac{\pi}{12}$$

$$\frac{\pi}{3} = \frac{\pi}{24} r^2$$

$$r^2 = \frac{\pi}{3} \cdot \frac{24}{\pi}$$

$$r^2 = 8$$

$$r = 2\sqrt{2} \text{ centimeters}$$

Note that we need only use the positive square root of 8, since we know our radius must be measured with positive units.

EXAMPLE 8   A lawn sprinkler located at the corner of a yard is set to rotate through 90° and project water out 30 feet. To three significant digits, what area of lawn is watered by the sprinkler?

Solution We have $\theta = 90° = \frac{\pi}{2}$ radians, and $r = 30$ feet.

$$A = \frac{1}{2} r^2 \theta$$

$$\approx \frac{1}{2} (30)^2 \left(\frac{\pi}{2} \right)$$

$$= 707 \text{ feet}^2$$

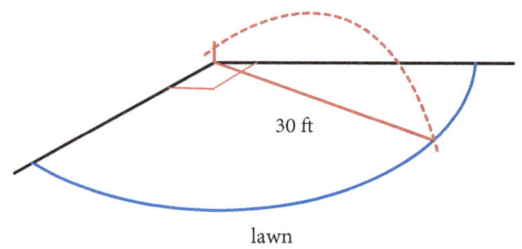

30 ft

lawn

FIGURE 8

Problem Set 3.4

Vocabulary

Use the vocabulary words and formulas below to fill in the blanks in the sentences.

$$\text{sector} \qquad A = \frac{1}{2}r^2\theta \qquad s = 2\left(\frac{\pi}{2}\right) \qquad s = r\theta \qquad A = \frac{1}{2} \cdot 4 \cdot \frac{\pi}{2}$$

For Problems 1 and 2, suppose a central angle of θ radians cuts off an arc of length s in a circle of radius r, then

1. The relationship between s, r, and θ is _____ .

2. The formula for the area A of the _____ formed by angle θ is
   _____ .

3. If the minute hand of a clock is 2 cm long, then the formula we would use to find the distance traveled by the tip of the minute hand in 15 minutes would be _____ .

4. To find the area of the sector swept out by the minute hand of the clock in Problem 3, we use the formula _____ .

Unless otherwise stated, all answers in this problem set that need to be rounded should be rounded to three significant digits.

For each problem below, θ is a central angle in a circle of radius r. In each case, find the length of arc s cut off by θ.

1. $\theta = 2$, $r = 3$ inches

2. $\theta = 3$, $r = 2$ inches

3. $\theta = 1.5$, $r = 1.5$ feet

4. $\theta = 2.4$, $r = 1.8$ feet

5. $\theta = \frac{\pi}{6}$, $r = 12$ centimeters

6. $\theta = \frac{\pi}{3}$, $r = 12$ centimeters

7. $\theta = 60°$, $r = 4$ millimeters

8. $\theta = 30°$, $r = 4$ millimeters

9. $\theta = 240°$, $r = 10$ inches

10. $\theta = 315°$, $r = 5$ inches

11. **Arc Length** The minute hand of a clock is 2.4 centimeters long. How far does the tip of the minute hand travel in 20 minutes?

12. **Arc Length** The minute hand of a clock is 1.2 centimeters long. How far does the tip of the minute hand travel in 40 minutes?

13. **Arc Length** A space shuttle 200 miles above the earth is orbiting the earth once every 6 hours. How far does the shuttle travel in 1 hour? (Assume the radius of the earth is 4,000 miles.) Give both the exact value and a three significant digit approximation for your answer.

14. **Arc Length** How long, in hours, does it take the space shuttle in Problem 13 to travel 8,400 miles? Give both the exact value and an approximate value for your answer.

15. **Arc Length** The pendulum on a grandfather clock swings from side to side once every second. If the length of the pendulum is 4 feet and the angle through which it swings is 20°, how far does the tip of the pendulum travel in one second?

16. **Arc Length** Find the total distance traveled in one minute by the tip of the pendulum on the grandfather clock in Problem 15.

17. **Diameter of the Moon** From the earth, the moon subtends an angle of approximately 0.5°. If the distance to the moon is approximately 240,000 miles, find an approximation for the diameter of the moon accurate to the nearest hundred miles. (See Example 4 and the discussion that precedes it.)

18. **Diameter of the Sun** If the distance to the sun is approximately 93 million miles, and, from the earth, the sun subtends an angle of approximately 0.5°, estimate the diameter of the sun to the nearest ten thousand miles.

In each problem below, θ is a central angle that cuts off an arc of length s. In each case, find the radius of the circle.

19. $\theta = 6, s = 3$ feet

20. $\theta = 1, s = 2$ feet

21. $\theta = 1.4, s = 4.2$ inches

22. $\theta = 5.1, s = 10.2$ inches

23. $\theta = \dfrac{\pi}{4}, s = \pi$ centimeters

24. $\theta = \dfrac{3\pi}{4}, s = \pi$ centimeters

25. $\theta = 90°, s = \dfrac{\pi}{2}$ meters

26. $\theta = 180°, s = \dfrac{\pi}{2}$ meters

27. **Altitude of a Plane** From the ground, a 747 Jumbo Jet flying overhead subtends an angle of 0.42°. If the length of the jet is 230 feet, find its altitude to the nearest thousand feet.

28. **Diameter to Mountain Top** From a point on the ground a person notices that a 100 foot antenna on the top of a mountain subtends an angle of 0.35°. If the angle of elevation to the top of the antenna is 35.35°, find the height of the mountain to the nearest hundred feet. First find the distance from the person on the ground to the top of the mountain, then use right triangle trigonometry to find the height of the mountain. Figure 9 illustrates.

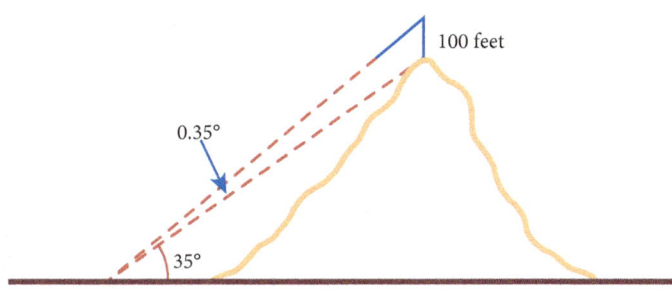

FIGURE 9

Find the area of the sector formed by central angle θ in a circle of radius r if

29. $\theta = 2$, $r = 3$ centimeters

30. $\theta = 3$, $r = 2$ centimeters

31. $\theta = 2.4$, $r = 4$ inches

32. $\theta = 1.8$, $r = 2$ inches

33. $\theta = \dfrac{\pi}{5}$, $r = 3$ meters

34. $\theta = \dfrac{2\pi}{5}$, $r = 3$ meters

35. $\theta = 15°$, $r = 5$ meters

36. $\theta = 15°$, $r = 10$ meters

37. A central angle of 2 radians cuts off an arc of length 4 inches. Find the area of the sector formed.

38. An arc of length 3 feet is cut off by a central angle of $\pi/4$ radians. Find the area of the sector formed.

39. If the sector formed by a central angle of 30° has an area of $\pi/3$ centimeters², find the radius of the circle.

40. What is the length of the arc cut off by angle θ in Problem 39?

41. A sector of area $2\pi/3$ inches² is formed by a central angle of 45°. What is the radius of the circle?

42. A sector of area 25 inches² is formed by a central angle of 4 radians. Find the radius of the circle.

43. **Lawn Sprinkler** A lawn sprinkler is located at the corner of a yard. The sprinkler is set to rotate through 90° and project water out 60 feet. What is the area of the yard watered by the sprinkler?

44. **Lawn Sprinkler** A lawn sprinkler is located at the edge of a lawn. The sprinkler is set to rotate through 110° and project water out 40 feet. What is the area of the lawn watered by the sprinkler?

45. Windshield Wiper An automobile windshield wiper 10 inches long rotates through an angle of 60°. If the rubber part of the blade covers only the last 9 inches of the wiper, find the area of the windshield cleaned by the windshield wiper.

46. Windshield Wiper A truck windshield wiper 24 inches long rotates through an angle of 80°. If there is no rubber on the first 6 inches of the wiper, find the area of the windshield cleaned by the windshield wiper.

47. Find the shaded area.

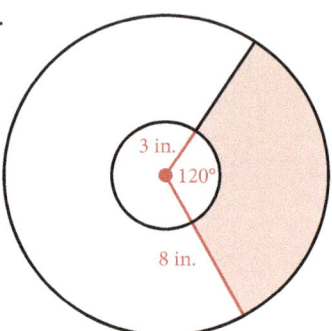

FIGURE 10

48. Find the shaded area.

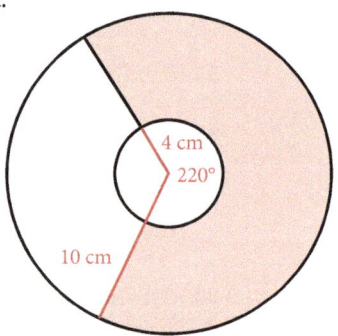

FIGURE 11

Getting Ready for the Next Section

Simplify. Leave answers in terms of π.

49. $\dfrac{2\pi}{30}$

50. $\dfrac{3\pi/4}{3}$

51. $\dfrac{\pi/2}{3/2}$

52. $\dfrac{5\pi/4}{5}$

Simplify. Round to three significant digits.

53. 2340/12

54. 13(180)

55. 270(3.14)

56. 848/12

57. $\dfrac{3(3.14)(60)}{5,280}$

58. $\dfrac{5(3.14)(60)}{4(5,280)}$

VELOCITIES

KEY WORDS

linear velocity

angular velocity

Learning Objectives

A   Find the linear velocity of a point moving on a circle.

B   Find the angular velocity of a point moving on a circle.

C   Find the relationship between linear velocity and angular velocity moving on a circle.

There are two kinds of velocities associated with a point moving on the circumference of a circle. One is called *linear velocity* and is a measure of distance traveled per unit time. The other is called *angular velocity*. Angular velocity is the central angle swept out by the point moving on the circle, divided by time.

A Linear Velocity

Definition

If P is a point on a circle of radius r, and P moves a distance s on the circumference of the circle, in an amount of time t, then the *linear velocity*, v, of P is given by the formula

$$v = \frac{s}{t}$$

EXAMPLE 1   A point on a circle travels 5 centimeters in 2 seconds. Find the linear velocity of the point.

Solution Substituting $s = 5$ and $t = 2$ into the equation $v = \frac{s}{t}$ gives us

$$v = \frac{5 \text{ centimeters}}{2 \text{ seconds}}$$

$$= 2.5 \text{ centimeters per second}$$

Note In all the examples and problems in this section, we are assuming that the point on the circle moves with uniform circular motion. That is, the velocity of the point is constant.

B Angular Velocity

Definition

If P is a point moving with uniform circular motion on a circle of radius r, and the line from the center of the circle through P sweeps out a central angle θ, in an amount of time t, then the *angular velocity*, ω, of P is given by the equation

$$\omega = \frac{\theta}{t} \qquad \text{where } \theta \text{ is measured in radians}$$

EXAMPLE 2   A point on a circle rotates through $\frac{3\pi}{4}$ radians in 3 seconds. Give the angular velocity of P.

Solution Substituting $\theta = \frac{3\pi}{4}$ and $t = 3$ into the equation $\omega = \frac{\theta}{t}$ gives us

$$\omega = \frac{3\pi/4 \text{ radians}}{3 \text{ seconds}}$$

$$= \frac{\pi}{4} \text{ radians per second}$$

EXAMPLE 3   A bicycle wheel with a radius of 13 inches turns with an angular velocity of 3 radians per second. Find the distance traveled by a point on the bicycle tire in 1 minute.

Solution   We have $\omega = 3$ radians per second, $r = 13$ inches, and $t = 60$ seconds. First we find θ using $\omega = \frac{\theta}{t}$.

$$\text{If}\quad \omega = \frac{\theta}{t}$$
$$\text{then } \theta = \omega t$$
$$= 3(60)$$
$$= 180 \text{ radians}$$

To find the distance traveled by the point in 60 seconds, we use the formula $s = r\theta$ from Section 3.4, with $r = 13$ inches, and $\theta = 180$.

$$s = 13(180)$$
$$= 2{,}340 \text{ inches}$$

If we want this result expressed in feet, we divide by 12.

$$s = \frac{2340}{12} \text{ feet}$$
$$= 195 \text{ feet}$$

A point on the tire of the bicycle will travel 195 feet in one minute. If the bicycle was being ridden under these conditions, the rider would travel 195 feet in one minute.

EXAMPLE 4   Figure 1 shows a fire truck parked on the shoulder of a freeway next to a long block wall. The red light on the top of the truck is 10 feet from the wall and rotates through one complete revolution every 2 seconds. Find the equations that give the lengths d and l in terms of time t.

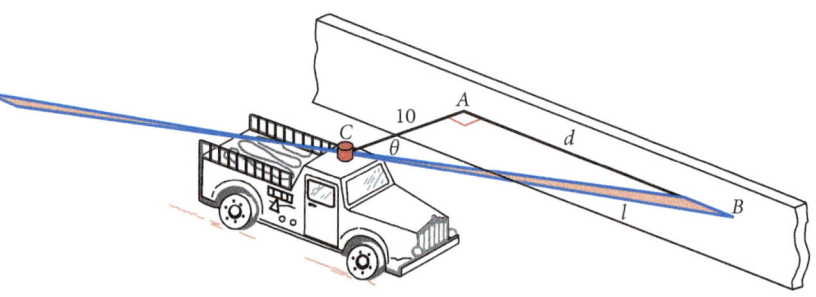

FIGURE 1

Solution   The angular velocity of the rotating red light is

$$\omega = \frac{\theta}{t} = \frac{2\pi \text{ radians}}{2 \text{ seconds}} = \pi \text{ radians per second}$$

From right triangle ABC we have the following relationships

$$\tan \theta = \frac{d}{10} \qquad \text{and} \qquad \sec \theta = \frac{l}{10}$$
$$d = 10 \tan \theta \qquad\qquad\qquad l = 10 \sec \theta$$

Now, these equations give us d and l in terms of θ. To write d and l in terms of t, we solve $\omega = \frac{\theta}{t}$ for θ to obtain $\theta = \omega t = \pi t$. Substituting this for θ in each equation we have d and l expressed in terms of t.

$$d = 10 \tan \pi t \qquad l = 10 \sec \pi t$$

C The Relationship between the Two Velocities

To find the relationship between the two kinds of velocities we have developed so far, we can take the equation that relates arc length and central angle measure, $s = r\theta$, and divide both sides by time, t.

$$\text{If} \quad s = r\theta$$

$$\text{then } \frac{s}{t} = \frac{r\theta}{t}$$

$$\frac{s}{t} = r\frac{\theta}{t}$$

$$v = r\omega$$

Linear velocity is the product of the radius and the angular velocity.

EXAMPLE 5 A phonograph record is turning at 45 rpm (revolutions per minute). If the distance from the center of the record to a point on the edge of the record is 3 inches, find the angular velocity and the linear velocity, in feet per minute, of the point.

Solution The quantity 45 rpm is another way of expressing the rate at which the point on the record is moving. We can obtain the angular velocity from it by remembering that one complete revolution is equivalent to 2π radians. Therefore,

$$\omega = 45(2\pi) \text{ radians per minute}$$

$$= 90\pi \text{ radians per minute}$$

To find the linear velocity, we multiply ω by the radius.

$$v = r\omega$$

$$= 3(90\pi)$$

$$= 270\pi \text{ inches per minute} \qquad \textit{Exact value}$$

$$\approx 848 \text{ inches per minute} \qquad \textit{To three significant digits}$$

If we want this last quantity expressed in feet per minute, we divide by 12.

$$v = \frac{848}{12} \text{ feet per minute}$$

$$= 70.7 \text{ feet per minute}$$

EXAMPLE 6 The London Eye, pictured below, is a Ferris wheel that opened in London in the year 2000. The diameter of the wheel is 120 feet. It has 32 capsules, each holding 25 passengers. The wheel rotates through one complete revolution every 30 minutes. (It does not stop for passenger loading and unloading.) Find the angular and linear velocity of one of the passengers as they ride the wheel.

iStockphoto.com © CO7

Solution To find the angular velocity, we use the fact that one revolution takes 30 minutes. In one trip around the wheel a rider travels through an angle of 2π radians. Therefore, the angular velocity is

$$\omega = \frac{2\pi \text{ radians}}{30 \text{ minutes}} = \frac{\pi}{15} \text{ radians/minute}$$

To find the linear velocity we use the formula $v = r\omega$ with the result from above and the fact that the radius is 60 feet to obtain

$$v = 60 \cdot \frac{\pi}{15} = 4\pi \text{ feet/minute}$$

We can change to units that are a little more intuitive to use by converting feet/minute to miles/hour. To do so we use the fact that there are 5,280 feet in 1 mile, and 60 minutes in 1 hour.

$$v = 4\pi \text{ feet/minute}$$

$$= \frac{4\pi \text{ feet}}{1 \text{ minute}}$$

$$= \frac{4\pi \text{ feet}}{1 \text{ minute}} \cdot \frac{60 \text{ minutes}}{1 \text{ hour}} \cdot \frac{1 \text{ mile}}{5,280 \text{ feet}}$$

$$\approx 0.14 \text{ miles/hour}$$

Problem Set 3.5

Vocabulary

Use the vocabulary words and formulas below to fill in the blanks in the sentences.

revolutions   linear   angular $v = \dfrac{s}{t}$ $\omega = \dfrac{\theta}{t}$

For Problems 1 and 2, suppose an object is moving around a circle with radius r, so that it rotates through an angle of θ radians every t seconds cutting off an arc of length s.

1. The _____ velocity of the object is _____.

2. The _____ velocity of the object is _____.

3. 45 rpm stands for 45 _____ per minute.

In this problem set, round any answers that need rounding to three significant digits.

Find the linear velocity of a point moving with uniform circular motion, if the point covers a distance s in an amount of time t, where

1. $s = 3$ feet and $t = 2$ minutes

2. $s = 10$ feet and $t = 2$ minutes

3. $s = 12$ centimeters and $t = 4$ seconds

4. $s = 12$ centimeters and $t = 2$ seconds

5. $s = 30$ miles and $t = 2$ hours

6. $s = 100$ miles and $t = 4$ hours

Find the distance s covered by a point moving with linear velocity v for a time t if

7. $v = 20$ feet per second and $t = 4$ seconds

8. $v = 10$ feet per second and $t = 4$ seconds

9. $v = 45$ miles per hour and $t = \frac{1}{2}$ hour

10. $v = 55$ miles per hour and $t = \frac{1}{2}$ hour

11. $v = 21$ miles per hour and $t = 20$ minutes

12. $v = 63$ miles per hour and $t = 10$ seconds

Point P sweeps out central angle θ as it rotates on a circle of radius r as given below. In each case, find the angular velocity of point P.

13. $\theta = \dfrac{2\pi}{3}$, $t = 5$ seconds

14. $\theta = \dfrac{3\pi}{4}$, $t = 5$ seconds

15. $\theta = 12$, $t = 3$ minutes

16. $\theta = 24$, $t = 6$ minutes

17. $\theta = 8\pi$, $t = 3\pi$ seconds

18. $\theta = 12\pi$, $t = 5\pi$ seconds

19. $\theta = 45\pi$, $t = 1.2$ hours

20. $\theta = 24\pi$, $t = 1.8$ hours

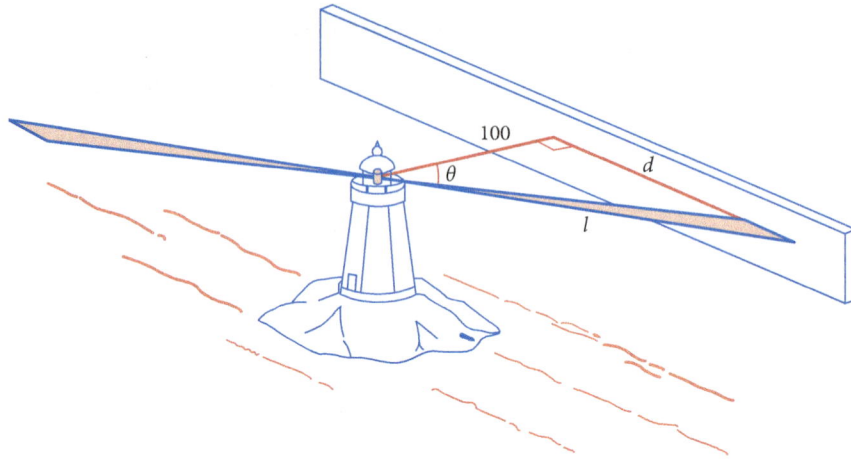

FIGURE 2

21. **Rotating Light** Figure 2 shows a lighthouse that is 100 feet from a long straight wall on the beach. The light in the lighthouse rotates through one complete rotation once every 4 seconds. Find an equation that gives the distance d in terms of time t, then find d when t is $\frac{1}{2}$ second and $\frac{3}{2}$ seconds. What happens when you try $t = 1$ second in the equation? How do you interpret this?

22. **Rotating Light** Using the diagram in Figure 2, find an equation that expresses l in terms of time t. Find l when t is 0.5 second, 1.0 second, and 1.5 seconds.

In the problems that follow, point P moves with angular velocity ω on a circle of radius r. In each case, find the distance s traveled by the point in time t.

23. $\omega = 4$ radians per second, $r = 2$ inches, $t = 5$ seconds

24. $\omega = 2$ radians per second, $r = 4$ inches, $t = 5$ seconds

25. $\omega = \frac{3\pi}{2}$ radians per second, $r = 4$ meters, $t = 30$ seconds

26. $\omega = \frac{4\pi}{3}$ radians per second, $r = 8$ meters, $t = 20$ seconds

27. $\omega = 15$ radians per second, $r = 5$ feet, $t = 1$ minute

28. $\omega = 10$ radians per second, $r = 6$ feet, $t = 2$ minutes

For each of the following problems, find the angular velocity associated with the given rpm's.

29. 10 rpm   30. 20 rpm

31. $33\frac{1}{3}$ rpm   32. $16\frac{2}{3}$ rpm

33. 5.8 rpm   34. 7.2 rpm

For each problem below, a point is rotating with uniform circular motion on a circle of radius r.

35. Find v if $r = 2$ inches and $\omega = 5$ radians per second

36. Find v if $r = 8$ inches and $\omega = 4$ radians per second

37. Find ω if $r = 6$ centimeters and $v = 3$ centimeters per second

38. Find ω if $r = 3$ centimeters and $v = 8$ centimeters per second

39. Find v if $r = 4$ feet and the point rotates at 10 rpm

40. Find v if $r = 1$ foot and the point rotates at 20 rpm

41. Velocity at the Equator The earth rotates through one complete revolution every 24 hours. Since the axis of rotation is perpendicular to the equator, you can think of a person standing on the equator as standing on the edge of a disc that is rotating through one complete revolution every 24 hours. Find the angular velocity of a person standing on the equator.

42. Velocity at the Equator Assuming the radius of the earth is 4,000 miles, use the information from Problem 41 to find the linear velocity of a person standing on the equator.

43. Distance A boy is twirling a model airplane on a string 5 feet long. If he twirls the plane at 0.5 rpm, how far does the plane travel in 2 minutes?

44. Velocity of a Mixer Blade A mixing blade on a food processor extends out 3 inches from its center. If the blade is turning at 600 rpm, what is the linear velocity of the tip of the blade in feet per minute?

45. Velocity of a Lawnmower Blade A gasoline driven lawnmower has a blade that extends out 1 foot from its center. The tip of the blade is traveling at the speed of sound, which is 1,100 feet per second. Through how many rpm is the blade turning?

46. Velocity of a Diskette An 8 inch floppy disk in a computer (radius = 4 inches) rotates at 300 rpm. Find the linear velocity of a point on the edge of the disk.

47. Distance How far does the tip of a 12 centimeter minute hand on a clock travel in 1 day?

48. Distance How far does the tip of a 10 centimeter hour hand on a clock travel in 1 day?

49. Velocity of a Bike Wheel A woman rides a bicycle for 1 hour and travels 16 kilometers (about 10 miles). Find the angular velocity of the wheel if the radius is 30 centimeters.

50. Velocity of a Bike Wheel Find the number of rpm for the wheel in Problem 49.

51. Ferris Wheel The photo below is the Ferris wheel known as the Reisenrad. The diameter of the wheel is 197 feet, and one complete revolution takes 16 minutes. The bottom of the wheel is 12 feet above the ground. In the movie The Third Man, Orson Welles takes one trip around the Reisenrad. Find his angular velocity and linear velocity as he does so.

52. Ferris Wheel The photo below is the Ferris wheel known as the High Roller that is located in Las Vegas. The diameter of the wheel is 550 feet, and one complete revolution takes 30 minutes. The bottom of the wheel is 30 feet above the ground. Find the linear and angular velocity of a person riding this wheel.

Maintaining Your Skills

The problems that follow review material we covered in Chapter 2.

53. Find sin A, cos A, and tan A in right triangle ABC, with $C = 90°$, if $b = 3$ and $c = 6$.

Simplify each expression as much as possible. Do not use tables.

54. $\sin^2 45° + \cos^2 30°$

55. $\dfrac{1}{\sec 30°}$

56. Convert 73.2° to degrees and minutes.

57. Convert 79° 30′ to decimal degrees.

58. Give the number of significant digits in each number 0.0028.

The following problems refer to right triangle ABC with $C = 90°$. In each case, find all the missing parts.

59. $a = 68.0$ and $b = 104$

60. $b = 305$ and $B = 24.9°$

61. Angle of Elevation If a 75 foot flag pole casts a shadow 43 feet long, what is the angle of elevation of the sun from the tip of the shadow?

62. Angle of Elevation A road up a hill makes an angle of 5° with the horizontal. If the road from the bottom of the hill to the top of the hill is 2.5 miles long, how high is the hill?

63. Dimensions of a Mirror A person standing 5 feet from a mirror notices that the angle of depression from his eyes to the bottom of the mirror is 13°, while the angle of elevation to the top of the mirror is 12°. Find the vertical dimensions of the mirror.

64. Bearing A boat travels on a course of bearing S 63° 50′ E for 100 miles. How many miles south and how many miles east has the boat traveled?

Examples

1. 30° is the reference angle for 30°, 150°, 210°, and 330°.

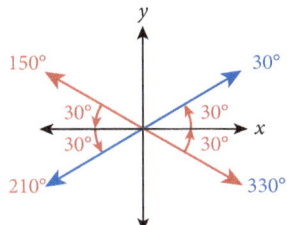

Reference Angle [3.1]

The *reference angle* $\hat{\theta}$ for any angle θ in standard position is the positive acute angle between the terminal side of θ and the *x*-axis.

2. $\sin 150° = \sin 30° = \dfrac{1}{2}$

$\sin 210° = -\sin 30° = -\dfrac{1}{2}$

$\sin 330° = -\sin 30° = -\dfrac{1}{2}$

Reference Angle Theorem [3.1]

A trigonometric function of an angle and its reference angle differ at most in sign.

   We find trigonometric functions for angles between 0° and 360° by first finding the reference angle. We then find the value of the trigonometric function of the reference angle and use the quadrant in which the angle terminates to assign the correct sign.

3. angle θ has a measure of 1 radian

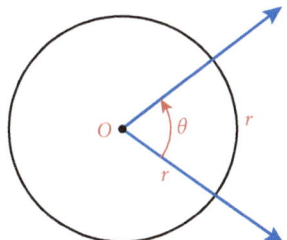

the vertex of θ is at the center of the circle; the arc cut off by θ is equal in length to the radius

Radian Measure [3.2]

In a circle, a central angle that cuts off an arc equal in length to the radius of the circle has a measure of 1 radian.

4. Radians to degrees

$$\frac{4\pi}{3}\text{ radians} = \frac{4\pi}{3}\left(\frac{180°}{\pi}\right) = 240°$$

Degrees to radians

$$450° = 450\left(\frac{\pi}{180°}\right)$$
$$= \frac{5\pi}{2}\text{ radians}$$

Radians and Degrees [3.2]

Changing from degrees to radians and radians to degrees is simply a matter of multiplying by the appropriate conversion factor.

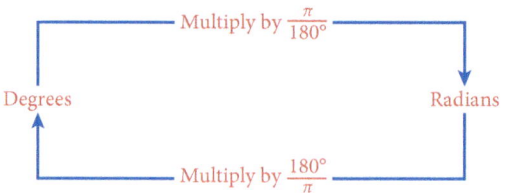

The Unit Circle [3.3]

5.

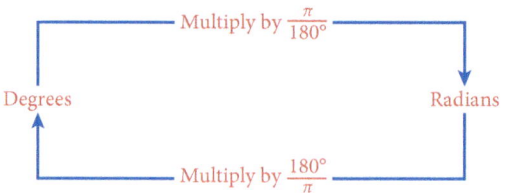

The unit circle is the circle with its center at the origin and a radius of 1. The equation of the unit circle is $x^2 + y^2 = 1$. Because the radius of the unit circle is 1, any point (x, y) on the circle is such that

$$x = \cos\theta \quad\text{and}\quad y = \sin\theta$$

Even and Odd Functions [3.3]

An *even function* is a function for which

$$f(-x) = f(x)\text{ for all }x\text{ in the domain of }f$$

and an *odd function* is a function for which

$$f(-x) = -f(x)\text{ for all }x\text{ in the domain of }f$$

Cosine is an even function, and sine is an odd function. That is,

$$\cos(-\theta) = \cos\theta \qquad\text{Cosine is an even function}$$

and

$$\sin(-\theta) = -\sin\theta \qquad\text{Sine is an odd function}$$

6. $\tan\theta$ is an odd function.

$$\tan(-\theta) = \frac{\sin(-\theta)}{\cos(-\theta)}$$
$$= \frac{-\sin\theta}{\cos\theta}$$
$$= -\frac{\sin\theta}{\cos\theta}$$
$$= -\tan\theta$$

Arc Length [3.4]

If s is an arc cut off by a central angle θ, measured in radians, in a circle of radius r, then

$$s = r\theta$$

7. The arc cut off by 2.5 radians in a circle of radius 4 inches is

$$s = 4(2.5)$$
$$= 10.0\text{ inches}$$

8. The area of the sector formed by a central angle of 2.5 radians in a circle of radius 4 inches is

$$A = \frac{1}{2}(4)^2(2.5)$$

$$= 20.0 \text{ inches}^2$$

Area of a Sector [3.4]

The area of the sector formed by a central angle θ in a circle of radius r is

$$A = \frac{1}{2}r^2\theta$$

where θ is measured in radians.

9. If a point moving at a uniform speed on a circle travels 12 centimeters every 3 seconds, then the linear velocity of the point is

$$v = \frac{12 \text{ centimeters}}{3 \text{ seconds}}$$

$$= 4 \text{ centimeters per second}$$

Linear Velocity [3.5]

If P is a point on a circle of radius r, and P moves a distance s on the circumference of the circle, in an amount of time t, then the *linear velocity*, v, of P is given by the formula

$$v = \frac{s}{t}$$

10. If a point moving at uniform speed on a circle of radius 4 inches rotates through $\frac{3\pi}{4}$ radians every 3 seconds, then the angular velocity of the point is

$$\omega = \frac{\frac{3\pi}{4} \text{ radians}}{3 \text{ seconds}}$$

$$= \frac{\pi}{4} \text{ radians per second}$$

The linear velocity of the same point is given by

$$v = 4\left(\frac{\pi}{4}\right)$$

$$= \pi \text{ inches per second}$$

Angular Velocity [3.5]

If P is a point moving with uniform circular motion on a circle of radius r, and the line from the center of the circle through P sweeps out a central angle θ, in an amount of time t, then the *angular velocity*, ω, of P is given by the equation

$$\omega = \frac{\theta}{r} \quad \text{where } \theta \text{ is measured in radians}$$

The relationship between linear velocity and angular velocity is given by the formula

$$v = r\omega$$

Draw each of the following angles in standard position and then name the reference angle:

1. 235°   **2.** 117.8°   **3.** 410° 20′   **4.** −225°

Use a calculator to find each of the following:

5. tan 320°   **6.** tan (−25°)   **7.** cos (−236.7°)

8. sin 322.3°   **9.** sec 140° 20′   **10.** csc 188° 50′

Use a calculator to find θ, if θ is between 0° and 360° and

11. $\sin \theta = 0.1045$ with θ in QII   **12.** $\cos \theta = -0.4772$ with θ in QIII

13. $\cot \theta = 0.9659$ with θ in QIII   **14.** $\sec \theta = 1.545$ with θ in QIV

Give the exact value of each of the following:

15. sin 225°   **16.** cos 135°   **17.** tan 330°   **18.** sec 390°

Convert each of the following to radian measure. Write each answer as an exact value.

19. 250°   **20.** −390°

Convert each of the following to degree measure:

21. $\dfrac{4\pi}{3}$ **22.** $\dfrac{7\pi}{12}$

Give the exact value of each of the following:

23. $\sin \dfrac{2\pi}{3}$ **24.** $\cos \dfrac{2\pi}{3}$ **25.** $4 \cos\left(-\dfrac{3\pi}{4}\right)$

26. $2 \cos\left(-\dfrac{5\pi}{3}\right)$ **27.** $\sec \dfrac{5\pi}{6}$ **28.** $\csc \dfrac{5\pi}{6}$

29. Evaluate $2 \cos\left(3x - \dfrac{\pi}{2}\right)$ when x is $\dfrac{\pi}{3}$.

30. Evaluate $4 \sin\left(2x + \dfrac{\pi}{4}\right)$ when x is $\dfrac{\pi}{4}$.

31. Show that cotangent is an odd function.

32. Prove the identity $\sin(-\theta) \sec(-\theta) \cot(-\theta) = 1$.

For each problem below, θ is a central angle in a circle of radius r. In each case, find the length of arc s cut off by θ.

33. $\theta = \dfrac{\pi}{6}$, $r = 12$ meters   **34.** $\theta = 60°$, $r = 6$ feet

In each problem below, θ is a central angle that cuts off an arc of length s. In each case, find the radius of the circle.

35. $\theta = \dfrac{\pi}{4}$, $s = \pi$ centimeters   **36.** $\theta = \dfrac{2\pi}{3}$, $s = \dfrac{\pi}{4}$ centimeters

Find the area of the sector formed by central angle θ in a circle of radius r if

37. $\theta = 90°$, $r = 4$ inches

38. $\theta = 2.4$, $r = 3$ centimeters

39. The minute hand of a clock is 2 centimeters long. How far does the tip of the minute hand travel in 30 minutes?

40. A central angle of 4 radians cuts off an arc of length 8 inches. Find the area of the sector formed.

Find the distance s covered by a point moving with linear velocity v for a time t if

41. $v = 30$ feet per second and $t = 3$ seconds

42. $v = 66$ feet per second and $t = 1$ minute

In the problems that follow, point P moves with angular velocity ω on a circle of radius r. In each case, find the distance s traveled by the point in time t.

43. $\omega = 4$ radians per second, $r = 3$ inches, $t = 6$ seconds

44. $\omega = \dfrac{3\pi}{4}$ radians per second, $r = 8$ feet, $t = 20$ seconds

For each of the following problems, find the angular velocity associated with the given rpm's.

45. 6 rpm

46. 2 rpm

For each problem below, a point is rotating with uniform circular motion on a circle of radius r.

47. Find ω if $r = 10$ centimeters and $v = 5$ centimeters per second

48. Find ω if $r = 3$ centimeters and $v = 5$ centimeters per second

49. Find v if $r = 2$ feet and the point rotates at 20 rpm

50. Find v if $r = 1$ foot and the point rotates at 10 rpm

51. Pulley Speed A belt connects a pulley of radius 8 centimeters to a pulley of radius 6 centimeters. Each point on the belt is traveling at 24 centimeters per second. Find the angular velocity of each pulley (see Figure 1).

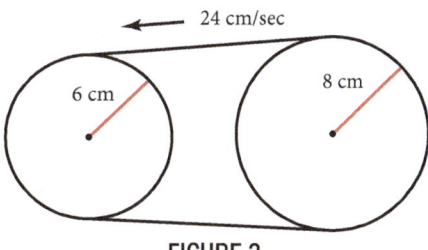

24 cm/sec

6 cm

8 cm

FIGURE 3

52. Propeller Speed A propeller with radius 1.5 feet is rotating at 900 rpm. Find the linear velocity of the tip of the propeller. Give the exact value and a three significant digit approximation.

Spotlight on Success

Stefanie, Student Instructor

Never confuse a single defeat with a final defeat.—F. Scott Fitzgerald

The idea that has worked best for my success in college, and more specifically in my math courses, is to stay positive and be resilient. I have learned that a 'bad' grade doesn't make me a failure; if anything it makes me strive to do better. That is why I never let a bad grade on a test or even in a class get in the way of my overall success.

By sticking with this positive attitude, I have been able to achieve my goals. My grades have never represented how well I know the material. This is because I have struggled with test anxiety and it has consistently lowered my test scores in a number of courses. However, I have not let it defeat me. When I applied to graduate school, I did not meet the grade requirements for my top two schools, but that did not stop me from applying.

One school asked that I convince them that my knowledge of mathematics was more than my grades indicated. If I had let my grades stand in the way of my goals, I wouldn't have been accepted to both of my top two schools, and will be attending one of them in the Fall, on my way to becoming a mathematics teacher.

GRAPHING and INVERSE FUNCTIONS

4

© iStockPhoto.com/Woodkern

WHAT ARE GRAPHING AND INVERSE FUNCTIONS?   Graphing and inverse functions are fundamental concepts that allow us to visualize and analyze relationships between angles and trigonometric ratios.

WHY ARE GRAPHING AND INVERSE FUNCTIONS IMPORTANT?   Graphing functions allows us to visualize the behavior of trigonometric functions, while inverse functions provide a way to find angles given trigonometric ratios or to solve equations involving trigonometric functions. These concepts are fundamental in trigonometry and have applications in various fields such as physics, engineering, and geometry.

WHERE MIGHT YOU SEE GRAPHING AND INVERSE FUNCTIONS?   Here are some examples of where you may see graphing and inverse functions in the real world:

MEDICAL IMAGING   Trigonometric functions are used in medical imaging techniques such as MRI (Magnetic Resonance Imaging) and CT (Computed Tomography) scans to reconstruct images of internal body structures. By graphing these functions, radiologists can visualize the spatial distribution of tissue densities and abnormalities within the body. Inverse trigonometric functions are used to convert raw data collected from imaging sensors into two- or three-dimensional images that can be interpreted by medical professionals.

ROBOTICS AND MOTION PLANNING   Trigonometric functions are used in robotics and motion planning algorithms to control the movements of robotic manipulators and mobile robots. By graphing these functions, roboticists can visualize the trajectories and configurations of robot end-effectors and joints in workspace and configuration spaces. Inverse trigonometric functions are used to calculate the joint angles and positions required to achieve desired end-effector positions and orientations during robotic tasks.

Scenario: Rotation, Height, and Ferris Wheels

The High Roller ferris wheel in Las Vegas is the tallest observation wheel in the world, with a diameter of 520 feet. Imagine you are riding the High Roller, which completes one full rotation every 30 minutes. Let's solve a problem to find your height above the ground at a specific point in time.

The High Roller starts its rotation from the lowest point (point A), which is 10 feet above the ground. Assume the wheel rotates counterclockwise. After 15 minutes, you are at point B on the wheel.

a. Calculate your height above the ground at point B.

b. Determine your height above the ground after any time t (in minutes).

Solution

a. Height above the ground at point B

First, we need to calculate the angle of rotation after 15 minutes. Since the wheel makes one full rotation (360 degrees) in 30 minutes, the angle rotated in 15 minutes is:

$$\theta = \frac{360°}{30} \times 15 = 180°$$

Next we calculate the height of point B above the center of the wheel. Point B is directly opposite point A (at the topmost point of the wheel). The radius r of the wheel is half its diameter:

$$r = \frac{520}{2} = 260 \text{ feet}$$

When you are at the top of the wheel, your height above the center is: $r = 260$ feet.

Since point A is 10 feet above the ground, the center of the wheel is: $260 + 10 = 270$ feet.

So, your height above the ground at point B is: $270 + 260 = 530$ feet.

b. Height above ground at time t

Let θ be the central angle you have rotated from the start. Then

$$\theta = \frac{t}{30} \cdot 2\pi = \frac{\pi t}{15}$$

Let x represent the vertical height you are below the center of the wheel. Then

$$\cos \theta = \frac{x}{260}$$
$$x = 260 \cos \theta = 260 \cos \frac{\pi t}{15}$$

Let h represent your height above the ground. Then

$$h = 270 - x = 270 - 260 \cos \frac{\pi t}{15}$$

This represents your height at any time t. As a check, when $t = 15$ from part a:

$$h = 270 - 260 \cos \left(\frac{\pi \cdot 15}{15} \right) = 270 - 260 \cos (\pi) = 270 + 260 = 530 \text{ feet}$$

Learning Objectives

A Graph sine and cosine functions.

B Graph tangent functions.

C Graph reciprocal trigonometric functions.

We will begin this section with a look at the graph of $y = \sin x$. As was the case when we first began graphing equations in algebra, we set up a table of values of x and y that satisfy the equation and then use the information in Table 1 to sketch the graph. To make it easy, we will let x take on values that are multiples of $\frac{\pi}{4}$. As an aid in sketching graphs, we will approximate $\frac{1}{\sqrt{2}}$ with 0.7.

Table 1

x	$y = \sin x$	(x, y)
0	$y = \sin 0 = 0$	$(0, 0)$
$\frac{\pi}{4}$	$y = \sin \frac{\pi}{4} = \frac{1}{\sqrt{2}} \approx 0.7$	$\left(\frac{\pi}{4}, 0.7\right)$
$\frac{\pi}{2}$	$y = \sin \frac{\pi}{2} = 1$	$\left(\frac{\pi}{2}, 1\right)$
$\frac{3\pi}{4}$	$y = \sin \frac{3\pi}{4} = \frac{1}{\sqrt{2}} \approx 0.7$	$\left(\frac{3\pi}{4}, 0.7\right)$
π	$y = \sin \pi = 0$	$(\pi, 0)$
$\frac{5\pi}{4}$	$y = \sin \frac{5\pi}{4} = -\frac{1}{\sqrt{2}} \approx -0.7$	$\left(\frac{5\pi}{4}, -0.7\right)$
$\frac{3\pi}{2}$	$y = \sin \frac{3\pi}{2} = -1$	$\left(\frac{3\pi}{2}, -1\right)$
$\frac{7\pi}{4}$	$y = \sin \frac{7\pi}{4} = -\frac{1}{\sqrt{2}} \approx -0.7$	$\left(\frac{7\pi}{4}, -0.7\right)$
2π	$y = \sin 2\pi = 0$	$(2\pi, 0)$

A Graphing $y = \sin x$ and $y = \cos x$

Graphing each ordered pair and then connecting them with a smooth curve, we obtain the graph shown in Figure 1.

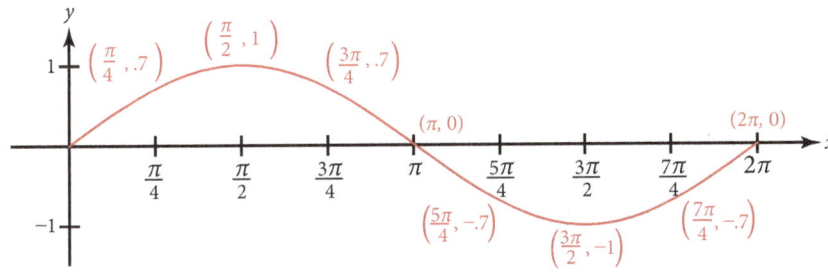

FIGURE 1

To further justify the graph in Figure 1, we could find additional ordered pairs that satisfy the equation. For example, we could continue our table by letting x take on multiples of $\frac{\pi}{6}$ and $\frac{\pi}{3}$. If we were to do so, we would find that any new ordered pair that satisfied the equation $y = \sin x$ would be such that its graph would lie on the curve in Figure 1. Figure 2 shows the curve in Figure 1 again, but this time with all the ordered pairs with x-coordinates that are multiples of $\frac{\pi}{6}$ or $\frac{\pi}{3}$ and the corresponding y-coordinates that satisfy the equation $y = \sin x$.

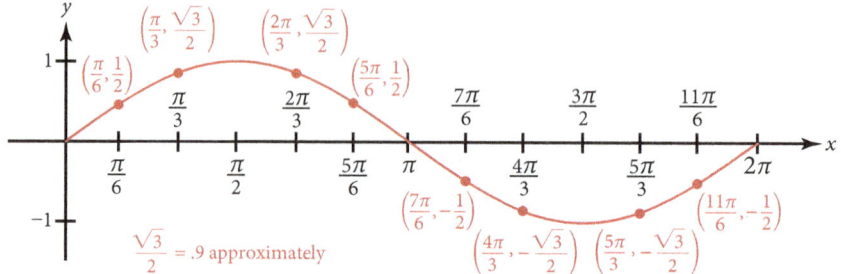

FIGURE 2

EXAMPLE 1   Use the graph of $y = \sin x$ in Figure 2 to find all values of x, between 0 and 2π, for which $\sin x = \frac{1}{2}$.

Solution   We locate 1/2 on the y-axis and draw a horizontal line through it. We follow this line to the points where it intersects the graph of $y \ \sin x$. The values of x just below these points of intersection are the values of x for which $\sin x = \frac{1}{2}$. As Figure 3 indicates, they are $\frac{\pi}{6}$ and $\frac{5\pi}{6}$.

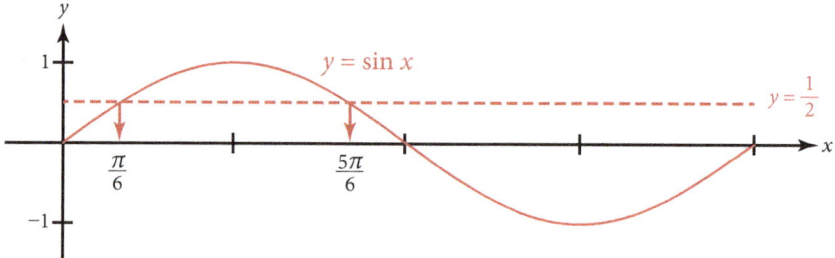

FIGURE 3

Graphing y = sin x Using the Unit Circle

We can also obtain the graph of $y = \sin x$ by using the unit circle. If we start at $(1, 0)$ and rotate once around the unit circle (through 2π) we can find the value of y in the equation $y = \sin t$ by simply keeping track of the y-coordinates of the points on the unit circle through which the terminal side of our angle t passes. (In this case, we use the variable t instead of the variable x to represent our angle so as not to confuse it with the x-coordinate of the point (x, y) on the unit circle.)

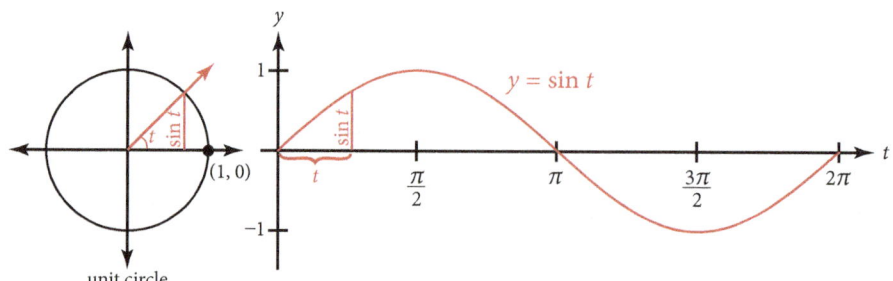

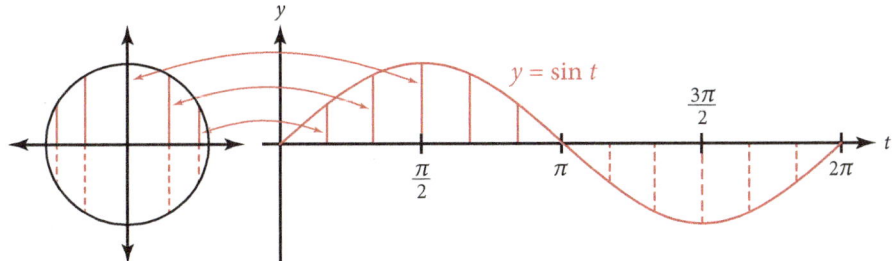

FIGURE 4

Extending the Sine Graph

Figure 1 through Figure 4 shows one complete cycle of $y = \sin x$. We can extend the graph of $y = \sin x$ to the left of $x = 0$ and to the right of $x = 2\pi$ by finding additional values for x and y. We don't need to go to all that trouble, however, since we know that, once we go past $x = 2\pi$, we will begin to name angles that are coterminal with the angles between 0 and 2π. Because of this, the values of $\sin x$ will start to repeat. Likewise, if we let x take on values below $x = 0$, we will simply get the values of $\sin x$ between 0 and 2π in the reverse order. Figure 4 shows the graph of $y = \sin x$ extended beyond the interval from $x = 0$ to $x = 2\pi$.

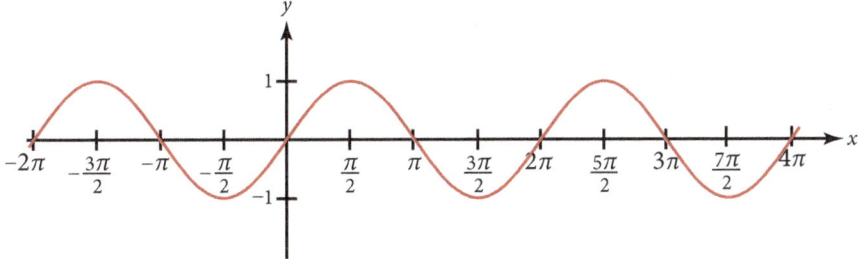

FIGURE 5

Our graph of $y = \sin x$ never goes above 1 or below -1 on the y-axis and it repeats itself every 2π units on the x-axis. This gives rise to the following two definitions:

> **Definition**
> The length of the smallest segment on the x-axis that it takes for a graph to go through one complete cycle is called the **period** of that graph or the equation it came from. In symbols, if p is the smallest positive number for which
> $$f(x + p) = f(x) \qquad \text{for all } x$$
> then we say the **period** of $f(x)$ is p.
>   In the case of $y = \sin x$, the period is 2π since 2π is the smallest positive number for p which $\sin (x + p) = \sin x$ for all x. See Figure 6.

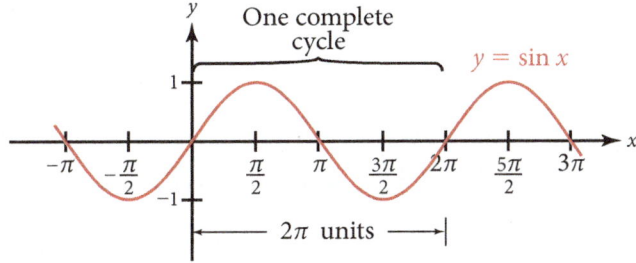

FIGURE 6

Definition

If the greatest value of y is M and the least value of y is m, then the **amplitude** of the graph of y is defined to be

$$A = \frac{1}{2}|M - m|$$

In the case of $y = \sin x$, the amplitude is 1 because $\frac{1}{2}|1 - (-1)| = \frac{1}{2}(2) = 1$. See Figure 6.

Using Technology: Graphing Calculators

Radian Mode

To graph $y = \sin x$ on a graphing calculator, first check to see that the calculator is in radian mode.

Mode: Radian

Then enter Y = sin X into your function list and set up the graph window as follows:

Window: X from -2π to 2π, Y from -4 to 4 (ZTrig on a TI-83/84)

Now trace along the graph and notice where the values of x are approximations to some multiples of $\pi/4$:

$\pi/4 \approx .785$ $\pi/2 \approx 1.57$ $\pi \approx 3.14$ $3\pi/2 \approx 4.71$ $2\pi \approx 6.28$

Using Technology: Graphing Calculators continued

Degree Mode

Set up your calculator as follows, then enter Y = sin X into your function list (if it is not there already):

Mode: Degree

Window: X from -360 to 360, Y from -4 to 4 (ZTrig on a TI-83/84)

Now when you trace along the graph, you will notice that the special points on the graph occur at multiples of 45°.

The graph of $y = \cos x$ has the same general shape as the graph of $y = \sin x$.

EXAMPLE 2 Sketch the graph of $y = \cos x$ for $0 \le x \le 2\pi$.

Solution We can arrive at the graph (Figure 7) by setting up a table of convenient values of x and y, as we did for $y = \sin x$ (Table 2).

Table 2

x	$y = \cos x$	(x, y)
0	$y = \cos 0 = 1$	$(0, 1)$
$\dfrac{\pi}{4}$	$y = \cos \dfrac{\pi}{4} = \dfrac{1}{\sqrt{2}}$	$\left(\dfrac{\pi}{4}, \dfrac{1}{\sqrt{2}}\right)$
$\dfrac{\pi}{2}$	$y = \cos \dfrac{\pi}{2} = 0$	$\left(\dfrac{\pi}{2}, 0\right)$
$\dfrac{3\pi}{4}$	$y = \cos \dfrac{3\pi}{4} = -\dfrac{1}{\sqrt{2}}$	$\left(\dfrac{3\pi}{4}, -\dfrac{1}{\sqrt{2}}\right)$
π	$y = \cos \pi = -1$	$(\pi, -1)$

Table 2 continued

x	$y = \cos x$	(x, y)
$\dfrac{5\pi}{4}$	$y = \cos \dfrac{5\pi}{4} = -\dfrac{1}{\sqrt{2}}$	$\left(\dfrac{5\pi}{4}, -\dfrac{1}{\sqrt{2}}\right)$
$\dfrac{3\pi}{2}$	$y = \cos \dfrac{3\pi}{2} = 0$	$\left(\dfrac{3\pi}{2}, 0\right)$
$\dfrac{7\pi}{4}$	$y = \cos \dfrac{7\pi}{4} = \dfrac{1}{\sqrt{2}}$	$\left(\dfrac{7\pi}{4}, \dfrac{1}{\sqrt{2}}\right)$
2π	$y = \cos 2\pi = 1$	$(2\pi, 1)$

As with $y = \sin t$, we can use the unit circle and the fact that $x = \cos t$ to plot x-coordinates to represent cosine values. See Figure 8.

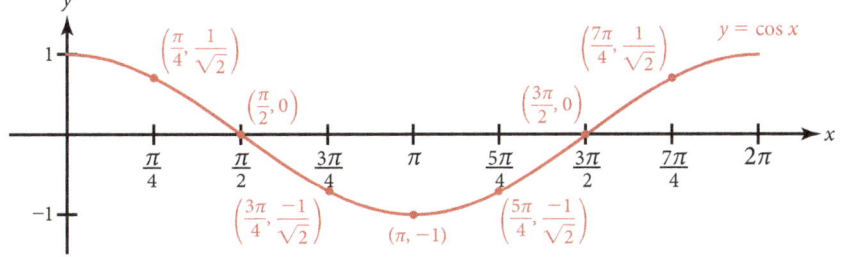

FIGURE 7

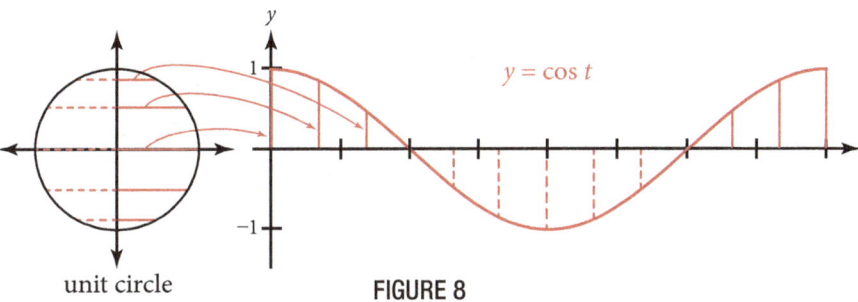

unit circle   **FIGURE 8**

EXAMPLE 5 Find all values of x for which $\cos x = -1$.

Solution We draw a horizontal line through $y = -1$ and notice where it intersects the graph of $y = \cos x$. The x-coordinates of those points are solutions to $\cos x = -1$.

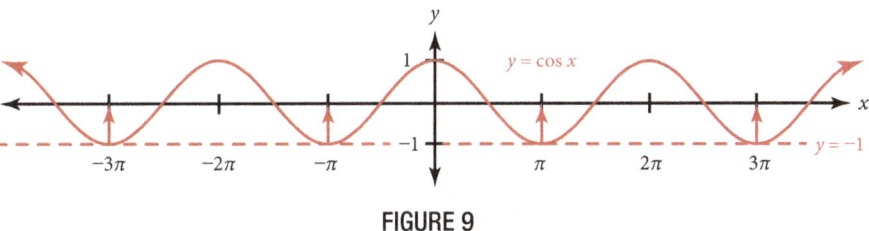

FIGURE 9

Figure 9 indicates that all solutions to $\cos x = -1$ are $x = \ldots -3\pi, -\pi, \pi, 3\pi, \ldots$. Since each pair of consecutive solutions differ by 2π, we can write the solutions in a more compact form as

$$x = \pi + 2k\pi \qquad \text{where } k \text{ is any integer}$$

B The Tangent Graph

Table 3 lists some solutions to $y = \tan x$ between $x = 0$ and $x = 2\pi$. Note that at $\frac{\pi}{2}$ and $\frac{3\pi}{2}$ the tangent is undefined.

The entries in Table 3 for which $y = \tan x$ is undefined correspond to values of x for which there is no corresponding value of y. That is, there will be no point on the graph with an x-coordinate of $\frac{\pi}{2}$ or $\frac{3\pi}{2}$. To help us remember this, we have drawn dashed vertical lines through $x = \frac{\pi}{2}$ and $x = \frac{3\pi}{2}$ in Figure 10. These vertical lines are **asymptotes**. Our graph will never cross or touch these lines.

Table 3

x	$y = \tan x$	x	$y = \tan x$
0	0	$\frac{5\pi}{4}$	1
$\frac{\pi}{4}$	1	$\frac{4\pi}{3}$	$\sqrt{3} \approx 1.7$
$\frac{\pi}{3}$	$\sqrt{3} \approx 1.7$	$\frac{3\pi}{2}$	Undefined
$\frac{\pi}{2}$	Undefined	$\frac{5\pi}{3}$	$-\sqrt{3} \approx -1.7$
$\frac{2\pi}{3}$	$-\sqrt{3} \approx -1.7$	$\frac{7\pi}{4}$	-1
$\frac{3\pi}{4}$	-1	2π	0
π	0		

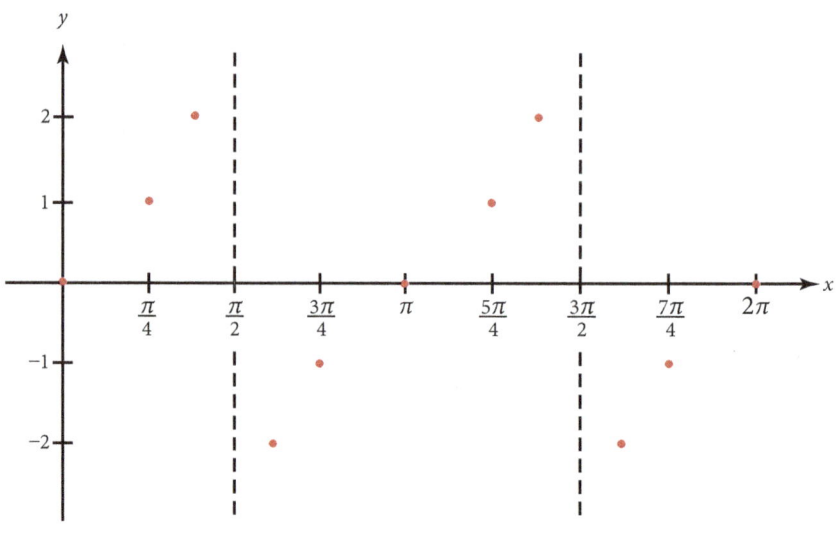

FIGURE 10

Figure 10 shows the information we have so far. If we were to use a calculator to find other values of tan x close to the asymptotes in Figure 11, we would find that tan x would become very large as we got close to the left side of an asymptote and very small as we got to the right side of the asymptote. For example, if we were to think of the numbers of the x-axis as degrees rather than radians, $x = 85°$ would be found just to the left of the asymptote at $\frac{\pi}{2}$ and tan 85° would be approximately 11. Likewise tan 89° would be approximately 57, and tan 89.9° would be about 573. As you can see, as x moves closer to $\frac{\pi}{2}$ from the left side of $\frac{\pi}{2}$, tan x gets larger and larger.

In Figure 11 we connect the points from Table 3 and then extend this graph to the right of 2π and to the left of 0 as we did with the sine and cosine curves.

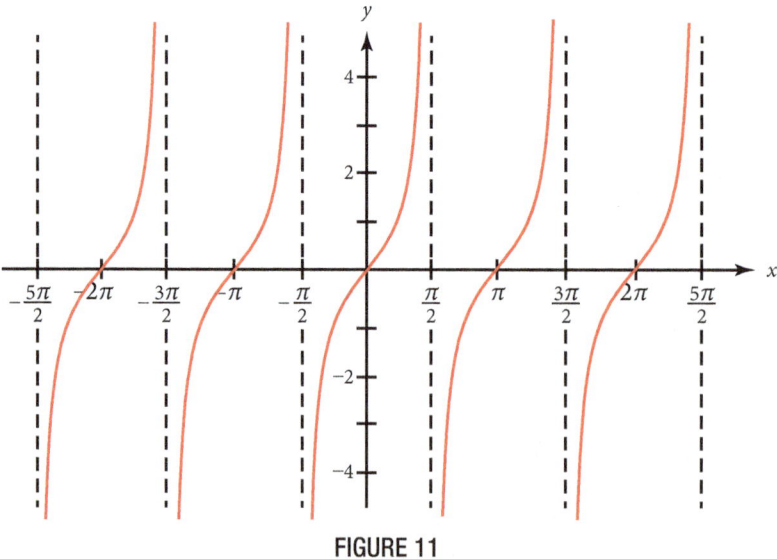

FIGURE 11

As Figure 11 indicates, the period of $y = \tan x$ is π. The tangent function has no amplitude since there is no largest or smallest value of y on the graph of $y = \tan x$.

EXAMPLE 4 Find the values of x between $x = -\frac{\pi}{2}$ and $x = \frac{3\pi}{2}$ that satisfy the equation $\tan x = -1$.

Solution

As Figure 12 shows, the values of x between $x = -\frac{\pi}{2}$ and $x = \frac{3\pi}{2}$ that satisfy the equation $\tan x = -1$ are $x = -\frac{\pi}{4}$ and $x = \frac{3\pi}{4}$.

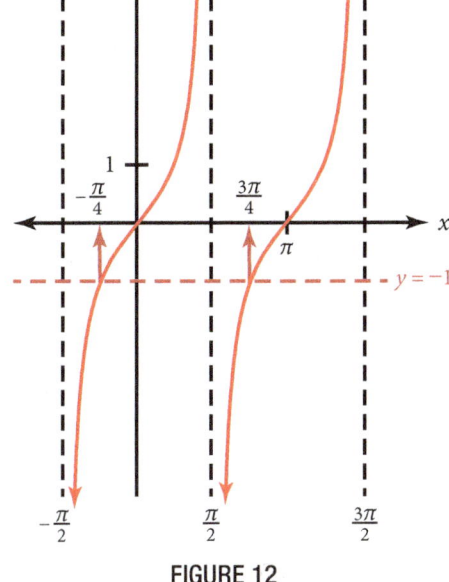

FIGURE 12

> **Using Technology: Graphing Calculators**
>
> Graph $y = \tan x$ on your graphing calculator in degree mode, and again in radian mode. (If you have a TI-83/84, use the ZTrig option under the Zoom key to set your graphing window.) If your calculator shows vertical lines connecting the sections of the tangent curve, remember that these lines are not part of the tangent graph—your calculator is showing them because it is finding individual points and then connecting them. If you trace over the points at which the asymptotes on the tangent graph occur, your calculator will not show corresponding y vales, because none exist.

C The Cosecant Graph

EXAMPLE 5 Sketch the graph of $y = \csc x$.

Solution Instead of using a table of values to help us graph $y = \csc x$, we can use the fact that $\csc x$ and $\sin x$ are reciprocals:

When sine is:	csc x will be:
1	1
$\frac{1}{2}$	2
$\frac{1}{3}$	3
$\frac{1}{4}$	4
0	Undefined
$-\frac{1}{4}$	-4
$-\frac{1}{3}$	-3
$-\frac{1}{2}$	-2
-1	-1

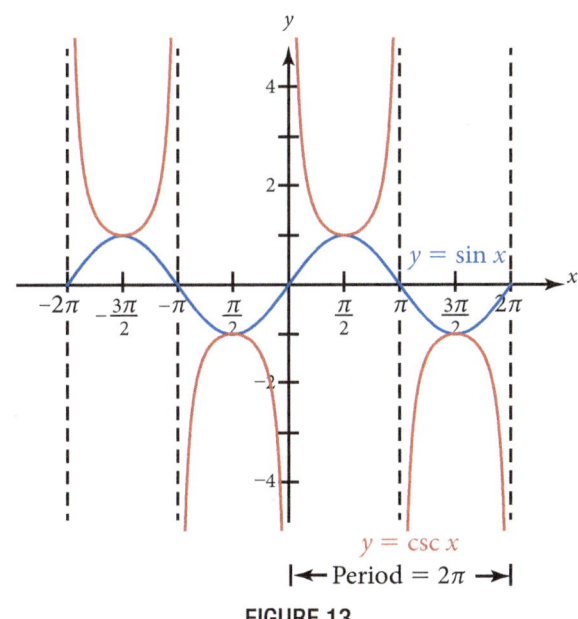

FIGURE 13

Figure 13 shows how this looks graphically. From the graph, we see that the period of $y = \csc x$ is 2π and, as was the case with $y = \tan x$, there is no amplitude.

> ### Using Technology: Graphing Calculators
>
> To graph $y = \csc x$ on your graphing calculator, you will have to enter
> Y = 1/sin X in your function list. You may want to enter both Y = sin X and Y = 1/sin X in the function list so you can see how the two graphs are related. Also, if you see vertical lines drawn on the graph, remember that they are not part of the graph (as mentioned on the preceding page).

The Cotangent and Secant Graphs

In Problem Set 4.1, you will be asked to graph $y = \cot x$ and $y = \sec x$. These graphs are shown in Figures 14 and 15 for reference.

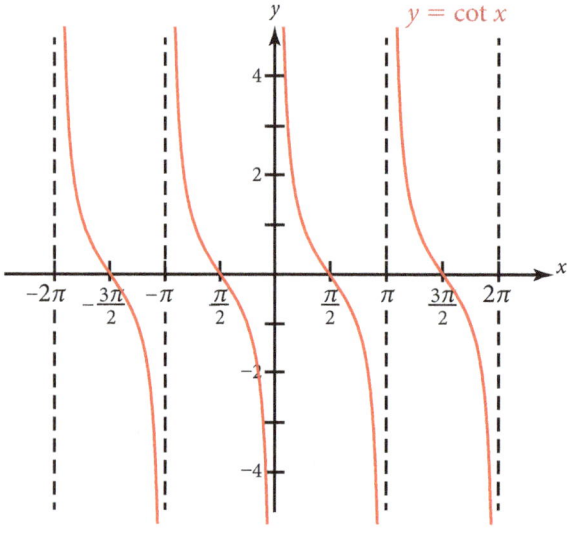

FIGURE 14

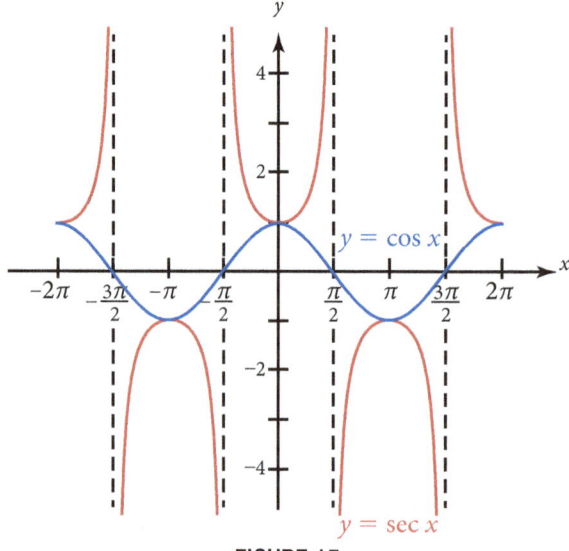

FIGURE 15

To end this section, we list some of the important facts we know about the graphs of four of the trigonometric functions. (Recall from algebra that the domain of a function $y = f(x)$ is the set of values that the variable x can assume, while the range is the set of values that y assumes.)

Functions	Domain	Range	Amplitude	Period
$y = \sin x$ and $y = \cos x$	all real numbers	$-1 \leq y \leq 1$	1	2π
$y = \tan x$	all real numbers except $x = \frac{\pi}{2} + k\pi$ where k is an integer	all real numbers	none	π
$y = \cot x$	all real numbers except $x = k\pi$ where k is an integer	all real numbers	none	π
$y = \sec x$	all real numbers except $x = \frac{\pi}{2} + k\pi$ where k is an integer	$y \leq -1$ or $y \geq 1$	none	2π
$y = \csc x$	all real numbers except $x = k\pi$ where k is an integer	$y \leq -1$ or $y \geq 1$	none	2π

Problem Set 4.1

Vocabulary

Use the vocabulary words below to fill in the blanks in the sentences.

asymptotes   amplitude   period   range   domain

1. The length of the smallest segment on the x-axis that it takes for a graph to go through one complete cycle is called the _____ of the graph.

2. The _____ of a graph is half the difference between the largest value of y and the smallest value of y.

3. For the tangent graph, the vertical lines that are associated with points on the x-axis for which the tangent is undefined are called _____ .

4. The _____ of the sine and cosine functions is $-1 \leq y \leq 1$.

5. The _____ for the function $y = \csc x$ is all real numbers except $x = k\pi$, where k is any integer.

Although we have worked a number of the problems below in the section previous to this, working them again yourself will help you become more familiar with the graphs of the six basic trigonometric functions. In each case, make a table of values for each function using multiples of $\frac{\pi}{4}$ for x. Then use the entries in the table to sketch the graph of each function for x between 0 and 2π.

1. $y = \cos x$

2. $y = \cot x$

3. $y = \csc x$

4. $y = \sin x$

5. $y = \tan x$

6. $y = \sec x$

Use the graphs you found in Problem 1 through 6 to find all values of x between 0 and 2π for which each of the following is true:

7. $\cos x = \frac{1}{2}$

8. $\sin x = \frac{1}{2}$

9. $\csc x = -1$

10. $\sec x = 1$

11. $\cos x = -\frac{1}{\sqrt{2}}$

12. $\cot x = -1$

13. $\tan x = 1$

14. $\sin x = -\frac{\sqrt{3}}{2}$

15. $\cos x = \frac{\sqrt{3}}{2}$

16. $\sin x = \frac{1}{\sqrt{2}}$

Sketch the graphs of each of the following between $x = -4\pi$ and $x = 4\pi$ by extending the graphs you made in Problems 1 through 6.

17. $y = \sin x$

18. $y = \cos x$

19. $y = \sec x$

20. $y = \csc x$

21. $y = \cot x$

22. $y = \tan x$

Use the graphs you found in Problems 17 through 22 to find all values of x between $x = 0$ and $x = 4\pi$ for which each of the following is true:

23. $\sin x = \dfrac{\sqrt{3}}{2}$

24. $\cos x = \dfrac{1}{\sqrt{2}}$

25. $\sec x = -1$

26. $\csc x = 1$

27. $\sin x = -\dfrac{1}{2}$

28. $\cos x = -\dfrac{1}{2}$

29. $\cot x = 1$

30. $\tan x = -1$

31. $\sin x = -\dfrac{1}{\sqrt{2}}$

32. $\cos x = -\dfrac{\sqrt{3}}{2}$

Find all the values of x for which the following are true:

33. $\cos x = 0$

34. $\sin x = 0$

35. $\sin x = 1$

36. $\cos x = 1$

37. $\tan x = 0$

38. $\cot x = 0$

39. Figure 16 is another diagram of the unit circle with the line segment corresponding to tan t showing. Make a diagram similar to the diagrams in Figures 4 and 8 from this section that shows how the unit circle can be used to obtain the graph of $y = \tan t$ from $t = -\frac{\pi}{2}$ to $t = \frac{\pi}{2}$.

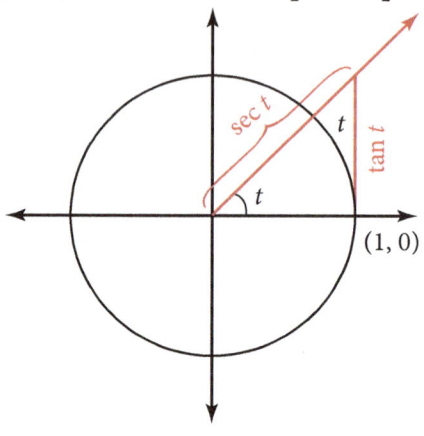

FIGURE 16

40. Following the instructions in Problem 39, use Figure 16 to sketch the graph of $y = \sec t$ from $t = 0$ to $t = \pi$.

Give the amplitude and period of each of the following graphs:

41.

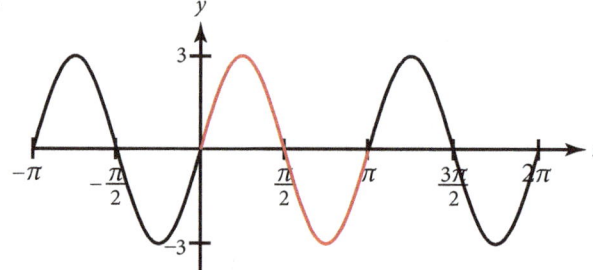

42.

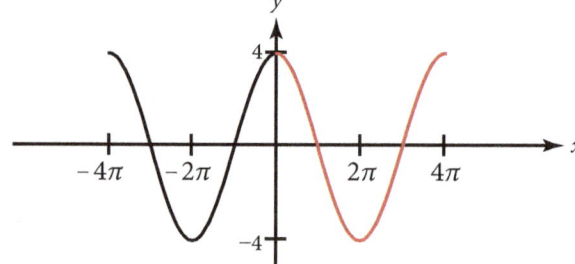

43.

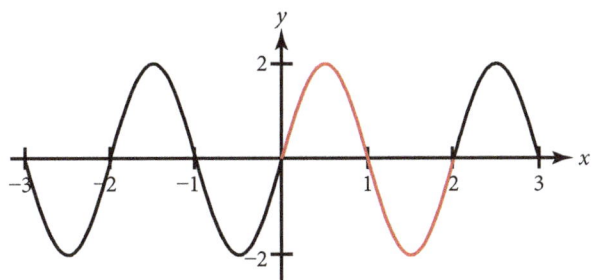

44.

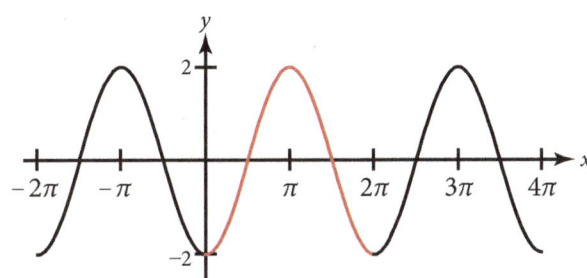

45.

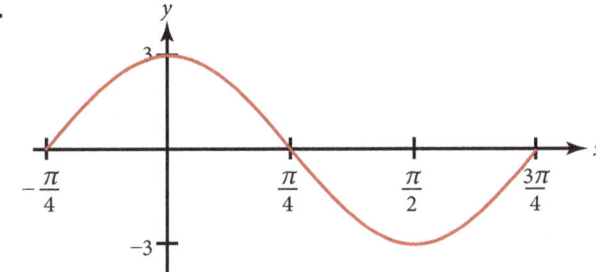

46.

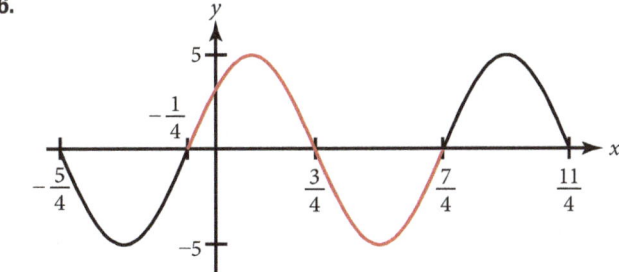

47. Sketch the graph of $y = 2 \sin x$ from $x = 0$ to $x = 2\pi$ by making a table using multiples of $\frac{\pi}{2}$ for x. What is the amplitude of the graph you obtain?

48. Sketch the graph of $y = \left(\frac{1}{2}\right)\cos x$ from $x = 0$ to $x = 2\pi$ by making a table using multiples of $\frac{\pi}{2}$ for x. What is the amplitude of the graph you obtain?

49. Make a table using multiples of $\frac{\pi}{4}$ for x to sketch the graph of $y = \sin 2x$ from $x = 0$ to $x = 2\pi$. After you have obtained the graph, state the number of complete cycles your graph goes through between 0 and 2π.

50. Make a table using multiples of $\frac{\pi}{6}$ for $\frac{x}{3}$ to sketch the graph of $y = \sin 3x$ from $x = 0$ to $x = 2\pi$. After you have obtained the graph, state the number of complete cycles your graph goes through between 0 and 2π.

Getting Ready for the Next Section

Simplify.

51. $\frac{1}{2}|2 - (-2)|$

52. $\frac{1}{2}|3 - (-3)|$

53. $\frac{2\pi}{1/2}$

54. $\frac{2\pi}{\pi/2}$

55. $3\left(\frac{2\pi}{3}\right)$

56. $4\left(\frac{3\pi}{2}\right)$

57. $2 \sin 0$

58. $2 \sin \pi$

59. $2 \sin \frac{\pi}{2}$

60. $2 \sin \frac{3\pi}{2}$

61. $\sin\left(2 \cdot \frac{3\pi}{4}\right)$

62. $\sin\left(2 \cdot \frac{5\pi}{4}\right)$

Learning Objectives

A   Find the amplitude and period of a given graph.

B   Graph trigonometric functions with negative coefficients.

In Section 4.1 the graphs of $y = \sin x$ and $y = \cos x$ were shown to have a period of 2π and an amplitude of 1. In this section we will extend our work with graphing to include a more detailed look at amplitude and period.

A Amplitude and Period

EXAMPLE 1   Sketch the graph of $y = 2 \sin x$, if $0 \leq x \leq 2\pi$.

Solution The coefficient 2 on the right side of the equation will multiply each value of $\sin x$ by a factor of 2. Therefore, the values of y in $y = 2 \sin x$ should all be twice the corresponding values of y in $y = \sin x$. Table 1 is a partial table of values for $y = 2 \sin x$, and Figure 1 shows the graphs of $y = \sin x$ and $y = 2 \sin x$. (We are including the graph of $y = \sin x$ on the same set of axes simply for reference. With both graphs to look at, it is easier to see what change is brought about by the coefficient 2.)

Table 1

x	$y = 2 \sin x$	(x, y)
0	$y = 2 \sin 0 = 2(0) = 0$	$(0, 0)$
$\dfrac{\pi}{2}$	$y = 2 \sin \dfrac{\pi}{2} = 2(1) = 2$	$\left(\dfrac{\pi}{2}, 2\right)$
π	$y = 2 \sin \pi = 2(0) = 0$	$(\pi, 0)$
$\dfrac{3\pi}{2}$	$y = 2 \sin \dfrac{3\pi}{2} = 2(-1) = -2$	$\left(\dfrac{3\pi}{2}, -2\right)$
2π	$y = 2 \sin 2\pi = 2(0) = 0$	$(2\pi, 0)$

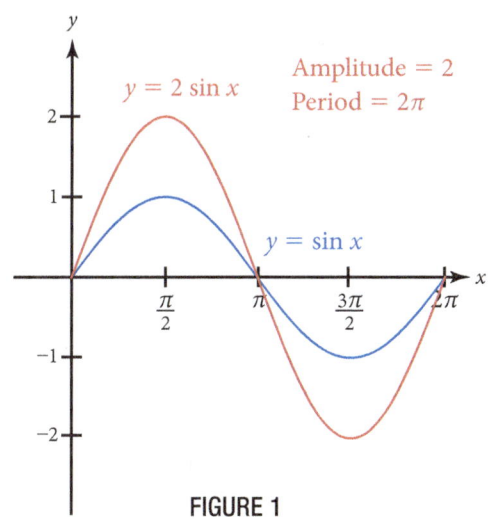

FIGURE 1

The coefficient 2 in $y = 2 \sin x$ changes the amplitude from 1 to 2: $\left[A = \dfrac{1}{2}|2 - (-2)| = \dfrac{1}{2}(4) = 2 \right]$.

EXAMPLE 2   Sketch one complete cycle of the graph of $y = \frac{1}{2} \cos x$.

Solution Again, we set up a partial table and then graph both $y = \frac{1}{2} \cos x$ and $y = \cos x$ on the same set of axes (Table 2, Figure 2).

The coefficient $\frac{1}{2}$ in $y = \frac{1}{2} \cos x$ is the amplitude of the graph.

Table 2

x	$y = \dfrac{1}{2} \cos x$	(x, y)
0	$y = \dfrac{1}{2} \cos 0 = \dfrac{1}{2}(1) = \dfrac{1}{2}$	$\left(0, \dfrac{1}{2}\right)$
$\dfrac{\pi}{2}$	$y = \dfrac{1}{2} \cos \dfrac{\pi}{2} = \dfrac{1}{2}(0) = 0$	$\left(\dfrac{\pi}{2}, 0\right)$
π	$y = \dfrac{1}{2} \cos \pi = \dfrac{1}{2}(-1) = -\dfrac{1}{2}$	$\left(\pi, -\dfrac{1}{2}\right)$
$\dfrac{3\pi}{2}$	$y = \dfrac{1}{2} \cos \dfrac{3\pi}{2} = \dfrac{1}{2}(0) = 0$	$\left(\dfrac{3\pi}{2}, 0\right)$
2π	$y = \dfrac{1}{2} \cos 2\pi = \dfrac{1}{2}(1) = \dfrac{1}{2}$	$\left(2\pi, \dfrac{1}{2}\right)$

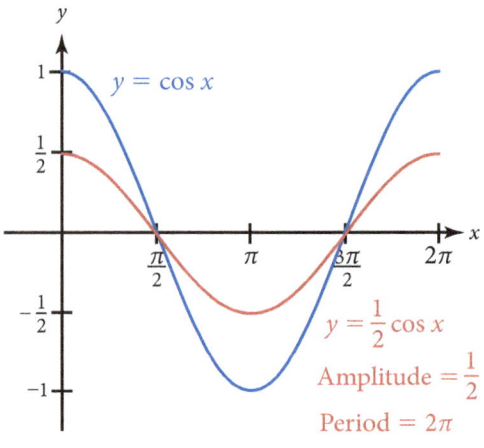

FIGURE 2

Generalizing the results of these last two examples, we can say:

If A is a positive number, then the graphs of $y = A \sin x$ and $y = A \cos x$ will have amplitude A.

EXAMPLE 3 Graph $y = 3 \cos x$ and $y = \dfrac{1}{4} \sin x$, if $0 \le x \le 2\pi$.

Solution The amplitude for $y = 3 \cos x$ is 3, while the amplitude for $y = \frac{1}{4} \sin x$ is $\frac{1}{4}$ (Figure 3).

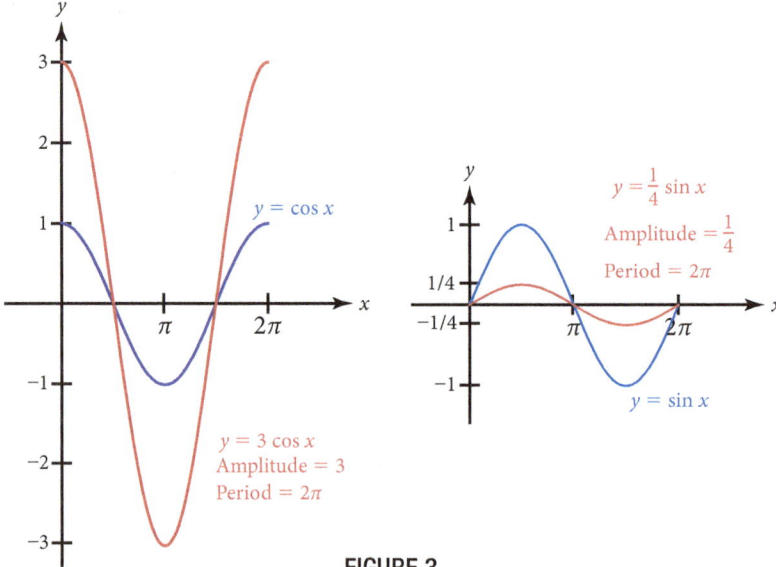

FIGURE 3

EXAMPLE 4   Graph $y = \sin 2x$ for $0 \le x \le 2\pi$.

Solution To see how the coefficient 2 in the equation $y = \sin 2x$ affects the graph, we will need a table in which the values of x are multiples of $\frac{\pi}{4}$, since the coefficient 2 divides the denominator 4 exactly. Table 3 is the table we need, and the graphs of $y = \sin x$ and $y = \sin 2x$ are shown in Figure 4.

Table 3

x	$y = \sin 2x$	(x, y)
0	$y = 2 \sin 2 \cdot 0 = 0$	$(0, 0)$
$\frac{\pi}{4}$	$y = \sin\left(2 \cdot \frac{\pi}{4}\right) = \sin \frac{\pi}{2} = 1$	$\left(\frac{\pi}{4}, 1\right)$
$\frac{\pi}{2}$	$y = \sin\left(2 \cdot \frac{\pi}{2}\right) = \sin \pi = 0$	$\left(\frac{\pi}{2}, 0\right)$
$\frac{3\pi}{4}$	$y = \sin\left(2 \cdot \frac{3\pi}{4}\right) = \sin \frac{3\pi}{2} = -1$	$\left(\frac{3\pi}{4}, -1\right)$
π	$y = \sin\left(2 \cdot \pi\right) = 0$	$(\pi, 0)$
$\frac{5\pi}{4}$	$y = \sin\left(2 \cdot \frac{5\pi}{4}\right) = \sin \frac{5\pi}{2} = 1$	$\left(\frac{5\pi}{4}, 1\right)$
$\frac{3\pi}{2}$	$y = \sin\left(2 \cdot \frac{3\pi}{2}\right) = \sin 3\pi = 0$	$\left(\frac{3\pi}{2}, 0\right)$
$\frac{7\pi}{4}$	$y = \sin\left(2 \cdot \frac{7\pi}{4}\right) = \sin \frac{7\pi}{2} = -1$	$\left(\frac{7\pi}{4}, -1\right)$
2π	$y = \sin\left(2 \cdot 2\pi\right) = \sin 4\pi = 0$	$(2\pi, 0)$

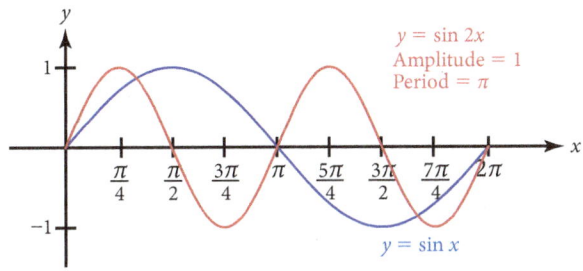

$y = \sin 2x$
Amplitude = 1
Period = π

$y = \sin x$

FIGURE 4

The graph of $y = \sin 2x$ has a period of π. It goes through two complete cycles in 2π units on the x-axis.

EXAMPLE 5   Graph $y = \sin 3x$, if $0 \le x \le 2\pi$.

Solution To see the effect of the coefficient 3 on the graph, we need a table in which the values of x are multiples of $\frac{\pi}{6}$, since the coefficient 3 divides the denominator 6 exactly.

The information in Table 4 indicates the period of $y = \sin 3x$ is $\frac{2\pi}{3}$. The graph will go through three complete cycles in 2π units on the x-axis. Figure 5 shows the graph of $y = \sin 3x$ and the graph of $y = \sin x$.

Table 4

x	$y = \sin 3x$	(x, y)
0	$y = \sin 3(0) = 3(0) = 0$	$(0, 0)$
$\frac{\pi}{6}$	$y = \sin 3\left(\frac{\pi}{6}\right) = \sin \frac{\pi}{2} = 1$	$\left(\frac{\pi}{6}, 1\right)$
$\frac{\pi}{3}$	$y = \sin 3\left(\frac{\pi}{3}\right) = \sin \pi = 0$	$\left(\frac{\pi}{3}, 0\right)$
$\frac{\pi}{2}$	$y = \sin 3\left(\frac{\pi}{2}\right) = \sin \frac{3\pi}{2} = -1$	$\left(\frac{\pi}{2}, -1\right)$
$\frac{2\pi}{3}$	$y = \sin 3\left(\frac{2\pi}{3}\right) = \sin 2\pi = 0$	$\left(\frac{2\pi}{3}, 0\right)$

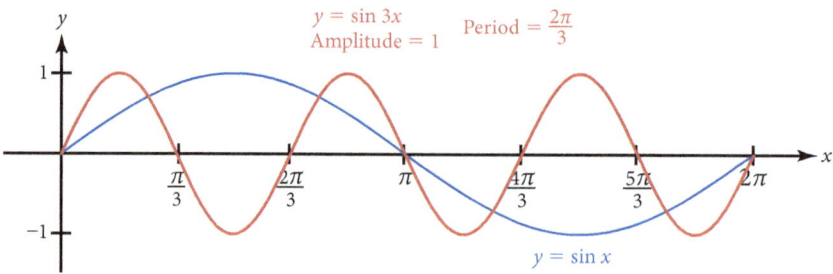

FIGURE 5

Table 5 summarizes the information obtained from Examples 4 and 5.

Table 5

Equation	Numbers of Cycles Every 2π Units	Period
$y = \sin x$	1	$\dfrac{2\pi}{1} = 2\pi$
$y = \sin 2x$	2	$\dfrac{2\pi}{2} = \pi$
$y = \sin 3x$	3	$\dfrac{2\pi}{3}$
$y = \sin Bx$	B	$\dfrac{2\pi}{B}$ *B is positive*

The following is a summary of everything we know so far about amplitude and period for sine and cosine.

> **Summary**
>
> The graphs of $y = A \sin Bx$ and $y = A \cos Bx$, where A *and* B are positive numbers, will have amplitude A and period $\frac{2\pi}{B}$.

EXAMPLE 6 Graph one complete cycle of $y = \cos \dfrac{1}{2} x$.

Solution The coefficient of x is $\frac{1}{2}$. The graph will go through $\frac{1}{2}$ of a complete cycle every 2π units. The period will be

$$\text{Period} = \frac{2\pi}{\frac{1}{2}} = 4\pi$$

Figure 6 shows the graph.

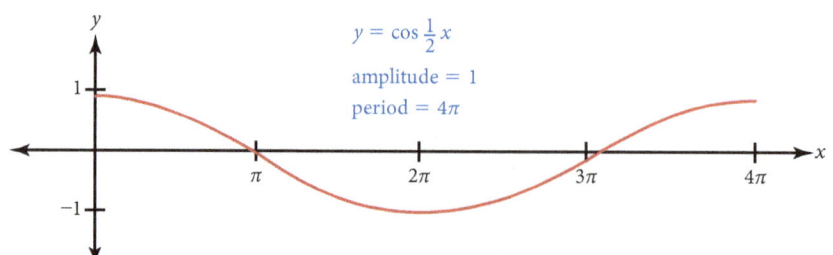

FIGURE 6

Continuing to generalize from the examples we have worked so far in this section, we can say that the graphs of $y = A \sin Bx$ and $y = A \cos Bx$, where A and B are positive numbers, will have amplitude A and period $2\pi/B$.

In the next three examples, we use this information about amplitude and period to graph one complete cycle of some sine and cosine curves, then we extend these graphs to cover more than one complete cycle.

EXAMPLE 7   Graph one complete cycle of each of the following equations and then extend the graph to cover the given interval.

a. $y = 3 \sin 2x,\ -\pi \leq x \leq 2\pi$

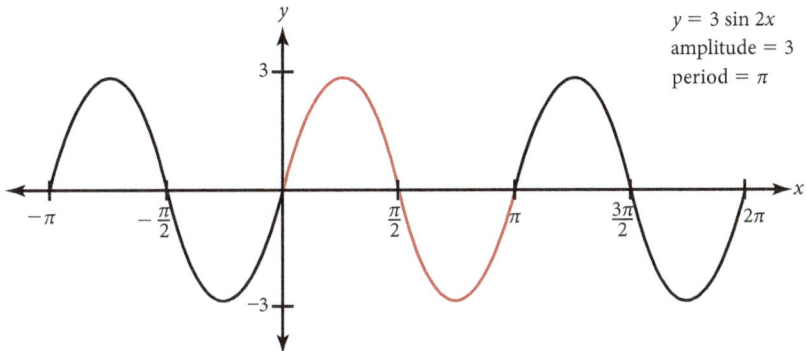

$y = 3 \sin 2x$
amplitude = 3
period = π

FIGURE 7

b. $y = 4 \cos \dfrac{1}{2}x,\ -4\pi \leq x \leq 4\pi$

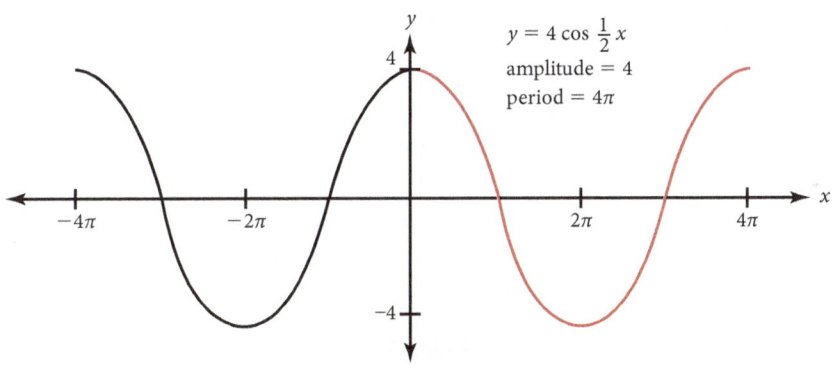

$y = 4 \cos \dfrac{1}{2}x$
amplitude = 4
period = 4π

FIGURE 8

c. $y = 2 \sin \pi x,\ -3 \leq x \leq 3$

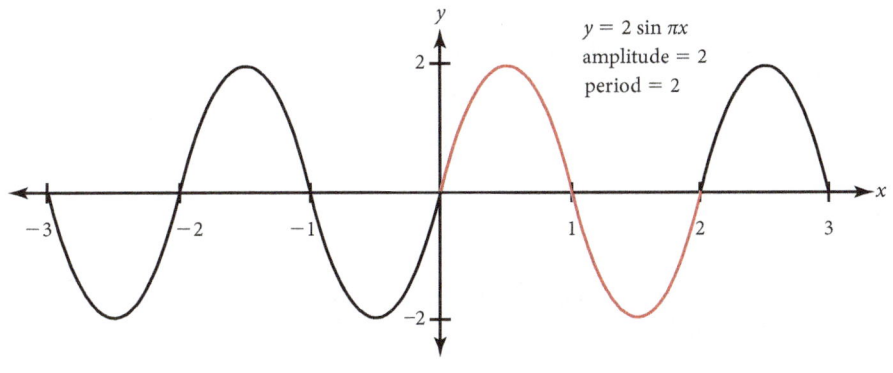

$y = 2 \sin \pi x$
amplitude = 2
period = 2

FIGURE 9

Note that, on the graphs in Figures 7, 8, and 9, the axes have not been labeled proportionally. Instead, they are labeled so that the amplitude and period are easy to read. As you can see, once we have the graph of one complete cycle of a curve, it is easy to extend the curve to any interval of interest.

B Negative Coefficients

So far in this section, all of the coefficients A and B we have encountered have been positive. If we are given an equation to graph in which B is negative, we can use the properties of even and odd functions to rewrite the equation with B positive. For example,

$$y = 3 \sin(-2x) \text{ is equivalent to } y = -3 \sin 2x$$
because sine is an odd function.

$$y = 3 \cos(-2x) \text{ is equivalent to } y = 3 \cos 2x$$
because cosine is an even function.

So we do not need to worry about negative values of B; we simply make them positive by using the properties of even and odd functions and then graph as usual. To see how a negative value of A affects graphing, we will graph $y = -2 \cos x$.

EXAMPLE 8   Graph $y = -2 \cos x$, from $x = -2\pi$ to $x = 4\pi$.

Solution Each value of y on the graph of $y = -2 \cos x$ will be the opposite of the corresponding value of y on the graph of $y = 2 \cos x$. The result is that the graph of $y = -2 \cos x$ is the reflection of the graph of $y = 2 \cos x$ about the x-axis. Figure 10 shows the extension of one complete cycle of $y = -2 \cos x$ to the interval $-2\pi \le x \le 4\pi$.

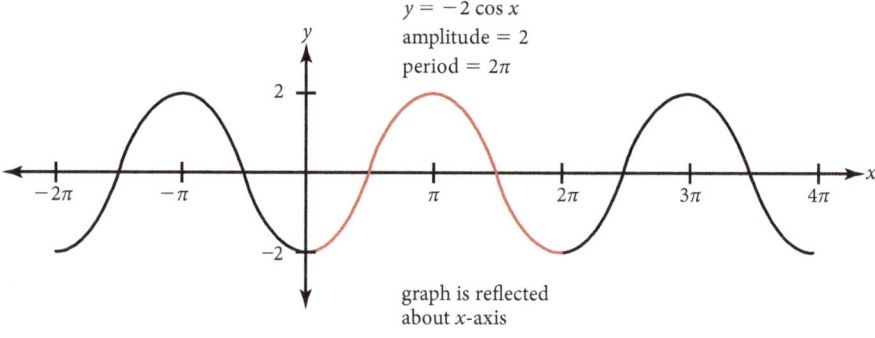

FIGURE 10

Summary

The graphs of $y = A \sin Bx$ and $y = A \cos Bx$, where B is a positive number, will have

$$\text{Amplitude} = |A|$$

$$\text{Period} = \frac{2\pi}{B}$$

They will be reflected about the x-axis if A is negative.

We conclude this section with a look at the graph of one of the equations we found in Example 4 of Section 3.5. Example 9 gives the main facts from that example.

EXAMPLE 9 Figure 11 shows a fire truck parked on the shoulder of a freeway next to a long block wall. The red light on the top of the truck is 10 feet from the wall and rotates through one complete revolution every 2 seconds. Graph the equation that gives the length d in terms of time t from $t = 0$ to $t = 2$.

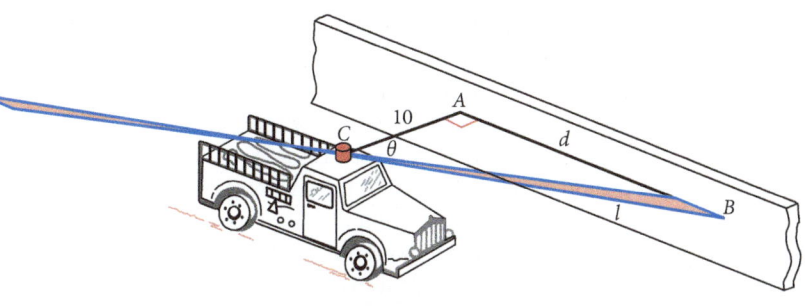

FIGURE 11

Solution From Example 4 in Section 3.5 we know that $d = 10 \tan \pi t$. To graph this equation between $t = 0$ and $t = 2$, we construct a table of values in which t assumes all multiples of $\frac{1}{4}$ from $t = 0$ to $t = 2$. Plotting the points given in Table 6 and then connecting them with a smooth tangent curve we have the graph in Figure 12.

Table 6

t	$d = 10 \tan \pi t$	d
0	$d = 10 \tan (\pi \cdot 0) = 10 \tan 0 = 0$	0
$\dfrac{1}{4}$	$d = 10 \tan \left(\pi \cdot \dfrac{1}{4} \right) = 10 \tan \dfrac{\pi}{4} = 10$	10
$\dfrac{1}{2}$	$d = 10 \tan \left(\pi \cdot \dfrac{1}{2} \right) = 10 \tan \dfrac{\pi}{2}$	undefined
$\dfrac{3}{4}$	$d = 10 \tan \left(\pi \cdot \dfrac{3}{4} \right) = 10 \tan \dfrac{3\pi}{4} = -10$	-10
1	$d = 10 \tan (\pi \cdot 1) = 10 \tan \pi = 0$	0
$\dfrac{5}{4}$	$d = 10 \tan \left(\pi \cdot \dfrac{5}{4} \right) = 10 \tan \dfrac{5\pi}{4} = 10$	10
$\dfrac{3}{2}$	$d = 10 \tan \left(\pi \cdot \dfrac{3}{2} \right) = 10 \tan \dfrac{3\pi}{2}$	undefined
$\dfrac{7}{4}$	$d = 10 \tan \left(\pi \cdot \dfrac{7}{4} \right) = 10 \tan \dfrac{7\pi}{4} = -10$	-10
2	$d = 10 \tan (\pi \cdot 2) = 10 \tan 2\pi = 0$	0

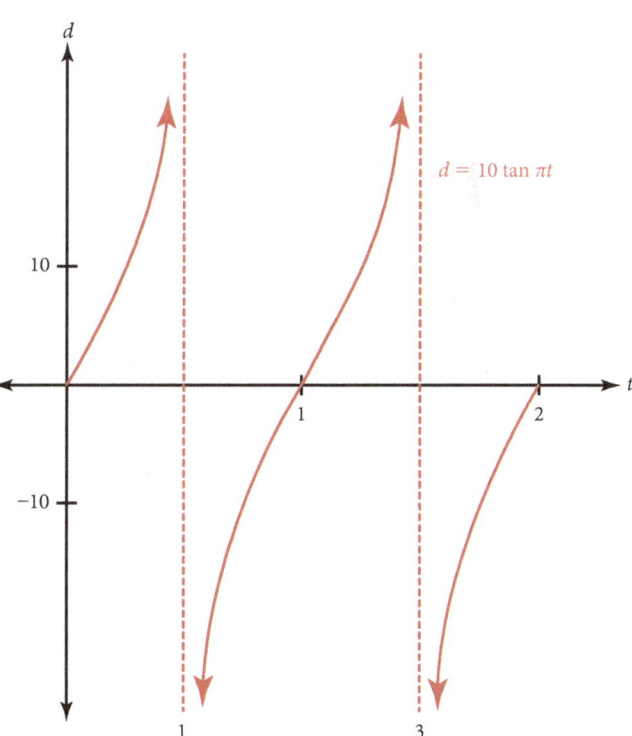

FIGURE 12

Note that the period of the function $d = 10 \tan \pi t$ is 1. Since the period of $y = \tan x$ is π, we can conclude that the new period 1 must come from dividing the normal period π by π to get 1. Also, the number 10 in the equation $d = 10 \tan \pi t$ causes the tangent graph to rise faster above the horizontal axis and fall faster below it.

We can generalize the results of Example 9 to conclude that, if A and B are positive numbers, the graphs of $y = A \tan Bx$ and $y = A \cot Bx$ will have period π/B. Each graph will rise and fall at a faster rate than the corresponding graphs of $y = \tan x$ and $y = \cot x$ if A is greater than 1 and at a slower rate if A is between 0 and 1.

In our next example, we'll look at the effect of A and B on the graph of $y = A \cot Bx$. First, recall from Section 4.1 (Figure 13) one cycle of the graph of $y = \cot x$:

Note that one cycle is completed in π units, so the period is π. As we would expect, then, the period of $y = A \cot Bx$ is $\frac{\pi}{B}$. You can think of B as compressing the period by a factor of B. In the sine and cosine functions, A is the amplitude, but since the cotangent does not have any amplitude, what is the effect of A on the cotangent graph? Remember that amplitude is a new term used to describe the vertical stretching, so A simply stretches the curve vertically. In the figure above, $\left(\frac{\pi}{4}, 1\right)$ is a point on the graph. The corresponding graph of $y = A \cot x$ would pass through the point $\left(\frac{\pi}{4}, A\right)$. We will illustrate this in the first example.

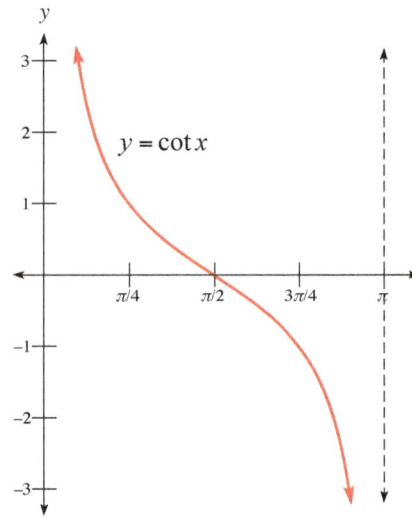

FIGURE 13

EXAMPLE 10 Graph one complete cycle of $y = 3 \cot 2x$.

Solution First note the period of the graph is $\frac{\pi}{B} = \frac{\pi}{2}$, so an entire cycle will be completed in the interval $0 < x < \frac{\pi}{2}$. At the quarter of this interval is $x = \frac{\pi}{8}$, where we would expect the point $\left(\frac{\pi}{8}, 1\right)$. However, since there is a vertical stretch by $A = 3$, the corresponding point is $\left(\frac{\pi}{8}, 3\right)$. Both curves are graphed on the same axes to illustrate this stretch.

Note that the red curve at the first quarter is triple the height of the original black curve.

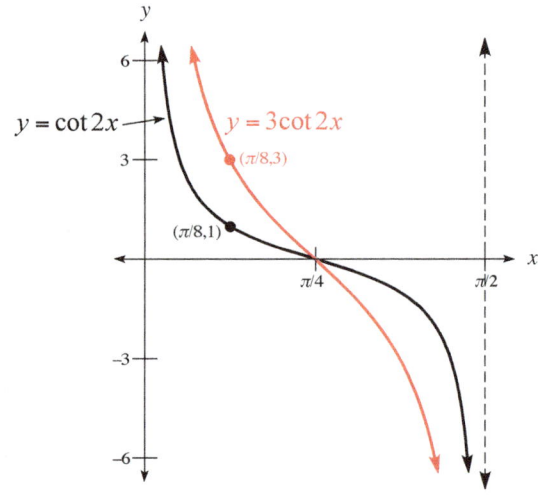

FIGURE 14

Next, we'll look at the effect of A and B on the graph of $y = A \sec Bx$. First, recall from Section 4.1 (Figure 15) one cycle of the graph of $y = \sec x$.

Note that one cycle is completed in 2π units, so the period is 2π. As we would expect, then, the period of $y = A \sec Bx$ is $\frac{2\pi}{B}$. You can think of B as compressing the period by a factor of B. As with the cotangent graph, A simply stretches the curve vertically. In the figure above, $(0, 1)$ is a point on the graph. The corresponding graph of $y = A \sec x$ would pass through the point $(0, A)$. We will illustrate this in the second example.

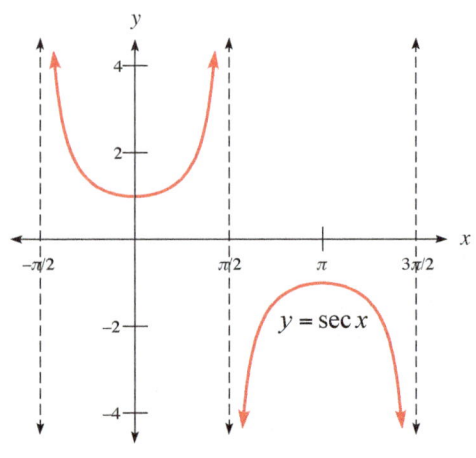

FIGURE 15

EXAMPLE 11 Graph one complete cycle of $y = 3 \sec 2x$.

Solution First note the period of the graph is $\frac{2\pi}{B} = \frac{2\pi}{2}$, so an entire cycle will be completed in the interval $-\frac{\pi}{4} < x < \frac{3\pi}{4}$. At the quarter of this interval is $x = 0$, where we would expect the point $(0, 1)$. However, since there is a vertical stretch by $A = 3$, the corresponding point is $(0, 3)$. Both curves are graphed on the same axes to illustrate this stretch:

Note that the red curve at the first quarter is triple the height of the original black curve, as well as the point at the third quarter.

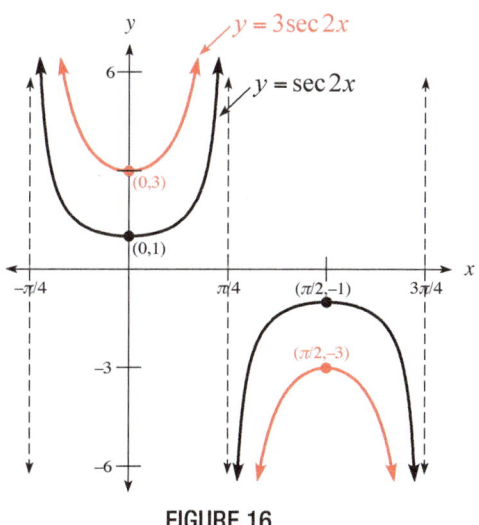

FIGURE 16

Vertical Translations

Adding a constant k to an equation that contains a trigonometric function will shift the graph of the equation up or down. That is, the graph of $y = k + 3 \cot 2x$, will be the graph of $y = 3 \cot 2x$ shifted k vertically. If k is positive the vertical shift is up; if k is negative the shift will be down.

EXAMPLE 12 Graph one complete cycle of $y = 2 + 3 \cot 2x$.

Solution The 2 will shift the graph of $y = 3 \cot 2x$ up 2 units, as shown in Figure 17.

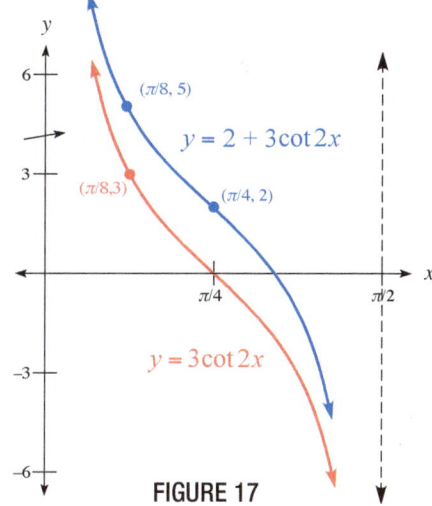

FIGURE 17

EXAMPLE 13 Graph one complete cycle of $y = -3 + 3 \sec 2x$.

Solution The only thing different about this equation from the equation in Example 11 is the -3 added to the equation in Example 2. From our previous work with vertical translations we know this will shift the graph of $y = 3 \sec 2x$ down 3 units, as shown in Figure 18.

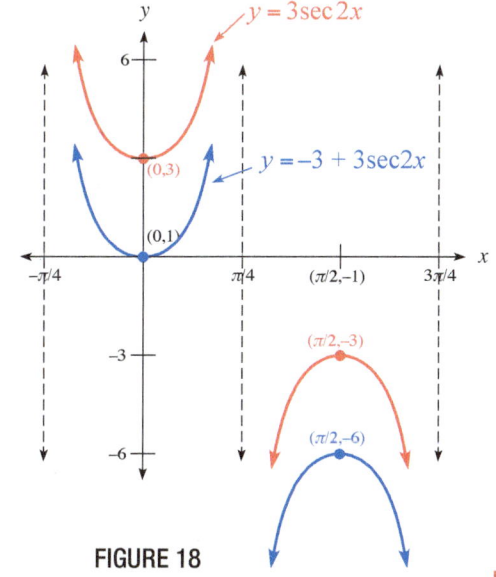

FIGURE 18

Vocabulary

Use the vocabulary words and formulas below to fill in the blanks in the sentences.

negative   amplitude   period   range   domain

1. The graph of $y = A \sin x$ will have an _____ equal to $|A|$.

2. If B is a positive number, then the graph of $y = \sin Bx$ will have a _____ equal to $2\pi/B$.

3. The graph of $y = A \sin x$ will have a reflection about the x-axis when A is a _____ number.

4. The graph of $y = 3 \sin 2x$ will have a _____ of π, and an _____ of 3.

5. The graph of $y = 3 \sin \pi x$ will have a _____ of 2; the _____ will be all real numbers, and the _____ will be $-3 \leq y \leq 3$.

Graph one complete cycle of each of the following. In each case label the axes accurately and identify the amplitude and period for each graph.

1. $y = 6 \sin x$ 2. $y = 6 \cos x$

3. $y = \sin 2x$ 4. $y = \sin \dfrac{1}{2} x$

5. $y = \cos \dfrac{1}{3} x$ 6. $y = \cos 3x$

7. $y = \dfrac{1}{3} \sin x$ 8. $y = \dfrac{1}{2} \cos x$

9. $y = \cos \pi x$ 10. $y = \sin \pi x$

11. $y = \cos \dfrac{\pi}{2} x$ 12. $y = \sin \dfrac{\pi}{2} x$

Graph one complete cycle for each of the following. In each case label the axes so that the amplitude and period are easy to read.

13. $y = 4 \sin 2x$ 14. $y = 2 \sin 4x$

15. $y = 2 \cos 4x$ 16. $y = 3 \cos 2x$

17. $y = 3 \sin \dfrac{1}{2} x$ 18. $y = 2 \sin \dfrac{1}{3} x$

19. $y = \dfrac{1}{2} \cos 3x$ 20. $y = \dfrac{1}{2} \sin 3x$

21. $y = \dfrac{1}{2} \sin \dfrac{\pi}{2} x$ 22. $y = 2 \sin \dfrac{\pi}{2} x$

Graph each of the following over the given interval. Label the axes so that the amplitude and period are easy to read.

23. $y = 2 \sin \pi x$, $-4 \le x \le 4$

24. $y = 3 \cos \pi x$, $-2 \le x \le 4$

25. $y = 3 \sin 2x$, $-\pi \le x \le 2\pi$

26. $y = -3 \sin 2x$, $-2\pi \le x \le 2\pi$

27. $y = -3 \cos \dfrac{1}{2} x$, $-2\pi \le x \le 6\pi$

28. $y = 3 \cos \dfrac{1}{2} x$, $-4\pi \le x \le 4\pi$

29. $y = -2 \sin(-3x)$, $0 \le x \le 2\pi$

30. $y = -2 \cos(-3x)$, $0 \le x \le 2\pi$

31. Electrical Circuits The current in an alternating circuit varies in intensity with time. If I represents the intensity of the current and t represents time, then the relationship between I and t is given by

$$I = 20 \sin 120\pi t$$

where I is measured in amperes and t is measured in seconds. Find the maximum value of I and the time it takes for I to go through one complete cycle.

32. Motion of a Spring A weight is hung from a spring and set in motion so that it moves up and down continuously. The velocity v of the weight at any time t is given by the equation

$$v = 3.5 \cos 2\pi t$$

where v is measured in meters and t is measured in seconds. Find the maximum velocity of the weight and the amount of time it takes for the weight to move from its lowest position to its highest position.

33. Since the period for $y = \tan x$ is π, the graph of $y = \tan x$ will go through one complete cycle every π units. Through how many cycles will the graph of $y = \tan 2x$ go every π units? What is the period of $y = \tan 2x$? Sketch the graph of $y = \tan 2x$, from $x = -\frac{\pi}{4}$ to $x = \frac{3\pi}{4}$.

34. Through how many complete cycles will the graph of $y = \tan \frac{1}{2}x$ go every π units? What is the period of this graph? Sketch the graph from $x = -\pi$ to $x = 3\pi$.

35. Rotating Light Figure 19 shows a lighthouse that is 100 feet from a long straight wall on the beach. The light in the lighthouse rotates through one complete rotation once every 4 seconds. In Problem 21 of Problem Set 3.5 you found the equation that gives d in terms of t to be $d = 100 \tan \frac{\pi}{2}t$. Graph this equation by making a table in which t assumes all multiples of $\frac{1}{2}$ from $t = 0$ to $t = 4$.

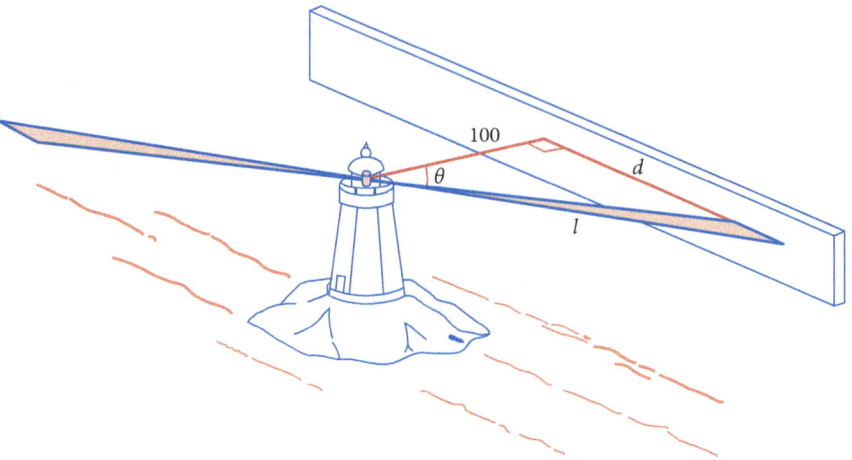

FIGURE 19

36. Sketch the graph of $y = -\tan x$, for $-\dfrac{\pi}{2} \le x \le \dfrac{3\pi}{2}$.

37. Give the period of $y = \cot 3x$. Sketch the graph from $x = 0$ to $x = \pi$.

38. Sketch the graph of $y = \cot \pi x$ from $x = -1$ to $x = 2$.

39. Graph $y = 2 \csc x$ by making a table in which x assumes multiples of $\frac{\pi}{4}$ starting at $x = 0$ and ending at $x = 2\pi$. (Use 1.4 as an approximation for $\sqrt{2}$.) Does the graph obtained by connecting the points from the table agree with the fact that $2 \csc x = 2\left(\frac{1}{\sin x}\right)$? That is, does your graph seem reasonable when viewed in terms of reciprocal functions?

40. Graph $y = \dfrac{1}{2}\sec x$ between $x = -\dfrac{\pi}{2}$ and $x = \dfrac{3\pi}{2}$.

41. Referring to Example 9 and Figure 11 from this section, the equation that gives l in terms of time t is $l = 10 \sec \pi t$. Graph this equation by making a table of values in which t takes on multiples of $\frac{1}{4}$ starting at $t = 0$ and ending at $t = 2$.

42. In Figure 13 above, the equation that gives l in terms of t is $l = 100 \sec \frac{\pi}{2}t$. Graph this equation from $t = 0$ to $t = 4$.

43. The period of $y = \csc 2x$ is $\frac{2\pi}{2} = \pi$. Graph $y = \csc 2x$ from $x = 0$ to $x = 2\pi$.

44. Graph $y = \sec 3x$ from $x = 0$ to $x = 2\pi$.

45. Graph one complete cycle of $y = 3 \csc 2x$.

46. Graph one complete cycle of $y = 2 \sec 3x$.

Sketch one complete cycle of each graph. Pay particular attention to the labeling of the x-axis. For problems 51 and 52, remember the negative has the effect of reflecting the curve across the x-axis.

47. $y = 2 \cot 4x$ **48.** $y = 3 \cot 4x$

49. $y = 2 \cot \pi x$ **50.** $y = 3 \cot \frac{\pi x}{2}$

51. $y = -4 \cot 3x$ **52.** $y = -5 \cot \left(\frac{1}{2} x \right)$

Sketch one complete cycle of each graph. Pay particular attention to the labeling of the x-axis. For problems 57 and 58, remember the negative has the effect of reflecting the curve across the x-axis.

53. $y = 3 \sec 4x$ **54.** $y = 4 \sec 3x$

55. $y = 5 \sec \frac{\pi x}{2}$ **56.** $y = 3 \sec \pi x$

57. $y = -4 \sec 4x$ **58.** $y = -5 \sec \left(\frac{1}{3} x \right)$

Sketch one complete cycle of each graph. Pay particular attention to the labeling of the x-axis.

59. $y = 1 + 2 \cot \pi x$ **60.** $y = -2 + 3 \cot \frac{\pi x}{2}$

61. $y = -2 + 3 \sec 4x$ **62.** $y = 2 + 4 \sec 3x$

63. $y = -1 + 5 \sec \frac{\pi x}{2}$ **64.** $y = -3 + 3 \sec \pi x$

Getting Ready for the Next Section

Evaluate each of the following if x is $\frac{\pi}{2}$ and y is $\frac{\pi}{6}$.

65. $\sin\left(x + \frac{\pi}{2}\right)$ **66.** $\sin\left(x - \frac{\pi}{2}\right)$

67. $\cos\left(y - \frac{\pi}{6}\right)$ **68.** $\cos\left(y + \frac{\pi}{6}\right)$

69. $\sin(x + y)$ **70.** $\cos(x + y)$

71. $\sin x + \sin y$ **72.** $\cos x + \cos y$

Simplify.

73. $2\pi - \frac{\pi}{2}$ **74.** $\frac{3\pi}{2} - \frac{\pi}{2}$

75. $\frac{-\pi/2}{2}$ **76.** $\frac{-\left(-\frac{\pi}{3}\right)}{2}$

77. $\frac{1}{2}\left(\frac{\pi}{6} + \frac{2\pi}{3}\right)$ **78.** $\frac{1}{2}\left(\frac{2\pi}{3} + \frac{7\pi}{6}\right)$

4.3 Videos

KEY WORDS

phase shift

vertical translation

Learning Objectives

A   Graph a trigonometric equation with a phase shift.

B   Graph a complete cycle of a trigonometric equation.

C   Graph a trigonometric equation with a vertical translation.

In this section we will consider equations of the form

$$y = A \sin(Bx + C) \quad \text{and}$$
$$y = A \cos(Bx + C)$$
$$\text{where } B > 0$$

We already know how the coefficients A and B affect the graphs of these equations. The only thing we have left to do is discover what effect C has on the graphs. We will start our investigation with a couple of equations in which A and B are equal to 1.

A Phase Shift

EXAMPLE 1   Graph $y = \sin\left(x + \frac{\pi}{2}\right)$, if $-\frac{\pi}{2} \leq x \leq \frac{3\pi}{2}$.

Solution   Since we have not graphed an equation of this form before, it is a good idea to begin by setting up a table (Table 1). In this case, multiples of $\frac{\pi}{2}$ will be the most convenient replacements for x in the table. Also, if we start with $x = -\frac{\pi}{2}$, our first value of y will be 0.

Graphing these points and then drawing the sine curve that connects them gives us the graph of $y = \sin\left(x + \frac{\pi}{2}\right)$ shown in Figure 1, Figure 1 also includes the graph of $y = \sin x$ for reference, since we are trying to discover how the graphs of $y = \sin\left(x + \frac{\pi}{2}\right)$ and $y = \sin x$ differ.

It seems that the graph of

$$y = \sin\left(x + \frac{\pi}{2}\right)$$

is shifted $\frac{\pi}{2}$ units to the left of the graph of $y = \sin x$. We say the graph of $y = \sin\left(x + \frac{\pi}{2}\right)$ has a **phase shift** of $-\frac{\pi}{2}$, or a horizontal translation of $\pi/2$ units to the left, where the negative sign indicates the shift is to the left (negative direction).

Table 1

x	$y = \sin\left(x + \dfrac{\pi}{2}\right)$	(x, y)
$-\dfrac{\pi}{2}$	$y = \sin\left(-\dfrac{\pi}{2} + \dfrac{\pi}{2}\right) = \sin 0 = 0$	$\left(-\dfrac{\pi}{2}, 0\right)$
0	$y = \sin\left(0 + \dfrac{\pi}{2}\right) = \sin\dfrac{\pi}{2} = 1$	$(0, 1)$
$\dfrac{\pi}{2}$	$y = \sin\left(\dfrac{\pi}{2} + \dfrac{\pi}{2}\right) = \sin \pi = 0$	$\left(\dfrac{\pi}{2}, 0\right)$
π	$y = \sin\left(\pi + \dfrac{\pi}{2}\right) = \sin\dfrac{3\pi}{2} = -1$	$(\pi, -1)$
$\dfrac{3\pi}{2}$	$y = \sin\left(\dfrac{3\pi}{2} + \dfrac{\pi}{2}\right) = \sin 2\pi = 0$	$\left(\dfrac{3\pi}{2}, 0\right)$

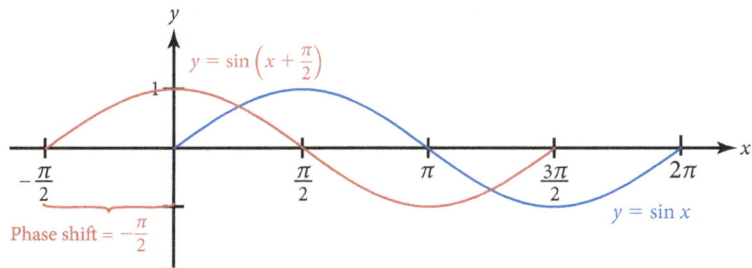

FIGURE 1

From the results in Example 1, we would expect the graph of $y = \sin\left(x - \frac{\pi}{2}\right)$ to have a phase shift of $+\frac{\pi}{2}$. That is, we expect the graph of $y = \sin\left(x - \frac{\pi}{2}\right)$ to be shifted $\frac{\pi}{2}$ units to the *right* of the graph of $y = \sin x$.

EXAMPLE 2   Graph one complete cycle of $y = \sin\left(x - \frac{\pi}{2}\right)$.

Solution Proceeding as we did in Example 1, we set up Table 2 using multiples of $\frac{\pi}{2}$ for x, and then use the information in the table to sketch the graph. In this example, we start with $x = \frac{\pi}{2}$, since the value of x will give us $y = 0$.

Table 2

x	$y = \sin\left(x - \dfrac{\pi}{2}\right)$	(x, y)
$\dfrac{\pi}{2}$	$y = \sin\left(\dfrac{\pi}{2} - \dfrac{\pi}{2}\right) = \sin 0 = 0$	$\left(\dfrac{\pi}{2}, 0\right)$
π	$y = \sin\left(\pi - \dfrac{\pi}{2}\right) = \sin\dfrac{\pi}{2} = 1$	$(\pi, 1)$
$\dfrac{3\pi}{2}$	$y = \sin\left(\dfrac{3\pi}{2} - \dfrac{\pi}{2}\right) = \sin \pi = 0$	$\left(\dfrac{3\pi}{2}, 0\right)$
2π	$y = \sin\left(2\pi - \dfrac{\pi}{2}\right) = \sin\dfrac{3\pi}{2} = -1$	$(2\pi, -1)$
$\dfrac{5\pi}{2}$	$y = \sin\left(\dfrac{5\pi}{2} - \dfrac{\pi}{2}\right) = \sin 2\pi = 0$	$\left(\dfrac{5\pi}{2}, 0\right)$

The graph of $y = \sin\left(x - \frac{\pi}{2}\right)$, as we expected, is a sine curved shifted $\frac{\pi}{2}$ units to the right of the graph of $y = \sin x$ (Figure 2). The phase shift in this case is $+\frac{\pi}{2}$.

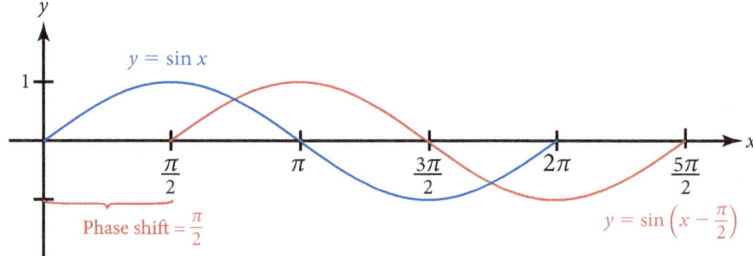

FIGURE 2

Before we write any conclusions about phase shift, we should look at another example in which A and B are not 1.

EXAMPLE 3   Graph $y = 3 \sin\left(2x + \frac{\pi}{2}\right)$, if $-\frac{\pi}{4} \le x \le \frac{3\pi}{4}$.

Solution We know the coefficient $B = 2$ will change the period from 2π to π. Because the period is smaller (π instead of 2π), we use values of x in Table 3 that are closer together—that is, multiples of $\frac{\pi}{4}$ instead of $\frac{\pi}{2}$.

Table 3

x	$y = 3 \sin\left(2x - \dfrac{\pi}{2}\right)$	(x, y)
$-\dfrac{\pi}{4}$	$y = 3 \sin\left[2\left(-\dfrac{\pi}{4}\right) + \dfrac{\pi}{2}\right] = 3 \sin 0 = 0$	$\left(-\dfrac{\pi}{4}, 0\right)$
0	$y = 3 \sin\left(2 \cdot 0 + \dfrac{\pi}{2}\right) = 3 \sin\dfrac{\pi}{2} = 3$	$(0, 3)$
$\dfrac{\pi}{4}$	$y = 3 \sin\left(2 \cdot \dfrac{\pi}{4} + \dfrac{\pi}{2}\right) = 3 \sin \pi = 0$	$\left(\dfrac{\pi}{4}, 0\right)$
$\dfrac{\pi}{2}$	$y = 3 \sin\left(2 \cdot \dfrac{\pi}{2} + \dfrac{\pi}{2}\right) = 3 \sin\dfrac{3\pi}{2} = -3$	$\left(\dfrac{\pi}{2}, -3\right)$
$\dfrac{3\pi}{4}$	$y = 3 \sin\left(2 \cdot \dfrac{3\pi}{4} + \dfrac{\pi}{2}\right) = 3 \sin 2\pi = 0$	$\left(\dfrac{3\pi}{4}, 0\right)$

The amplitude and period are as we would expect. The phase shift, however, is half of $-\frac{\pi}{2}$. The phase shift, $-\frac{\pi}{4}$, comes from the ratio $-\frac{C}{B}$ or, in this case,

$$\frac{-\pi/2}{2} = -\frac{\pi}{4}$$

The phase shift in the equation $y = A \sin(Bx + C)$ depends on both B and C. In this example, the period is half of the period of $y = \sin x$, and the phase shift is half of the phase shift of $y = \sin\left(x + \frac{\pi}{2}\right)$ that was found in Example 1 (see Figure 3).

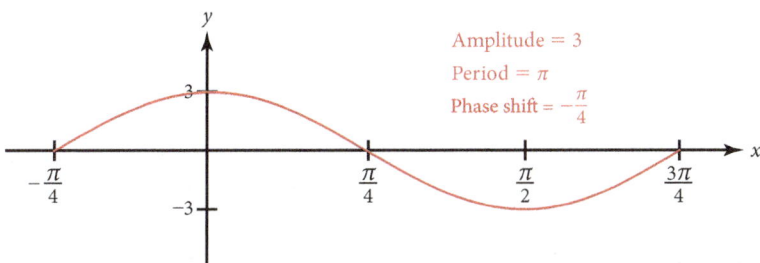

Amplitude = 3
Period = π
Phase shift = $-\frac{\pi}{4}$

FIGURE 3

Although all of the examples we have completed so far in this section have been sine curves, the results also apply to cosine curves. Here is a summary of what we know about the amplitude, period, and phase shift for since and cosine curves.

Amplitude, Period, and Phase Shift
The graphs of $y = A \sin(Bx + C)$ and $y = A \cos(Bx + C)$, where $B > 0$, have the following characteristics.

1. Amplitude $= |A|$ **2.** Period $= \dfrac{2\pi}{B}$ **3.** Phase Shift $= -\dfrac{C}{B}$

The information on amplitude, period, and phase shift allows us to sketch sine and cosine curves without having to prepare tables.

B Graphing Complete Cycles

EXAMPLE 4 Graph one complete cycle of $y = 2 \sin(3x + \pi)$.

Solution Here is a detailed list of steps to use in graphing sine and cosine curves for which B is positive:

Step 1: Use A, B, and C to find the amplitude, period, and phase shift.

Amplitude $= |A| = 2$

Period $= \dfrac{2\pi}{B} = \dfrac{2\pi}{3}$

Phase Shift $= -\dfrac{C}{B} = -\dfrac{\pi}{3}$

Step 2: On the x-axis, label the starting point, ending point, and the point halfway between them for each cycle of the curve in question. The starting point is the phase shift. The ending point it the phase shift plus the period. It is also a good idea to label the points one-fourth and three-fourths of the way between the starting point and ending point. See Figure 4.

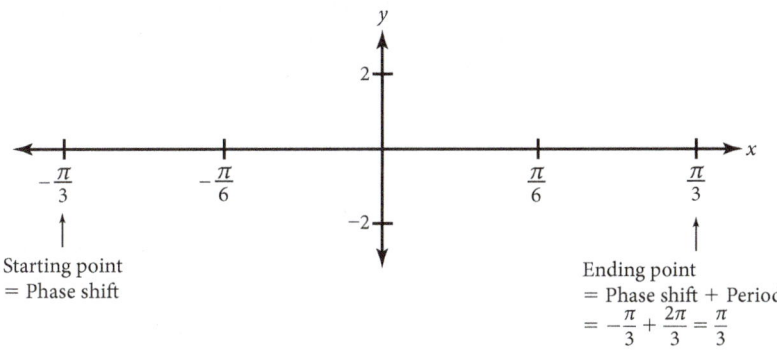

Starting point
= Phase shift

Ending point
= Phase shift + Period
$= -\dfrac{\pi}{3} + \dfrac{2\pi}{3} = \dfrac{\pi}{3}$

FIGURE 4

Step 3: Label the y-axis with the amplitude and the opposite of the amplitude. It is okay if the units on the x-axis and the y-axis are not proportional. That is, one unit on the y-axis can be a different length than one unit on the x-axis. The idea is to make the graph easy to read.

Step 4: Sketch in the curve in question. In this case, we want a sine curve that will be 0 at the starting point and 0 at the ending point. See Figure 5.

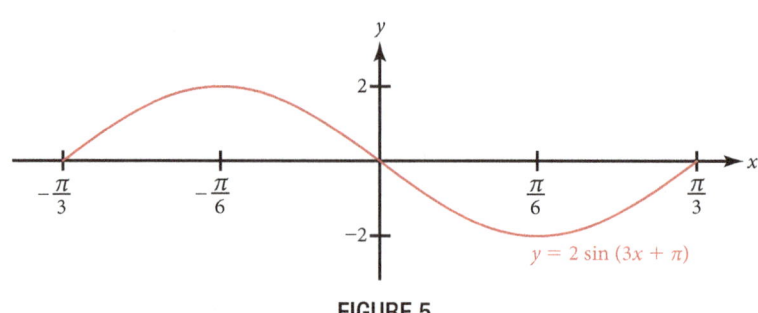

FIGURE 5

The steps listed in Example 4 may seem complicated at first. But with a little practice they do not take much time at all, especially when compared with the time it would take to prepare a table. Also, once we have graphed one complete cycle of the curve, it would be fairly easy to extend the graph in either direction.

EXAMPLE 5 Graph $y = 2\cos(3x + \pi)$ from $x = -\dfrac{2\pi}{3}$ to $x = \dfrac{2\pi}{3}$.

Solution A, B, and C are the same here as they were in Example 4. We use the same labeling on the axes as we used in Example 4, but we draw in a cosine curve instead of a sine curve and extend it to cover the interval $-\frac{2\pi}{3} \le x \le \frac{2\pi}{3}$, as shown in Figure 6.

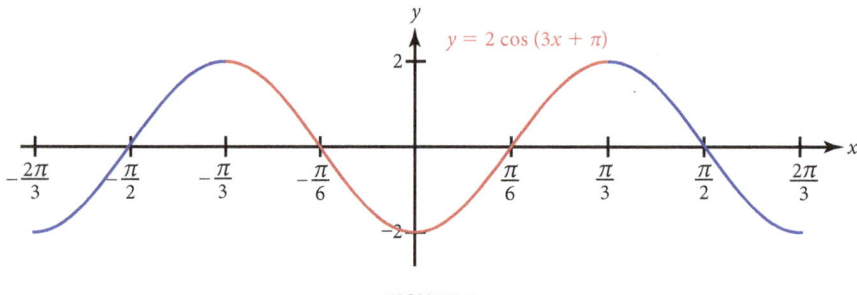

FIGURE 6

More on Labeling the x-Axis

Now that we have been through a few examples in detail we can be more specific about how we label the x-axis. Suppose the line in Figure 7 represents the x-axis, and the five points we have labeled with the letters a through e are the five points we will use to graph one complete cycle of a sine or cosine curve. Because these points are equally spaced along the x-axis, we can find their coordinates in the following manner (and the following order).

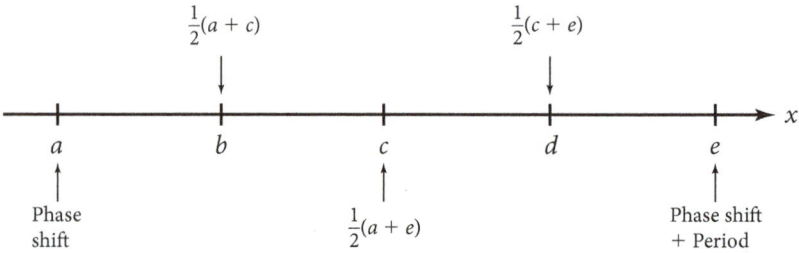

FIGURE 7

1. The coordinate of the first point, a, is always the phase shift.

2. The coordinate of the last point, e, is the sum of the phase shift and the period.

3. The center point, c, is the average of the first point and the last point. That is,

$$c = \frac{a + e}{2} = \frac{1}{2}(a + e)$$

4. The point b is the average of the first point and the center point. In symbols,

$$b = \frac{a + c}{2} = \frac{1}{2}(a + c)$$

5. The point d is the average of the center point and the last point. In symbols,

$$d = \frac{c + e}{2} = \frac{1}{2}(c + e)$$

EXAMPLE 6   Graph one complete cycle of $y = 4 \sin\left(2x - \frac{\pi}{3}\right)$.

Solution In this case $A = 4$, $B = 2$, and $C = -\frac{\pi}{3}$. This gives us a sine curve with

$$\text{Amplitude} = 4$$

$$\text{Period} = \frac{2\pi}{2} = \pi$$

$$\text{Phase Shift} = \frac{-(-\pi/3)}{2} = \frac{\pi}{6}$$

Because the amplitude is 4, we label the y-axis with 4 and -4. The x-axis is labeled according to the discussion that preceded this example, as shown in Figure 8.

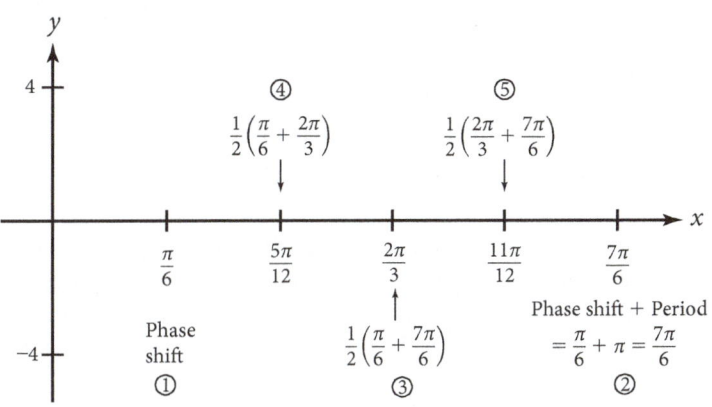

FIGURE 8

We finish the graph by drawing in a sine curve that starts at $x = \frac{\pi}{6}$ and ends at $x = \frac{7\pi}{6}$, as shown in Figure 9.

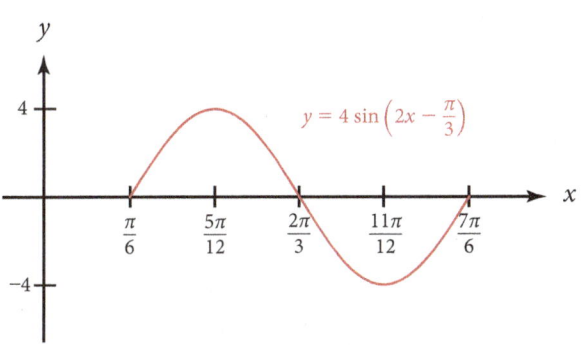

FIGURE 9

EXAMPLE 7 Graph one complete cycle of
$y = 5 \sin\left(\pi x + \frac{\pi}{4}\right)$.

Solution In this example, $A = 5$, $B = \pi$, and $C = \frac{\pi}{4}$.
The graph will have the following characteristics:

$$\text{Amplitude} = 5$$

$$\text{Period} = \frac{2\pi}{\pi} = 2$$

$$\text{Phase Shift} = -\frac{\pi/4}{\pi} = -\frac{1}{4}$$

The graph is shown in Figure 10.

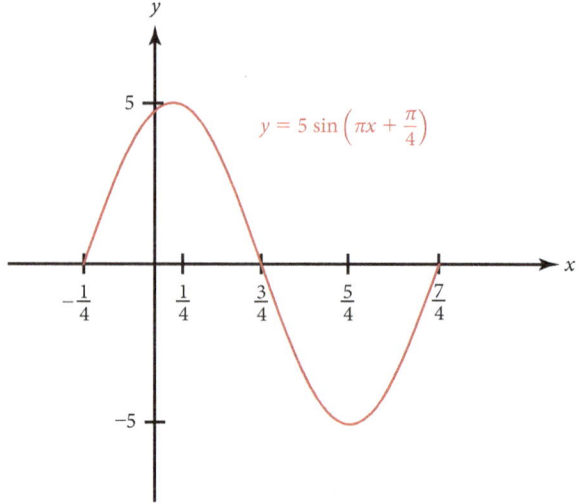

FIGURE 10

C Vertical Translations

Recall that adding a constant k to an equation containing trigonometric functions translates the graph vertically up or down. That is, the graph of $y = k + A \sin(Bx + C)$ will have the same shape (amplitude, period, phase shift, and reflection, if indicated) as the graph of $y = A \sin(Bx + C)$ but will be translated k units vertically from the graph of $y = A \sin(Bx + C)$.

EXAMPLE 8 Use the graph shown in Example 7 to graph one complete cycle of each of the following:

a. $y = 3 + 5 \sin\left(\pi x + \frac{\pi}{4}\right)$ **b.** $y = -5 + 5 \sin\left(\pi x + \frac{\pi}{4}\right)$

Solution The amplitude, period, and phase shift for each graph will be the same as those shown on the graph in Figure 10. The graph of the first equation is shifted up 3 units from the graph shown in Figure 10 (shown in blue in the graphs below), while the graph of the second equation is shifted down 5 units from the graph shown in Figure 10. One complete cycle of each graph will start at $x = -\frac{1}{4}$ and end 2 units later at $x = \frac{7}{4}$. The graphs are shown in Figures 11 and 12.

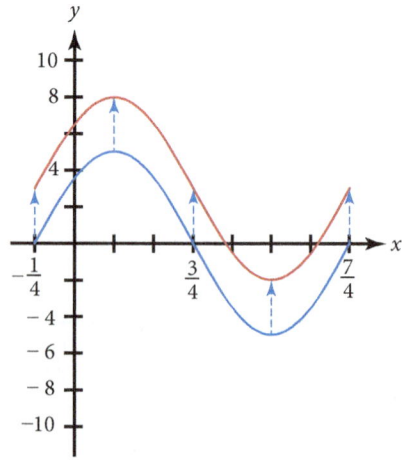

FIGURE 11

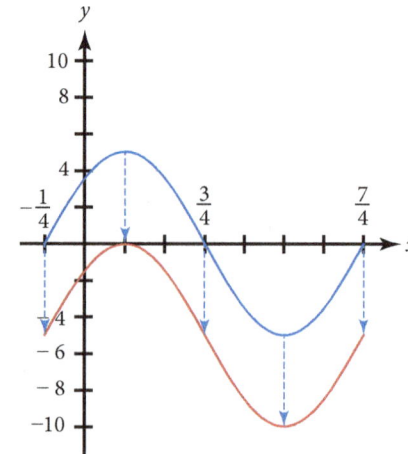

FIGURE 12

In our next example, we'll look at the effect of C on the graph of

$$y = A \tan(Bx + C)$$

First, recall from Section 4.1 (Figure 11) one cycle of the graph of $y = \tan x$:

Note that one cycle is completed in π units, so the period is π. As we would expect, then, the period of $y = A \tan(Bx + C)$ is $\frac{\pi}{B}$. The value of A stretches the curve vertically, as with the cotangent in the previous worksheet. The phase shift, as with sine and cosine, is $-\frac{C}{B}$. We'll illustrate these in the following example.

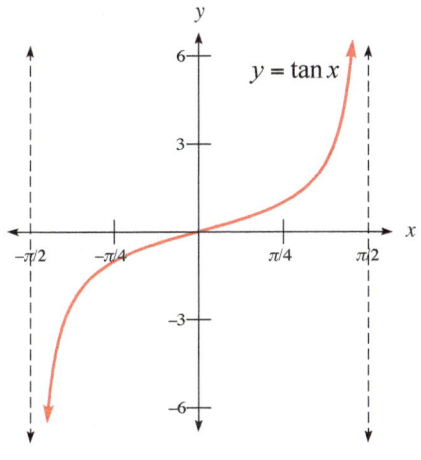

FIGURE 13

EXAMPLE 9 Graph one complete cycle of $y = 2 \tan\left(2x + \frac{\pi}{2}\right)$.

Solution First note the period of the graph is $\frac{\pi}{B} = \frac{\pi}{2}$, so an entire cycle will be completed in the interval $-\frac{\pi}{4} < x < \frac{\pi}{4}$. At the quarter of this interval is $x = -\frac{\pi}{8}$, where we would expect the point $\left(-\frac{\pi}{8}, -1\right)$. However, since there is a vertical stretch by $A = 2$, the corresponding point is $\left(-\frac{\pi}{8}, -2\right)$.

Finally, the phase shift is

$$-\frac{C}{B} = \frac{\frac{\pi}{2}}{2} = -\frac{\pi}{4}$$

which means one cycle of the graph is completed in the interval $-\frac{\pi}{2} < x < 0$. Here is the completed graph:

Note the points $\left(-\frac{3\pi}{8}, -2\right)$ and $\left(-\frac{\pi}{8}, -2\right)$ indicating the vertical stretch by a factor of 2. It is often helpful to first find the x-axis interval where the final cycle will be graphed, then fill in the corresponding points before graphing.

Original interval for tangent: $-\frac{\pi}{2} < x < \frac{\pi}{2}$

Period factor of 2: $-\frac{\pi}{4} < x < \frac{\pi}{4}$

Phase shift of $-\frac{\pi}{4}$: $-\frac{\pi}{2} < x < 0$

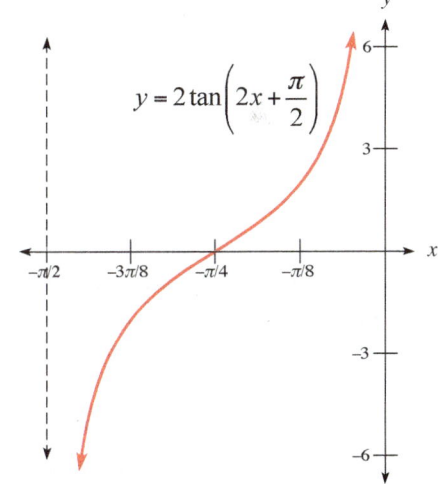

FIGURE 14

In the following example, we'll look at the effect of C on the graph of $y = A \csc(Bx + C)$. First, recall from Section 4.1 (Figure 15) one cycle of the graph of $y = \csc x$:

Note that one cycle is completed in 2π units, so the period is π. As we would expect, then, the period of $y = A \csc(Bx + C)$ is $\frac{2\pi}{B}$. The value of A stretches the curve vertically, as with the secant in the previous worksheet. The phase shift, as with sine and cosine, is $-\frac{C}{B}$. We'll illustrate these in the following example.

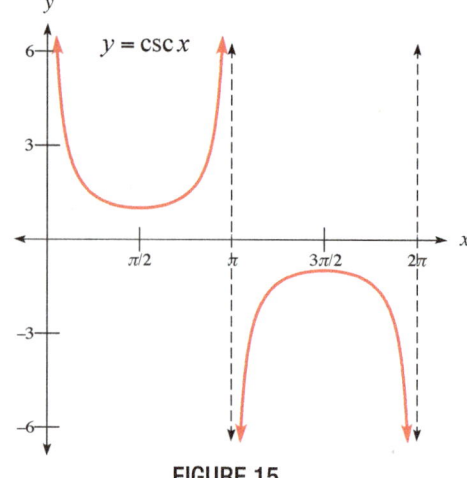

FIGURE 15

EXAMPLE 10 Graph one complete cycle of $y = 2 \csc(2x - \pi)$.

Solution First note the period of the graph is $\frac{2\pi}{B} = \frac{2\pi}{2} = \pi$, so an entire cycle will be completed in the interval $0 < x < \pi$. The value $A = 2$ stretches the graph by a factor of 2. The phase shift is $-\frac{C}{B} = \frac{-\pi}{2} = \frac{\pi}{2}$, so the final interval for the graph is $\frac{\pi}{2} < x < \frac{3\pi}{2}$. The final graph is sketched to the right:

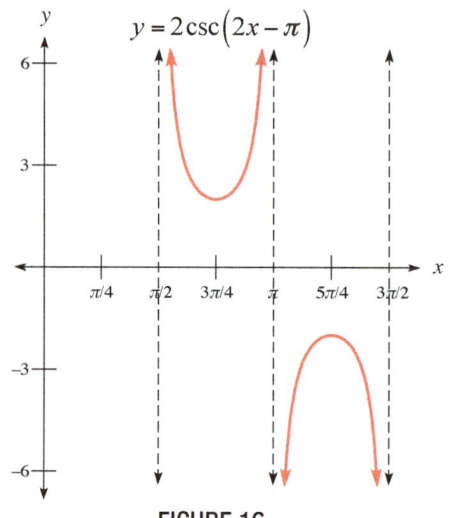

FIGURE 16

Vertical Translations of Tangent

As we have noted previously, a constant term added to a trigonometric function will translate the graph of the original function up or down vertically, depending on whether the constant term is positive or negative.

EXAMPLE 11 Graph one complete cycle of $y = 3 + 2 \tan\left(2x + \frac{\pi}{2}\right)$.

Solution: If you look back to Example 1, you will see that the graph of this function, will be the graph of $y = 3 + 2 \tan\left(2x + \frac{\pi}{2}\right)$ translated vertically up 3 units. Figure 17 shows the graph.

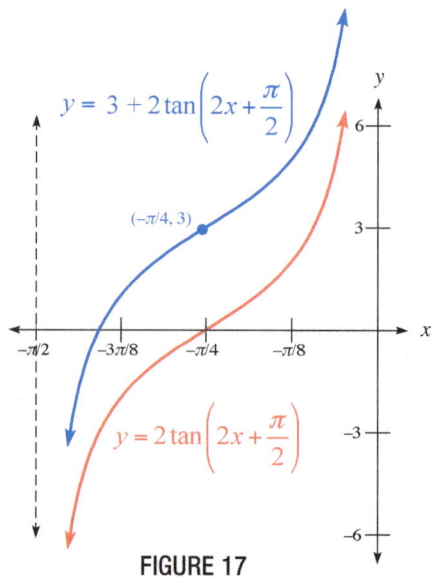

FIGURE 17

Problem Set 4.3

Vocabulary

Use the vocabulary words and formulas below to fill in the blanks in the sentences.

vertical   amplitude   horizontal   phase   cycle

1. The phase shift of a graph can also be thought of as a _____ translation.

2. The graph of $y = \sin\left(x + \frac{\pi}{3}\right)$ will have a _____ shift equal to $-\frac{\pi}{3}$.

3. The graph of $y = 2 + \sin x$ will have a _____ translation equal to 2.

4. The graph of $y = 3\sin\left(2x + \frac{\pi}{3}\right)$ will go through one complete _____ in π units on the x-axis. The _____ of the graph will be 3 and the _____ shift is $-\frac{\pi}{6}$.

For each equation, first identify the phase shift and then sketch one complete cycle of the graph. In each case, graph $y = \sin x$ on the same coordinate system.

1. $y = \sin\left(x + \frac{\pi}{4}\right)$

2. $y = \sin\left(x + \frac{\pi}{6}\right)$

3. $y = \sin\left(x - \frac{\pi}{4}\right)$

4. $y = \sin\left(x - \frac{\pi}{6}\right)$

5. $y = \sin\left(x + \frac{\pi}{3}\right)$

6. $y = \sin\left(x - \frac{\pi}{3}\right)$

For each equation, identify the phase shift and then sketch one complete cycle of the graph. In each case, graph one complete cycle of $y = \cos x$ on the same set of axes.

7. $y = \cos\left(x - \frac{\pi}{2}\right)$

8. $y = \cos\left(x + \frac{\pi}{2}\right)$

9. $y = \cos\left(x + \frac{\pi}{3}\right)$

10. $y = \cos\left(x - \frac{\pi}{4}\right)$

For each equation, identify the amplitude, period, and phase shift. Then label the axes accordingly and sketch one complete cycle of the curve.

11. $y = \sin(2x - \pi)$

12. $y = \sin(2x + \pi)$

13. $y = \cos\left(2x + \dfrac{\pi}{2}\right)$

14. $y = \cos\left(2x - \dfrac{\pi}{2}\right)$

15. $y = 2\sin\left(\dfrac{1}{2}x + \dfrac{\pi}{2}\right)$

16. $y = 3\cos\left(\dfrac{1}{2}x + \dfrac{\pi}{3}\right)$

17. $y = 3\sin\left(2x - \dfrac{\pi}{2}\right)$

18. $y = 3\cos\left(2x - \dfrac{\pi}{3}\right)$

19. $y = 4\cos\left(2x - \dfrac{\pi}{2}\right)$

20. $y = 3\sin\left(2x - \dfrac{\pi}{3}\right)$

21. $y = \dfrac{1}{2}\cos\left(3x - \dfrac{\pi}{2}\right)$

22. $y = \dfrac{4}{3}\cos\left(3x + \dfrac{\pi}{2}\right)$

23. $y = \dfrac{2}{3}\sin\left(3x + \dfrac{\pi}{2}\right)$

24. $y = \dfrac{3}{4}\sin\left(3x - \dfrac{\pi}{2}\right)$

Use your answers to Problems 11-16 for reference, and graph one complete cycle of each of the following equations.

25. $y = 1 + \sin(2x - \pi)$

26. $y = -1 + \sin(2x + \pi)$

27. $y = 2 - \cos\left(2x + \dfrac{\pi}{2}\right)$

28. $y = -2 - \cos\left(2x - \dfrac{\pi}{2}\right)$

29. $y = -2 + 2\sin\left(\dfrac{1}{2}x + \dfrac{\pi}{2}\right)$

30. $y = 3 + 3\cos\left(\dfrac{1}{2}x + \dfrac{\pi}{3}\right)$

Sketch one complete cycle of each graph. Pay particular attention to the labeling of the x-axis. For problems 35 and 36, remember the negative has the effect of reflecting the curve across the x-axis.

31. $y = \tan\left(2x + \dfrac{\pi}{4}\right)$

32. $y = \tan(3x + \pi)$

33. $y = 3\tan\left(\pi x - \dfrac{\pi}{2}\right)$

34. $y = 2\tan\left(\dfrac{\pi}{2}x - \dfrac{\pi}{4}\right)$

35. $y = -4\tan\left(3x + \dfrac{\pi}{2}\right)$

36. $y = -3\tan\left(2x - \dfrac{\pi}{2}\right)$

Sketch one complete cycle of each graph. Pay particular attention to the labeling of the x-axis. For problems 41 and 42, remember the negative has the effect of reflecting the curve across the x-axis.

37. $y = \csc\left(2x + \dfrac{\pi}{4}\right)$

38. $y = \csc\left(3x + \pi\right)$

39. $y = 3\csc\left(\pi x - \dfrac{\pi}{2}\right)$

40. $y = 2\csc\left(\dfrac{\pi}{2}x - \dfrac{\pi}{4}\right)$

41. $y = -4\csc\left(3x + \dfrac{\pi}{2}\right)$

42. $y = -3\csc\left(2x - \dfrac{\pi}{2}\right)$

Sketch one complete cycle of each graph.

43. $y = 3 + \tan\left(2x + \dfrac{\pi}{4}\right)$

44. $y = -3 + \tan\left(3x + \pi\right)$

45. $y = -1 + 3\tan\left(\pi x - \dfrac{\pi}{2}\right)$

46. $y = 1 + 2\tan\left(\dfrac{\pi}{2}x - \dfrac{\pi}{4}\right)$

47. $y = 3 + \csc\left(2x + \dfrac{\pi}{4}\right)$

48. $y = -3 + \csc\left(3x + \pi\right)$

Getting Ready for the Next Section

For each of the following lines, names the slope and y-intercept. Then write the equation of the line in slope-intercept form.

49.
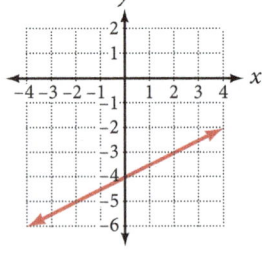

50.
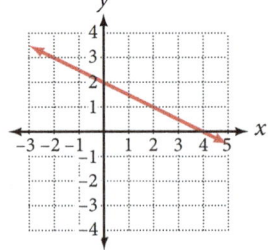

51.
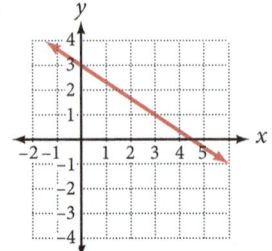

52.
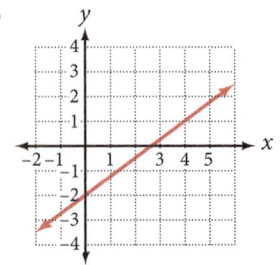

4.4 Videos

Learning Objectives

A Find the equation of a given graph.

In this section, we will reverse what we have done in previous sections of this chapter and produce an equation that describes a graph, rather than a graph that describes an equation. Let's start with an example from algebra.

EXAMPLE 1 Find the equation of the line shown in Figure 1.

Solution From algebra we know that the equation of any straight line (except a vertical one) can be written in slope-intercept form as

$$y = mx + b$$

where m is the slope of the line, and b is its y-intercept.

Since the line in Figure 1 crosses the y-axis at 3, we know the y-intercept b is 3. To find the slope of the line, we find the ratio of the vertical change to the horizontal change between any two points on the line (sometimes called rise/run). From Figure 1 we see that this ratio is $-\frac{1}{2}$. Therefore, $m = -\frac{1}{2}$. The equation of our line must be

$$y = -\frac{1}{2}x + 3$$

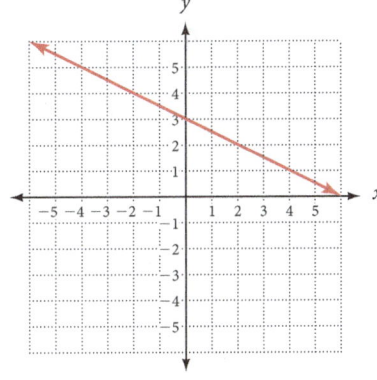

FIGURE 1

EXAMPLE 2 One cycle of the graph of a trigonometric function is shown in Figure 2. Find an equation to match the graph.

Solution The graph is a sine curve with an amplitude of 3, period 2π, and no phase shift. The equation is

$$y = 3 \sin x \quad \text{for} \quad 0 \le x \le 2\pi$$

Note that we have specified the domain for each equation. Doing so gives us a way of indicating how much of a curve is showing. In Figure 2, the curve starts at 0 and ends at 2π. We can indicate this by restricting x to the interval $0 \le x \le 2\pi$.

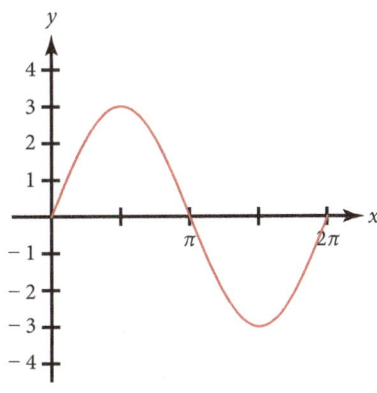

FIGURE 2

EXAMPLE 3 Find the equation of the graph shown in Figure 3.

Solution Again, we have a sine curve, so we know the equation will have the form

$$y = k + A \sin (Bx + C)$$

From Figure 3 we see that the amplitude is 3, which means that $A = 3$. There is no phase shift, nor is there any vertical translation of the graph. Therefore, both C and k are 0.

To find B, we notice that the period is π. Since the formula for the period is $\frac{2\pi}{B}$, we have

$$\pi = \frac{2\pi}{B}$$

which means that B is 2. Our equation must be

$$y = 0 + 3 \sin (2x + 0)$$

which simplifies to

$$y = 3 \sin 2x \quad \text{for} \quad 0 \le x \le 2\pi$$

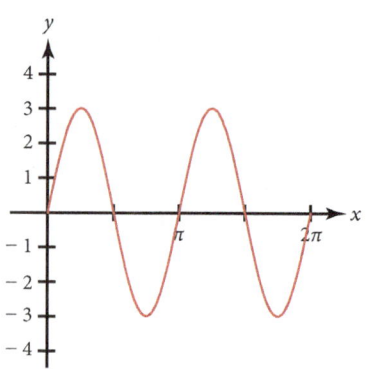

FIGURE 3

EXAMPLE 4 Find the equation of the graph shown in Figure 4.

Solution The graph in Figure 4 has the same shape (amplitude, period, and phase shift) as the graph shown in Figure 3. In addition, it has undergone a vertical shift up of 2 units; therefore, the equation is

$$y = 2 + 3 \sin 2x \quad \text{for} \quad 0 \le x \le 2\pi$$

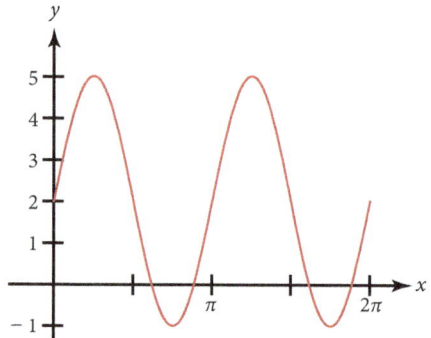

FIGURE 4

EXAMPLE 5 Find the equation of the graph shown in Figure 5.

Solution If we look at the graph from $x = 0$ to $x = 2$, it looks like a cosine curve that has been reflected about the x-axis. The general form for a cosine curve is

$$y = k + A \cos (Bx + C)$$

From Figure 5 we see that the amplitude is 5. Since the graph has been reflected about the x-axis, $A = -5$. The period is 2, giving us an equation to solve for B:

$$\text{Period} = \frac{2\pi}{B} = 2 \quad \Rightarrow \quad B = \pi$$

There is no phase shift or vertical translation of the curve (if we assume it is a cosine curve), so C and k are both 0. The equation that describes this graph is

$$y = -5 \cos \pi x \quad \text{for} \quad -0.5 \le x \le 2.5$$

Notice that we could have done this same problem using a sine curve and the result would be the equation

$$y = -5 \sin \left(\pi x + \frac{\pi}{2} \right)$$

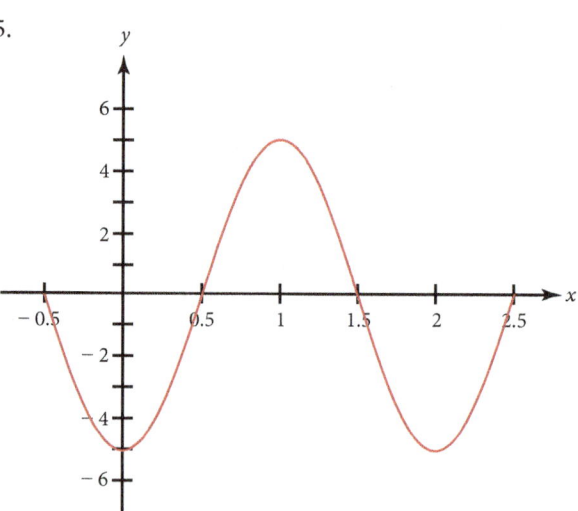

FIGURE 5

EXAMPLE 6 Find the equation of the curve shown in Figure 6.

Solution The graph has the same shape as the graph shown in Figure 5 except it has been moved to the right 0.5 unit. Using the results from Example 5, we know that $B = \pi$. To find C, we use the formula for phase shift:

$$\text{Phase Shift} = -\frac{C}{B}$$

$$0.5 = -\frac{C}{\pi}$$

Solving for C, we have

$$C = -\frac{\pi}{2}$$

The equation is

$$y = -5 \cos\left(\pi x - \frac{\pi}{2}\right) \quad \text{for} \quad 0 \leq x \leq 3$$

Try this problem again assuming it is a sine curve. You will see that the resulting equation is actually simpler:

$$y = -5 \sin \pi x$$

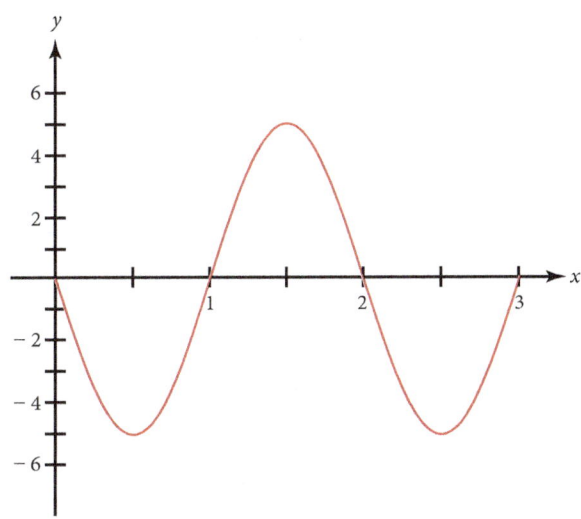

FIGURE 6

In our next example, we use a table of values for two variables to obtain a graph. From the graph, we find the equation.

EXAMPLE 7 A model of the Ferris wheel built by George Ferris that we mentioned previously is shown in Figure 7. Recall that the diameter is 250 feet and it rotates through one complete revolution every 20 minutes.

First, set up a table that shows the rider's height, h, above the ground for each of the nine values of t shown in Figure 7. Then graph the ordered pairs indicated by the table. Finally, use the graph to find an equation that gives h as a function of t.

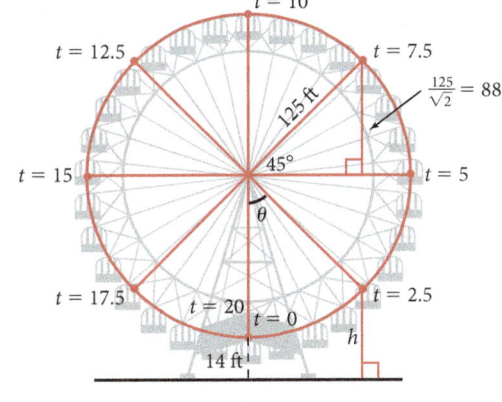

FIGURE 7

Solution Here is the reasoning we use to find a value of h for each of the given values of t: When $t = 0$, the rider is at the start of the ride, 14 feet above the ground. When $t = 10$, the rider is at the top of the wheel, which is 264 feet above the ground. At times $t = 5$ and $t = 15$, the rider is even with the center of the wheel, which is 139 feet above the ground. The other four values of h are found using a 45°–45°–90° triangle with a hypotenuse of 125 feet. Table 1 lists the corresponding values of t and h.

Table 1

t	h
0 min	14 ft
2.5 min	51 ft
5 min	139 ft
7.5 min	227 ft
10 min	264 ft
12.5 min	227 ft
15 min	139 ft
17.5 min	51 ft
20 min	14 ft

If we graph the points (t, h) on a rectangular coordinate system and then connect them with a smooth curve, we produce the diagram shown in Figure 8. The curve is a cosine curve that has been reflected about the t-axis and then shifted up vertically.

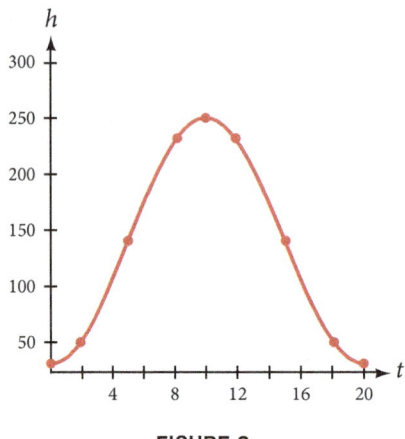

FIGURE 8

Since the curve is a cosine curve, the equation will have the form

$$h = k + A \cos(Bx + C)$$

To find the equation for the curve in Figure 8, we must find values for k, A, B, and C. We begin by finding C.

Since the curve starts at $t = 0$, the phase shift is 0, which gives us

$$C = 0$$

The amplitude, half the difference between the highest point and the lowest point on the graph, must be 125 (also equal to the radius of the wheel). However, since the graph is a *reflected* cosine curve, we have

$$A = -125$$

The period is 20 minutes, so we find B by using the equation

$$20 = \frac{2\pi}{B} \quad \Rightarrow \quad B = \frac{\pi}{10}$$

The amount of vertical translation, k, is the distance the center of the wheel is off the ground. Therefore,

$$k = 139$$

The equation that gives the height of the rider at any time t during the ride is

$$h = 139 - 125 \cos\frac{\pi}{10} t$$

Find the equation of each of the following lines. Write your answers in slope-intercept form, $y = mx + b$.

1.

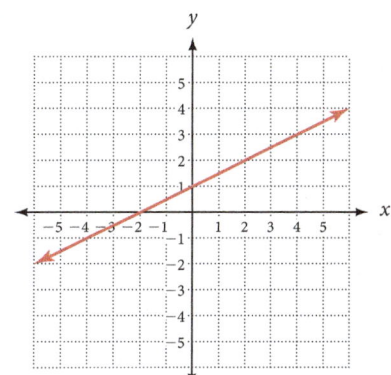

2.

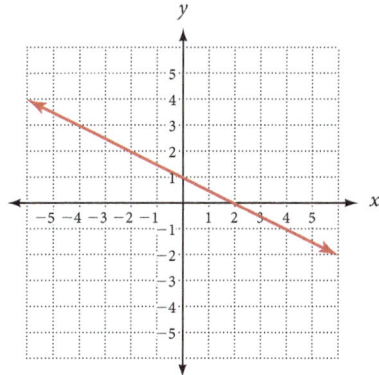

3.

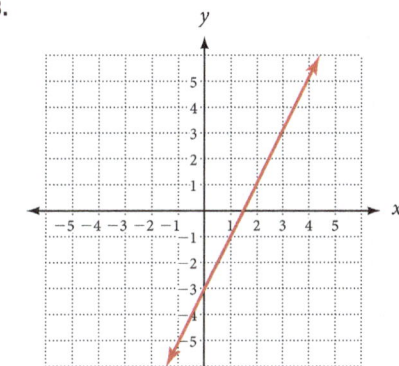

4.

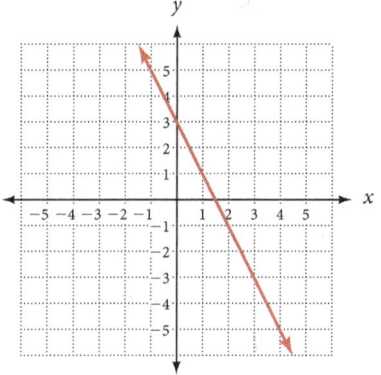

Each graph below is one complete cycle of the graph of an equation containing a trigonometric function. In each case, find an equation to match the graph.

5.

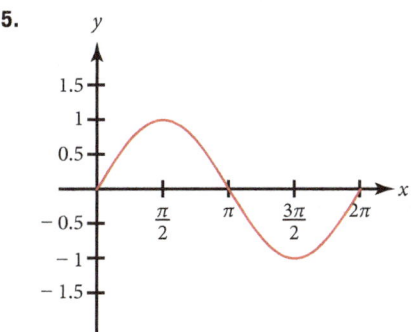

6.

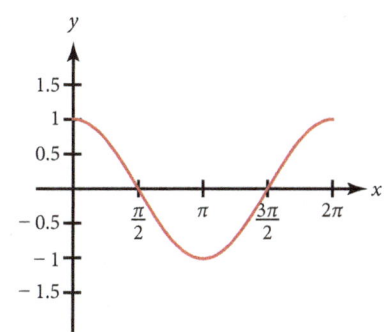

7.

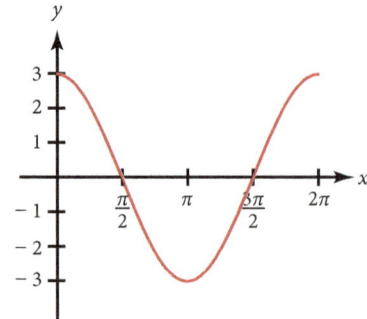

8.

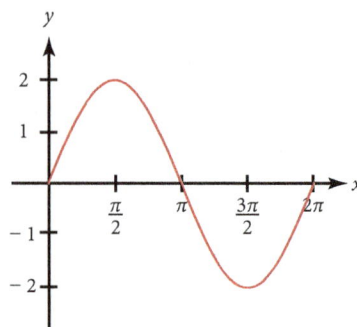

9.

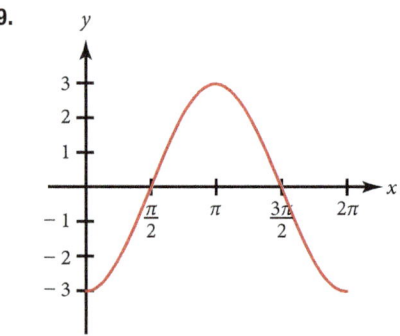

10.

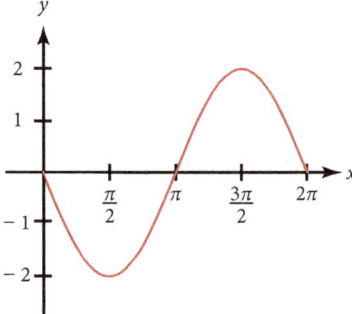

11.

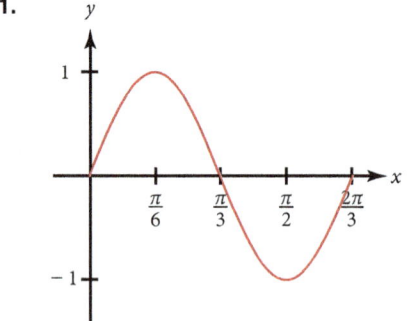

12.

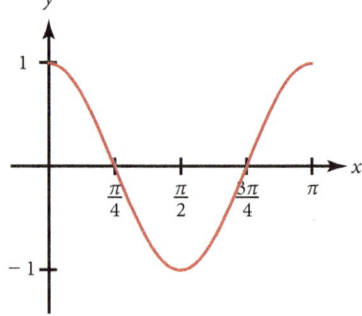

13.

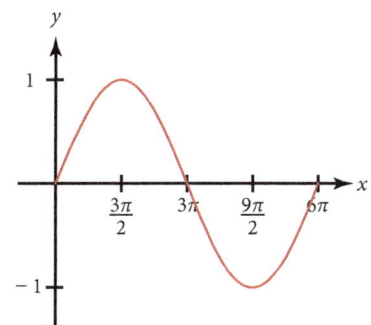

14.

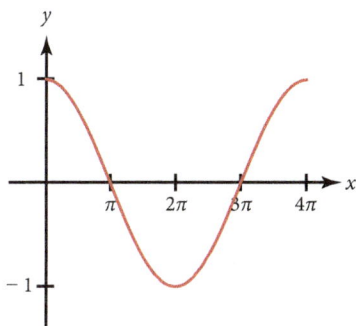

15.

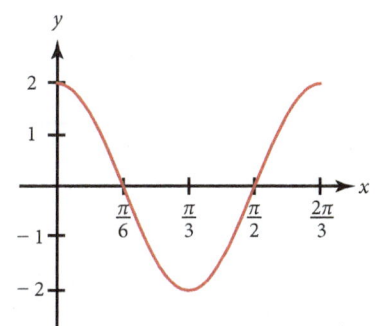

16.

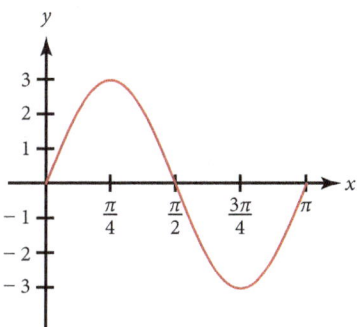

17.

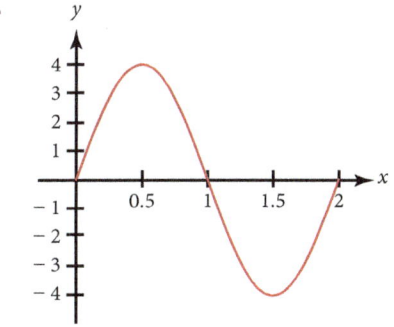

18.

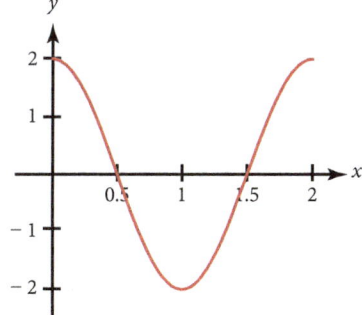

19.

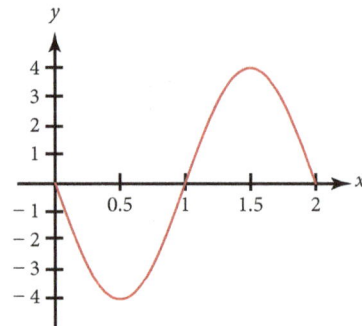

20.

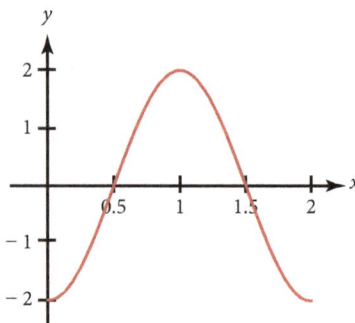

21.

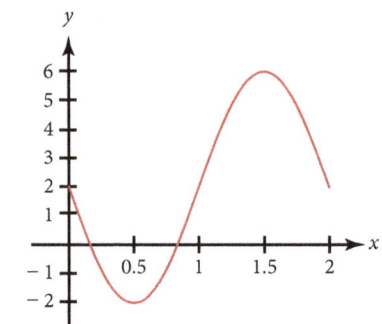

22.

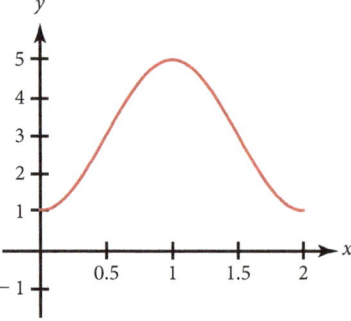

23.

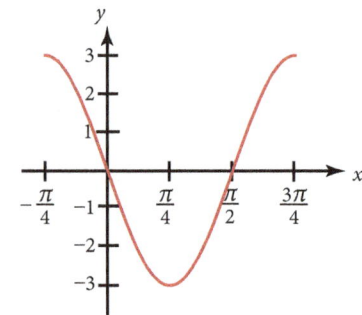

24.

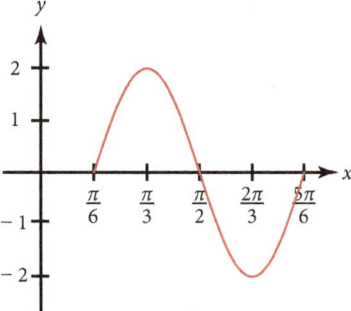

25.

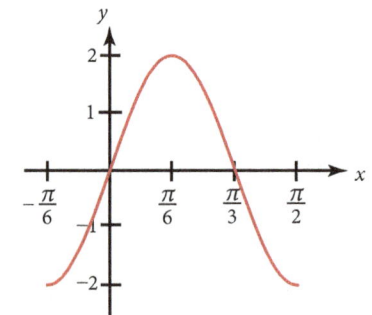

26.

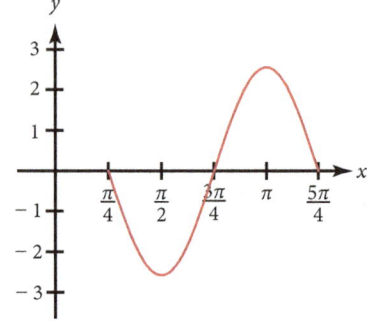

Applying the Concepts

27. Ferris Wheel   Figure 9 is a model of the Reisenrad Ferris wheel. The diameter of the wheel is 197 feet, and one complete revolution takes 15 minutes. The bottom of the wheel is 12 feet above the ground. Refer to Figure 9 to complete Table 2. The plot the points (t, h) from the table. Finally, connect the points with a smooth curve, and use the curve to find an equation that will give a passenger's height above the ground at any time t during the ride.

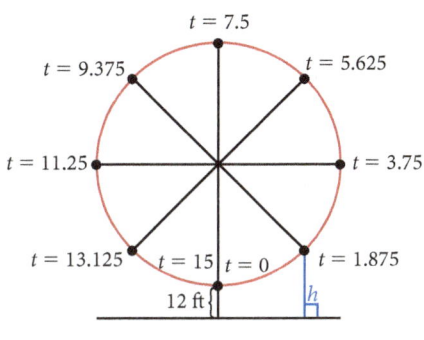

FIGURE 9

Table 2

t	h
0 min	
1.875 min	
3.75 min	
5.625 min	
7.5 min	
9.375 min	
11.25 min	
13.125 min	
15 min	

28. Ferris Wheel   A Ferris wheel called Colossus was built in St. Louis in 1986. The diameter of the wheel is 165 feet, it rotates at 1.5 revolutions per minute, and the bottom of the wheel is 9 feet above the ground. Find an equation that gives a passenger's height above the ground at any time t during the ride. Assume the passenger starts the ride at the bottom of the wheel.

Getting Ready for the Next Section

Evaluate the following. Round to nearest hundredth, if necessary.

29. $\dfrac{\pi}{2} + \sin \dfrac{\pi}{2}$

30. $\pi + \sin \pi$

31. $0 + \sin 0$

32. $\dfrac{3\pi}{2} + \sin \dfrac{3\pi}{2}$

33. $2\pi + \sin 2\pi$

34. $\dfrac{\pi}{4} + \sin \dfrac{\pi}{4}$

Learning Objectives

A   Graph algebraic and trigonometric functions of *x*.

In this section we will graph equations of the form $y = y_1 + y_2$ where y_1 and y_2 are algebraic or trigonometric functions of *x*. For instance, the equation $y = 1 + \sin x$ can be thought of as the sum of the two functions $y_1 = 1$ and $y_2 = \sin x$. That is,

$$\text{if} \quad y_1 = 1 \text{ and } y_2 = \sin x,$$

$$\text{then } y = y_1 + y_2$$

Using this kind of reasoning, the graph of $y = 1 + \sin x$ is obtained by adding each value of y_2 in $y_2 = \sin x$ to the corresponding value of y_1 in $y_1 = 1$. Graphically, we can show this by adding the values of *y* from the graph of y_2 to the corresponding values of *y* from the graph of y_1 (Figure 1).

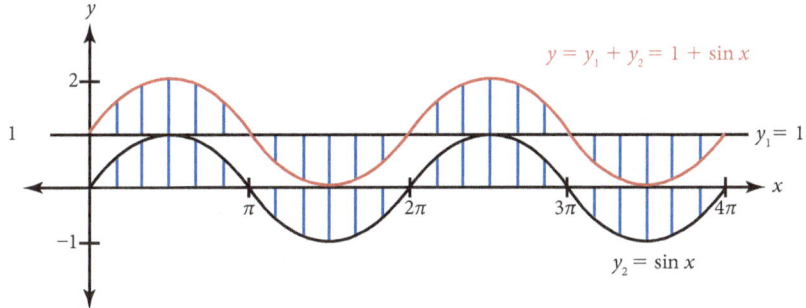

FIGURE 1

Although in actual practice you may not draw in the little vertical lines we have shown here, they do serve the purpose of allowing us to visualize the idea of adding the *y*-coordinates on one graph to the corresponding *y*-coordinates on another graph.

EXAMPLE 1   Graph $y = \frac{1}{3}x - \sin x$ between $x = 0$ and $x = 4\pi$.

Solution   We can think of the equation $y = \frac{1}{3}x - \sin x$ as the sum of the equations $y_1 = \frac{1}{3}x$ and $y_2 = -\sin x$. Graphing each of these two equations on the same set of axes and then adding the values of y_2 to the corresponding values of y_1, we have the graph shown in Figure 2.

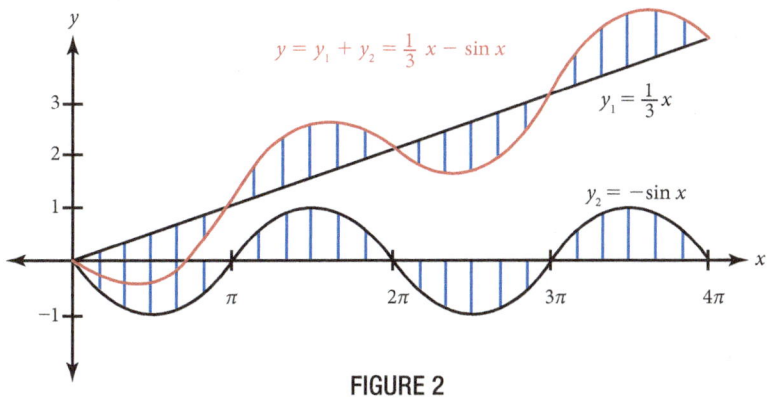

FIGURE 2

For the rest of the examples in this section, we will not show the vertical lines used to visualize the process of adding y-coordinates. Sometimes the graphs become too confusing to read when the vertical lines are included. It is the idea behind the vertical lines that is important, not the lines themselves. (And remember, the alternative to graphing these types of equations by adding y-coordinates is to make a table. If you try using a table on some of the examples that follow, you will see that the method being presented here is much faster.)

EXAMPLE 2 Graph $y = 2 \sin x + \cos 2x$ for x between 0 and 4π.

Solution We can think of y as the sum of y_1 and y_2, where

$$y_1 = 2 \sin x \text{ (amplitude 2, period } 2\pi) \quad \text{and} \quad y_2 = \cos 2x \text{ (amplitude 1, period } \pi)$$

The graphs of y_1, y_2 and $y = y_1 + y_2$ are shown in Figure 3.

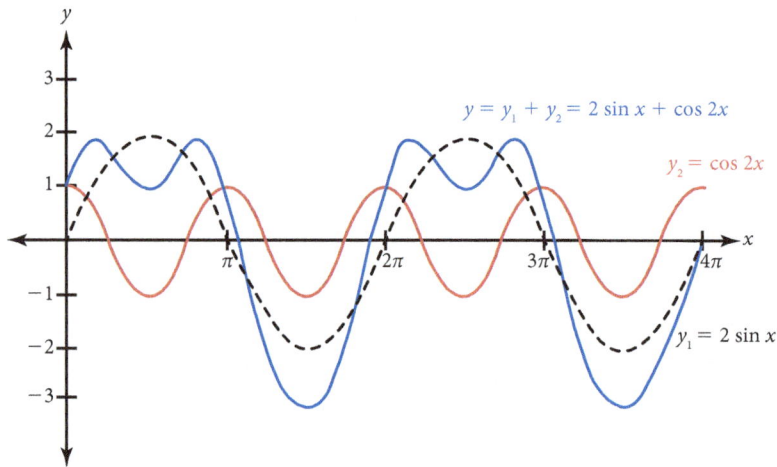

FIGURE 3

EXAMPLE 3 Graph $y = \cos x + \cos 2x$ for $0 \le x \le 4\pi$.

Solution We let $y = y_1 + y_2$, where

$$y_1 = \cos x \text{ (amplitude 1, period } 2\pi) \quad \text{and} \quad y_2 = \cos 2x \text{ (amplitude 1, period } \pi)$$

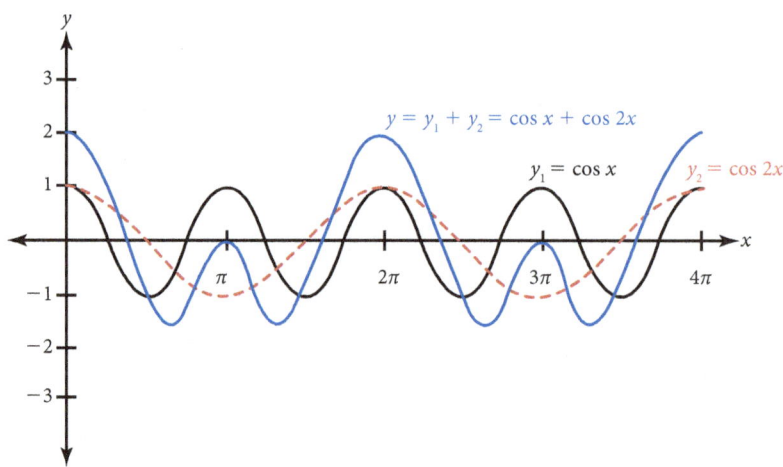

FIGURE 4

EXAMPLE 4   Graph $y = \sin x + \cos x$ for x between 0 and 4π.

Solution   We let $y_1 = \sin x$ and $y_2 = \cos x$ and graph y_1, y_2, and $y = y_1 + y_2$.

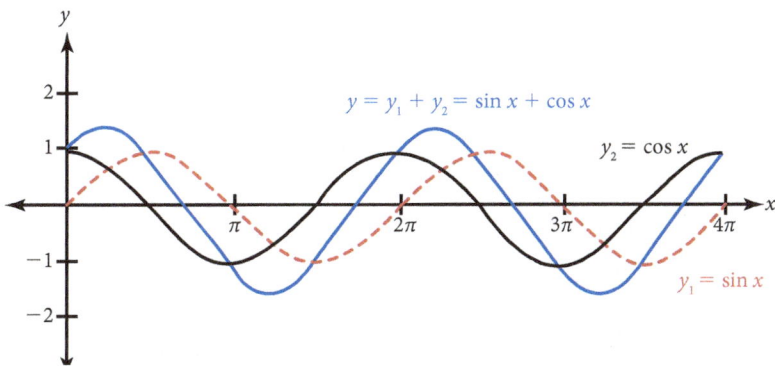

FIGURE 5

The graph of $y = \sin x + \cos x$ has amplitude $\sqrt{2}$. If we were to extend the graph to the left, we would find it crossed the x-axis at $-\pi/4$. It would then be apparent that the graph of $y = \sin x + \cos x$ is the same as the graph of $y = \sqrt{2}$ $\sin(x + \pi/4)$; both are the sine curves with amplitude $\sqrt{2}$ and phase shift $-\pi/4$. This connection between $y = \sin x + \cos x$ and $y = \sqrt{2} \sin(x + \pi/4)$ will be studied further in the next chapter.

Use addition of y-coordinates to sketch the graph of each of the following between $x = 0$ and $x = 4\pi$.

1. $y = 1 + \sin x$

2. $y = 1 + \cos x$

3. $y = 2 - \cos x$

4. $y = 2 - \sin x$

5. $y = 4 + 2 \sin x$

6. $y = 4 + 2 \cos x$

7. $y = \frac{1}{3} x - \cos x$

8. $y = \frac{1}{2} x - \sin x$

9. $y = \frac{1}{2} x - \cos x$

10. $y = \frac{1}{3} x + \cos x$

Sketch the graph of each equation from $x = 0$ to $x = 8$.

11. $y = x + \sin \pi x$

12. $y = x + \cos \pi x$

Sketch the graph from $x = 0$ to $x = 4\pi$.

13. $y = 3 \sin x + \cos 2x$

14. $y = 3 \cos x + \sin 2x$

15. $y = 2 \sin x - \cos 2x$

16. $y = 2 \cos x - \sin 2x$

17. $y = \sin x + \sin \frac{x}{2}$

18. $y = \cos x + \cos \frac{x}{2}$

19. $y = \sin x + \sin 2x$

20. $y = \cos x + \cos 2x$

21. $y = \cos x + \frac{1}{2} \sin 2x$

22. $y = \sin x + \frac{1}{2} \cos 2x$

23. $y = \sin x - \cos x$

24. $y = \cos x - \sin x$

25. Make a table using multiples of $\frac{\pi}{2}$ for x between 0 and 4π to help sketch the graph of $y = x \sin x$.

26. Make a table using multiples of $\frac{\pi}{2}$ for x between 0 and 4π to help sketch the graph of $y = x \cos x$.

Getting Ready for the Next Section

Solve for y.

27. $2y - 3 = x$

28. $x = 2y + 3$

29. $x = y^2 + 4$

30. $x = y^2 - 3$

Find in degrees. Round to the nearest tenth of a degree.

31. $\sin \theta = \frac{1}{\sqrt{2}}, 0° < \theta < 90°$

32. $\sin \theta = -\frac{1}{\sqrt{2}}, -90° < \theta < 0°$

33. $\sin \theta = 0.5075, 0° < \theta < 90°$

34. $\sin \theta = -0.5075, -90° < \theta < 0°$

35. $\cos \theta = -0.6428, 0° < \theta < 180°$

36. $\cos \theta = 0.6428, 0° < \theta < 90°$

Learning Objectives

A   Find the equation of the inverse of a function.

B   Sketch a graph of a function and its inverse.

C   Identify inverse functions.

D   Evaluate inverse trigonometric functions.

The following diagram (Figure 1) shows the route Justin takes to school. He leaves his home and drives 3 miles east, and then turns left and drives 2 miles north. When he leaves school to drive home, he drives the same two segments, but in the reverse order and the opposite direction; that is, he drives 2 miles south, turns right, and drives 3 miles west. When he arrives home from school, he is right where he started. His route home "undoes" his route to school, leaving him where he began.

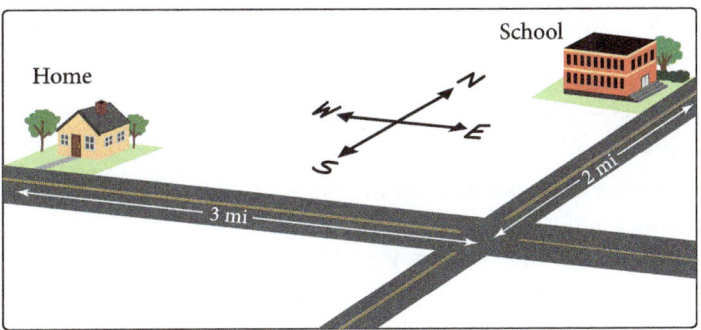

FIGURE 1

As you will see, the relationship between a function and its inverse function is similar to the relationship between Justin's route from home to school and his route from school to home.

Suppose the function f is given by

$$f = \{(1, 4), (2, 5), (3, 6), (4, 7)\}$$

The inverse of f is obtained by reversing the order of the coordinates in each ordered pair in f. The inverse of f is the relation given by

$$g = \{(4, 1), (5, 2), (6, 3), (7, 4)\}$$

It is obvious that the domain of f is now the range of g, and the range of f is now the domain of g. Every function (or relation) has an inverse that is obtained from the original function by interchanging the components of each ordered pair.

A Finding the Inverse Relation

Suppose a function f is defined with an equation instead of a list of ordered pairs. We can obtain the equation of the inverse of f by interchanging the role of x and y in the equation for f.

EXAMPLE 1   If the function f is defined by $f(x) = 2x - 3$, find the equation that represents the inverse of f.

Solution Because the inverse of f is obtained by interchanging the components of all the ordered pairs belonging to f, and each ordered pair in f satisfies the equation $y = 2x - 3$, we simply exchange x and y in the equation $y = 2x - 3$ to get the formula for the inverse of f:

$$x = 2y - 3$$

We now solve this equation for y in terms of x:

$$x + 3 = 2y$$

$$\frac{x + 3}{2} = y$$

$$y = \frac{x + 3}{2}$$

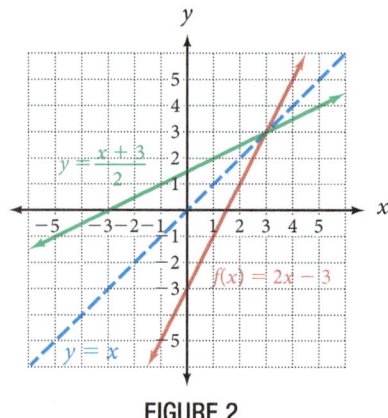

FIGURE 2

The last line gives the equation that defines the inverse of f. Let's compare the graphs of f and its inverse as given here. (See Figure 2.)

The graphs of f and its inverse have symmetry about the line $y = x$. This is a reasonable result since the one function was obtained from the other by interchanging x and y in the equation. The ordered pairs (a, b) and (b, a) always have symmetry about the line $y = x$.

B Graph a Function with Its Inverse

EXAMPLE 2   Graph the function $y = x^2 - 2$ and its inverse. Give the equation for the inverse.

Solution   We can obtain the graph of the inverse of $y = x^2 - 2$ by graphing $y = x^2 - 2$ by the usual methods, and then reflecting the graph about the line $y = x$.

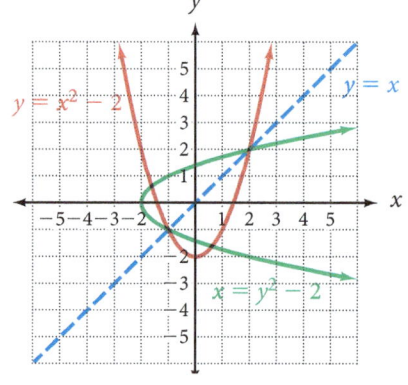

FIGURE 3

The equation that corresponds to the inverse of $y = x^2 - 2$ is obtained by interchanging x and y to get $x = y^2 - 2$.

We can solve the equation $x = y^2 - 2$ for y in terms of x as follows:

$$x = y^2 - 2$$

$$x + 2 = y^2$$

$$y = \pm\sqrt{x + 2}$$

Comparing the graphs from Examples 1 and 2, we observe that the inverse of a function is not always a function. In Example 1, both f and its inverse have graphs that are nonvertical straight lines and therefore both represent functions. In Example 2, the inverse of function f is not a function, since a vertical line crosses it in more than one place.

One-to-One Functions

We can distinguish between those functions with inverses that are also functions and those functions with inverses that are not functions with the following definition.

> **One-to-One Functions**
> A function is a *one-to-one function* if every element in the range comes from exactly one element in the domain.

This definition indicates that a one-to-one function will yield a set of ordered pairs in which no two different ordered pairs have the same second coordinates. For example, the function

$$f = \{(2, 3), (-1, 3), (5, 8)\}$$

is not one-to-one because the element 3 in the range comes from both 2 and -1 in the domain. On the other hand, the function

$$g = \{(5, 7), (3, -1), (4, 2)\}$$

is a one-to-one function because every element in the range comes from only one element in the domain.

Horizontal Line Test

If we have the graph of a function, we can determine if the function is one-to-one with the following test. If a horizontal line crosses the graph of a function in more than one place, then the function is not a one-to-one function because the points at which the horizontal line crosses the graph will be points with the same *y*-coordinates, but different *x*-coordinates. Therefore, the function will have an element in the range (the *y*-coordinate) that comes from more than one element in the domain (the *x*-coordinates).

Of the functions we have covered previously, all the linear functions and exponential functions are one-to-one functions because no horizontal lines can be found that will cross their graphs in more than one place.

C Functions Whose Inverses Are Also Functions

Because one-to-one functions do not repeat second coordinates, when we reverse the order of the ordered pairs in a one-to-one function, we obtain a relation in which no two ordered pairs have the same first coordinate — by definition, this relation must be a function. In other words, every one-to-one function has an inverse that is itself a function. Because of this, we can use function notation to represent that inverse.

Inverse Function Notation

If $y = f(x)$ is a one-to-one function, then the inverse of f is also a function and can be denoted by $y = f^{-1}(x)$.

To illustrate, in Example 1 we found that the inverse of $f(x) = 2x - 3$ was the function $y = \frac{x+3}{2}$. We can write this inverse function with inverse function notation as

$$f^{-1}(x) = \frac{x+3}{2}$$

On the other hand, the inverse of the function in Example 2 is not itself a function, so we do not use the notation $f^{-1}(x)$ to represent it.

Note The notation f^{-1} does not represent the reciprocal of f. That is, the -1 in this notation is not an exponent. The notation f^{-1} is defined as representing the inverse function for a one-to-one function.

Here are three important points about inverse functions:
1. The equation of the inverse is found by exchanging *x* and *y* in the equation of the function.
2. The graph of the inverse can be found by reflecting the graph of the function about the line $y = x$
3. A graph is not the graph of a function if a vertical line crosses it in more than one place.

The Inverse Sine Relation

To find the inverse of $y = \sin x$, we interchange *x* and *y* to obtain

$$x = \sin y$$

This is the equation of the inverse sine relation.

To graph $x = \sin, y$, we simply reflect the graph of $y = \sin x$ about the line $y = x$, as shown in figure 1.

As you can see from the graph, $x = \sin y$ is a relation but not a function. For every value of *x* in the domain, there are many values of *y*. The graph of $x = \sin y$ fails the vertical line test.

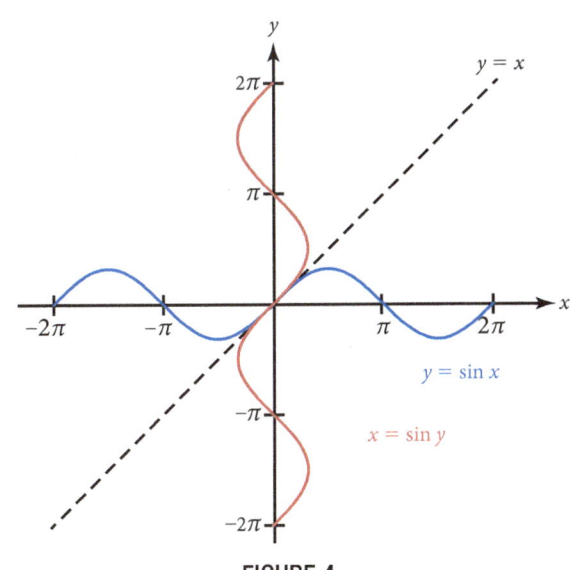

FIGURE 4

Inverse Trigonometric Functions

In order for the function $y = \sin x$ to have an inverse that is also a function, it is necessary to restrict the values that y can assume. The interval we restrict it to is the interval $-\frac{\pi}{2} \le y \le \frac{\pi}{2}$. Figure 2 contains the graph of $x = \sin y$ with the restricted interval.

It is apparent from Figure 5 that if $x = \sin y$ is restricted to the interval $-\frac{\pi}{2} \le y \le \frac{\pi}{2}$, then each value of x is associated with exactly one value of y and we have a function rather than just a relation. (That is, on the interval $-\frac{\pi}{2} \le y \le \frac{\pi}{2}$, the graph of $x = \sin y$ is such that no vertical line crosses it in more than one place.) The equation $x = \sin y$, together with the restriction $-\frac{\pi}{2} \le y \le \frac{\pi}{2}$, form the inverse sine function. To designate this function we use the following notation.

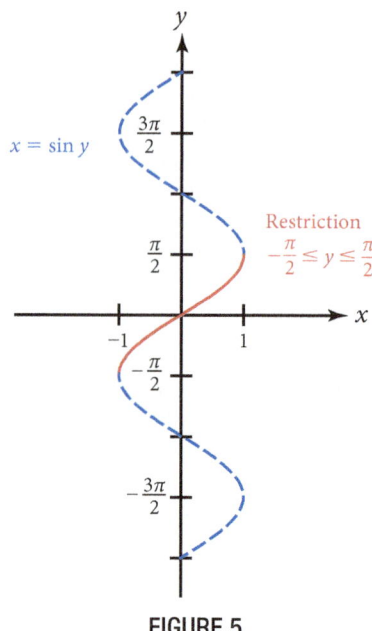

FIGURE 5

Notation

The notation used to indicate the **inverse sine function** is as follows:

Notation	Meaning
$y = \sin^{-1}x$ or $y = \arcsin x$	$x = \sin y$ and $-\frac{\pi}{2} \le y \le \frac{\pi}{2}$

In this section, we will limit our discussion of inverse trigonometric functions to the inverses of the three major functions: sine, cosine, and tangent. The other three inverse trigonometric functions can be handled with the use of reciprocals.

Figure 6 shows the graphs of $y = \cos^{-1}x$ and $y = \tan^{-1}x$ and the restrictions that allow them to become functions instead of just relations.

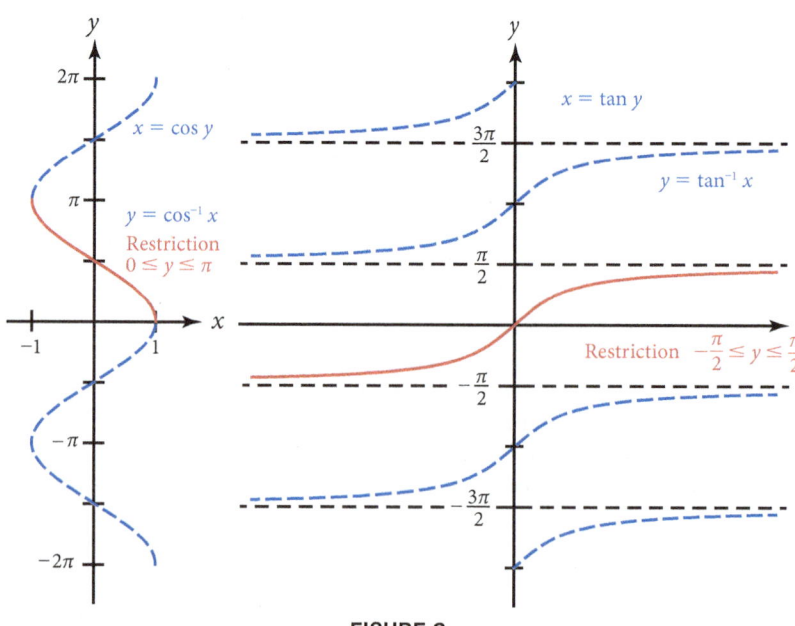

FIGURE 6

To summarize, here is the definition for the three major inverse trigonometic functions.

Definition
The inverse trigonometric functions for $y = \sin x$, $y = \cos x$, and $y = \tan x$ are as follows:

Inverse Function	Meaning
$y = \sin^{-1} x$ or $y = \arcsin x$	$x = \sin y$ and $-\dfrac{\pi}{2} \le y \le \dfrac{\pi}{2}$
	In Words: y is the angle between $-\dfrac{\pi}{2}$ and $\dfrac{\pi}{2}$, inclusive, whose sine is x.
$y = \cos^{-1} x$ or $y = \arccos x$	$x = \cos y$ and $0 \le y \le \pi$
	In Words: y is the angle between 0 and π, inclusive, whose cosine is x.
$y = \tan^{-1} x$ or $y = \arctan x$	$x = \tan y$ and $-\dfrac{\pi}{2} < y < \dfrac{\pi}{2}$
	In Words: y is the angle between $-\dfrac{\pi}{2}$ and $\dfrac{\pi}{2}$ whose tangent is x.

D Evaluating Trigonometric Functions

EXAMPLE 3   Evaluate in radians without using calculator.

a. $\sin^{-1} \dfrac{1}{2}$
b. $\arccos\left(-\dfrac{\sqrt{3}}{2}\right)$
c. $\tan^{-1}(-1)$

Solution

a. The angle between $-\dfrac{\pi}{2}$ and $\dfrac{\pi}{2}$ whose sine is $\dfrac{1}{2}$ is $\dfrac{\pi}{6}$.

$$\sin^{-1} \frac{1}{2} = \frac{\pi}{6}$$

b. The angle between 0 and π with a cosine of $-\dfrac{\sqrt{3}}{2}$ is $\dfrac{5\pi}{6}$.

$$\arccos\left(-\frac{\sqrt{3}}{2}\right) = \frac{5\pi}{6}$$

c. The angle between $-\dfrac{\pi}{2}$ and $\dfrac{\pi}{2}$ with a tangent of -1 is $-\dfrac{\pi}{4}$.

$$\tan^{-1}(-1) = -\frac{\pi}{4}$$

It would be incorrect to give the answer as $\frac{7\pi}{4}$. It is true that $\tan \frac{7\pi}{4} = -1$, but $\frac{7\pi}{4}$ is not between $-\frac{\pi}{2}$ and $\frac{\pi}{2}$.

Note   The notation $\sin^{-1}x$ is not to be interpreted as meaning the reciprocal of $\sin x$. That is,

$$\sin^{-1}x \ne \frac{1}{\sin x}$$

If we want the reciprocal of $\sin x$, we use $\csc x$ or $(\sin x)^{-1}$, but never $\sin^{-1}x$.

EXAMPLE 4 Use a calculator to evaluate each expression to the nearest tenth of a degree.

a. arcsin 0.5075

b. arcsin (-0.5075)

c. $\cos^{-1} 0.6428$

d. $\cos^{-1} (-0.6428)$

e. arctan 4.474

f. arctan (-4.474)

Solution The easiest method of evaluating these expressions is to use a calculator. Make sure the calculator is set to the degree mode, and then enter the number and push the appropriate keys. Scientific calculators are programmed so that the restrictions on the inverse trigonometric functions are automatic. We just push the keys and the rest is taken care of.

a. arcsin $0.5075 \approx 30.5°$
b. arcsin $(-0.5075) \approx -30.5°$ $\Big\}$ Reference angle 30.5°

c. $\cos^{-1} 0.6428 \approx 50.0°$
d. $\cos^{-1} (-0.6428) \approx 130.0°$ $\Big\}$ Reference angle 50°

e. arctan $4.474 \approx 77.4°$
f. arctan $(-4.474) \approx -77.4°$ $\Big\}$ Reference angle 77.4°

EXAMPLE 5 Evaluate $\sin\left(\tan^{-1}\dfrac{3}{4}\right)$ without using a calculator.

Solution We begin by letting $\theta = \tan^{-1} \frac{3}{4}$. (Remember, $\tan^{-1} x$ is the angle whose tangent is x.) Then we have:

$$\text{If } \theta = \tan^{-1}\frac{3}{4} \quad \text{then} \quad \tan\theta = \frac{3}{4} \quad \text{and} \quad 0° < \theta < 90°$$

We can draw a triangle in which one of the acute angles is θ. We imagine that this is the reference triangle for θ, if θ was in standard position. Since $\tan\theta = \frac{3}{4}$, we label the side opposite θ with 3 and the side adjacent to θ with 4. (If we imagine that θ is in standard position, the side opposite θ is the y-coordinate of a point on the terminal side of θ, while the side adjacent to θ is the x-coordinate.) The hypotenuse is found by applying the Pythagorean Theorem, as shown in Figure 7.

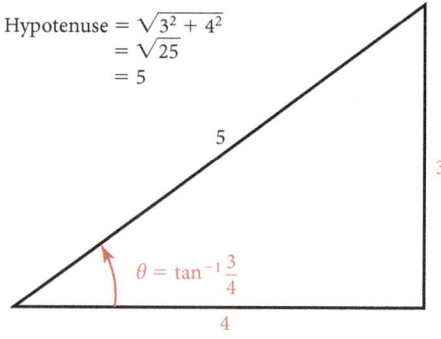

Hypotenuse $= \sqrt{3^2 + 4^2}$
$= \sqrt{25}$
$= 5$

$\theta = \tan^{-1}\dfrac{3}{4}$

FIGURE 7

From Figure 7 we find $\sin\theta$ using the ratio of the side opposite θ to the hypotenuse.

$$\sin\left(\tan^{-1}\frac{3}{4}\right) = \sin\theta = \frac{3}{5}$$

Calculator Note If we were to do the same problem with the aid of a calculator, the sequence would look like this:

Scientific Calculator

3 ÷ 4 = $\boxed{\tan^{-1}}$ $\boxed{\sin}$

Graphing Calculator

$\boxed{\sin}$ $\boxed{\tan^{-1}}$ (3 ÷ 4) $\boxed{\text{ENTER}}$

The display would read 0.6, which is $\frac{3}{5}$.

Although it is a lot easier to use a calculator on problems like the one in Example 5, solving it without a calculator will be of more use to you in the future.

EXAMPLE 6   Write the expression $\sin(\cos^{-1} x)$ as an equivalent expression in x only. (Assume x is positive.)

Solution   We let $\theta = \cos^{-1} x$, then $\cos \theta = x$. To help visualize the problem we draw a right triangle with an acute angle of θ and label it so that $\cos \theta = x$. This is accomplished by labeling the side adjacent to θ with x and the hypotenuse with 1. That way, the ratio of the side adjacent θ to the hypotenuse is $\frac{x}{1} = x$. The side opposite θ is found by applying the Pythagorean theorem.

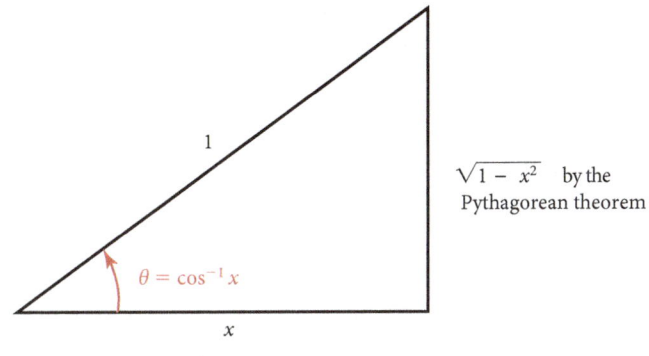

$$\sqrt{1 - x^2} \quad \text{by the Pythagorean theorem}$$

$\theta = \cos^{-1} x$

x

FIGURE 8

Finding $\sin \theta$ is simply a matter of taking the raio of the opposite to the hypotenuse.

$$\sin(\cos^{-1} x) = \sin \theta = \frac{\sqrt{1 - x^2}}{1} = \sqrt{1 - x^2}$$

Note that our assumption that x was positive isn't really needed, but it does assist us in drawing the triangle. If x was negative, then $\cos^{-1} x$ is between 90° and 180°, in which case the sine is still positive, so $\sin(\cos^{-1} x)$ is always positive.

Problem Set 4.6

Vocabulary

Use the vocabulary words and formulas below to fill in the blanks in the sentences.

horizontal   inverse   function   restrict

1. To test to see if a function is a one-to-one function, we can use the _____ line test.

2. If a function is a one-to-one function, then its inverse is also a _____ .

3. The graph of $y = \sin x$ is not a one-to-one function, therefore its _____ is not a function.

4. To insure that $y = \sin x$ has an inverse that is also a function, we _____ the domain in its inverse, $x = \sin y$, to $-\frac{\pi}{2} \le y \le \frac{\pi}{2}$.

For each equation below, write an equation for its inverse, and then sketch the graph of the equation and its inverse on the same coordinate system.

1. $y = x^2 + 4$ 2. $y = x^2 - 3$ 3. $y = 2x^2$ 4. $y = \frac{1}{2}x^2$

5. $y = 3x - 2$ 6. $y = 2x + 3$ 7. $x^2 + y^2 = 9$ 8. $x^2 + y^2 = 4$

9. $y = 3^x$ 10. $y = 4^x$

11. Graph $y = \cos x$ between -2π and 2π, and then reflect the graph about the line $y = x$ to obtain the graph of $x = \cos y$.

12. Graph $y = \sin x$ between $-\frac{\pi}{2}$ and $\frac{\pi}{2}$, and then reflect the graph about the line $y = x$ to obtain the graph of $y = \sin^{-1} x$ between $-\frac{\pi}{2}$ and $\frac{\pi}{2}$.

13. Graph $y = \tan x$ for x between $-\frac{3\pi}{2}$ and $\frac{3\pi}{2}$, and then reflect the graph about the line $y = x$ to obtain the graph of $x = \tan y$.

14. Graph $y = \cot x$ for x between 0 and 2π, and then reflect the graph about the line $y = x$ to obtain the graph of $x = \cot y$.

The domain of a function is the set of all first coordinates (the set of all values that x can assume). The range is the set of all second coordinates (all the values that y can assume).

15. What is the domain of the function $y = \sin^{-1} x$?

16. What is the range of the function $y = \sin^{-1} x$?

17. What is the range of the function $y = \arccos x$?

18. What is the domain of the function $y = \arccos x$?

Simplify.

19. $\sin^{-1} \dfrac{\sqrt{3}}{2}$ 20. $\cos^{-1} \dfrac{1}{2}$ 21. $\cos^{-1} (-1)$ 22. $\cos^{-1} 0$

23. $\tan^{-1} 1$ 24. $\tan^{-1} 0$ 25. $\arccos\left(-\dfrac{1}{\sqrt{2}}\right)$ 26. $\arccos 1$

27. $\sin^{-1}\left(-\dfrac{1}{2}\right)$ 28. $\sin^{-1} \dfrac{1}{\sqrt{2}}$ 29. $\arctan \sqrt{3}$ 30. $\arctan \dfrac{1}{\sqrt{3}}$

31. $\arcsin 0$ 32. $\arcsin\left(-\dfrac{\sqrt{3}}{2}\right)$ 33. $\tan^{-1}\left(-\dfrac{1}{\sqrt{3}}\right)$ 34. $\tan^{-1}(-\sqrt{3})$

35. $\cos^{-1}\left(-\dfrac{1}{2}\right)$ 36. $\sin^{-1} 1$ 37. $\arccos \dfrac{\sqrt{3}}{2}$ 38. $\arcsin(-1)$

Use a calculator to evaluate each expression to the nearest tenth of a degree.

39. $\sin^{-1} 0.1702$

40. $\sin^{-1}(-0.1702)$

41. $\cos^{-1}(-0.8425)$

42. $\cos^{-1} 0.8425$

43. $\tan^{-1} 0.3799$

44. $\tan^{-1}(-0.3799)$

45. $\arcsin 0.9627$

46. $\arccos 0.9627$

47. $\cos^{-1}(-0.4664)$

48. $\sin^{-1}(-0.4664)$

49. $\arctan(-2.748)$

50. $\arctan(-0.3640)$

51. $\sin^{-1}(-0.7660)$

52. $\cos^{-1}(-0.7660)$

Evaluate without using a calculator.

53. $\cos\left(\tan^{-1}\frac{3}{4}\right)$

54. $\csc\left(\tan^{-1}\frac{3}{4}\right)$

55. $\tan\left(\sin^{-1}\frac{3}{5}\right)$

56. $\tan\left(\cos^{-1}\frac{3}{5}\right)$

57. $\sec\left(\cos^{-1}\frac{1}{\sqrt{5}}\right)$

58. $\sin\left(\cos^{-1}\frac{1}{\sqrt{5}}\right)$

59. $\sin\left(\cos^{-1}\frac{1}{2}\right)$

60. $\cos\left(\sin^{-1}\frac{1}{2}\right)$

61. $\cot\left(\tan^{-1}\frac{1}{2}\right)$

62. $\cot\left(\tan^{-1}\frac{1}{3}\right)$

63. $\sin\left(\sin^{-1}\frac{3}{5}\right)$

64. $\cos\left(\cos^{-1}\frac{3}{5}\right)$

65. $\cos\left(\cos^{-1}\frac{1}{2}\right)$

66. $\sin\left(\sin^{-1}\frac{1}{\sqrt{2}}\right)$

67. $\tan\left(\tan^{-1}\frac{1}{2}\right)$

68. $\tan\left(\tan^{-1}\frac{3}{4}\right)$

For each expression below, write an equivalent expression that involves x only. (Assume x is positive.)

69. $\cos(\cos^{-1} x)$

70. $\sin(\sin^{-1} x)$

71. $\cos(\sin^{-1} x)$

72. $\tan(\cos^{-1} x)$

73. $\sin(\tan^{-1} x)$

74. $\cos(\tan^{-1} x)$

Maintaining Your Skills

Name the reference angle.

75. 235°

76. 117.8°

Convert each of the following to degree measure:

77. $\dfrac{4\pi}{3}$

78. $\dfrac{7\pi}{12}$

Give the exact value of each of the following:

79. $\cos\dfrac{2\pi}{3}$

80. $\csc\dfrac{5\pi}{6}$

81. $\theta = \frac{\pi}{6}$ is a central angle in a circle of radius 12 meters. Find the length of arc s cut off by θ.

82. $\theta = \frac{\pi}{4}$ is a central angle that cuts off an arc of length π. Find the radius of the circle.

83. A central angle of 4 radians cuts off an arc of length 8 inches. Find the area of the sector formed.

84. A point is rotating with uniform circular motion on a circle of radius r. Find ω if $r = 10$ centimeters and $v = 5$ centimeters per second.

Spotlight on Success

CJ, Student Instructor

We are what we repeatedly do. Excellence, then, is not an act, but a habit.
—Aristotle

Something that has worked for me in college, in addition to completing the assigned homework, is working on some extra problems from each section. Working on these extra problems is a great habit to get into because it helps further your understanding of the material, and you see the many different types of problems that can arise. If you have completed every problem that your book offers, and you still don't feel confident that you have a full grasp of the material, look for more problems. Many problems can be found online or in other books. Your professors may even have some problems that they would suggest doing for extra practice.

The biggest benefit to working all the problems in the course's assigned textbook is that often teachers will choose problems either straight from the book or ones similar to problems that were not assigned for tests. Doing this will ensure that you do your best in all your classes.

Examples

1. Since $\sin(x + 2\pi) = \sin x$, the function $y = \sin x$ is periodic with period 2π. Likewise, since $\tan(x + \pi) = \tan x$, the function $y = \tan x$ is periodic with period π.

Periodic Functions [4.1]

A function $y = f(x)$ is said to be periodic with period p if p is the smallest positive number such that $f(x + p) = f(x)$ for all x in the domain of f.

2. a.

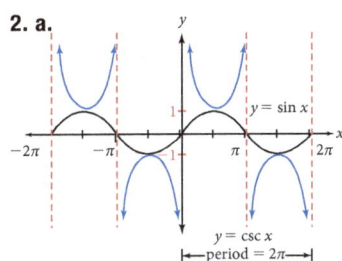

b.

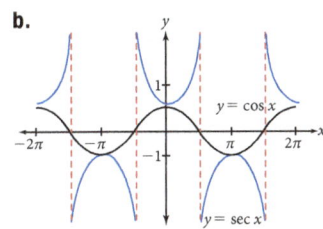

c.

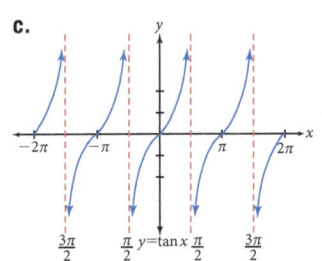

Basic Graphs [4.1]

The graphs of $y = \sin x$ and $y = \cos x$ are both periodic with period 2π. The amplitude of each graph is 1. The sine curve passes through 0 on the y-axis, while the cosine curve passes through 1 on the y-axis.

The graphs of $y = \csc x$ and $y = \sec x$ are also periodic with period 2π. We graph them by using the fact that they are reciprocals of sine and cosine. Since there is no largest or smallest value of y, we say the secant and cosecant curves have no amplitude.

The graphs of $y = \tan x$ and $y = \cot x$ are periodic with period π. The tangent curve passes through the origin, while the cotangent is undefined when x is 0. There is no amplitude for either graph.

d.

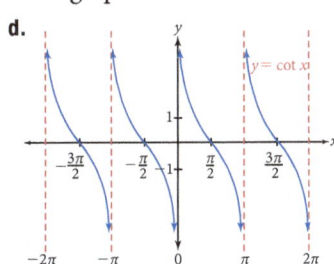

3.

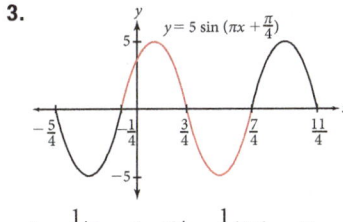

$$A = \frac{1}{2}|5 - (-5)| = \frac{1}{2}(10) = 5$$

Amplitude [4.2]

The *amplitude* A of a curve is half the absolute value of the difference between the largest value of y, denoted by M, and the smallest value of y, denoted by m.

$$A = \frac{1}{2}|M - m|$$

4. The phase shift for the graph in Example 3 is $-1/4$.

Phase Shift [4.3]

The *phase shift* for a sine or cosine curve, is the distance the curve has moved right or left from the curve $y = \sin x$ or $y = \cos x$. For example, we usually think of the graph of $y = \sin x$ as starting at the origin. If we graph another sine curve that starts at $\pi/4$, then we say this curve has a phase shift of $\pi/4$.

5. a.

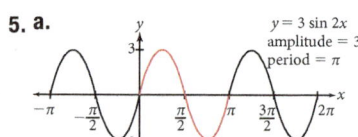

$y = 3 \sin 2x$
amplitude = 3
period = π

b.

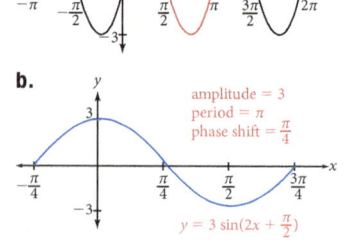

amplitude = 3
period = π
phase shift = $\frac{\pi}{4}$

$y = 3 \sin(2x + \frac{\pi}{2})$

Graphing Sine and Cosine Curves [4.2, 4.3]

The graphs of $y = A \sin(Bx + C)$ and $y = A \cos(Bx + C)$, where $B > 0$, will have the following characteristics:

$$\text{Amplitude} = |A|$$

$$\text{Period} = \frac{2\pi}{B}$$

$$\text{Phase shift} = -\frac{C}{B}$$

To graph one of these curves, we first find the phase shift and label that point on the x-axis (this will be our starting point). We then add the period to the phase shift and mark the result on the x-axis (this is our ending point). We mark the y-axis with the amplitude. Finally, we sketch in one complete cycle of the curve in question keeping in mind that, if A is negative, the graph must be reflected about the x-axis.

6.

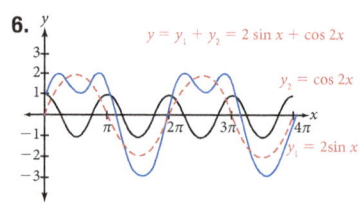

$y = y_1 + y_2 = 2 \sin x + \cos 2x$

$y_2 = \cos 2x$

$y_1 = 2\sin x$

Graphing by Addition of y-Coordinates [4.5]

To graph equations of the form $y = y_1 + y_2$, where y_1 and y_2 are algebraic or trigonometric functions of x, we graph y_1 and y_2 separately on the same coordinate system and then add the two graphs to obtain the graph of y.

7. Evaluate in radians without using a calculator.

a. $\sin^{-1} \frac{1}{2}$

The angle between $-\frac{\pi}{2}$ and $\frac{\pi}{2}$ whose sine is $\frac{1}{2}$ is $\frac{\pi}{6}$.

$$\sin^{-1} \frac{1}{2} = \frac{\pi}{6}$$

b. $\arccos\left(-\frac{\sqrt{3}}{2}\right)$

The angle between 0 and π with a cosine of $-\frac{\sqrt{3}}{2}$ is $\frac{5\pi}{6}$.

$$\arccos\left(-\frac{\sqrt{3}}{2}\right) = \frac{5\pi}{6}$$

Inverse Trigonometric Functions [4.6]

Inverse Function	Meaning
$y = \sin^{-1}x$ or $y = \arcsin x$	$x = \sin y$ *and* $-\frac{\pi}{2} \leq y \leq \frac{\pi}{2}$
	In Words: y is the angle between $-\frac{\pi}{2}$ and $\frac{\pi}{2}$, inclusive, whose sine is x.
$y = \cos^{-1}x$ or $y = \arccos x$	$x = \cos y$ and $0 \leq y \leq \pi$
	In Words: y is the angle between 0 and π, inclusive, whose cosine is x.
$y = \tan^{-1}x$ or $y = \arctan x$	$x = \tan y$ and $-\frac{\pi}{2} < y < \frac{\pi}{2}$
	In Words: y is the angle between $-\frac{\pi}{2}$ and $\frac{\pi}{2}$ whose tangent is x.

Graph each of the following between $x = -4\pi$ and $x = 4\pi$.

1. $y = \sin x$

2. $y = \cos x$

3. $y = \tan x$

4. $y = \sec x$

5. How many complete cycles of the graph of the equation $y = \sin x$ are shown in your answer to Problem 1?

6. How many complete cycles of the graph of the equation $y = \tan x$ are shown in your answer to Problem 3?

7. Use your answer to Problem 4 to find all values of x between -4π and 4π for which $\sec x = -1$.

8. Use your answer to Problem 2 to find all values of x between -4π and 4π for which $\cos x = \frac{1}{2}$.

For each equation below, first identify the amplitude and period and then use this information to sketch one complete cycle of the graph.

9. $y = \cos \pi x$

10. $y = -3 \cos x$

Graph each of the following on the given interval.

11. $y = 3 \sin 2x, \ -\pi \leq x \leq 2\pi$

12. $y = 2 \sin \pi x, \ -4 \leq x \leq 4$

For each equation below, identify the amplitude, period, and phase shift and then use this information to sketch one complete cycle of the graph.

13. $y = \sin\left(x + \dfrac{\pi}{4}\right)$

14. $y = \cos\left(x - \dfrac{\pi}{2}\right)$

15. $y = 3 \sin\left(2x - \dfrac{\pi}{3}\right)$

16. $y = 3 \sin\left(\dfrac{\pi}{3}x - \dfrac{\pi}{3}\right)$

17. $y = \csc\left(x + \dfrac{\pi}{4}\right)$

18. $y = \tan\left(2x - \dfrac{\pi}{2}\right)$

Graph each of the following on the given interval.

19. $y = 2 \sin(3x - \pi), \ -\dfrac{\pi}{3} \leq x \leq \dfrac{5\pi}{3}$

20. $y = 2 \sin\left(\dfrac{\pi}{2}x - \dfrac{\pi}{4}\right), \ -\dfrac{1}{2} \leq x \leq \dfrac{13}{2}$

Sketch the following between $x = 0$ and $x = 4\pi$.

21. $y = \dfrac{1}{2}x - \sin x$

22. $y = \sin x + \cos 2x$

Find all values of y, in degrees, that satisfy each of the following expressions:

23. $y = \sin^{-1} 0$

24. $y = \cos^{-1}(-1)$

25. Graph $y = \cos^{-1} x$

26. Graph $y = \arcsin x$

Evaluate each expression without using a calculator or tables and write your answer in radians.

27. $\sin^{-1}\left(\dfrac{1}{2}\right)$

28. $\cos^{-1}\left(-\dfrac{\sqrt{3}}{2}\right)$

29. $\arctan(-1)$

30. $\arcsin(1)$

Use a calculator or tables to evaluate each expression to the nearest tenth of a degree.

31. $\arcsin(0.5934)$

32. $\arctan(-0.8302)$

33. $\arccos(-0.6981)$

34. $\arcsin(-0.2164)$

Evaluate without using a calculator.

35. $\tan\left(\cos^{-1}\dfrac{2}{3}\right)$

36. $\cos\left(\tan^{-1}\dfrac{2}{3}\right)$

For each expression below, write an equivalent expression that involves x only. (Assume x is positive.)

37. $\sin(\cos^{-1} x)$

38. $\tan(\sin^{-1} x)$

Identities and Formulas

5

WHAT ARE IDENTITIES AND FORMULAS? Identities and formulas are mathematical expressions that establish relationships between the trigonometric functions and their various properties. These identities and formulas are used to simplify expressions, solve equations, and prove mathematical statements.

WHY ARE IDENTITIES AND FORMULAS IMPORTANT? Identities and formulas in trigonometry provide a framework for solving real-world problems and applications in various fields such as renewable energy, civil engineering, navigation, mechanical engineering, and astronomy.

WHERE MIGHT YOU SEE IDENTITIES AND FORMULAS? Here are some examples of where you may see identities and formulas in the real world:

BRIDGE CONSTRUCTION In the construction of a bridge, an engineer needs to determine the length of a diagonal support cable. If the vertical and horizontal distances between the anchor points of the cable are 50 meters and 70 meters respectively, find the length of the support cable using the Pythagorean trigonometric identity.

NAVIGATION A ship is navigating in a straight line at a constant speed. The ship's captain needs to adjust the course of the ship due to ocean currents. If the current makes the ship drift kilometers south for every 10 kilometers east it travels, determine the angle at which the ship should steer to maintain its intended course.

Scenario: Solar Efficiency

Solve a solar panels angles problem, given the following:

1. Angle of inclination of the solar panel is 30°.

2. Efficiency of the solar panel is $\cos^2(\theta)$, where θ is the angle between the sunlight and the normal to the panel.

© istockphoto.com Halfpoint

Solution We need to find the efficiency of the solar panel. First, let's understand the scenario. The angle of inclination of the solar panel refers to the angle between the panel and the horizontal surface. Let's call this angle $\alpha = 30°$.

The angle between the sunlight and the normal to the panel (θ) can be determined using the relationship between angles in a right triangle. Since the angle of inclination is 30°, the angle between the panel and the horizontal surface is also 30°. Therefore, $\theta = 90° - \alpha = 90° - 30° = 60°$.

Now, we can calculate the efficiency of the solar panel using $\cos^2(\theta)$: Efficiency = $\cos^2(60°)$.

Using the identity $\cos^2(\theta) = \dfrac{1 + \cos(2\theta)}{2}$:

$$\cos^2(60°) = \frac{1 + \cos(2 \cdot 60°)}{2} = \frac{1 + \cos(120°)}{2} = \frac{1 - \frac{1}{2}}{2} = \frac{1}{4}$$

So the efficiency of the solar panel is $\frac{1}{4}$ or 25%. Therefore, the efficiency of the solar panel when the angle of inclination is 30° is 25%.

Learning Objectives

A Prove trigonometric identities.

We began proving identities in Chapter 1. In this section, we will extend the work we did in Chapter 1 to include proving more complicated identities. For review, here are the basic identities and some of their more important equivalent forms.

Table 1

	Basic Identities	Common Equivalent Forms
Reciprocal	$\csc \theta = \dfrac{1}{\sin \theta}$	$\sin \theta = \dfrac{1}{\csc \theta}$
	$\sec \theta = \dfrac{1}{\cos \theta}$	$\cos \theta = \dfrac{1}{\sec \theta}$
	$\cot \theta = \dfrac{1}{\tan \theta}$	$\tan \theta = \dfrac{1}{\cot \theta}$
Ratio	$\tan \theta = \dfrac{\sin \theta}{\cos \theta}$	
	$\cot \theta = \dfrac{\cos \theta}{\sin \theta}$	
Pythagorean	$\cos^2 \theta + \sin^2 \theta = 1$ $1 + \tan^2 \theta = \sec^2 \theta$ $1 + \cot^2 \theta = \csc^2 \theta$	$\sin^2 \theta = 1 - \cos^2 \theta$ $\sin \theta = \pm \sqrt{1 - \cos^2 \theta}$ $\cos^2 \theta = 1 - \sin^2 \theta$ $\cos \theta = \pm \sqrt{1 - \sin^2 \theta}$

Next we want to use the eight basic identities and their equivalent forms to verify other trigonometric identities. To prove (or verify) that a trigonometric identity is true, we use trigonometric substitutions and algebraic manipulations to either

1. Transform the right side into the left side.

Or:

2. Transform the left side into the right side.

   The main thing to remember in proving identities is to work on each side of the identity separately. We do not want to use properties from algebra that involve both sides of the identity—such as the addition property of equality. We prove identities in order to develop the ability to transform one trigonometric expression into another. When we encounter problems in other courses that require the use of the techniques used to verify identities, we usually find that the solution to these problems hinges upon transforming an expression containing trigonometric functions into a less complicated expression. In these cases, we do not usually have an equal sign to work with.

EXAMPLE 1 Verify the identity: $\sin\theta\cot\theta = \cos\theta$

Proof To prove this identity we transform the left side into the right side:

$$\sin\theta\cot\theta = \sin\theta\cdot\frac{\cos\theta}{\sin\theta} \qquad \text{Ratio identity}$$

$$= \frac{\sin\theta\cos\theta}{\sin\theta} \qquad \text{Multiply.}$$

$$= \cos\theta \qquad \text{Divide out common factor } \sin\theta.$$

EXAMPLE 2 Prove: $\tan x + \cos x = \sin x(\sec x + \cot x)$.

Proof We begin by applying the distributive property to the right side of the identity. Then we change each expression on the right side to an equivalent expression involving only $\sin x$ and $\cos x$.

$$\sin x(\sec x + \cot x) = \sin x\sec x + \sin x\cot x \qquad \text{Multiply.}$$

$$= \sin x\cdot\frac{1}{\cos x} + \sin x\cdot\frac{\cos x}{\sin x} \qquad \text{Reciprocal and ratio identities}$$

$$= \frac{\sin x}{\cos x} + \cos x \qquad \text{Multiply and divide out common factor } \sin x.$$

$$= \tan x + \cos x \qquad \text{Ratio identity}$$

In this case, we transformed the right side into the left side.

Before we go on to the next example, let's list some guidelines that may be useful in learning how to prove identities. Probably the best advice is to remember that these are simply guidelines. The best way to become proficient at proving trigonometric identities is to practice. The more identities you prove, the more you will be able to prove and the more confident you will become. Don't be afraid to stop and start over if you don't seem to be getting anywhere. With most identities, there are a number of different proofs that will lead to the same result. Some of the proofs will be longer than others.

> **Guidelines for Proving Identities**
> 1. It is usually best to work on the more complicated side first.
> 2. Look for trigonometric substitutions involving the basic identities that may help simplify things.
> 3. Look for algebraic operations, such as adding fractions, the distributive property, or factoring, that may simplify the side you are working with or that will at least lead to an expression that will be easier to simplify.
> 4. If you cannot think of anything else to do, change everything to sines and cosines and see if that helps.
> 5. Always keep an eye on the side you are not working with to be sure you are working toward it. There is a certain sense of direction that accompanies a successful proof.

EXAMPLE 3 Prove: $\dfrac{\cos^4 t - \sin^4 t}{\cos^2 t} = 1 - \tan^2 t$.

Proof In this example, factoring the numerator on the left side will reduce the exponents there from 4 to 2.

$$\frac{\cos^4 t - \sin^4 t}{\cos^2 t} = \frac{(\cos^2 t + \sin^2 t)(\cos^2 t - \sin t)}{\cos^2 t} \qquad \text{Factor}$$

$$= \frac{1(\cos^2 t - \sin^2 t)}{\cos^2 t} \qquad \text{Pythagorean identity}$$

$$= \frac{\cos^2 t}{\cos^2 t} - \frac{\sin^2 t}{\cos^2 t} \qquad \text{Separate into two fractions.}$$

$$= 1 - \tan^2 t \qquad \text{Ratio identity}$$

EXAMPLE 4   Prove: $1 + \cos\theta = \dfrac{\sin^2\theta}{1 - \cos\theta}$.

Proof We begin by applying an alternative form of the Pythagorean identity to the right side to write $\sin^2\theta$ as $1 - \cos^2\theta$. Then we factor $1 - \cos^2\theta$ and reduce to lowest terms.

$$\dfrac{\sin^2\theta}{1 - \cos\theta} = \dfrac{1 - \cos^2\theta}{1 - \cos\theta} \qquad \text{Pythagorean identity}$$

$$= \dfrac{(1 - \cos\theta)(1 + \cos\theta)}{1 - \cos\theta} \qquad \text{Factor.}$$

$$= 1 + \cos\theta \qquad \text{Reduce.}$$

EXAMPLE 5   Prove $\tan x + \cot x = \sec x \csc x$.

Proof We begin by rewriting the left side in terms of $\sin x$ and $\cos x$. Then we simplify by finding a common denominator, changing to equivalent fractions, and adding.

$$\tan x + \cot x = \dfrac{\sin x}{\cos x} + \dfrac{\cos x}{\sin x} \qquad \text{Change to expressions in } \sin x \text{ and } \cos x.$$

$$= \dfrac{\sin x}{\cos x} \cdot \dfrac{\sin x}{\sin x} + \dfrac{\cos x}{\sin x} \cdot \dfrac{\cos x}{\cos x} \qquad \text{LCD}$$

$$= \dfrac{\sin^2 x + \cos^2 x}{\cos x \sin x} \qquad \text{Add fractions.}$$

$$= \dfrac{1}{\cos x \sin x} \qquad \text{Pythagorean identity}$$

$$= \dfrac{1}{\cos x} \cdot \dfrac{1}{\sin x} \qquad \text{Write as separate fractions.}$$

$$= \sec x \csc x \qquad \text{Reciprocal identities}$$

EXAMPLE 6   Prove: $\dfrac{\sin A}{1 + \cos A} + \dfrac{1 + \cos A}{\sin A} = 2\csc A$.

Proof The LCD for the left side is $\sin A(1 + \cos A)$.

$$\dfrac{\sin A}{1 + \cos A} + \dfrac{1 + \cos A}{\sin A} = \dfrac{\sin A}{\sin A} \cdot \dfrac{\sin A}{1 + \cos A} + \dfrac{1 + \cos A}{\sin A} \cdot \dfrac{1 + \cos A}{1 + \cos A} \qquad \text{LCD}$$

$$= \dfrac{\sin^2 A + (1 + \cos A)^2}{\sin A(1 + \cos A)} \qquad \text{Add fractions.}$$

$$= \dfrac{\sin^2 A + 1 + 2\cos A + \cos^2 A}{\sin A(1 + \cos A)} \qquad \text{Expand } (1 + \cos A)^2.$$

$$= \dfrac{2 + 2\cos A}{\sin A(1 + \cos A)} \qquad \text{Pythagorean identity}$$

$$= \dfrac{2(1 + \cos A)}{\sin A(1 + \cos A)} \qquad \text{Factor out 2.}$$

$$= \dfrac{2}{\sin A} \qquad \text{Reduce.}$$

$$= 2\csc A \qquad \text{Reciprocal identity}$$

EXAMPLE 7   Prove: $\dfrac{1 + \sin t}{\cos t} = \dfrac{\cos t}{1 - \sin t}$.

Proof The trick to proving this identity is to multiply the numerator and denominator on the right side by $1 + \sin t$.

$$\dfrac{\cos t}{1 - \sin t} = \dfrac{\cos t}{1 - \sin t} \cdot \dfrac{1 + \sin t}{1 + \sin t}$$ Multiply numerator and denominator by $1 + \sin t$.

$$= \dfrac{\cos t(1 + \sin t)}{1 - \sin^2 t}$$ Multiply out the denominator.

$$= \dfrac{\cos t(1 + \sin t)}{\cos^2 t}$$ Pythagorean identity

$$= \dfrac{1 + \sin t}{\cos t}$$ Reduce.

Note that it would have been just as easy for us to verify this identity by multiplying the numerator and denominator on the left side by $1 - \sin t$.

The reason why we chose $1 + \sin t$ (called the conjugate of $1 - \sin t$) is to create the expression $1 - \sin^2 t$, which is replaced with $\cos^2 t$ in the following step. Occasionally we make a trigonometric expression more complicated in order to allow us to apply an identity.

Problem Set 5.1

Vocabulary

Use the vocabulary words below to fill in the blanks in the sentences.

complicated   Pythagorean   identity   ratio   reciprocal

1. An _____ in mathematics is a statement that two expressions are equal for all values of the variable for which they are defined.
2. The basic identity $\csc \theta = \frac{1}{\sin \theta}$ is called a _____ identity.
3. The basic identity $\tan \theta = \frac{\sin \theta}{\cos \theta}$ is called a _____ identity.
4. The basic identity $\sin^2 \theta + \cos^2 \theta = 1$ is a _____ identity.
5. When we verify an identity, we usually try to transform the more _____ expression into the simpler expression.

Prove that each of the following identities is true:

1. $\cos \theta \tan \theta = \sin \theta$

2. $\sec \theta \cot \theta = \csc \theta$

3. $\csc \theta \tan \theta = \sec \theta$

4. $\tan \theta \cot \theta = 1$

5. $\dfrac{\tan A}{\sec A} = \sin A$

6. $\dfrac{\cot A}{\csc A} = \cos A$

7. $\sec \theta \cot \theta \sin \theta = 1$

8. $\tan \theta \csc \theta \cos \theta = 1$

9. $\cos x(\csc x + \tan x) = \cot x + \sin x$

10. $\sin x(\sec x + \csc x) = \tan x + 1$

11. $\csc x(\cot x - 1) = \cot x(\csc x - \sec x)$

12. $\tan x(\cos x + \cot x) = \sin x + 1$

13. $\cos^2 x(1 + \tan^2 x) = 1$

14. $\sin^2 x(\cot^2 x + 1) = 1$

15. $(1 - \sin x)(1 + \sin x) = \cos^2 x$

16. $(1 - \cos x)(1 + \cos x) = \sin^2 x$

17. $\dfrac{\cos^4 t - \sin^4 t}{\sin^2 t} = \cot^2 t - 1$

18. $\dfrac{\sin^4 t - \cos^4 t}{\sin^2 t \cos^2 t} = \sec^2 t - \csc^2 t$

19. $1 + \sin \theta = \dfrac{\cos^2 \theta}{1 - \sin \theta}$

20. $1 - \sin \theta = \dfrac{\cos^2 \theta}{1 + \sin \theta}$

21. $\dfrac{1 - \sin^4 \theta}{1 + \sin^2 \theta} = \cos^2 \theta$

22. $\dfrac{1 - \cos^4 \theta}{1 + \cos^2 \theta} = \sin^2 \theta$

23. $\sec^2 \theta - \tan^2 \theta = 1$

24. $\csc^2 \theta - \cot^2 \theta = 1$

25. $\sec^4 \theta - \tan^4 \theta = \dfrac{1 + \sin^2 \theta}{\cos^2 \theta}$

26. $\csc^4 \theta - \cot^4 \theta = \dfrac{1 + \cos^2 \theta}{\sin^2 \theta}$

27. $\tan \theta - \cot \theta = \dfrac{\sin^2 \theta - \cos^2 \theta}{\sin \theta \cos \theta}$

28. $\sec \theta - \csc \theta = \dfrac{\sin \theta - \cos \theta}{\sin \theta \cos \theta}$

29. $\csc B - \sin B = \cot B \cos B$

30. $\sec B - \cos B = \tan B \sin B$

31. $\cot \theta \cos \theta + \sin \theta = \csc \theta$

32. $\tan \theta \sin \theta + \cos \theta = \sec \theta$

33. $\dfrac{\cos x}{1 + \sin x} + \dfrac{1 + \sin x}{\cos x} = 2 \sec x$

34. $\dfrac{\cos x}{1 + \sin x} - \dfrac{1 - \sin x}{\cos x} = 0$

35. $\dfrac{1}{1 + \cos x} + \dfrac{1}{1 - \cos x} = 2 \csc^2 x$

36. $\dfrac{1}{1 - \sin x} + \dfrac{1}{1 + \sin x} = 2 \sec^2 x$

37. $\dfrac{1 - \sec x}{1 + \sec x} = \dfrac{\cos x - 1}{\cos x + 1}$

38. $\dfrac{\csc x - 1}{\csc x + 1} = \dfrac{1 - \sin x}{1 + \sin x}$

39. $\dfrac{\cos t}{1 + \sin t} = \dfrac{1 - \sin t}{\cos t}$

40. $\dfrac{\sin t}{1 + \cos t} = \dfrac{1 - \cos t}{\sin t}$

41. $\dfrac{(1 - \sin t)^2}{\cos^2 t} = \dfrac{1 - \sin t}{1 + \sin t}$

42. $\dfrac{\sin^2 t}{(1 - \cos t)^2} = \dfrac{1 + \cos t}{1 - \cos t}$

43. $\dfrac{\sec \theta + 1}{\tan \theta} = \dfrac{\tan \theta}{\sec \theta - 1}$

44. $\dfrac{\csc \theta - 1}{\cot \theta} = \dfrac{\cot \theta}{\csc \theta + 1}$

45. Show that $\sin(A + B)$ is, in general, not equal to $\sin A + \sin B$ by substituting 30° for A and 60° for B in both expressions and simplifying.

46. Show that $\sin 2x \neq 2 \sin x$ by substituting 30° for x and then simplifying both sides.

Getting Ready for the Next Section

47. If $\sin A = \frac{3}{5}$ and A terminates in quadrant I, find $\cos A$ and $\tan A$.

48. If $\cos B = -\frac{5}{13}$ with B in quadrant III, find $\sin B$ and $\tan B$.

Give the exact value of each of the following:

49. $\sin \dfrac{\pi}{3}$ **50.** $\cos \dfrac{\pi}{3}$ **51.** $\cos \dfrac{\pi}{6}$ **52.** $\sin \dfrac{\pi}{6}$

Convert to degrees.

53. $\dfrac{\pi}{12}$

54. $\dfrac{5\pi}{12}$

Graph one complete cycle.

55. $y = 3 \cos 2x$

56. $y = 2 \sin \dfrac{\pi}{2} x$

KEY WORDS

sum and difference formulas

expansion formula

Learning Objectives

A   Find the value of trigonometric functions using sum and difference formulas.

The expression $\sin(A + B)$ and $\cos(A + B)$ occur frequently enough in mathematics that it is necessary to find expressions equivalent to them that involve sines and cosines of single angles. The most obvious question to begin with is $\sin(A + B) = \sin A + \sin B$?

The answer is no. Substituting almost any pair of numbers for A and B in the formula will yield a false statement. As a counterexample, we can let $A = 30°$ and $B = 60°$ in the formula above then simplify each side.

$$\sin(30° + 60°) = \sin 30° + \sin 60°$$

$$\sin 90° \overset{?}{=} \frac{1}{2} + \frac{\sqrt{3}}{2}$$

$$1 \neq \frac{1 + \sqrt{3}}{2}$$

The formula just doesn't work. The next question is, what are the formulas for $\sin(A + B)$ and $\cos(A + B)$? The answer to that question is what this section is all about. Let's start by deriving the formula for $\cos(A + B)$.

We begin by drawing A in standard position and then adding B and $-B$ to it. These angles are shown in Figure 1 in relation to the unit circle. The unit circle is the circle with its center at the origin with a radius of 1. Since the radius of the unit circle is 1, the point through which the terminal side of A passes will have coordinates $(\cos A, \sin A)$. [If P_2 in Figure 1 has coordinates (x, y), then by definition of $\sin A$, $\cos A$, and the unit circle, $\cos A = \frac{x}{r} = \frac{x}{1} = x$ and $\sin A = \frac{y}{r} = \frac{y}{1} = y$. Therefore, $(x, y) = (\cos A, \sin A)$.] The points on the unit circle through which the terminal sides of the other angles in Figure 1 pass are found in the same manner.

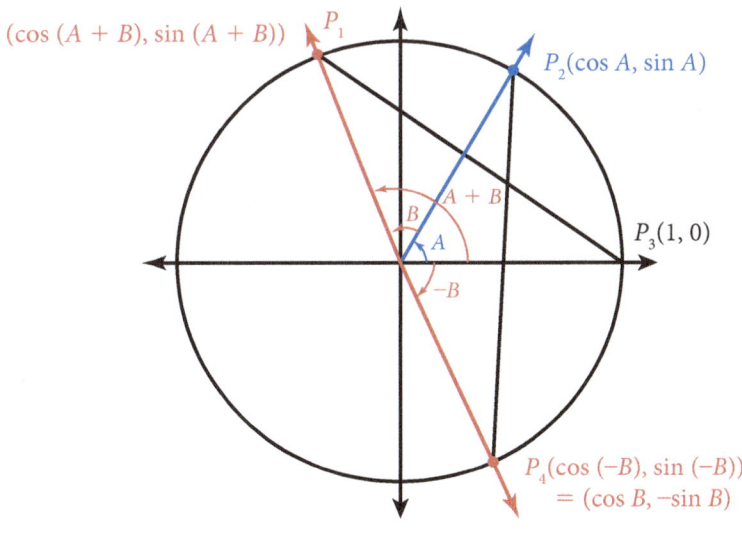

FIGURE 1

To derive the formula for $\cos(A + B)$, we simply have to see that line segment P_1P_3 is equal to line segment P_2P_4. (From geometry, they are chords cut off by equal central angles.)

$$\overline{P_1P_3} = \overline{P_2P_4}$$

Squaring both sides gives us

$$(\overline{P_1P_3})^2 = (\overline{P_2P_4})^2$$

Now, applying the distance formula, we have

$$[\cos(A + B) - 1]^2 + [\sin(A + B) - 0]^2 = (\cos A - \cos B)^2 + (\sin A + \sin B)^2$$

Let's call this Equation 1. Taking the left side of Equation 1, expanding it, and then simplifying by using the Pythagorean identity, gives us

Left side of Equation 1: $\cos^2(A + B) - 2\cos(A + B) + 1 + \sin^2(A + B)$ Expand squares.
$$= -2\cos(A + B) + 2$$ Pythagorean identity

Applying the same two steps to the right side of Equation 1 looks like this:

Right side of Equation 1: $\cos^2 A - 2\cos A \cos B + \cos^2 B + \sin^2 A + 2\sin A \sin B + \sin^2 B$
$$= -2\cos A \cos B + 2\sin A \sin B + 2$$

Equating the simplified versions of the left and right sides of Equation 1, we have

$$-2\cos(A + B) + 2 = -2\cos A \cos B + 2\sin A \sin B + 2$$

Adding -2 to both sides and then dividing both sides by -2 gives us the formula we are after.
This is the first formula in a series of formulas for trigonometric functions of the sum or difference of two angles. It must be memorized. Before we derive the others, let's look at some of the ways we can use our first formula.

$$\cos(A + B) = \cos A \cos B - \sin A \sin B$$

EXAMPLE 1   Find the exact value for $\cos 75°$.

Solution   We write $75°$ as $45° + 30°$ and then apply the formula for $\cos(A + B)$.

$$\cos 75° = \cos(45° + 30°)$$

$$= \cos 45° \cos 30° - \sin 45° \sin 30°$$

$$= \frac{\sqrt{2}}{2} \cdot \frac{\sqrt{3}}{2} - \frac{\sqrt{2}}{2} \cdot \frac{1}{2}$$

$$= \frac{\sqrt{6} - \sqrt{2}}{4}$$

EXAMPLE 2   Show that $\cos(x + 2\pi) = \cos x$.

Solution   Applying the formula for $\cos(A + B)$, we have

$$\cos(x + 2\pi) = \cos x \cos 2\pi - \sin x \sin 2\pi$$

$$= \cos x \cdot 1 - \sin x \cdot 0$$

$$= \cos x$$

Notice that this is not a new relationship. We already know that if two angles are coterminal, then their cosines are equal; and $x + 2\pi$ and x are coterminal. What we have done here is shown this to be true with a formula instead of the definition of cosine.

EXAMPLE 3   Write $\cos 3x \cos 2x - \sin 3x \sin 2x$ as a single cosine.

Solution We apply the formula for $\cos(A + B)$ in the reverse direction.

$$\cos 3x \cos 2x - \sin 3x \sin 2x = \cos(3x + 2x)$$

$$= \cos 5x$$

The formula for $\cos(A - B)$ is similar to the formula we have for $\cos(A + B)$. It looks like this:

$$\cos(A - B) = \cos A \cos B + \sin A \sin B$$

The only difference in the formulas for the expansion of $\cos(A + B)$ and $\cos(A - B)$ is the sign between the two terms. Here are both formulas again:

$$\cos(A + B) = \cos A \cos B - \sin A \sin B$$

$$\cos(A - B) = \cos A \cos B + \sin A \sin B$$

Again, both formulas are important and should be memorized.

EXAMPLE 4   Show that $\cos(90° - A) = \sin A$.

Solution We will need this formula when we derive the formula for $\sin(A + B)$.

$$\cos(90° - A) = \cos 90° \cos A + \sin 90° \sin A$$

$$= 0 \cdot \cos A + 1 \cdot \sin A$$

$$= \sin A$$

The angles $90° - A$ and A are complementary angles. This formula states that the sine of an angle is always equal to the cosine of its complement. We could also state it this way:

$$\sin(90° - A) = \cos A$$

We can use this information to derive the formula for $\sin(A + B)$. To understand this derivation, you must recognize that $A + B$ and $90° - (A + B)$ are complementary angles.

$$\sin(A + B) = \cos[90° - (A + B)] \qquad \text{The sine of an angle is the cosine of its complement.}$$

$$= \cos[90° - A - B] \qquad \text{Remove parentheses.}$$

$$= \cos[(90° - A) - B] \qquad \text{Regroup with brackets.}$$

Now we expand using the formula for the cosine of a difference:

$$= \cos(90° - A) \cos B + \sin(90° - A) \sin B$$

$$= \sin A \cos B + \cos A \sin B$$

This gives us an expansion formula for $\sin(A + B)$:

$$\sin(A + B) = \sin A \cos B + \cos A \sin B$$

This is the formula for the sine of a sum. The formula for $\sin(A - B)$ is similar and looks like this:

$$\sin(A - B) = \sin A \cos B - \cos A \sin B$$

EXAMPLE 5 Find the exact value of $\sin \dfrac{\pi}{12}$.

Solution We have to write $\frac{\pi}{12}$ in terms of two numbers, the exact values of which are known. The numbers $\frac{\pi}{3}$ and $\frac{\pi}{4}$ will work since their difference is $\frac{\pi}{12}$.

$$\sin \frac{\pi}{12} = \sin \left(\frac{\pi}{3} - \frac{\pi}{4} \right)$$

$$= \sin \frac{\pi}{3} \cos \frac{\pi}{4} - \cos \frac{\pi}{3} \sin \frac{\pi}{4}$$

$$= \frac{\sqrt{3}}{2} \cdot \frac{\sqrt{2}}{2} - \frac{1}{2} \cdot \frac{\sqrt{2}}{2}$$

$$= \frac{\sqrt{6} - \sqrt{2}}{4}$$

This is the same answer we obtained in Example 1 when we found the exact value of cos 75°. It should be, since $\frac{\pi}{12} = 15°$, which is the complement of 75°, and the cosine of an angle is equal to the sine of its complement.

EXAMPLE 6 If $\sin A = \frac{3}{5}$ with A in QI and $\cos B = -\frac{5}{13}$ with B in QIII, find $\sin (A + B)$, $\cos (A + B)$, and $\tan (A + B)$.

Solution We have $\sin A$ and $\cos B$. We need to find $\cos A$ and $\sin B$ before we can apply any of our formulas. Some equivalent forms of the Pythagorean identity will help here.

If $\sin A = \dfrac{3}{5}$ with A in QI, then

$$\cos A = \sqrt{1 - \sin^2 A}$$

$$= \sqrt{1 - \left(\frac{3}{5} \right)^2} = \frac{4}{5}$$

If $\cos B = -\dfrac{5}{13}$ with B in QIII, then

$$\sin B = -\sqrt{1 - \left(-\frac{5}{13} \right)^2}$$

$$= -\frac{12}{13}$$

Note that we chose the negative radical since $\sin B < 0$ when B is in QIII.

We have

$$\sin A = \frac{3}{5} \qquad \sin B = -\frac{12}{13}$$

$$\cos A = \frac{4}{5} \qquad \cos B = -\frac{5}{13}$$

Therefore,

$$\sin (A + B) = \sin A \cos B + \cos A \sin B$$

$$= \frac{3}{5} \left(-\frac{5}{13} \right) + \frac{4}{5} \left(-\frac{12}{13} \right)$$

$$= -\frac{63}{65}$$

$$\cos(A + B) = \cos A \cos B - \sin A \sin B$$

$$= \frac{4}{5}\left(-\frac{5}{13}\right) - \frac{3}{5}\left(-\frac{12}{13}\right)$$

$$= \frac{16}{65}$$

$$\tan(A + B) = \frac{\sin(A + B)}{\cos(A + B)}$$

$$= -\frac{\frac{63}{65}}{\frac{16}{65}}$$

$$= -\frac{63}{16}$$

Notice also that $A + B$ must terminate in quadrant IV because $\sin(A + B) < 0$ and $\cos(A + B) > 0$.

While working through the last part of Example 6, you may have wondered if there is a separate formula for $\tan(A + B)$. (More likely, you are hoping there isn't.) There is, and it is derived from the formulas we already have:

$$\tan(A + B) = \frac{\sin(A + B)}{\cos(A + B)}$$

$$= \frac{\sin A \cos B + \cos A \sin B}{\cos A \cos B - \sin A \sin B}$$

To be able to write this last line in terms of tangents only, we must divide the numerator and denominator by $\cos A \cos B$:

$$= \frac{\dfrac{\sin A \cos B}{\cos A \cos B} + \dfrac{\cos A \sin B}{\cos A \cos B}}{\dfrac{\cos A \cos B}{\cos A \cos B} - \dfrac{\sin A \sin B}{\cos A \cos B}}$$

$$= \frac{\tan A + \tan B}{1 - \tan A \tan B}$$

The formula for $\tan(A - B)$ can be derived in a similar manner:

$$\tan(A - B) = \frac{\tan A - \tan B}{1 + \tan A \tan B}$$

To summarize:

$$\tan(A + B) = \frac{\tan A + \tan B}{1 - \tan A \tan B}$$

$$\tan(A - B) = \frac{\tan A - \tan B}{1 + \tan A \tan B}$$

EXAMPLE 7 If $\sin A = \frac{3}{5}$ with A in QI and $\cos B = -\frac{5}{13}$ with B in QIII, find $\tan (A + B)$ by using the formula

$$\tan (A + B) = \frac{\tan A + \tan B}{1 - \tan A \tan B}$$

Solution The angles A and B as given here are the same ones used previously in Example 6. Looking over Example 6 again, we find that

$$\tan A = \frac{3}{4} \qquad \tan B = \frac{12}{5}$$

Therefore,

$$\tan(A + B) = \frac{\tan A + \tan B}{1 - \tan A \tan B)}$$

$$= \frac{\dfrac{3}{4} + \dfrac{12}{5}}{1 - \dfrac{3}{4} \cdot \dfrac{12}{5}}$$

$$= \frac{\dfrac{15}{20} + \dfrac{48}{20}}{1 - \dfrac{9}{5}}$$

$$= \frac{\dfrac{63}{20}}{-\dfrac{4}{5}}$$

$$= -\frac{63}{20} \cdot \left(-\frac{5}{4}\right)$$

$$= -\frac{63}{16}$$

Which is the same result we obtained previously.

EXAMPLE 8 Graph $y = 4 \sin 6x \cos 4x - 4 \cos 6x \sin 4x$ from $x = 0$ to $x = 2\pi$.

Solution To write the equation in the form $y = A \sin Bx$, we factor 4 from each term on the right and then apply the formula for $\sin (A - B)$ to the remaining expression to write it as a single trigonometric function.

$$y = 4 \sin 6x \cos 4x - 4 \cos 6x \sin 4x$$

$$= 4 \sin (6x - 4x)$$

$$= 4 \sin 2x$$

The graph is shown in Figure 2.

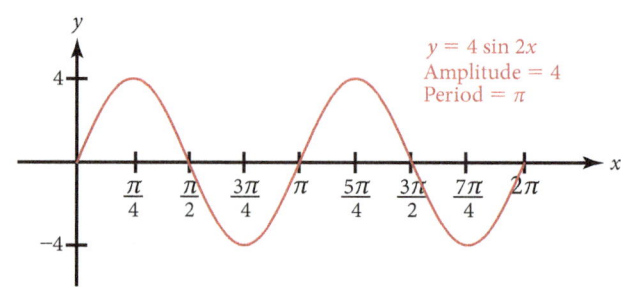

FIGURE 2

Problem Set 5.2

Vocabulary

Use the vocabulary words below to fill in the blanks in the sentences.

> not $\sin A$ $\sin B$ $\cos A$ $\cos B$

1. The statement $\sin (A + B) = \sin A + \sin B$ is _____ an identity because it is not true for all values of A and B.

2. $\sin (A + B) = \sin A \cos B +$ _____ $\sin B$

3. $\sin (A - B) =$ _____ $\cos B -$ _____ $\sin B$

4. $\cos (A + B) = \cos A \cos B -$ _____ $\sin B$

5. $\cos (A - B) =$ _____ $\cos B + \sin A$ _____

Find exact values for each of the following:

1. $\sin 15°$
2. $\sin 75°$
3. $\cos 15°$
4. $\cos 75°$
5. $\tan 15°$
6. $\tan 75°$
7. $\sin \dfrac{7\pi}{12}$ *Hint:* $\dfrac{7\pi}{12} = \dfrac{\pi}{3} + \dfrac{\pi}{4}$
8. $\cos \dfrac{7\pi}{12}$
9. $\cos 195°$ *Hint:* $195° = 150° + 45°$
10. $\sin 195°$

Use sum and difference formulas to prove that each statement is true.

11. $\sin (x + 2\pi) = \sin x$
12. $\cos (x - 2\pi) = \cos x$
13. $\cos \left(x - \dfrac{\pi}{2} \right) = \sin x$
14. $\sin \left(x - \dfrac{\pi}{2} \right) = -\cos x$
15. $\cos (180° - \theta) = -\cos \theta$
16. $\sin (180° - \theta) = \sin \theta$
17. $\sin (90° + \theta) = \cos \theta$
18. $\cos (90° + \theta) = -\sin \theta$
19. $\tan \left(x + \dfrac{\pi}{4} \right) = \dfrac{1 + \tan x}{1 - \tan x}$
20. $\tan \left(x - \dfrac{\pi}{4} \right) = \dfrac{\tan x - 1}{\tan x + 1}$

Write each expression as a single trigonometric function.

21. $\sin 3x \cos 2x + \cos 3x \sin 2x$
22. $\cos 3x \cos 2x + \sin 3x \sin 2x$
23. $\cos 5x \cos x - \sin 5x \sin x$
24. $\sin 8x \cos x - \cos 8x \sin x$
25. $\sin 45° \cos \theta + \cos 45° \sin \theta$
26. $\cos 60° \cos \theta - \sin 60° \sin \theta$
27. $\cos 30° \sin \theta + \sin 30° \cos \theta$
28. $\cos 60° \sin \theta + \sin 60° \cos \theta$
29. $\cos 15° \cos 75° - \sin 15° \sin 75°$
30. $\cos 15° \cos 75° + \sin 15° \sin 75°$

31. Let $\sin A = \dfrac{3}{5}$ with A in QII and $\sin B = -\dfrac{5}{13}$ with B in QIII. Find $\sin (A + B)$, $\cos (A + B)$, and $\tan (A + B)$.

32. Let $\cos A = -\dfrac{5}{13}$ with A in QII and $\sin B = \dfrac{3}{5}$ with B in QI.

Find $\sin (A - B)$, $\cos (A - B)$, and $\tan (A - B)$.

33. If $\sin A = \dfrac{1}{\sqrt{5}}$ with A in QI and $\tan B = \dfrac{3}{4}$ with B in QI, find

$\tan (A + B)$ and $\cot (A + B)$.

34. If $\sec A = \sqrt{5}$ with A in QI and $\sec B = \sqrt{10}$ with B in QI, find $\sec (A + B)$.
[First find $\cos (A + B)$.]

35. If $\tan (A + B) = 3$ and $\tan B = \dfrac{1}{2}$, find $\tan A$.

36. If $\tan(A + B) = 2$ and $\tan B = \dfrac{1}{3}$, find $\tan A$.

37. Write the formula for $\sin 2x$ by writing $\sin 2x$ as $\sin (x + x)$ and using the formula for the sine of a sum.

38. Write a formula for $\cos 2x$ by writing $\cos 2x$ as $\cos (x + x)$ and using the formula for the cosine of a sum.

39. Graph one complete cycle of $y = \sin x \cos \dfrac{\pi}{4} + \cos x \sin \dfrac{\pi}{4}$ by first rewriting the right side in the form $\sin (A + B)$.

40. Graph one complete cycle of $y = \sin x \cos \dfrac{\pi}{6} - \cos x \sin \dfrac{\pi}{6}$ by first rewriting the right side in the form $\sin (A - B)$.

41. Graph one complete cycle of $y = 2 \sin x \cos \dfrac{\pi}{3} + 2 \cos x \sin \dfrac{\pi}{3}$ by first rewriting the right side in the form $2 \sin (A + B)$.

42. Graph one complete cycle of $y = 2 \sin x \cos \dfrac{\pi}{3} - 2 \cos x \sin \dfrac{\pi}{3}$ by first rewriting the right side in the form $2 \sin (A - B)$.

Prove each identity.

43. $\sin (90° + x) + \sin (90° - x) = 2 \cos x$

44. $\sin (90° + x) - \sin (90° - x) = 0$

45. $\cos (x - 90°) - \cos (x + 90°) = 2 \sin x$

46. $\cos (x + 90°) + \cos (x - 90°) = 0$

Getting Ready for the Next Section

47. If $\sin A = \dfrac{3}{5}$ with A in Q2, find $\cos A$.

48. If $\cos A = \dfrac{5}{13}$ with A in Q1, find $\sin A$.

Simplify.

49. $\sqrt{1 - \left(\dfrac{3}{5}\right)^2}$

50. $1 - 2\left(\dfrac{1}{\sqrt{5}}\right)^2$

51. $1 + 2\left(\dfrac{2}{\sqrt{5}}\right)^2$

52. $\sqrt{1 - \left(\dfrac{5}{13}\right)^2}$

5.3 Videos

KEY WORDS

double-angle formula

Learning Objectives

A   Find the value of trigonometric functions using double-angle formulas.

B   Prove trigonometric identities using double-angle formulas.

We will begin this section by deriving the formulas for $\sin 2A$ and $\cos 2A$ using the formulas for $\sin (A + B)$ and $\cos (A + B)$. The formulas we derive for $\sin 2A$ and $\cos 2A$ are called **double-angle formulas**. Here is the derivation of the formula for $\sin 2A$.

$$\sin 2A = \sin (A + A) \qquad \text{Write } 2A \text{ as } A + A.$$
$$= \sin A \cos A + \cos A \sin A \qquad \text{Sum formula}$$
$$= \sin A \cos A + \sin A \cos A \qquad \text{Commutative property}$$
$$= 2 \sin A \cos A$$

The last line gives us our first double-angle formula.

The first thing to notice about this formula is that it indicates that the 2 in $\sin 2A$ cannot be factored out and written as a coefficient. That is,

$$\sin 2A \neq 2 \sin A$$

$$\sin 2A = 2 \sin A \cos A$$

A Solving Trigonometric Functions

Here are some examples of how we can apply the double-angle formula $\sin 2A = 2 \sin A \cos A$.

EXAMPLE 1   If $\sin A = \dfrac{3}{5}$ and A terminates in QII, find $\sin 2A$.

Solution   In order to apply the formula for $\sin 2A$, we must first find $\cos A$.

$$\cos A = -\sqrt{1 - \sin^2 A} = -\sqrt{1 - \left(\frac{3}{5}\right)^2} = -\sqrt{\frac{16}{25}} = -\frac{4}{5}$$

Note that we chose the negative radical since $\cos A < 0$ when A is in QII.

Now we can apply the formula for $\sin 2A$:

$$\sin 2A = 2 \sin A \cos A$$
$$= 2\left(\frac{3}{5}\right)\left(-\frac{4}{5}\right) = -\frac{24}{25}$$

We can also use our new formula to expand the work we did previously with identities.

EXAMPLE 2   Prove: $(\sin \theta + \cos \theta)^2 = 1 + \sin 2\theta$.

Proof $(\sin \theta + \cos \theta)^2 = \sin^2 \theta + 2 \sin \theta \cos \theta + \cos^2 \theta \qquad$ Expand

$$= 1 + 2 \sin \theta \cos \theta \qquad \text{Pythagorean identity}$$
$$= 1 + \sin 2\theta \qquad \text{Double-angle identity}$$

There are three forms of the double-angle formula for cos 2A. The first involves both sine and cosine, the second involves only cosine, and the third, just sine. Here is how we obtain the first of these three formulas.

$$\cos 2A = \cos(A + A) \qquad \textit{Write 2A as A + A}$$

$$= \cos A \cos A - \sin A \sin A \qquad \textit{Sum formula}$$

$$= \cos^2 A - \sin^2 A$$

To write this last formula in terms of cos A only, we substitute $1 - \cos^2 A$ for $\sin^2 A$:

$$\cos 2A = \cos^2 A - (1 - \cos^2 A)$$

$$= \cos^2 A - 1 + \cos^2 A$$

$$= 2 \cos^2 A - 1$$

To write the formula for cos 2A in terms of sin A only, we substitute $1 - \sin^2 A$ for $\cos^2 A$ in the last line above:

$$\cos 2A = 2 \cos^2 A - 1$$

$$= 2(1 - \sin^2 A) - 1$$

$$= 2 - 2 \sin^2 A - 1$$

$$= 1 - 2 \sin^2 A$$

Here are the three forms of the double-angle formula for cos 2A:

$$\cos 2A = \cos^2 A - \sin^2 A \quad \textit{First form}$$

$$= 2 \cos^2 A - 1 \quad \textit{Second form}$$

$$= 1 - 2 \sin^2 A \quad \textit{Third form}$$

EXAMPLE 3   If $\sin A = \dfrac{1}{\sqrt{5}}$, find cos 2A.

Solution   In this example, since we are given sin A, applying the third form of the formula for cos 2A will give us the answer more quickly than applying either of the other two forms.

$$\cos 2A = 1 - 2 \sin^2 A$$

$$= 1 - 2\left(\frac{1}{\sqrt{5}}\right)^2$$

$$= 1 - \frac{2}{5}$$

$$= \frac{3}{5}$$

B Proving Identities

EXAMPLE 4 Prove $\sin 2x = \dfrac{2 \cot x}{1 + \cot^2 x}$.

Solution

$$\frac{2 \cot x}{1 + \cot^2 x} \cdot \frac{\sin^2 x}{\sin^2 x} = \frac{2 \cdot \dfrac{\cos x}{\sin x}}{1 + \dfrac{\cos^2 x}{\sin^2 x}} \cdot \frac{\sin^2 x}{\sin^2 x}$$

$$= \frac{2 \sin x \cos x}{\sin2 x + \cos^2 x}$$

$$= 2 \sin x \cos x$$

$$= \sin 2x$$

EXAMPLE 5 Prove $\cos 4x = 8 \cos^4 x - 8 \cos^2 x + 1$.

Solution We can write $\cos 4x$ as $\cos 2 \cdot 2x$ and apply our double-angle formula. Since the right side is written in terms of $\cos x$ only, we will choose the second form of our double-angle formula for $\cos 2A$.

$$\cos 4x = \cos (2 \cdot 2x)$$

$= 2 \cos^2 2x - 1$	Double-angle formula
$= 2(2 \cos^2 x - 1)^2 - 1$	Double-angle formula
$= 2(4 \cos^4 x - 4 \cos^2 x + 1) - 1$	Square
$= 8 \cos^4 x - 8 \cos^2 x + 2 - 1$	Distribute
$= 8 \cos^4 x - 8 \cos^2 x + 1$	Simplify

EXAMPLE 6 Prove $\tan \theta = \dfrac{1 - \cos 2\theta}{\sin 2\theta}$.

Solution

$\dfrac{1 - \cos 2\theta}{\sin 2\theta} = \dfrac{1 - (1 - 2 \sin^2 \theta)}{2 \sin \theta \cos \theta}$	Double-angle formulas
$= \dfrac{2 \sin^2 \theta}{2 \sin \theta \cos \theta}$	Simplify numerator
$= \dfrac{\sin \theta}{\cos \theta}$	Divide out common factor $2 \sin \theta$
$= \tan \theta$	Ratio identity

EXAMPLE 7   Graph $y = 4 \cos^2 \dfrac{x}{4} - 4 \sin^2 \dfrac{x}{4}$ from $x = -4\pi$ to $x = 4\pi$.

Solution To write the equation in the form $y = A \cos Bx$, we factor 4 from each term on the right side and then apply the formula for cos $2A$ to the remaining expression to write it as a single trigonometric function.

$$y = 4 \cos^2 \frac{x}{4} - 4 \sin^2 \frac{x}{4}$$

$$= 4 \left(\cos^2 \frac{x}{4} - \sin^2 \frac{x}{4} \right) \qquad \text{Factor 4 from each term}$$

$$= 4 \cos \left(2 \cdot \frac{x}{4} \right) \qquad \text{Double-angle formula}$$

$$= 4 \cos \frac{1}{2}x$$

The graph will have an amplitude of 4 and a period of $\frac{2\pi}{1/2} = 4\pi$. The graph is shown in Figure 1.

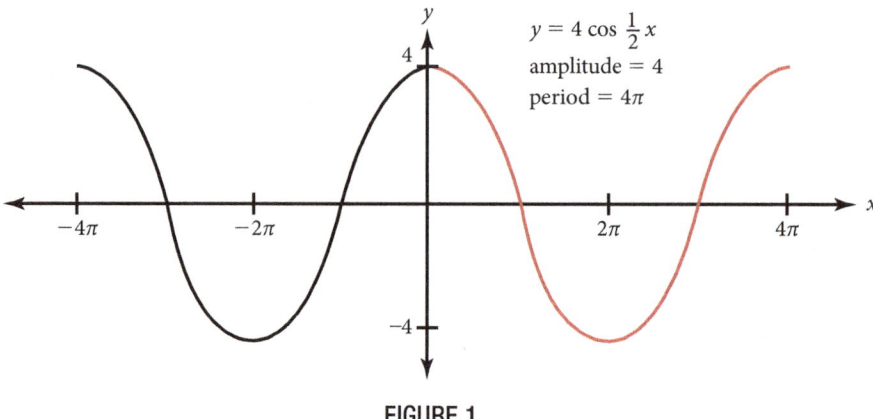

$y = 4 \cos \frac{1}{2}x$
amplitude = 4
period = 4π

FIGURE 1

Problem Set 5.3

Vocabulary

Use the vocabulary words below to fill in the blanks in the sentences.

is   is not   true   false

1. This statement _____ an identity: $\sin 2A = 2 \sin A$

2. This statement _____ an identity: $\cos 2\theta = \cos^2 \theta - \sin^2 \theta$

3. This statement is _____ : $\cos 2\theta = 1 - \sin^2 \theta$

4. This statement is _____ : $\cos 2\theta = 2 \cos^2 \theta - 1$

Let $\sin A = -\dfrac{3}{5}$ with $180° < A < 270°$, and find

1. $\sin 2A$ **2.** $\cos 2A$

3. $\tan 2A$ **4.** $\cot 2A$

Let $\cos x = \dfrac{1}{\sqrt{10}}$ with $270° < x < 360°$, and find

5. $\cos 2x$ **6.** $\sin 2x$

7. $\cot 2x$ **8.** $\tan 2x$

Let $\tan \theta = \frac{5}{12}$ with θ in QI and find

9. $\sin 2\theta$ **10.** $\cos 2\theta$ **11.** $\csc 2\theta$ **12.** $\sec 2\theta$

Let $\csc t = \sqrt{5}$ with t in QII and find

13. $\cos 2t$ **14.** $\sin 2t$ **15.** $\sec 2t$ **16.** $\csc 2t$

17. Express $\cos 100°$ in terms of $\sin 50°$.

18. Express $\cos 100°$ in terms of $\cos 50°$.

19. Express $\sin \frac{\pi}{5}$ in terms of $\sin \frac{\pi}{10}$ and $\cos \frac{\pi}{10}$.

20. Express $\cos \frac{\pi}{4}$ in terms of $\cos \frac{\pi}{8}$.

Use exact values to show that each of the following is true:

21. $\sin 60° = 2 \sin 30° \cos 30°$ **22.** $\cos 60° = 1 - 2 \sin^2 30°$

23. $\cos 120° = \cos^2 260° - \sin^2 260°$ **24.** $\sin 90° = 2 \sin 45° \cos 45°$

25. $\cos \pi = 2 \cos^2 \dfrac{\pi}{2} - 1$ **26.** $\sin \pi = 2 \sin \dfrac{\pi}{2} \cos \dfrac{\pi}{2}$

27. $\sin \dfrac{2\pi}{3} = 2 \sin \dfrac{\pi}{3} \cos \dfrac{\pi}{3}$ **28.** $\cos \dfrac{2\pi}{3} = \cos^2 \dfrac{\pi}{3} - \sin^2 \dfrac{\pi}{3}$

29. Find the exact value of $\sin 15°$ from the expression $\cos 30° = 1 - 2 \sin^2 15°$.

30. Find the exact value of $\cos 15°$ from the expression $\cos 30° = 2 \cos^2 15° - 1$.

Prove each of the following identities:

31. $(\sin x - \cos x)^2 = 1 - \sin 2x$

32. $(\cos x - \sin x)(\cos x + \sin x) = \cos 2x$

33. $\cos^2 \theta = \dfrac{1 + \cos 2\theta}{2}$ **34.** $\sin^2 \theta = \dfrac{1 - \cos 2\theta}{2}$

35. $\cot \theta = \dfrac{\sin 2\theta}{1 - \cos 2\theta}$ **36.** $\cos 2\theta = \dfrac{1 - \tan^2 \theta}{1 + \tan^2 \theta}$

37. $\cos^4 x - \sin^4 x = \cos 2x$ **38.** $\sin^4 x - \cos^4 x = -\cos 2x$

39. $\sin 3\theta = 3 \sin \theta - 4 \sin^3 \theta$ **40.** $\cos 3\theta = 4 \cos^3 \theta - 3 \cos \theta$

41. $2 \csc 2x = \tan x + \cot x$ **42.** $2 \cot 2x = \cot x - \tan x$

43. $\cot \theta - \tan \theta = \dfrac{\cos 2\theta}{\sin \theta \cos \theta}$ **44.** $\csc \theta - 2 \sin \theta = \dfrac{\cos 2\theta}{\sin \theta}$

Graph each of the following from $x = 0$ to $x = 2\pi$.

45. $y = 8 \sin x \cos x$ **46.** $y = 4 \sin 2x \cos 2x$

47. $y = 2 \cos^2 2x - 2 \sin^2 2x$ **48.** $y = 3 \cos^2 x - 3 \sin^2 x$

49. $y = 6 - 12 \sin^2 \dfrac{x}{2}$ **50.** $y = 2 \cos^2 \dfrac{3x}{2} - 1$

Getting Ready for the Next Section

Simplify. Leave your answer in terms of square roots.

51. $\sqrt{1 - \left(-\dfrac{12}{13}\right)^2}$ **52.** $\sqrt{1 - \left(\dfrac{3}{5}\right)^2}$

53. $\sqrt{\dfrac{1 - \dfrac{3}{5}}{2}}$ **54.** $\sqrt{\dfrac{1 - \dfrac{5}{13}}{2}}$

55. $\sqrt{\dfrac{1 - \dfrac{1}{2}}{2}}$ **56.** $\sqrt{\dfrac{1 - \dfrac{\sqrt{3}}{2}}{2}}$

5.4 Videos

KEY WORDS

half-angle formula

Learning Objectives

A   Find the value of trigonometric functions using half-angle formulas.

We are now ready to derive the formulas for

$$\sin \frac{A}{2} \quad \text{and} \quad \cos \frac{A}{2}$$

These formulas are called **half-angle formulas** and are derived from the double-angle formulas for $\cos 2A$.

We can derive the formula for $\sin \frac{A}{2}$ by first solving the double-angle formula $\cos 2x = 1 - 2 \sin^2 x$ for $\sin x$, and then applying a simple substitution.

$$1 - 2 \sin^2 x = \cos 2x \qquad \text{Double-angle formula}$$

$$-2 \sin^2 x = -1 + \cos 2x \qquad \text{Add } -1 \text{ to both sides.}$$

$$\sin^2 x = \frac{1 - \cos 2x}{2} \qquad \text{Divide both sides by } -2.$$

$$\sin x = \pm \sqrt{\frac{1 - \cos 2x}{2}} \qquad \text{Take the square root of both sides.}$$

Since every value of x can be written as $\frac{1}{2}$ of some other number A, we can replace x with $\frac{A}{2}$. This is equivalent to saying $2x = A$.

$$\sin \frac{A}{2} = \pm \sqrt{\frac{1 - \cos A}{2}}$$

This last expression is the half-angle formula for $\sin \frac{A}{2}$. To find the half-angle formula for $\cos \frac{A}{2}$, we solve $\cos 2x = 2 \cos^2 x - 1$ for $\cos x$ and then replace x with $\frac{A}{2}$ (and $2x$ with A). Without showing the steps involved in this process, here is the result:

$$\cos \frac{A}{2} = \pm \sqrt{\frac{1 + \cos A}{2}}$$

In both half-angle formulas the sign, $+$ or $-$, in front of the radical is determined by the quadrant in which $\frac{A}{2}$ terminates.

EXAMPLE 1   If $\cos A = \dfrac{3}{5}$ with $270° < A < 360°$, find $\sin \dfrac{A}{2}$, $\cos \dfrac{A}{2}$, and $\tan \dfrac{A}{2}$.

Solution   First of all, $\frac{A}{2}$ terminates in QII. Here is why.

$$270° < A < 360° \Rightarrow \frac{270°}{2} < \frac{A}{2} < \frac{360°}{2}$$

or

$$135° < \frac{A}{2} < 180° \Rightarrow \frac{A}{2} \in \text{QII}$$

In quadrant II, sine is positive and cosine is negative.

$$\sin \frac{A}{2} = \sqrt{\frac{1 - \cos A}{2}} \qquad\qquad \cos \frac{A}{2} = -\sqrt{\frac{1 + \cos A}{2}}$$

$$= \sqrt{\frac{1 - \frac{3}{5}}{2}} \qquad\qquad = -\sqrt{\frac{1 + \frac{3}{5}}{2}}$$

$$= \sqrt{\frac{1}{5}} \qquad\qquad\qquad = -\sqrt{\frac{4}{5}}$$

$$= \frac{1}{\sqrt{5}} \qquad\qquad\qquad = -\frac{2}{\sqrt{5}}$$

$$\tan \frac{A}{2} = \frac{\sin \frac{A}{2}}{\cos \frac{A}{2}} = \frac{\frac{1}{\sqrt{5}}}{-\frac{2}{\sqrt{5}}} = -\frac{1}{2}$$

EXAMPLE 2   If $\sin A = -\dfrac{12}{13}$ and $180° < A < 270°$, find $\sin \dfrac{A}{2}$, $\cos \dfrac{A}{2}$, and $\tan \dfrac{A}{2}$.

Solution   To use the half-angle formulas, we need to find $\cos A$.

$$\cos A = -\sqrt{1 - \sin^2 A} = -\sqrt{1 - \left(-\frac{12}{13}\right)^2} = -\sqrt{\frac{25}{169}} = -\frac{5}{13}$$

Also, $\frac{A}{2} \in$ QII because

$$180° < A < 270° \Rightarrow \frac{180°}{2} < \frac{A}{2} < \frac{270°}{2}$$

$$90° < \frac{A}{2} < 135° \Rightarrow \frac{A}{2} \in \text{QII}$$

In quadrant II, sine is positive and cosine is negative.

$$\sin \frac{A}{2} = \sqrt{\frac{1 - \left(-\frac{5}{13}\right)}{2}} \qquad\qquad \cos \frac{A}{2} = -\sqrt{\frac{1 + \left(-\frac{5}{13}\right)}{2}}$$

$$= \sqrt{\frac{9}{13}} \qquad\qquad\qquad = -\sqrt{\frac{4}{13}}$$

$$= \frac{3}{\sqrt{13}} \qquad\qquad\qquad = -\frac{2}{\sqrt{13}}$$

$$\tan \frac{A}{2} = \frac{\sin \frac{A}{2}}{\cos \frac{A}{2}} = \frac{\frac{3}{\sqrt{13}}}{-\frac{2}{\sqrt{13}}} = -\frac{3}{2}$$

EXAMPLE 3 Use a half-angle formula to find the exact value of sin 15°.

Solution Since $15° = \frac{30°}{2}$ we have

$$\sin 15° = \sin \frac{30°}{2} = \sqrt{\frac{1 - \cos 30°}{2}}$$

$$= \sqrt{\frac{1 - \frac{\sqrt{3}}{2}}{2}}$$

$$= \sqrt{\frac{2 - \sqrt{3}}{4}}$$

$$= \frac{\sqrt{2 - \sqrt{3}}}{2}$$

Note We found the exact value of sin 15° previously in Example 5 of Section 5.2. (At that time, we were working in radians and thus called it $\frac{\pi}{12}$. But $\frac{\pi}{12} = 15°$.) The value of sin 15° from Example 5, Section 5.2 was

$$\sin 15° = \frac{\sqrt{6} - \sqrt{2}}{4}$$

Can you show that the two answers are the same?

EXAMPLE 4 Prove $\sin^2 \frac{x}{2} = \frac{\tan x - \sin x}{2 \tan x}$.

Proof We can use a half-angle formula on the left side. In this case, since we have $\sin^2 \frac{x}{2}$ we write the half-angle formula without the square root sign. After that, we multiply the numerator and denominator on the left side by tan x because the right side has tan x in both the numerator and denominator.

$$\sin^2 \frac{x}{2} = \frac{1 - \cos x}{2} \qquad \textcolor{green}{\text{Square of half-angle formula}}$$

$$= \frac{\tan x}{\tan x} \cdot \frac{1 - \cos x}{2} \qquad \textcolor{green}{\text{Multiply numerator and denominator by tan x}}$$

$$= \frac{\tan x - \tan x \cos x}{2 \tan x} \qquad \textcolor{green}{\text{Distributive property}}$$

$$= \frac{\tan x - \frac{\sin x}{\cos x} \cdot \cos x}{2 \tan x}$$

$$= \frac{\tan x - \sin x}{2 \tan x} \qquad \textcolor{green}{\text{tan x cos x is sin x}}$$

Problem Set 5.4

Vocabulary

Use the vocabulary words below to fill in the blanks in the sentences.

     is   is not   true   false

1. This statement _____ an identity: $\sin \frac{A}{2} = \pm\sqrt{\frac{1 - \cos A}{2}}$.

2. This statement _____ an identity: $\sin \frac{A}{2} = \pm\sqrt{\frac{1 + \cos A}{2}}$.

3. This statement is _____: $\cos \frac{A}{2} = \pm\sqrt{\frac{1 - \cos A}{2}}$.

4. This statement is _____ : $\sin 15° = \sqrt{\frac{1 - \cos 30°}{2}}$.

If $\cos A = \dfrac{1}{2}$ and $270° < A < 360°$, find:

1. $\sin \dfrac{A}{2}$ **2.** $\cos \dfrac{A}{2}$

3. $\csc \dfrac{A}{2}$ **4.** $\sec \dfrac{A}{2}$

If $\sin A = -\dfrac{3}{5}$ and $180° < A < 270°$, find:

5. $\cos \dfrac{A}{2}$ **6.** $\sin \dfrac{A}{2}$

7. $\cot \dfrac{A}{2}$ **8.** $\tan \dfrac{A}{2}$

9. $\sec \dfrac{A}{2}$ **10.** $\csc \dfrac{A}{2}$

If $\sin A = \dfrac{4}{5}$ with $90° < A < 180°$ and $\sin B = \dfrac{3}{5}$ with $0° < B < 90°$, find:

11. $\sin \dfrac{A}{2}$ **12.** $\cos \dfrac{A}{2}$

13. $\cos 2A$ **14.** $\sin 2A$

15. $\sec 2A$ **16.** $\csc 2A$

17. $\cos \dfrac{B}{2}$ **18.** $\sin \dfrac{B}{2}$

19. $\sin(A + B)$ **20.** $\cos(A + B)$

21. $\cos(A - B)$ **22.** $\sin(A - B)$

Use half-angle formulas to find exact values for each of the following:

23. $\cos 15°$ **24.** $\tan 15°$

25. $\sin 75°$ **26.** $\cos 75°$

27. $\cos 105°$ **28.** $\sin 105°$

Prove the following identities.

29. $\sin^2 \dfrac{\theta}{2} = \dfrac{\csc \theta - \cot \theta}{2 \csc \theta}$ **30.** $2 \cos^2 \dfrac{\theta}{2} = \dfrac{\sin^2 \theta}{1 - \cos \theta}$

31. $\cos^2 \dfrac{\theta}{2} = \dfrac{\tan \theta + \sin \theta}{2 \tan \theta}$ **32.** $2 \sin^2 \dfrac{\theta}{2} = \dfrac{\sin^2 \theta}{1 + \cos \theta}$

33. $\tan \dfrac{x}{2} = \dfrac{1 - \cos x}{\sin x}$ **34.** $\tan \dfrac{x}{2} = \dfrac{\sin x}{1 + \cos x}$

35. $\cos^4 \theta = \dfrac{1}{4} + \dfrac{\cos 2\theta}{2} + \dfrac{\cos^2 2\theta}{4}$ **36.** $4 \sin^4 \theta = 1 - 2 \cos 2\theta + \cos^2 2\theta$

37. $\sin \dfrac{t}{2} \cos \dfrac{t}{2} = \dfrac{1}{2} \sin t$ **38.** $\left(\cos \dfrac{t}{2} - \sin \dfrac{t}{2} \right)^2 = 1 - \sin t$

Getting Ready for the Next Section

Evaluate without using a calculator or tables.

39. $\sin\left(\arcsin\dfrac{3}{5}\right)$

40. $\cos\left(\arcsin\dfrac{3}{5}\right)$

41. $\cos(\arctan 2)$

42. $\sin(\arctan 2)$

Write an equivalent expression that involves x only. (Assume x is positive.)

43. $\sin(\tan^{-1} x)$

44. $\cos(\tan^{-1} x)$

45. $\tan(\sin^{-1} x)$

46. $\tan(\cos^{-1} x)$

5.5 Videos

Learning Objectives

A   Evaluate identities and formulas involving inverse functions.

B   Find the value of trigonometric functions using the product to sum formula.

C   Find the value of trigonometric functions using the sum to product formula.

There are two main parts to this section, both of which rely on the work we have done previously with identities and formulas. In the first part of this section, we will extend our work on identities to include problems that involve inverse trigonometric functions. In the second part, we will use the formulas we obtained for the sine and cosine of a sum or difference to write some new formulas involving sums and products.

KEY WORDS

product-to-sum formulas

sum-to-product formulas

A Identities and Formulas Involving Inverse Functions

The solution to our first example combines our knowledge of inverse trigonometric functions with our formula for $\sin(A + B)$.

EXAMPLE 1   Evaluate $\sin(\arcsin \frac{3}{5} + \arctan 2)$ without using a calculator or tables.

Solution   We can simplify things somewhat if we let $\alpha = \arcsin \frac{3}{5}$ and $\beta = \arctan 2$.

$$\sin\left(\arcsin \frac{3}{5} + \arctan 2\right)$$

$$= \sin(\alpha + \beta)$$

$$= \sin \alpha \cos \beta + \cos \alpha \sin \beta$$

Drawing and labeling a triangle for α and another for β we have

From the triangles in Figure 1 we have

$$\sin \alpha = \frac{3}{5} \qquad \sin \beta = \frac{2}{\sqrt{5}}$$

$$\cos \alpha = \frac{4}{5} \qquad \cos \beta = \frac{1}{\sqrt{5}}$$

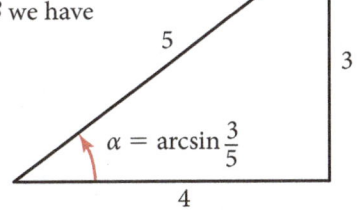
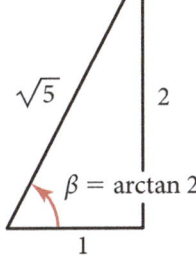

FIGURE 1

Substituting these numbers into

$$\sin \alpha \cos \beta + \cos \alpha \sin \beta$$

gives us,

$$\frac{3}{5} \cdot \frac{1}{\sqrt{5}} + \frac{4}{5} \cdot \frac{2}{\sqrt{5}} = \frac{11}{5\sqrt{5}}$$

Note: If we were to work Example 1 on a calculator, the display would show 0.9839 to four decimal places, which is the decimal approximation of $\frac{11}{5\sqrt{5}}$. It is appropriate to check your work on problems like this by using your calculator. The concepts are best understood, however, by working through the problems without using a calculator.

Here is a similar example involving inverse trigonometric functions and a double-angle identity.

EXAMPLE 2 Write $\sin(2\tan^{-1}x)$ as an equivalent expression involving only x. (Assume x is positive.)

Solution We begin by letting $\theta = \tan^{-1}x$ and then drawing a right triangle with an acute angle of θ. If we label the opposite side with x and the adjacent side with 1, the ratio of the side opposite θ to the side adjacent θ is $\frac{x}{1} = x$.

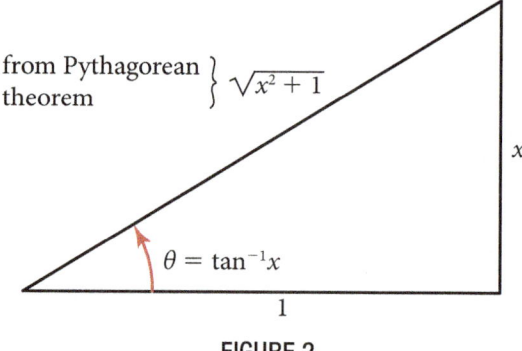

from Pythagorean theorem $\Big\}$ $\sqrt{x^2+1}$

$\theta = \tan^{-1}x$

FIGURE 2

From Figure 2, we have $\sin\theta = \dfrac{x}{\sqrt{x^2+1}}$ and $\cos\theta = \dfrac{1}{\sqrt{x^2+1}}$. Therefore,

$$\sin(2\tan^{-1}x) = \sin 2\theta \qquad \text{Substitute } \theta \text{ for } \tan^{-1}x$$

$$= 2\sin\theta\cos\theta \qquad \text{Double-angle identity}$$

$$= 2 \cdot \frac{x}{\sqrt{x^2+1}} \cdot \frac{1}{\sqrt{x^2+1}} \qquad \text{From Figure 2}$$

$$= \frac{2x}{x^2+1} \qquad \text{Multiplication}$$

To conclude our work with identities in this chapter, we will derive some additional formulas that contain sums and products of sines and cosines.

B Product-to-Sum Formulas

If we add the formula for $\sin(A-B)$ to the formula for $\sin(A+B)$, we will eventually arrive at a formula for the product $\sin A \cos B$.

$$\sin A \cos B + \cos A \sin B = \sin(A+B)$$
$$\underline{\sin A \cos B - \cos A \sin B = \sin(A-B)}$$
$$2\sin A \cos B \qquad\qquad = \sin(A+B) + \sin(A-B)$$

Dividing both sides of this result by 2 gives us

$$\sin A \cos B = \frac{1}{2}[\sin(A+B) + \sin(A-B)] \tag{1}$$

By similar methods, we can derive the formulas that follow.

$$\cos A \sin B = \frac{1}{2}[\sin(A+B) - \sin(A-B)] \tag{2}$$

$$\cos A \cos B = \frac{1}{2}[\cos(A+B) + \cos(A-B)] \tag{3}$$

$$\sin A \sin B = \frac{1}{2}[\cos(A-B) - \cos(A+B)] \tag{4}$$

These four product formulas are of use in calculus. The reason they are useful is that they indicate how we can convert a product into a sum. In calculus, it is sometimes much easier to work with sums of trigonometric functions than it is to work with products.

EXAMPLE 3   Verify product formula (3) for $A = 30°$ and $B = 120°$.

Solution   Substituting $A = 30°$ and $B = 120°$ into

$$\cos A \cos B = \frac{1}{2}[\cos(A + B) + \cos(A - B)]$$

we have

$$\cos 30° \cos 120° = \frac{1}{2}[\cos 150° + \cos(-90°)]$$

$$\frac{\sqrt{3}}{2} \cdot \left(-\frac{1}{2}\right) = \frac{1}{2}\left(-\frac{\sqrt{3}}{2} + 0\right)$$

$$-\frac{\sqrt{3}}{4} = -\frac{\sqrt{3}}{4} \quad \textit{A true statement}$$

EXAMPLE 4   Write $10 \cos 5x \sin 3x$ as a sum or difference.

Solution

$$10 \cos 5x \sin 3x = 10 \cdot \frac{1}{2}[\sin(5x + 3x) - \sin(5x - 3x)]$$

$$= 5(\sin 8x - \sin 2x)$$

C Sum-to-Product Formulas

By some simple manipulations we can change our product formulas into sum formulas. If we take the formula for $\sin A \cos B$ and exchange sides and then multiply through by 2 we have

$$\sin(A + B) + \sin(A - B) = 2 \sin A \cos B$$

If we let $\alpha = A + B$ and $\beta = A - B$, then we can solve for A by adding the left sides and the right sides.

$$A + B = \alpha$$
$$\underline{A - B = \beta}$$
$$2A \quad\quad = \alpha + \beta$$

$$A = \frac{\alpha + \beta}{2}$$

By subtracting the expression for β from the expression for α, we have

$$B = \frac{\alpha - \beta}{2}$$

Writing the equation $\sin(A + B) + \sin(A - B) = 2 \sin A \cos B$ in terms of α and β gives us our first sum formula

$$\sin \alpha + \sin \beta = 2 \sin \frac{\alpha + \beta}{2} \cos \frac{\alpha - \beta}{2} \tag{5}$$

Similarly, the following sum formulas can be derived from the other product formulas

$$\sin \alpha - \sin \beta = 2 \cos \frac{\alpha + \beta}{2} \sin \frac{\alpha - \beta}{2} \qquad (6)$$

$$\cos \alpha + \cos \beta = 2 \cos \frac{\alpha + \beta}{2} \cos \frac{\alpha - \beta}{2} \qquad (7)$$

$$\cos \alpha - \cos \beta = -2 \sin \frac{\alpha + \beta}{2} \sin \frac{\alpha - \beta}{2} \qquad (8)$$

EXAMPLE 5   Verify sum formula (7) for $\alpha = 30°$ and $\beta = 90°$.

Solution   We substitute $\alpha = 30°$ and $\beta = 90°$ into sum formula (7) and simplify each side of the resulting equation

$$\cos 30° + \cos 90° = 2 \cos \frac{30° + 90°}{2} \cos \frac{30° - 90°}{2}$$

$$\cos 30° + \cos 90° = 2 \cos 60° \cos(-30°)$$

$$\frac{\sqrt{3}}{2} + 0 = 2\left(\frac{1}{2}\right)\left(\frac{\sqrt{3}}{2}\right)$$

$$\frac{\sqrt{3}}{2} = \frac{\sqrt{3}}{2} \qquad \textcolor{teal}{\textit{A true statement}}$$

EXAMPLE 6   Verify the identity.

$$-\tan x = \frac{\cos 3x - \cos x}{\sin 3x + \sin x}$$

Proof   Applying the formulas for $\cos \alpha - \cos \beta$ and $\sin \alpha + \sin \beta$ to the right side and then simplifying, will get us to $-\tan x$.

$$\frac{\cos 3x - \cos x}{\sin 3x + \sin x} = \frac{-2 \sin \dfrac{3x + x}{2} \sin \dfrac{3x - x}{2}}{2 \sin \dfrac{3x + x}{2} \cos \dfrac{3x - x}{2}} \qquad \textcolor{teal}{\textit{Sum to product formulas}}$$

$$= \frac{-2 \sin 2x \sin x}{2 \sin 2x \cos x} \qquad \textcolor{teal}{\textit{Simplify}}$$

$$= -\frac{\sin x}{\cos x} \qquad \textcolor{teal}{\textit{Divide out common factors}}$$

$$= -\tan x \qquad \textcolor{teal}{\textit{Ratio identity}}$$

Evaluate each expression below without using a calculator or tables. (Assume any variables represent positive numbers.)

1. $\sin\left(\arcsin\dfrac{3}{5} - \arctan 2\right)$

2. $\cos\left(\arcsin\dfrac{3}{5} - \arctan 2\right)$

3. $\cos\left(\tan^{-1}\dfrac{1}{2} + \sin^{-1}\dfrac{1}{2}\right)$

4. $\sin\left(\tan^{-1}\dfrac{1}{2} - \sin^{-1}\dfrac{1}{2}\right)$

5. $\sin\left(2\cos^{-1}\dfrac{1}{\sqrt{5}}\right)$

6. $\sin\left(2\tan^{-1}\dfrac{3}{4}\right)$

7. $\tan(\sin^{-1}x)$

8. $\tan(\cos^{-1}x)$

9. $\sin(2\sin^{-1}x)$

10. $\sin(2\cos^{-1}x)$

11. $\cos(2\cos^{-1}x)$

12. $\cos(2\sin^{-1}x)$

13. Verify product formula (4) for $A = 30°$ and $B = 120°$.

14. Verify product formula (1) for $A = 120°$ and $B = 30°$.

Rewrite each expression as a sum or difference, then simplify if possible.

15. $10\sin 5x\cos 3x$

16. $10\sin 5x\sin 3x$

17. $\cos 8x\cos 2x$

18. $\cos 2x\sin 8x$

19. $\sin 60°\cos 30°$

20. $\cos 90°\cos 180°$

21. $\sin 4\pi\sin 2\pi$

22. $\cos 3\pi\sin\pi$

23. Verify sum formula (6) for $\alpha = 30°$ and $\beta = 90°$.

24. Verify sum formula (8) for $\alpha = 90°$ and $\beta = 30°$.

Rewrite each expression as a product. Simplify if possible.

25. $\sin 7x + \sin 3x$

26. $\cos 5x - \cos 3x$

27. $\cos 45° + \cos 15°$

28. $\sin 75° - \sin 15°$

29. $\sin\dfrac{7\pi}{12} - \sin\dfrac{\pi}{12}$

30. $\cos\dfrac{\pi}{12} + \cos\dfrac{7\pi}{12}$

Verify each identity.

31. $-\cot x = \dfrac{\sin 3x + \sin x}{\cos 3x - \cos x}$

32. $\tan x = \dfrac{\cos 3x + \cos x}{\sin 3x - \sin x}$

33. $\cot x = \dfrac{\sin 4x + \sin 6x}{\cos 4x - \cos 6x}$

34. $-\tan 4x = \dfrac{\cos 3x - \cos 5x}{\sin 3x - \sin 5x}$

35. $\tan 4x = \dfrac{\sin 5x + \sin 3x}{\cos 3x + \cos 5x}$

36. $\cot 2x = \dfrac{\sin 3x - \sin x}{\cos x - \cos 3x}$

Maintaining Your Skills

The problems below review material we covered in Chapter 4.

37. Graph $y = 1 + \sin x$ for x between 0 and 4π.

38. Graph $y = x + \sin \pi x$ for x between 0 and 8.

39. Graph $y = \cos x + \dfrac{1}{2} \sin 2x$ for x between 0 and 4π.

40. Graph $y = \sqrt{3} \sin x + \cos x$, from $x = 0$ to $x = 4\pi$.

Graph one complete cycle.

41. $y = \sin\left(x + \dfrac{\pi}{4}\right)$ **42.** $y = \sin\left(x - \dfrac{\pi}{4}\right)$

43. $y = 3 \sin\left(2x - \dfrac{\pi}{2}\right)$ **44.** $y = 3 \sin\left(2x + \dfrac{\pi}{2}\right)$

Basic Identities [5.1]

	Basic Identities	Common Equivalent Forms
Reciprocal	$\csc \theta = \dfrac{1}{\sin \theta}$	$\sin \theta = \dfrac{1}{\csc \theta}$
	$\sec \theta = \dfrac{1}{\cos \theta}$	$\cos \theta = \dfrac{1}{\sec \theta}$
	$\cot \theta = \dfrac{1}{\tan \theta}$	$\tan \theta = \dfrac{1}{\cot \theta}$
Ratio	$\tan \theta = \dfrac{\sin \theta}{\cos \theta}$	
	$\cot \theta = \dfrac{\cos \theta}{\sin \theta}$	
Pythagorean	$\cos^2\theta + \sin^2\theta = 1$	$\sin^2\theta = 1 - \cos^2\theta$
	$1 + \tan^2\theta = \sec^2\theta$	$\sin \theta = \pm \sqrt{1 - \cos^2 - \theta}$
	$1 + \cot^2\theta = \csc^2\theta$	$\cos^2\theta = 1 - \sin^2\theta$
		$\cos \theta = \pm \sqrt{1 - \sin^2\theta}$

Examples

1. To prove $\tan x + \cos x$
$= \sin x(\sec x + \cot x)$

we apply the distributive property to the right side and then change to sines and cosines:

$\sin x(\sec x + \cot x)$
$= \sin x \sec x + \sin x \cot x$

$= \sin x \cdot \dfrac{1}{\cos x} + \sin x \cdot \dfrac{\cos x}{\sin x}$

$= \tan x + \cos x$

Proving Identities [5.1]

An identity in mathematics is a statement that two expressions are equal for all replacements of the variable for which each statement is defined. To prove a trigonometric identity, we use trigonometric substitutions and algebraic manipulations to either:

1. Transform the right side into the left.
2. Transform the left side into the right.

Remember to work on each side separately. We do not want to use properties from algebra that involve both sides of the identity—such as the addition property of equality.

2. To find the exact value for $\cos 75°$, we write $75°$ as $45° + 30°$ and then apply the formula for $\cos (A + B)$.

$\cos 75° = \cos (45° + 30°)$

$= \cos 45° \cos 30° - \sin 45° \sin 30°$

$= \dfrac{\sqrt{2}}{2} \cdot \dfrac{\sqrt{3}}{2} - \dfrac{\sqrt{2}}{2} \cdot \dfrac{1}{2}$

$= \dfrac{\sqrt{6} - \sqrt{2}}{4}$

Sum and Difference Formulas [5.2]

$$\sin(A + B) = \sin A \cos B + \cos A \sin B$$

$$\sin(A - B) = \sin A \cos B - \cos A \sin B$$

$$\cos(A + B) = \cos A \cos B - \sin A \sin B$$

$$\cos(A - B) = \cos A \cos B + \sin A \sin B$$

$$\tan(A + B) = \dfrac{\tan A + \tan B}{1 - \tan A \tan B}$$

$$\tan(A - B) = \dfrac{\tan A - \tan B}{1 + \tan A \tan B}$$

3. If $\sin A = \dfrac{3}{5}$ with A in QII, then

$$\cos 2A = 1 - 2\sin^2 A$$

$$= 1 - 2\left(\frac{3}{5}\right)^2$$

$$= \frac{7}{25}$$

Double-Angle Formulas [5.3]

$$\sin 2A = 2\sin A \cos A$$

$$\cos 2A = \cos^2 A - \sin^2 A \qquad \text{First form}$$

$$= 2\cos^2 A - 1 \qquad \text{Second form}$$

$$= 1 - 2\sin^2 A \qquad \text{Third form}$$

4. If $\cos\theta = \dfrac{1}{2}$ and $0° < \theta < 90°$, then

$$\cos\frac{\theta}{2} = \sqrt{\frac{1 + \frac{1}{2}}{2}}$$

$$= \sqrt{\frac{3}{4}}$$

$$= \frac{\sqrt{3}}{2}$$

Half-Angle Formulas [5.4]

$$\sin\frac{A}{2} = \pm\sqrt{\frac{1 - \cos A}{2}}$$

$$\cos\frac{A}{2} = \pm\sqrt{\frac{1 + \cos A}{2}}$$

5. We can write the product
$$10\cos 5x \sin 3x$$
as the difference by applying the second product to sum formula

$$10\cos 5x \sin 3x$$

$$= 10 \cdot \frac{1}{2}[\sin(5x+3x) - \sin(5x-3x)]$$

$$= 5(\sin 8x - \sin 2x)$$

Product to Sum Formulas [5.5]

$$\sin A \cos B = \frac{1}{2}[\sin(A + B) + \sin(A - B)]$$

$$\cos A \sin B = \frac{1}{2}[\sin(A + B) - \sin(A - B)]$$

$$\cos A \cos B = \frac{1}{2}[\cos(A + B) + \cos(A - B)]$$

$$\sin A \sin B = \frac{1}{2}[\cos(A - B) - \cos(A + B)]$$

6. Prove $-\tan x = \dfrac{\cos 3x - \cos x}{\sin 3x + \sin x}$

Proof
$$\frac{\cos 3x - \cos x}{\sin 3x + \sin x} = \frac{-2\sin\frac{3x+x}{2}\sin\frac{3x-x}{2}}{2\sin\frac{3x+x}{2}\cos\frac{3x-x}{2}}$$

$$= \frac{-2\sin 2x \sin x}{2\sin 2x \cos x}$$

$$= -\frac{\sin x}{\cos x}$$

$$= -\tan x$$

Sum to Product Formulas [5.5]

$$\sin\alpha + \sin\beta = 2\sin\frac{\alpha + \beta}{2}\cos\frac{\alpha - \beta}{2}$$

$$\sin\alpha - \sin\beta = 2\cos\frac{\alpha + \beta}{2}\sin\frac{\alpha - \beta}{2}$$

$$\cos\alpha + \cos\beta = 2\cos\frac{\alpha + \beta}{2}\cos\frac{\alpha - \beta}{2}$$

$$\cos\alpha - \cos\beta = -2\sin\frac{\alpha + \beta}{2}\sin\frac{\alpha - \beta}{2}$$

Prove each identity.

1. $\tan \theta = \sin \theta \sec \theta$

2. $\dfrac{\cot \theta}{\csc \theta} = \cos \theta$

3. $(\sec x - 1)(\sec x + 1) = \tan^2 x$

4. $\sec \theta - \cos \theta = \tan \theta \sin \theta$

5. $\dfrac{\cos t}{1 - \sin t} = \dfrac{1 + \sin t}{\cos t}$

6. $\dfrac{1}{1 - \sin t} + \dfrac{1}{1 + \sin t} = 2 \sec^2 t$

7. $\sin(\theta - 90) = -\cos \theta$

8. $\cos\left(\dfrac{\pi}{2} + \theta\right) = -\sin \theta$

9. $\cos^4 A - \sin^4 A = \cos 2A$

10. $\cot A = \dfrac{\sin 2A}{1 - \cos 2A}$

11. $\cot x - \tan x = \dfrac{\cos 2x}{\sin x \cos x}$

12. $\tan \dfrac{x}{2} = \dfrac{\sin x}{1 + \cos x}$

Let $\sin A = -\frac{3}{5}$ with A in QIV and $\sin B = \frac{12}{13}$ with B in QII and find

13. $\sin (A + B)$

14. $\cos (A - B)$

15. $\cos 2B$

16. $\sin 2B$

17. $\sin \dfrac{A}{2}$

18. $\cos \dfrac{A}{2}$

Find exact values for each of the following:

19. $\sin 75°$

20. $\cos 15°$

21. $\tan \dfrac{\pi}{12}$

22. $\cot \dfrac{\pi}{12}$

Write each expression as a single trigonometric function.

23. $\cos 4x \cos 5x - \sin 4x \sin 5x$

24. $\sin 15° \cos 75° + \cos 15° \sin 75°$

25. If $\sin A = -\dfrac{1}{\sqrt{5}}$ with A in QIII, find $\cos 2A$ and $\cos \dfrac{A}{2}$.

26. If $\sec A = \sqrt{10}$ with A in QI, find $\sin 2A$ and $\sin \dfrac{A}{2}$.

27. Find $\tan A$ if $\tan B = \dfrac{1}{2}$ and $\tan (A + B) = 3$.

28. Find $\cos x$ if $\cos 2x = \dfrac{1}{2}$.

Evaluate each expression below without using a calculator or tables. (Assume any variables represent positive numbers.)

29. $\cos \left(\arcsin \dfrac{4}{5} - \arctan 2\right)$

30. $\sin \left(\arccos \dfrac{4}{5} + \arctan 2\right)$

31. $\cos (2 \sin^{-1} x)$

32. $\sin (2\cos^{-1} x)$

33. Rewrite the product $\sin 6x \sin 4x$ as a sum or difference.

34. Rewrite the sum $\cos 15° + \cos 75°$ as a product and simplify.

Spotlight on Success

Gordon, Student Instructor

Math takes time. This fact holds true in the smallest of math problems as much as it does in the most math intensive careers. I see proof in each video I make. My videos get progressively better with each take, though I still make mistakes and find aspects I can improve on with each new video. In order to keep trying to improve in spite of any failures or lack of improvement, something else is needed. For me it is the sense of a specific goal in sight, to help me maintain the desire to put in continued time and effort.

When I decided on the number one university I wanted to attend, I wrote the name of that school in bold block letters on my door, written to remind myself daily of my ultimate goal. Stuck in the back of my head, this end result pushed me little by little to succeed and meet all of the requirements for the university I had in mind. And now I can say I'm at my dream school bringing with me that skill.

I recognize that others may have much more difficult circumstances than my own to endure, with the goal of improving or escaping those circumstances, and I deeply respect that. But that fact demonstrates to me how easy but effective it is, in comparison, to "stay with the problems longer" with a goal in mind of something much more easily realized, like a good grade on a test. I've learned to set goals, small or big, and to stick with them until they are realized.

EQUATIONS

WHAT ARE TRIGONOMETRIC EQUATIONS? Solving trigonometric equations involves using algebraic techniques, trigonometric identities, and properties of trigonometric functions. Depending on the complexity of the equation, solutions may involve finding exact values, using numerical methods, or applying calculus techniques.

WHY ARE TRIGONOMETRIC EQUATIONS IMPORTANT? Learning about trigonometric equations is important not only for understanding fundamental mathematical principles but also for their wide-ranging applications in science, technology, engineering, and everyday life.

WHERE MIGHT YOU SEE TRIGONOMETRIC EQUATIONS? Here are some examples of where you may see trigonometric equations in the real world:

SATELLITE COMMUNICATION Trigonometry is essential in satellite communication systems for determining the angles at which satellite dishes must be pointed to establish a connection. The direction of a satellite dish can be calculated using trigonometric functions based on the satellite's position in the sky and the observer's location on Earth.

OCEANOGRAPHY AND TIDES Trigonometry helps scientists study ocean tides, which are influenced by the gravitational forces of the moon and sun. The variation in water level over time at a particular location can be modeled using trigonometric functions, allowing researchers to predict high and low tides accurately.

Scenario: Tidal Heights

The height of the tide in a particular coastal area can be modeled by the function

$$h(t) = 2 \sin\left(\frac{\pi}{6}t - \frac{\pi}{3}\right) + 3$$

where $h(t)$ represents the height of the tide in meters and t is the time in hours after midnight.

Determine the height of the tide at 6 AM.

Solution

Since 6 AM is 6 hours after midnight, we have $t = 6$.

$$h(6) = 2 \sin\left(\frac{\pi}{6}(6) - \frac{\pi}{3}\right) + 3$$

$$= 2 \sin\left(\pi - \frac{\pi}{3}\right) + 3$$

$$= 2 \sin\left(\frac{2\pi}{3}\right) + 3$$

If $\sin\left(\frac{2\pi}{3}\right) = \sin\left(\pi - \frac{\pi}{3}\right) = \sin\left(\frac{\pi}{3}\right) = \frac{\sqrt{3}}{2}$, then $h(6) = 2\left(\frac{\sqrt{3}}{2}\right) + 3 = \sqrt{3} + 3$.

So the height of the tide at 6 AM is $\sqrt{3} + 3$ meters ≈ 4.73 meters.

Learning Objectives

A   Solve trigonometric equations using the properties of equality.

The solution set for an equation is the set of all numbers which, when used in place of the variable, make the equation a true statement. For example, the solution set for the equation $4x^2 - 9 = 0$ is $\left\{ -\frac{3}{2}, \frac{3}{2} \right\}$ since these are the only two numbers that, when used in place of x, turn the equation into a true statement.

In algebra, the first kind of equations you learned to solve were linear (or first-degree) equations in one variable. Solving these equations was accomplished by applying two important properties: the *addition property of equality* and the *multiplication property of equality*. These two properties were stated as follows:

Addition Property of Equality

For any three algebraic expressions A, B, and C

$$\text{If } A = B$$
$$\text{then } A + C = B + C$$

In Words: Adding the same quantity to both sides of an equation will not change the solution set.

Multiplication Property of Equality

For any three algebraic expressions A, B, and C, with $C \neq 0$,

$$\text{If } A = B$$
$$\text{then } AC = BC$$

In Words: Multiplying both sides of an equation by the same nonzero quantity will not change the solution set.

Here is an example that shows how we use these two properties to solve a linear equation in one variable.

EXAMPLE 1   Solve for x: $5x + 7 = 2x - 5$.

Solution

$$5x + 7 = 2x - 5 \qquad \text{Add } -2x \text{ to each side.}$$

$$3x + 7 = -5 \qquad \text{Add } -7 \text{ to each side.}$$

$$3x = -12 \qquad \text{Multiply each side by } \tfrac{1}{3}.$$

$$x = -4$$

Notice in the last step, we could just as easily have divided both sides by 3 instead of multiplying both sides by $\frac{1}{3}$. Division by a number and multiplication by its reciprocal are equivalent operations.

The process of solving trigonometric equations is very similar to the process of solving algebraic equations. With trigonometric equations, we look for values of an *angle* that will make the equation into a true statement. We usually begin by solving for a specific trigonometric function of that angle and then use the concepts we have developed earlier to find the angle. Here are some examples that illustrate this procedure.

EXAMPLE 2   Solve for x: $2 \sin x - 1 = 0$.

Solution   We can solve for $\sin x$ using our methods from algebra. We then use our knowledge of trigonometry to find x.

$$2 \sin x - 1 = 0$$

$$2 \sin x = 1$$

$$\sin x = \frac{1}{2}$$

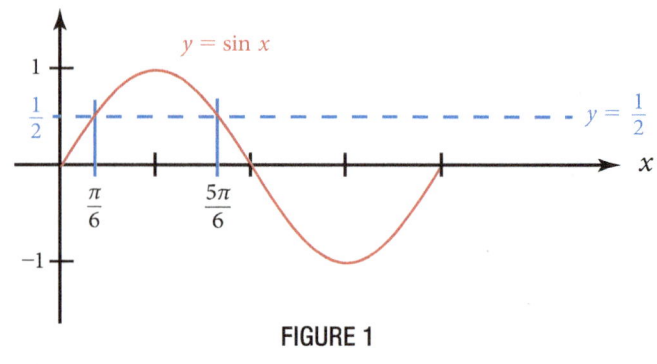

FIGURE 1

From Figure 1 we can see that if we are looking for radian solutions between 0 and 2π, then x is either $\pi/6$ or $5\pi/6$. On the other hand, if we want degree solutions between 0° and 360°, then our solutions will be 30° and 150°. Without the aid of Figure 1, we would reason that, since $\sin x = \frac{1}{2}$, the reference angle for x is 30°. Then, since $\frac{1}{2}$ is a positive number and the sine function is positive in quadrants I and II, x must be 30° or 150°.

Since the sine function is periodic with period 2π (or 360°), adding multiples of 2π (or 360°) will give us all solutions.

Solutions Between 0° and 360° or 0° and 2π	
In Degrees	**In Radians**
$x = 30°$ or $x = 150°$	$x = \frac{\pi}{6}$ or $x = \frac{5\pi}{6}$

All Solutions (k is any integer)	
In Degrees	**In Radians**
$x = 30° + 360°k$	$x = \frac{\pi}{6} + 2k\pi$
or $x = 150° + 360°k$	or $x = \frac{5\pi}{6} + 2k\pi$

EXAMPLE 3   Find all values of x for which $2 \cos x - \sqrt{3} = 0$, if $0° \leq x < 360°$.

Solution   We can solve for $\cos x$ using methods from algebra. Then we use our knowledge of trigonometry to find x.

$$2 \cos x - \sqrt{3} = 0$$

$$2 \cos x = \sqrt{3} \qquad \text{Add } \sqrt{3} \text{ to each side.}$$

$$\cos x = \frac{\sqrt{3}}{2} \qquad \text{Divide each side by 2.}$$

From Section 1.3 we know that if x is between 0° and 360° and $\cos x = \frac{\sqrt{3}}{2}$, then x is either 30° or 330°.

$$x = 30° \quad \text{or} \quad x = 330°$$

Note   Remember, $\cos x$ is positive when x terminates in quadrant I or IV, and the reference angle whose cosine is $\frac{\sqrt{3}}{2}$ is 30°.

EXAMPLE 4   Solve $2 \sin \theta - 3 = 0$, if $0° \leq \theta < 360°$.

Solution   We begin by solving for $\sin \theta$:

$$2 \sin \theta - 3 = 0$$

$$2 \sin \theta = 3 \qquad \text{Add 3 to each side.}$$

$$\sin \theta = \frac{3}{2} \qquad \text{Divide each side by 2.}$$

Since $\sin \theta$ is between -1 and 1 for all values of θ, it can never be $\frac{3}{2}$. Therefore, there is no solution to our equation.

EXAMPLE 5   Solve $3 \sin \theta - 2 = 7 \sin \theta - 1$ for θ, if $0° \le \theta < 360°$.

Solution We can solve for $\sin \theta$ by collecting all the variable terms on the left side and all the constant terms on the right side.

$$3 \sin \theta - 2 = 7 \sin \theta - 1$$

$$-4 \sin \theta - 2 = -1 \qquad \text{Add } -7 \sin \theta \text{ to each side.}$$

$$-4 \sin \theta = 1 \qquad \text{Add 2 to each side.}$$

$$\sin \theta = -\frac{1}{4} \qquad \text{Divide each side by } -4.$$

Since we have not memorized the angle whose sine is $-\frac{1}{4}$, we must convert $-\frac{1}{4}$ to a decimal and use a calculator to find the reference angle:

$$\sin \theta = -\frac{1}{4} = -0.2500$$

We find that the angle whose sine is nearest to 0.2500 is 14.5°. Therefore, the reference angle is 14.5°. Since $\sin \theta$ is negative, θ will terminate in quadrant III or IV.

In Quadrant III We Have:	**In Quadrant IV We Have:**
$\theta \approx 180° + 14.5°$	$\theta \approx 360° - 14.5°$
$= 194.5°$	$= 345.5°$

Calculator Note Remember, because of the restricted values on the inverse sine function, if you enter -0.2500 and press the $\sin^{-1}$ key, your calculator will display approximately $-14.5°$, which is incorrect. The best way to proceed is to use your calculator to find the reference angle by entering 0.2500 and pressing the $\sin^{-1}$ key. Then do the rest of the calculations as we have above.

The next trigonometric equation we will solve is quadratic in form. Let's review how to solve quadratic equations in algebra.

EXAMPLE 6   Solve $2x^2 - 9x = 5$ for x.

Solution We begin by writing the equation in standard form (0 on one side—decreasing powers of the variable on the other). We then factor the left side and set each factor equal to 0.

$$2x^2 - 9x = 5$$

$$2x^2 - 9x - 5 = 0 \qquad \text{Standard form}$$

$$(2x + 1)(x - 5) = 0 \qquad \text{Factor}$$

$$2x + 1 = 0 \quad \text{or} \quad x - 5 = 0 \qquad \text{Set each factor to 0}$$

$$x = -\frac{1}{2} \qquad\qquad x = 5 \qquad \text{Solving resulting equations}$$

The two solutions, $x = -\frac{1}{2}$ and $x = 5$, are the only two numbers that satisfy the original equation.

EXAMPLE 7   Solve $2 \cos^2 t - 9 \cos t = 5$, if $0 \le t < 2\pi$.

Solution The fact that t is between 0 and 2π indicates that we are to write out solutions in radians.

$$2 \cos^2 t - 9 \cos t = 5$$

$$2 \cos^2 t - 9 \cos t - 5 = 0 \qquad \text{Standard form}$$

$$(2 \cos t + 1)(\cos t - 5) = 0 \qquad \text{Factor.}$$

$$2 \cos t + 1 = 0 \quad \text{or} \quad \cos t - 5 = 0 \qquad \text{Set each factor to 0.}$$

$$\cos t = -\frac{1}{2} \qquad\qquad \cos t = 5$$

The first result, $\cos t = -\frac{1}{2}$, gives us $t = \frac{2\pi}{3}$ or $t = \frac{4\pi}{3}$. The second result, $\cos t = 5$, is meaningless. For any value of t, $\cos t$ must be between -1 and 1. It can never be 5.

If you look at the algebraic steps in the above two examples you'll notice they are identical. If $x = \cos t$ in Example 6, the equation becomes that of Example 7. The only difference is in the interpretation of the final answer.

EXAMPLE 8   Solve $2 \sin^2 \theta + 2 \sin \theta - 1 = 0$, if $0° \leq \theta < 360°$.

Solution The equation is already in standard form. However, if we try to factor the left side, we find it does not factor. We must use the quadratic formula.

The coefficients a, b, and c are

$$a = 2 \qquad b = 2 \qquad c = -1$$

Using these numbers, we solve for $\sin \theta$ as follows:

$$\sin \theta = \frac{-2 \pm \sqrt{4 - 4(2)(-1)}}{2(2)}$$

$$= \frac{-2 \pm \sqrt{12}}{4}$$

$$= \frac{-2 \pm 2\sqrt{3}}{4}$$

$$= \frac{-1 \pm \sqrt{3}}{2}$$

Using the approximation $\sqrt{3} \approx 1.7321$, we arrive at the following decimal approximations for $\sin \theta$:

$$\sin \theta \approx \frac{-1 + \sqrt{3}}{2} \qquad \text{or} \qquad \sin \theta \approx \frac{-1 - \sqrt{3}}{2}$$

$$\sin \theta \approx 0.3660 \qquad \text{or} \qquad \sin \theta \approx -1.3660$$

We will not obtain any solutions from the second expression, $\sin \theta = -1.3660$, since $\sin \theta$ must be between -1 and 1. For $\sin \theta = 0.3660$, using a calculator to find the reference angle to be $\hat{\theta} = \sin^{-1}(0.3660) \approx 21.5°$. Since $\sin \theta$ is positive, θ must terminate in quadrant I or II. Therefore,

$$\theta \approx 21.5° \qquad \text{or} \qquad \theta \approx 180° - 21.5° = 158.5°$$

Vocabulary

Use the vocabulary words below to fill in the blanks in the sentences.

one   multiples   nonzero   no   two

1. Multiplying both sides of an equation by the same _____ quantity will not change the solution set.

2. When we solve a trigonometric equation for all values of the variable that make the equation true, we sometimes find solutions between 0° and 360° and then add on _____ of 360°.

3. If we solve the equation $\sin \theta = -0.2500$, for $0° < \theta < 360°$ we know there are _____ solutions, and our calculator will give us just _____ of those solutions. We find the other one ourselves.

4. Because $\sin \theta$ is always a number between -1 and 1, we know the equation $\sin \theta = \frac{3}{2}$ has _____ solution.

Solve each equation for θ if $0° \leq \theta \leq 360°$. Do not use a calculator.

1. $2 \sin \theta = 1$

2. $2 \cos \theta = 1$

3. $2 \cos \theta - \sqrt{3} = 0$

4. $2 \cos \theta + \sqrt{3} = 0$

5. $2 \tan \theta + 2 = 0$

6. $\sqrt{3} \cot \theta - 1 = 0$

Solve each equation for all solutions between 0 and 2π (including 0 and 2π, if they are solutions). Give all answers as exact values in radians. Do not use a calculator.

7. $4 \sin x - \sqrt{3} = 2 \sin x$

8. $\sqrt{3} + 5 \sin x = 3 \sin x$

9. $2 \cos t = 6 \cos t - 2\sqrt{3}$

10. $5 \cos t + 2\sqrt{3} = \cos t$

11. $3 \sin x - 5 = -2 \sin x$

12. $3 \sin x + 4 = 4$

Find all solutions between 0° and 360° (including 0° and 360°, if they are solutions). Use a calculator on the last step and write all answers to the nearest tenth of a degree.

13. $4 \sin \theta - 3 = 0$ **14.** $4 \sin \theta + 3 = 0$

15. $2 \cos \theta - 5 = 3 \cos \theta - 2$ **16.** $4 \cos \theta - 1 = 3 \cos \theta + 4$

17. $\sin \theta - 3 = 5 \sin \theta$ **18.** $\sin \theta - 4 = -2 \sin \theta$

Solve for x, if $0 \leq x < 2\pi$. Write your answers in exact values only.

19. $(\sin x - 1)(2 \sin x - 1) = 0$ **20.** $(\cos x - 1)(2 \cos x - 1) = 0$

21. $\tan x \, (\tan x - 1) = 0$ **22.** $\tan x \, (\tan x + 1) = 0$

23. $\sin x + 2 \sin x \cos x = 0$ **24.** $\cos x - 2 \sin x \cos x = 0$

25. $2 \sin^2 x - \sin x - 1 = 0$ **26.** $2 \cos^2 x + \cos x - 1 = 0$

Solve for θ, if $0° \leq \theta < 360°$.

27. $(2 \cos \theta + \sqrt{3})(2 \cos \theta + 1) = 0$ **28.** $(2 \sin \theta - \sqrt{3})(2 \sin \theta - 1) = 0$

29. $\sqrt{3} \tan \theta - 2 \sin \theta \tan \theta = 0$ **30.** $\tan \theta - 2 \cos \theta \tan \theta = 0$

31. $2 \cos^2 \theta + 11 \cos \theta = -5$ **32.** $2 \sin^2 \theta - 7 \sin \theta = -3$

Use the quadratic formula to find all solutions between 0° and 360° to the nearest tenth of a degree.

33. $2 \sin^2 \theta - 2 \sin \theta - 1 = 0$ **34.** $2 \cos^2 \theta - 2 \cos \theta - 1 = 0$

35. $\cos^2 \theta + \cos \theta - 1 = 0$ **36.** $\sin^2 \theta + \sin \theta - 1 = 0$

37. $4 \sin^2 \theta + 1 = 4 \sin \theta$ **38.** $1 - 4 \cos \theta = -4 \cos^2 \theta$

Write expressions representing all solutions to the equations you solved in the problems below.

39. Problem 1   **40.** Problem 2

41. Problem 7   **42.** Problem 8

43. Problem 11   **44.** Problem 12

45. Problem 13   **46.** Problem 14

If a projectile (such as a bullet) is fired into the air with an initial velocity of v at an angle of inclination θ, then the height h of the projectile at time t is given by

$$h = -16t^2 + vt \sin \theta$$

47. Projectile Motion Give the equation for the height, if v is 1,500 feet per second and θ is 30°.

48. Projectile Motion Give the equation h, if v is 600 feet per second and θ is 45°. (Leave your answer in exact value form.)

49. Projectile Motion Use the equation found in Problem 47 to find the height of the object after 2 seconds.

50. Projectile Motion Use the equation found in Problem 48 to find the height of the object after $\sqrt{3}$ seconds.

51. Projectile Motion Find the angle of inclination θ of a rifle barrel, if a bullet fired at 1500 feet per second takes 2 seconds to reach a height of 750 feet. Give your answer to the nearest tenth of a degree.

52. Projectile Motion Find the angle of inclination θ of a rifle barrel, if a bullet fired at 1500 feet per second takes 3 seconds to reach a height of 750 feet. Give your answer to the nearest tenth of a degree.

Getting Ready for the Next Section

53. Write the double-angle formula for $\sin 2A$.

54. Write $\cos 2A$ in terms of $\sin A$ only.

55. Write $\cos 2A$ in terms of $\cos A$ only.

56. Write $\cos 2A$ in terms of $\sin A$ and $\cos A$.

57. Expand $\sin(\theta + 45°)$ and then simplify.

58. Expand $\sin(\theta + 30°)$ and then simplify.

59. Find the exact value of $\sin 15°$ by using the formula for $\sin(45° - 30°)$.

60. Find the exact value of $\cos 75°$ by using the formula for $\cos(45° + 30°)$.

Learning Objectives

A Solve trigonometric equations using trigonometric identities.

Now we will extend the process of solving trigonometric equations to include the use of trigonometric identities. That is, we will use our knowledge of identities to replace some parts of the equations we are solving with equivalent expressions that will make the equations easier to solve. Here are some examples.

EXAMPLE 1 Solve $2 \cos x - 1 = \sec x$, if $0 \leq x < 2\pi$.

Solution To solve the equation as we have solved the equations in the previous examples, we must write each term using the same trigonometric functions. To do so, we can use a reciprocal identity to write $\sec x$ in terms of $\cos x$:

$$2 \cos x - 1 = \frac{1}{\cos x}$$

To clear the equation of fractions, we multiply both sides by $\cos x$:

$$\cos x(2 \cos x - 1) = \frac{1}{\cos x} \cdot \cos x$$

$$2 \cos^2 x - \cos x = 1$$

We are left with a quadratic equation, which we write in standard form and then solve.

$2 \cos^2 x - \cos x - 1 = 0$	Standard form
$(2 \cos x + 1)(\cos x - 1) = 0$	Factor.
$2 \cos x + 1 = 0$ or $\cos x - 1 = 0$	Set each factor to 0.

$$\cos x = -\frac{1}{2} \qquad\qquad \cos x = 1$$

$$x = \frac{2\pi}{3}, \frac{4\pi}{3} \qquad\qquad x = 0$$

EXAMPLE 2 Solve $\sin 2\theta + \sqrt{2} \cos \theta = 0$, if $0° \leq \theta < 360°$.

Solution In order to solve this equation, both trigonometric functions must be functions of the same angle. As the equation stands now, one angle is 2θ, while the other is θ. We can write everything as a function of θ by using the double-angle identity $\sin 2\theta = 2 \sin \theta \cos \theta$.

$\sin 2\theta + \sqrt{2} \cos \theta = 0$	
$2 \sin \theta \cos \theta + \sqrt{2} \cos \theta = 0$	Identity
$\cos \theta(2 \sin \theta + \sqrt{2}) = 0$	Factor out $\cos \theta$.
$\cos \theta = 0$ or $2 \sin \theta + \sqrt{2} = 0$	Set each factor to 0.

$$\sin \theta = -\frac{\sqrt{2}}{2}$$

$$\theta = 90°, 270° \qquad\qquad \theta = 225°, 315°$$

EXAMPLE 3   Solve $\cos 2\theta + 3 \sin \theta - 2 = 0$, if $0° \leq \theta \leq 360°$.

Solution   We have the same problem with this equation as we did with the equation in Example 2. We must rewrite $\cos 2\theta$ in terms of functions of just θ. Recall that there are three forms of the double-angle identity for $\cos 2\theta$. We choose the double-angle identity that involves $\sin \theta$ only, since the middle term of our equation contains $\sin \theta$ and we want all the terms to involve the same trigonometric function.

$$\cos 2\theta + 3 \sin \theta - 2 = 0$$

$1 - 2 \sin^2 \theta + 3 \sin \theta - 2 = 0$	Identity
$-2 \sin^2 \theta + 3 \sin \theta - 1 = 0$	Simplify.
$2 \sin^2 \theta - 3 \sin \theta + 1 = 0$	Multiply each side by -1.
$(2 \sin \theta - 1)(\sin \theta - 1) = 0$	Factor.
$2 \sin \theta - 1 = 0 \quad \text{or} \quad \sin \theta - 1 = 0$	Set factors to 0.

$$\sin \theta = \frac{1}{2} \qquad\qquad \sin \theta = 1$$

$$\theta = 30°, 150° \qquad \theta = 90°$$

EXAMPLE 4   Solve $4 \cos^2 x + 4 \sin x - 5 = 0$, if $0 \leq x < 2\pi$.

Solution   We cannot factor and solve this quadratic equation until each term involves the same trigonometric function. If we replace the $\cos^2 x$ in the first term with $1 - \sin^2 x$, we will obtain an equation that involves the sine function only.

$$4 \cos^2 x + 4 \sin x - 5 = 0$$

$4(1 - \sin^2 x) + 4 \sin x - 5 = 0$	$\cos^2 x = 1 - \sin^2 x$
$4 - 4 \sin^2 x + 4 \sin x - 5 = 0$	Distributive property
$-4 \sin^2 x + 4 \sin x - 1 = 0$	Simplify.
$4 \sin^2 x - 4 \sin x + 1 = 0$	Multiply each side by -1.
$(2 \sin x - 1)^2 = 0$	Factor.
$2 \sin x - 1 = 0$	Set factor to 0.

$$\sin x = \frac{1}{2}$$

$$x = \frac{\pi}{6}, \frac{5\pi}{6}$$

EXAMPLE 5   Solve $\sin \theta - \cos \theta = 1$, if $0° \leq \theta \leq 360°$.

Solution   If we separate $\sin \theta$ and $\cos \theta$ on opposite sides of the equal sign, and then square both sides of the equation, we will be able to use an identity to write the equation in terms of one trigonometric function only.

$$\sin \theta - \cos \theta = 1$$

$$\sin \theta = 1 + \cos \theta \qquad \textit{Add } \cos \theta \textit{ to each side}$$

$$\sin^2 \theta = (1 + \cos \theta)^2 \qquad \textit{Square each side}$$

$$\sin^2 \theta = 1 + 2 \cos \theta + \cos^2 \theta \qquad \textit{Expand } (1 + \cos \theta)^2$$

$$1 - \cos^2 \theta = 1 + 2 \cos \theta + \cos^2 \theta \qquad \sin^2 \theta = 1 - \cos^2 \theta$$

$$0 = 2 \cos \theta + 2 \cos^2 \theta \qquad \textit{Standard form}$$

$$0 = 2 \cos \theta (1 + \cos \theta) \qquad \textit{Factor}$$

$$2 \cos \theta = 0 \quad \text{or} \quad 1 + \cos \theta = 0 \qquad \textit{Set factors to 0}$$

$$\cos \theta = 0 \qquad\qquad \cos \theta = -1$$

$$\theta = 90°, 270° \qquad\qquad \theta = 180°$$

We have three possible solutions, some of which may be extraneous since we squared both sides of the equation in Step 2. Any time we raise both sides of an equation to an even power, we have the possibility of introducing extraneous solutions. We must check each possible solution in our original equation.

Checking $\theta = 90°$	*Checking* $\theta = 180°$	*Checking* $\theta = 270°$
$\sin 90° - \cos 90° \overset{?}{=} 1$	$\sin 180° - \cos 180° \overset{?}{=} 1$	$\sin 270° - \cos 270° \overset{?}{=} 1$
$1 - 0 \overset{?}{=} 1$	$0 - (-1) \overset{?}{=} 1$	$-1 - 0 \overset{?}{=} 1$
$1 = 1$	$1 = 1$	$-1 \neq 1$
$\theta = 90°$ is a solution	$\theta = 180°$ is a solution	$\theta = 270°$ is not a solution

All possible solutions, except $\theta = 270°$, produce true statements when used in place of the variable in the original equation. $\theta = 270°$ is an extraneous solution produced by squaring both sides of the equation. Our solution set is $\{90°, 180°\}$.

Problem Set 6.2

Vocabulary

Use the vocabulary words below to fill in the blanks in the sentences.

degrees   double angle   extraneous   radians   reciprocal

1. To solve the equation $2 \cos x - 1 = \sec x$ we first use a
_____ identity to write $\sec x$ in terms of $\cos x$.

2. Generally if we are solving a trigonometric equation for x we give the
solutions in _____ , and when we solve a trigonometric equation
of θ we give the solutions in _____ .

3. To solve the equation $\sin 2\theta + \sqrt{2} \cos \theta = 0$ we first use a _____
identity to write all trigonometric functions in terms of just θ and not 2θ.

4. When we square both sides of an equation in the process of solving
that equation, we have the possibility that we have introduced
_____ solutions. And that is why we need to check each possible
solution in the original equation.

For all equations written in terms of θ, find all degree solutions from $\theta = 0°$ to
$\theta = 360°$. If the equation is written in terms of x, write your solutions in radians
using exact values for $0 \leq x \leq 2\pi$.

1. $\sqrt{3} \sec \theta = 2$

2. $\sqrt{2} \csc \theta = 2$

3. $\sqrt{2} \csc x + 5 = 3$

4. $2\sqrt{3} \sec x + 7 = 3$

5. $4 \sin x - 2 \csc x = 0$

6. $4 \cos x - 3 \sec x = 0$

7. $\sec \theta - 2 \tan \theta = 0$

8. $\csc \theta + 2 \cot \theta = 0$

9. $\sin 2\theta - \cos \theta = 0$

10. $2 \sin \theta + \sin 2\theta = 0$

11. $2 \cos \theta + 1 = \sec \theta$

12. $2 \sin \theta - 1 = \csc \theta$

13. $\cos 2x - 3 \sin x - 2 = 0$

14. $\cos 2x - \cos x - 2 = 0$

15. $\cos \theta - \cos 2\theta = 0$

16. $\sin \theta = -\cos 2\theta$

17. $2\cos^2 x + \sin x - 1 = 0$

18. $2\sin^2 x - \cos x - 1 = 0$

19. $4\sin^2 x + 4\cos x - 5 = 0$

20. $4\cos^2 x - 4\sin x - 5 = 0$

21. $2\sin x + \cot x - \csc x = 0$

22. $2\cos x + \tan x = \sec x$

23. $\sin\theta + \cos\theta = \sqrt{2}$

24. $\sin\theta - \cos\theta = \sqrt{2}$

25. $\sqrt{3}\sin\theta + \cos\theta = \sqrt{3}$

26. $\sin\theta - \sqrt{3}\cos\theta = \sqrt{3}$

27. $\sqrt{3}\sin\theta - \cos\theta = 1$

28. $\sin\theta - \sqrt{3}\cos\theta = 1$

29. $\sin\dfrac{x}{2} - \cos x = 0$

30. $\sin\dfrac{x}{2} + \cos x = 1$

31. $\cos\dfrac{x}{2} - \cos x = 1$

32. $\cos\dfrac{x}{2} - \cos x = 0$

Write expressions that give all solutions to the equations you solved in the problems given below.

33. Problem 3

34. Problem 4

35. Problem 23

36. Problem 24

37. Problem 31

38. Problem 32

39. Physiology In the human body, the value of θ that makes the expression below equal to zero is the angle at which an artery of radius r will branch off from a larger artery of radius R in order to minimize the energy loss due to friction. Show that the expression below is 0 when $\cos\theta = \dfrac{r^4}{R^4}$.

$$r^4 \csc^2\theta - R^4 \csc\theta \cot\theta$$

40. Physiology Find the value of θ that makes the equation in Problem 39 equal to 0, if $r = 2$ millimeters and $R = 4$ millimeters. (Give your answer to the nearest tenth of a degree.)

Getting Ready for the Next Section

41. Find all values of θ between 0° and 720° for which $\cos \theta = \dfrac{\sqrt{3}}{2}$.

42. Find all values of x between 0 and 4π for which $\sin x = \dfrac{1}{\sqrt{2}}$.

43. Find all values of x between 0 and 4π for which $\tan x = 1$.

44. Find all values of θ between 0° and 720° for which $\sin \theta = 0$.

Learning Objectives

A   Solve trigonometric equations involving multiple angles.

In this section, we will consider equations that contain multiple angles. We will use most of the same techniques to solve these equations that we have used in the past. We have to be careful at the last step, however, when our equations contain multiple angles. Here is an example.

EXAMPLE 1   Solve $\cos 2\theta = \frac{\sqrt{3}}{2}$, if $0° \leq \theta \leq 360°$.

Solution   The equation cannot be simplified further. Since we are looking for θ between $0°$ and $360°$, we must first find all values of 2θ between $0°$ and $720°$ that satisfy the equation, because

$$\text{if } 0° \leq \theta \leq 360°, \text{ then } 2(0°) \leq 2\theta \leq 2(360°)$$

$$\text{or } 0° \leq 2\theta \leq 720°$$

To find all values of 2θ between $0°$ and $720°$ that satisfy our original equation, we first find all solutions to $\cos 2\theta = \frac{\sqrt{3}}{2}$ between $0°$ and $360°$. Then we add $360°$ to each of these solutions to obtain all solutions between $0°$ and $720°$. That is,

$$\text{If } \cos 2\theta = \frac{\sqrt{3}}{2}$$

$$\overbrace{30° + 360°}^{} \quad \overbrace{330° + 360°}^{}$$

$$\text{then } 2\theta = 30° \text{ or } 2\theta = 330° \text{ or } 2\theta = 390° \text{ or } 2\theta = 690°$$

Dividing both sides of each equation by 2 gives us all θ between $0°$ and $360°$ that satisfy $\cos 2\theta = \frac{\sqrt{3}}{2}$.

$2\theta = 30°$	$2\theta = 330°$	$2\theta = 390°$	$2\theta = 690°$
$\theta = 15°$	$\theta = 165°$	$\theta = 195°$	$\theta = 345°$

If we had originally been asked to find *all solutions* to the equation in Example 1, instead of only those between $0°$ and $360°$, our work would have been a little less complicated. To find all solutions, we would first find all values of 2θ between $0°$ and $360°$ that satisfy the equation and then we would add on multiples of $360°$, since the period of the cosine function is $360°$. After that, it is just a matter of dividing everything by 2.

$$\text{If } \cos 2\theta = \frac{\sqrt{3}}{2}$$

then $2\theta = 30° + 360°k$ or $2\theta = 330° + 360°k$ *k any integer*
          $\theta = 15° + 180°k$ or $\theta = 165° + 180°k$ *Divide by 2*

The last line gives us *all values* of θ that satisfy $\cos 2\theta = \frac{\sqrt{3}}{2}$. Note that $k = 0$ and $k = 1$ will give us the four solutions between $0°$ and $360°$ that we found in Example 1. Note also that we are including negative angles as solutions since k can assume values of $-1, -2, -3$, and so forth.

EXAMPLE 2   Find all solutions to tan $3x = 1$, if x is measured in radians with exact values.

Solution First we find all values of $3x$ between 0 and π that satisfy tan $3x = 1$, and then we add on multiples of π because the period of the tangent function is π. After that, we simply divide by 3 to solve for x.

$$\text{If tan } 3x = 1$$

$$\text{then } 3x = \frac{\pi}{4} + k\pi$$

$$x = \frac{\pi}{12} + \frac{k\pi}{3} \qquad \textit{Divide by 3}$$

Note that $k = 0$, 1, and 2 will give us all values of x between 0 and π that satisfy tan $3x = 1$.

EXAMPLE 3   Solve sin $2x$ cos x + cos $2x$ sin $x = \frac{1}{\sqrt{2}}$, if $0 \le x \le 2\pi$.

Solution We can simplify the left side by using the formula for $\sin(A + B)$.

$$\sin 2x \cos x + \cos 2x \sin x = \frac{1}{\sqrt{2}}$$

$$\sin(2x + x) = \frac{1}{\sqrt{2}}$$

$$\sin 3x = \frac{1}{\sqrt{2}}$$

Finding all solutions looks like this

$$3x = \frac{\pi}{4} + 2k\pi \quad \text{or} \quad 3x = \frac{3\pi}{4} + 2k\pi \qquad \textit{k any integer}$$

$$x = \frac{\pi}{12} + \frac{2k\pi}{3} \quad \text{or} \quad x = \frac{\pi}{4} + \frac{2k\pi}{3} \qquad \textit{Divide by 3}$$

To find only those solutions that lie between 0 and 2π, we let k take on values of 0, 1, and 2. Doing so results in the following solutions:

$$x = \frac{\pi}{12}, \ \frac{\pi}{4}, \ \frac{9\pi}{12}, \ \frac{11\pi}{12}, \ \frac{17\pi}{12}, \ \text{ and } \ \frac{19\pi}{12}$$

EXAMPLE 4   Find all solutions to $2\sin^2 3\theta - \sin 3\theta - 1 = 0$, if θ is measured in degrees.

Solution We have an equation that is quadratic in sin 3θ. We factor and solve as usual.

$$2\sin^2 3\theta - \sin 3\theta - 1 = 0 \qquad \textit{Standard form}$$

$$(2\sin 3\theta + 1)(\sin 3\theta - 1) = 0 \qquad \textit{Factor}$$

$$2\sin 3\theta + 1 = 0 \quad \text{or} \quad \sin 3\theta - 1 = 0 \qquad \textit{Set factors to 0}$$

$$\sin 3\theta = -\frac{1}{2} \quad \text{or} \qquad \sin 3\theta = 1$$

$$3\theta = 210° + 360°k \quad \text{or} \quad 3\theta = 330° + 360°k \quad \text{or} \quad 3\theta = 90° + 360°k$$

$$\theta = 70° + 120°k \quad \text{or} \quad \theta = 110° + 120°k \quad \text{or} \quad \theta = 30° + 120°k$$

EXAMPLE 5   Find all solutions, in radians, for $\tan^2 3x = 1$.

Solution Taking the square root of both sides we have

$$\tan^2 3x = 1$$
$$\tan 3x = \pm 1 \qquad \textcolor{green}{\textit{Square root of both sides}}$$

Since the period of the tangent function is π, we find all solutions to $\tan 3x = 1$ and $\tan 3x = -1$ between 0 and π, and then we add multiples of π to these solutions. Finally, we divide each side of the resulting equations by 3.

$$3x = \frac{\pi}{4} + k\pi \qquad \text{or} \qquad 3x = \frac{3\pi}{4} + k\pi$$

$$x = \frac{\pi}{12} + \frac{k\pi}{3} \qquad \text{or} \qquad x = \frac{\pi}{4} + \frac{k\pi}{3}$$

EXAMPLE 6   Solve $\sin\theta - \cos\theta = 1$ if $0° \leq \theta \leq 360°$.

Solution We have solved this equation previously in Section 6.2. This time we will simply square both sides.

$$\sin\theta - \cos\theta = 1$$
$$(\sin\theta - \cos\theta)^2 = 1^2 \qquad \textcolor{green}{\textit{Square both sides}}$$
$$\sin^2\theta - 2\sin\theta\cos\theta + \cos^2\theta = 1 \qquad \textcolor{green}{\textit{Expand left side}}$$
$$-2\sin\theta\cos\theta + 1 = 1 \qquad \textcolor{green}{\sin^2\theta + \cos^2\theta = 1}$$
$$-2\sin\theta\cos\theta = 0 \qquad \textcolor{green}{\textit{Add} -1 \textit{ to both sides}}$$
$$-\sin 2\theta = 0 \qquad \textcolor{green}{\textit{Double-angle identity}}$$
$$2\theta = 0°, 180°, 360°, 540°, 720°$$
$$\theta = 0°, 90°, 180°, 270°, 360°$$

Since we squared both sides of the equation in Step 2, we must check all the possible solutions to see if they satisfy the original equation. Doing so gives us solutions $x = 90°$ and $180°$. The others are all extraneous.

Problem Set 6.3

Find all solutions if $0° \leq \theta \leq 360°$.

1. $\sin 2\theta = \dfrac{\sqrt{3}}{2}$

2. $\sin 2\theta = -\dfrac{\sqrt{3}}{2}$

3. $\tan 2\theta = -1$

4. $\cot 2\theta = 1$

5. $\cos 3\theta = -1$

6. $\sin 3\theta = -1$

Find all solutions if $0 \leq x \leq 2\pi$. Use exact values only.

7. $\sin 2x = \dfrac{1}{\sqrt{2}}$

8. $\cos 2x = \dfrac{1}{\sqrt{2}}$

9. $\sec 3x = -1$

10. $\csc 3x = 1$

11. $\tan 2x = \sqrt{3}$

12. $\tan 2x = -\sqrt{3}$

Find all degree solutions for each of the following:

13. $\sin 2\theta = \dfrac{1}{2}$

14. $\sin 2\theta = -\dfrac{\sqrt{3}}{2}$

15. $\cos 3\theta = 0$

16. $\cos 3\theta = -1$

17. $\sin 10\theta = \dfrac{\sqrt{3}}{2}$

18. $\cos 8\theta = \dfrac{1}{2}$

Find all solutions if $0 \leq x \leq 2\pi$. Use exact values only.

19. $\sin 2x \cos x + \cos 2x \sin x = \dfrac{1}{2}$

20. $\sin 2x \cos x + \cos 2x \sin x = -\dfrac{1}{2}$

21. $\cos 2x \cos x - \sin 2x \sin x = -\dfrac{\sqrt{3}}{2}$

22. $\cos 2x \cos x - \sin 2x \sin x = \dfrac{1}{\sqrt{2}}$

Find all solutions in radians using exact values only.

23. $\sin 3x \cos 2x + \cos 3x \sin 2x = 1$

24. $\sin 2x \cos 3x + \cos 2x \sin 3x = -1$

25. $\sin^2 4x = 1$

26. $\cos^2 4x = 1$

27. $\cos^3 5x = -1$

28. $\sin^3 5x = -1$

Find all degree solutions.

29. $2 \sin^2 3\theta + \sin 3\theta - 1 = 0$

30. $2 \sin^2 3\theta + 3 \sin 3\theta + 1 = 0$

31. $2 \cos^2 2\theta + 3 \cos 2\theta + 1 = 0$

32. $2 \cos^2 2\theta - \cos 2\theta - 1 = 0$

33. $\tan^2 3\theta = 3$

34. $\cot^2 3\theta = 1$

Find all solutions if $0° \le \theta \le 360°$.

35. $\cos \theta - \sin \theta = 1$

36. $\sin \theta - \cos \theta = 1$

37. $\sin \theta + \cos \theta = -1$

38. $\cos \theta - \sin \theta = -1$

39. Geometry The formula below gives the relationship between the number of sides, n, the radius, r, and the length of each side, l, in a regular polygon. Find n if $l = r$.

$$l = 2r \sin \frac{180°}{n}$$

40. Geometry If central angle θ cuts off a chord of length c in a circle of radius r, then the relationship between θ, c, and r is given by

$$2r \sin \frac{\theta}{2} = c$$

Find θ, if $c = \sqrt{3}\,r$.

41. Rotating Light Previously, we found the equation that gives d in terms of t in Figure 1 to be $d = 10 \tan \pi t$. If a person is standing against the wall, 10 feet from point A, how long after the light is at point A will the person see the light? (You must find t when d is 10.)

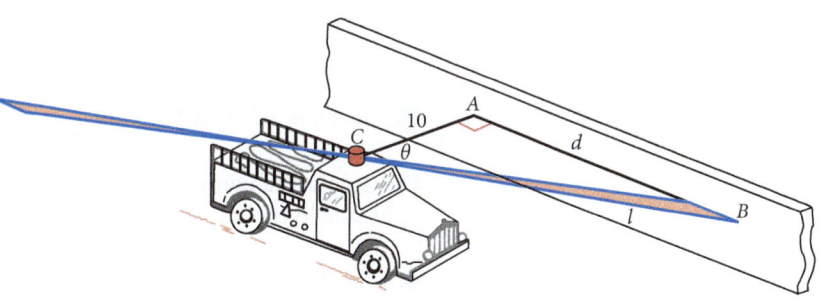

FIGURE 1

42. Rotating Beacon Previously, you found the equation that gives d in terms of t in Figure 2 to be $d = 100 \tan \frac{\pi}{2} t$. Two people are sitting on the wall. One of them is directly opposite the lighthouse, while the other person is 100 feet further down the wall. How long after one of them sees the light does the other one see the light? (There are two solutions depending on who sees the light first.)

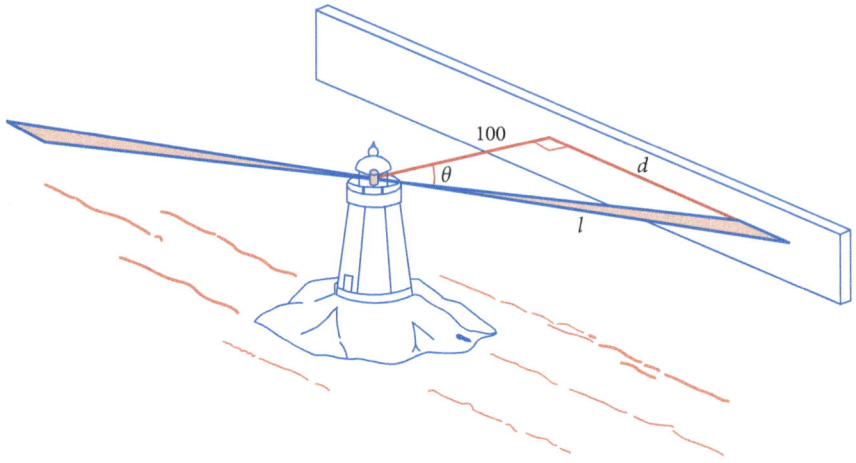

FIGURE 2

43. Find the smallest positive value of t for which $\sin 2\pi t = \dfrac{1}{2}$.

44. Find the smallest positive value of t for which $\sin 2\pi t = \dfrac{1}{\sqrt{2}}$.

Getting Ready for the Next Section

Graph each equation between 0 and 4π.

45. $y = \sin x + \cos x$ **46.** $y = \sin x - \cos x$

Write as a single trigonometric function.

47. $\sin x \cos 45° + \cos x \sin 45°$

48. $\sin x \cos 30° + \cos x \sin 30°$

49. Graph one complete cycle of $y = \sin x \cos \dfrac{\pi}{4} + \cos x \sin \dfrac{\pi}{4}$ by rewriting the right side in the form $\sin(A + B)$.

50. Graph one complete cycle of $y = 2\left(\sin x \cos \dfrac{\pi}{3} - \cos x \sin \dfrac{\pi}{3} \right)$ by rewriting the right side in the form $2 \sin(A - B)$.

PARAMETRIC EQUATIONS AND FURTHER GRAPHING

6.4

6.4 Videos

KEY WORDS

parametric equations

parameter

Learning Objectives

A   Solve problems involving parametric equations.

B   Graph trigonometric equations of the form $y = a \sin x + b \cos x$.

Many times in mathematics a set of points (x, y) in the plane is described by a pair of equations rather than one equation. For example, $x = \cos t$ and $y = \sin t$, where t is any real number, are a pair of equations that describe a certain curve in the xy-plane. The equations are called **parametric equations** and the variable t is called the **parameter**. The variable t does not appear as part of the graph, but rather, produces values of x and y that do appear as ordered pairs (x, y) on the graph.

Table 1 (see next page) shows the values of x and y produced by substituting convenient values of t into each equation.

Plotting each ordered pair (x, y) from the last column of Table 1 on a coordinate system, we see that the set of points form the unit circle, starting at $(1, 0)$ when $t = 0$ and ending at $(1, 0)$ when $t = 2\pi$. (See Figure 1.)

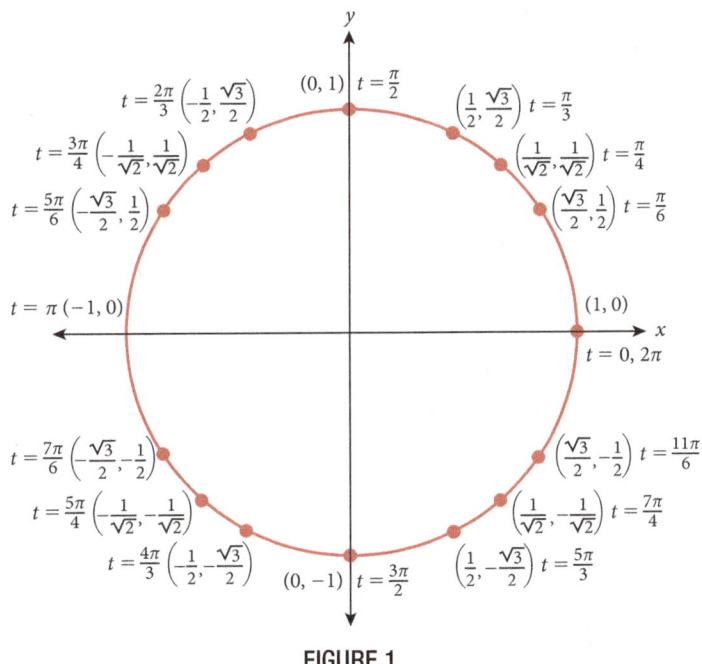

FIGURE 1

We could have found this graph without going to as much work by using the Pythagorean identity $\cos^2 t + \sin^2 t = 1$. Substituting $x = \cos t$ and $y = \sin t$ into

$$\cos^2 t + \sin^2 t = 1$$

we have

$$x^2 + y^2 = 1$$

which is the equation of the unit circle. The process of going directly to an equation that contains x and y but not t is called *eliminating the parameter*.

Table 1			
t	$x = \cos t$	$y = \sin t$	(x, y)
0	$x = \cos 0 = 1$	$y = \sin 0 = 0$	$(1, 0)$
$\dfrac{\pi}{6}$	$x = \cos \dfrac{\pi}{6} = \dfrac{\sqrt{3}}{2}$	$y = \sin \dfrac{\pi}{6} = \dfrac{1}{2}$	$\left(\dfrac{\sqrt{3}}{2}, \dfrac{1}{2} \right)$
$\dfrac{\pi}{4}$	$x = \cos \dfrac{\pi}{4} = \dfrac{1}{\sqrt{2}}$	$y = \sin \dfrac{\pi}{4} = \dfrac{1}{\sqrt{2}}$	$\left(\dfrac{1}{\sqrt{2}}, \dfrac{1}{\sqrt{2}} \right)$
$\dfrac{\pi}{3}$	$x = \cos \dfrac{\pi}{3} = \dfrac{1}{2}$	$y = \sin \dfrac{\pi}{3} = \dfrac{\sqrt{3}}{2}$	$\left(\dfrac{1}{2}, \dfrac{\sqrt{3}}{2} \right)$
$\dfrac{\pi}{2}$	$x = \cos \dfrac{\pi}{2} = 0$	$y = \sin \dfrac{\pi}{2} = 1$	$(0, 1)$
$\dfrac{2\pi}{3}$	$x = \cos \dfrac{2\pi}{3} = -\dfrac{1}{2}$	$y = \sin \dfrac{2\pi}{3} = \dfrac{\sqrt{3}}{2}$	$\left(-\dfrac{1}{2}, \dfrac{\sqrt{3}}{2} \right)$
$\dfrac{3\pi}{4}$	$x = \cos \dfrac{3\pi}{4} = -\dfrac{1}{\sqrt{2}}$	$y = \sin \dfrac{3\pi}{4} = \dfrac{1}{\sqrt{2}}$	$\left(-\dfrac{1}{\sqrt{2}}, \dfrac{1}{\sqrt{2}} \right)$
$\dfrac{5\pi}{6}$	$x = \cos \dfrac{5\pi}{6} = -\dfrac{\sqrt{3}}{2}$	$y = \sin \dfrac{5\pi}{6} = \dfrac{1}{2}$	$\left(-\dfrac{\sqrt{3}}{2}, \dfrac{1}{2} \right)$
π	$x = \cos \pi = -1$	$y = \sin \pi = 0$	$(-1, 0)$
$\dfrac{7\pi}{6}$	$x = \cos \dfrac{7\pi}{6} = -\dfrac{\sqrt{3}}{2}$	$y = \sin \dfrac{7\pi}{6} = -\dfrac{1}{2}$	$\left(-\dfrac{\sqrt{3}}{2}, -\dfrac{1}{2} \right)$
$\dfrac{5\pi}{4}$	$x = \cos \dfrac{5\pi}{4} = -\dfrac{1}{\sqrt{2}}$	$y = \sin \dfrac{5\pi}{4} = -\dfrac{1}{\sqrt{2}}$	$\left(-\dfrac{1}{\sqrt{2}}, -\dfrac{1}{\sqrt{2}} \right)$
$\dfrac{4\pi}{3}$	$x = \cos \dfrac{4\pi}{3} = -\dfrac{1}{2}$	$y = \sin \dfrac{4\pi}{3} = -\dfrac{\sqrt{3}}{2}$	$\left(-\dfrac{1}{2}, -\dfrac{\sqrt{3}}{2} \right)$
$\dfrac{3\pi}{2}$	$x = \cos \dfrac{3\pi}{2} = 0$	$y = \sin \dfrac{3\pi}{2} = -1$	$(0, -1)$
$\dfrac{5\pi}{3}$	$x = \cos \dfrac{5\pi}{3} = \dfrac{1}{2}$	$y = \sin \dfrac{5\pi}{3} = -\dfrac{\sqrt{3}}{2}$	$\left(\dfrac{1}{2}, -\dfrac{\sqrt{3}}{2} \right)$
$\dfrac{7\pi}{4}$	$x = \cos \dfrac{7\pi}{4} = \dfrac{1}{\sqrt{2}}$	$y = \sin \dfrac{7\pi}{4} = -\dfrac{1}{\sqrt{2}}$	$\left(\dfrac{1}{\sqrt{2}}, -\dfrac{1}{\sqrt{2}} \right)$
$\dfrac{11\pi}{6}$	$x = \cos \dfrac{11\pi}{6} = \dfrac{\sqrt{3}}{2}$	$y = \sin \dfrac{11\pi}{6} = -\dfrac{1}{2}$	$\left(\dfrac{\sqrt{3}}{2}, -\dfrac{1}{2} \right)$
2π	$x = \cos 2\pi = 1$	$y = \sin 2\pi = 0$	$(1, 0)$

A Eliminating the Parameter

EXAMPLE 1 Eliminate the parameter t from the parametric equations

$$x = 3 \cos t \qquad y = 2 \sin t$$

Solution Again, we will use the identity $\cos^2 t + \sin^2 t = 1$. Before we do so, however, we must solve the first equation for $\cos t$ and the second equation for $\sin t$.

$$x = 3 \cos t \Rightarrow \cos t = \frac{x}{3}$$

$$y = 2 \sin t \Rightarrow \sin t = \frac{y}{2}$$

Substituting $\frac{x}{3}$ and $\frac{y}{2}$ for $\cos t$ and $\sin t$ into the Pythagorean identity gives us

$$\cos^2 t + \sin^2 t = 1$$

$$\left(\frac{x}{3}\right)^2 + \left(\frac{y}{2}\right)^2 = 1$$

$$\frac{x^2}{9} + \frac{y^2}{4} = 1$$

which is the equation of an ellipse. The center is at the origin, the x-intercepts are 3 and -3, and the y-intercepts are 2 and -2. Figure 2 shows the graph.

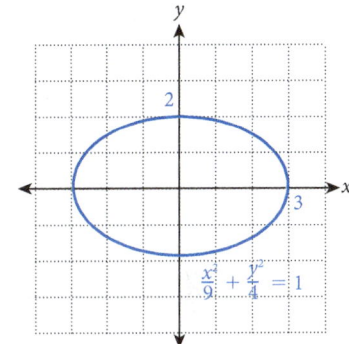

FIGURE 2

EXAMPLE 2 Eliminate the parameter t from the equations

$$x = 3 + \sin t \qquad y = \cos t - 2$$

Solution Solving the first equation for $\sin t$ and the second equation for $\cos t$, we have

$$\sin t = x - 3 \quad \text{and} \quad \cos t = y + 2$$

Substituting these expressions into the Pythagorean identity for $\sin t$ and $\cos t$ gives us

$$\cos^2 t + \sin^2 t = 1$$

$$(x - 3)^2 + (y + 2)^2 = 1$$

which is the equation of a circle with a radius of 1 and center at $(3, -2)$.

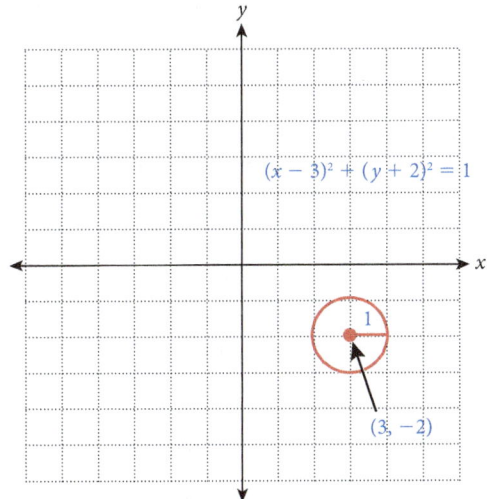

FIGURE 3

EXAMPLE 3   Eliminate the parameter t.

$$x = 3 + 2 \sec t \qquad y = 2 + 4 \tan t$$

Solution In this case, we solve for $\sec t$ and $\tan t$ and then use the identity $1 + \tan^2 t = \sec^2 t$.

$$x = 3 + 2 \sec t \Rightarrow \sec t = \frac{x - 3}{2}$$

$$y = 2 + 4 \tan t \Rightarrow \tan t = \frac{y - 2}{4}$$

so

$$1 + \tan^2 t = \sec^2 t$$

$$1 + \left(\frac{y - 2}{4}\right)^2 = \left(\frac{x - 3}{2}\right)^2$$

or

$$\frac{(x - 3)^2}{4} - \frac{(y - 2)^2}{16} = 1$$

This is the equation of a hyperbola.

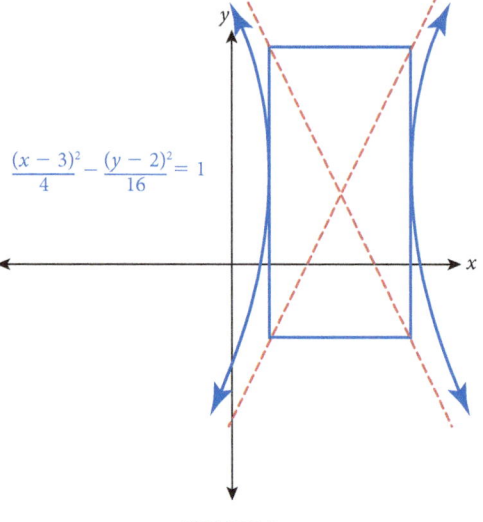

$$\frac{(x - 3)^2}{4} - \frac{(y - 2)^2}{16} = 1$$

FIGURE 4

B Graphing More Trigonometric Equations

We will end this chapter by looking at one more type of graphing. Recall that in Chapter 4 we graphed the equation $y = \sin x + \cos x$ by graphing $y = \sin x$ and $y = \cos x$ separately and then adding y-coordinates. There is another method of graphing equations of the form $y = a \sin x + b \cos x$ that does not require two separate graphs. Instead, we rewrite the equation so it has the form $y = A \sin(Bx + C)$ so that we can identify the amplitude, period, and phase shift and sketch the graph from them. The key to this new method is multiplying and dividing the right side of $y = a \sin x + b \cos x$ by $\sqrt{a^2 + b^2}$ (we did something similar to this in Example 6 of Section 6.2). That is, we write

$$y = a \sin x + b \cos x$$

as

$$y = \sqrt{a^2 + b^2} \left(\frac{a}{\sqrt{a^2 + b^2}} \sin x + \frac{b}{\sqrt{a^2 + b^2}} \cos x \right)$$

Doing so does not change the equation, because we have simply multiplied by

$$\frac{\sqrt{a^2 + b^2}}{\sqrt{a^2 + b^2}}$$

which is 1.

EXAMPLE 4   Graph one complete cycle of $y = \sin x + \cos x$.

Solution $y = \sin x + \cos x$ has the form $y = a \sin x + b \cos x$, where a and b are both 1.
$$\sqrt{a^2 + b^2} = \sqrt{1^2 + 1^2} = \sqrt{2}$$

Multiplying and dividing the right side of our original equation by $\sqrt{2}$, gives us

$$y = \sqrt{2} \left(\frac{1}{\sqrt{2}} \sin x + \frac{1}{\sqrt{2}} \cos x \right)$$

Next we substitute $\cos \frac{\pi}{4}$ for the first $\frac{1}{\sqrt{2}}$ and $\sin \frac{\pi}{4}$ for the second $1/\sqrt{2}$.

$$y = \sqrt{2} \left(\cos \frac{\pi}{4} \sin x + \sin \frac{\pi}{4} \cos x \right)$$

Exchanging the order of the products in each term gives us an expression on the right side that we recognize as an expanded form of the sum formula for $\sin(A + B)$

$$y = \sqrt{2}\left(\sin x \cos \frac{\pi}{4} + \cos x \sin \frac{\pi}{4}\right)$$

Writing the right side in the form $\sin(A + B)$ we have

$$y = \sqrt{2}\sin\left(x + \frac{\pi}{4}\right)$$

the graph of which is a sine curve with amplitude $\sqrt{2}$, period 2π, and phase shift $-\pi/4$.

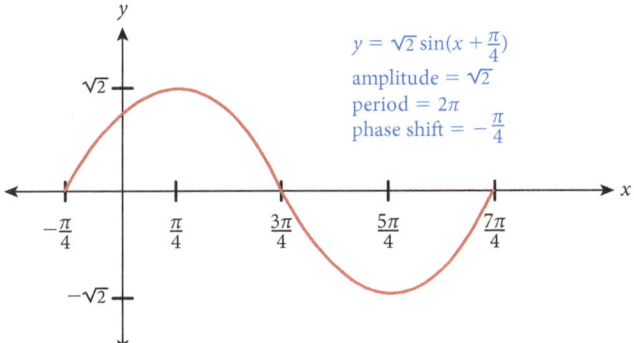

FIGURE 5

EXAMPLE 5 The London Eye, pictured below, is a Ferris wheel that we have worked with previously. Recall that the diameter of the wheel is 120 feet and the bottom of the wheel is 15 feet above the ground. The wheel rotates through one complete revolution every 30 minutes. Figure 6 is a simplified model of the path of a rider who gets on the wheel at Point A and then rides to Point B, forming central angle θ. Use triangle DBC to derive parametric equations for x and y in terms of the parameter θ. (Doing so will give you the position of the rider at any time t after they start their ride, which is different than anything we have done previously.)

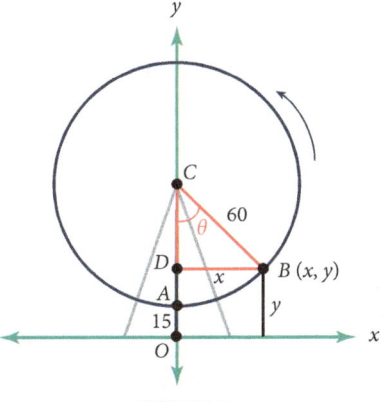

FIGURE 6

Solution Let's take triangle DBC out of the diagram to simplify our work. Notice that the horizontal side of our triangle is simply x, while the vertical side is the difference between the distance from the origin of our coordinate system, 75, and y, the riders height above the ground at Point B.

The triangle containing angle θ in Figure 6 is reproduced in Figure 7. Note that the horizontal side of the triangle is simply x. The vertical side is the difference between 75 (the distance from the origin to the center of the wheel) and the y-coordinate of point P_1.

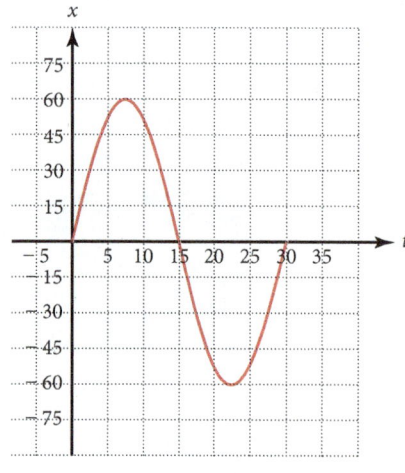

FIGURE 7

Using right-triangle trigonometry, we write the following relationships:

$$\sin \theta = \frac{x}{60}$$

$$\cos \theta = \frac{75 - y}{60}$$

Solving these equations for x and y, we have

$$x = 60 \sin \theta$$

$$y = 75 - 60 \cos \theta$$

These are parametric equations that give us any point (x, y) on the circumference of the wheel as a function of θ.

To rewrite our equations in terms of time t, we use the fact that a rider will travel once around the wheel every 30 minutes.

$$\frac{2\pi}{30 \text{ min}} = \frac{\theta}{t}$$

Solving for θ, we have

$$\theta = \frac{\pi}{15} t$$

We substitute this expression for θ into our parametric equations to obtain

$$\left. \begin{array}{l} x = 60 \sin \dfrac{\pi}{15}t \\[2em] y = 75 - 60 \cos \dfrac{\pi}{15}t \end{array} \right\} \quad \text{Parametric equations with } x \text{ and } y \text{ as functions of } t$$

If we graph all the ordered pairs (x, y) using either of our two sets of parametric equations, we produce the Ferris wheel shown earlier in Figure 6. On the other hand, if we graph each of the parametric equations above separately, as if they were two separate functions of t, we have the two curves shown in Figure 8 and 9. Notice that the horizontal axes are the t axes, while the vertical axes are labeled x and y, respectively.

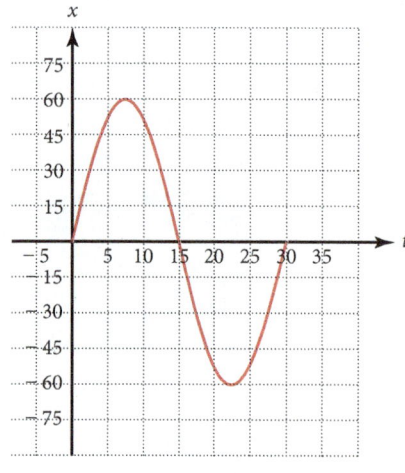

FIGURE 8

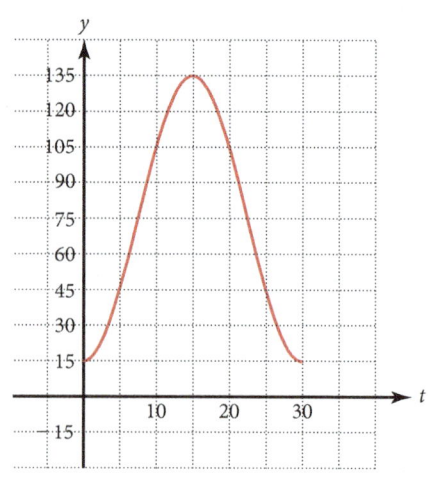

FIGURE 9

Figure 8 shows the distance the rider is from the y-axis at any time t during the ride. (If the sun were directly overhead during the ride, this curve would describe the motion of the shadow of the rider.) Figure 9 gives us the height y the rider is above the ground at time t.

EXAMPLE 6   A human cannonball leaves the cannon 88 feet per second (approximately 60 miles per hour). If θ is the angle the cannon makes with the horizontal, then the position of the cannonball, (x, y), can be given with the parametric equations

$$x = (88 \cos \theta)t$$
$$y = (88 \sin \theta)t - 16t^2$$

The first equation gives the horizontal distance the cannon ball has traveled after t seconds. The second equation gives the vertical position of the cannonball after t seconds.

a. If θ is 30°, how long will the cannonball be in the air, before landing in the net?

b. Graph the second equation when θ is 30°, when θ is 45°, and when θ is 60°. Use using t for the horizontal axes and y for the vertical axes.

Solution Here is a diagram of the situation when θ is 30°:

Human Cannonball Diagram

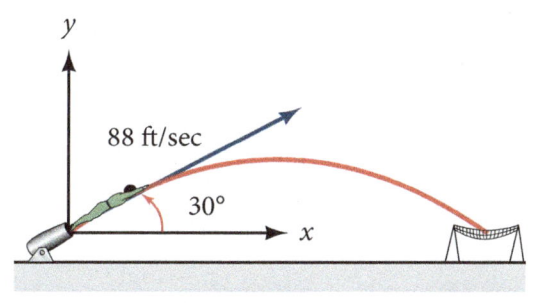

Simplified Mathematical Model

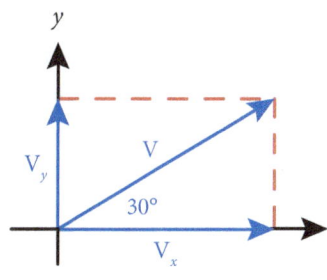

FIGURE 10

a. Substituting $\theta = 30°$ into our second equation we have the height y of the cannonball t seconds after leaving the cannon.

$$y = (88 \sin 30°)t - 16t^2 = 44t - 16t^2$$

The trip will end when when he lands in the net. The height is 0 at the beginning of the trip and at the end of the trip. To find out which times correspond to $y = 0$, we solve this equation

$$0 = 44t - 16t^2$$
$$0 = 4t(11 - 4t)$$
$$4t = 0 \quad \text{or} \quad 11 - 4t = 0$$
$$t = 0 \qquad\qquad t = \frac{11}{4} = 2.75$$

The cannonball leaves the cannon at 0 seconds, and then lands in the net 2.75 seconds later.

b. To graph the second equation for the different values of θ, we first substitute for θ to find the equations. Then we graph each equation.

When $\theta = 30°$, the equation is $y = 44t - 16t^2$.
When $\theta = 45°$, the equation is $y = (88 \sin 45°)t - 16t^2 = 62t - 16t^2$.
When $\theta = 60°$, the equation is $y = (88 \sin 60°)t - 16t^2 = 76t - 16t^2$.

Here are the three graphs:

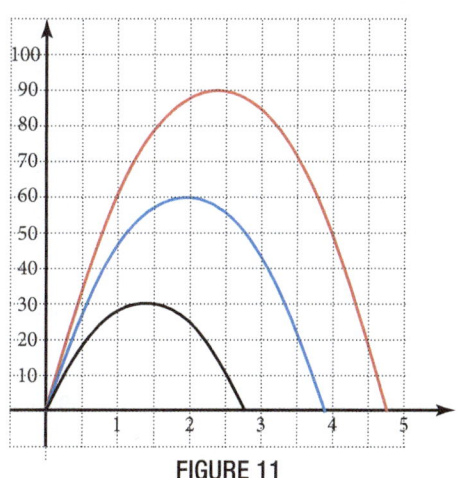

FIGURE 11

Problem Set 6.4

Vocabulary

Use the vocabulary words and formulas below to fill in the blanks in the sentences.

parameter   eliminating   parametric   identities

1. In this section we worked with pairs of equations, such as $x = \cos t$ and $y = \sin t$ that describe a curve in the xy-plane. These equations are called _____ equations and the variable t is called the _____ .

2. Sometimes we can write a pair of parametric equations in terms of just the variables x and y, by _____ the parameter t.

3. Many times, when we are eliminating the parameter t from a pair of parametric equations, we use trigonometric _____ .

Eliminate the parameter t from each of the following and then sketch the graph of each:

1. $x = \sin t$
$\quad y = \cos t$

2. $x = -\sin t$
$\quad y = \cos t$

3. $x = 3\cos t$
$\quad y = 3\sin t$

4. $x = 2\cos t$
$\quad y = 2\sin t$

5. $x = 2\sin t$
$\quad y = 4\cos t$

6. $x = 3\sin t$
$\quad y = 4\cos t$

7. $x = 2 + \sin t$
$\quad y = 3 + \cos t$

8. $x = 3 + \sin t$
$\quad y = 2 + \cos t$

9. $x = \sin t - 2$
$\quad y = \cos t - 3$

10. $x = \cos t - 3$
$\quad y = \sin t + 2$

11. $x = 3 + 2\sin t$
$\quad y = 1 + 2\cos t$

12. $x = 2 + 3\sin t$
$\quad y = 1 + 3\cos t$

13. $x = 3\cos t - 3$
$\quad y = 3\sin t + 1$

14. $x = 4\sin t - 5$
$\quad y = 4\cos t - 3$

Eliminate the parameter t in each of the following:

15. $x = \sec t$
$\quad y = \tan t$

16. $x = \tan t$
$\quad y = \sec t$

17. $x = 3\sec t$
$\quad y = 3\tan t$

18. $x = 3\cot t$
$\quad y = 3\csc t$

19. $x = 2 + 3\tan t$
$\quad y = 4 + 3\sec t$

20. $x = 3 + 5\tan t$
$\quad y = 2 + 5\sec t$

302

21. $x = \cos 2t$
   $y = \sin t$

22. $x = \cos 2t$
   $y = \cos t$

23. $x = \sin t$
   $y = \sin t$

24. $x = \cos t$
   $y = \cos t$

25. $x = 3 \sin t$
   $y = 2 \sin t$

26. $x = 2 \sin t$
   $y = 3 \sin t$

Graph one complete cycle of each of the following by first changing to a single sine function and then using amplitude, period, and phase shift. (See Example 4.)

27. $y = \sin x - \cos x$

28. $y = \sqrt{2} \sin x + \sqrt{2} \cos x$

29. $y = \sin x + \sqrt{3} \cos x$

30. $y = \sin x - \sqrt{3} \cos x$

31. $y = \sqrt{3} \sin x + \cos x$

32. $y = \sqrt{3} \sin x - \cos x$

Applying the Concepts

33. Ferris Wheel Previously we worked some problems involving the Ferris wheel known as the High Roller that is located in Las Vegas. Recall that the diameter of the wheel is 550 feet, and one complete revolution takes 30 minutes, and the bottom of the wheel is 30 feet above the ground. The position of a rider t seconds after the ride starts is given by (x, y). Using Example 5 as a model, find parametric equations for x and y in terms of t.

34. Ferris Wheel The Ferris wheel known as the Reisenrad has a diameter of the wheel is 197 feet, one complete revolution takes 16 minutes, and the bottom of the wheel is 12 feet above the ground. The position of a rider t seconds after the ride starts is given by (x, y). Using Example 5 as a model, find parametric equations for x and y in terms of t.

35. Human Cannonball Gemma "The Jet" Kirby was the youngest human cannonball when she started working for Ringling Bros. and Barnum & Bailey Circus. In a demonstration on the David Letterman Show she leaves the cannon at 74 feet per second (approximately 50 miles per hour). If her cannon makes an angle of 45° with the horizontal, then her position t seconds after leaving the cannon can be given with the parametric equations

$$x = (74 \cos 45°)t$$

$$y = (74 \sin 45°)t - 16t^2$$

a. How long is she in the air before landing in the net. (To the nearest tenth of a second.)

b. Graph the second equation using t for the horizontal axes and y for the vertical axes.

36. Human Cannonball A human cannonball leaves the cannon 64 feet per second (approximately 44 miles per hour). If θ is the angle the cannon makes with the horizontal, then the position of the cannonball, (x, y), can be given with the parametric equation

$$x = (64 \cos \theta)t$$
$$y = (64 \sin \theta)t - 16t^2$$

a. If θ is 30°, how long will the cannonball be in the air, before landing in the net?

b. Graph the second equation when θ is 30°, when θ is 45°, and when θ is 60°.

Review Problems

The problems that follow review material we covered in Chapter 5.

Prove each identity.

37. $\sec A - \cos A = \tan A \sin A$ **38.** $\dfrac{\sin x}{1 + \cos x} = \dfrac{1 - \cos x}{\sin x}$

39. $\dfrac{1}{1 + \cos t} + \dfrac{1}{1 - \cos t} = 2 \csc^2 t$ **40.** $\tan \dfrac{A}{2} = \dfrac{\sin A}{1 + \cos A}$

Let $\sin A = -\frac{3}{5}$ with A in QIV and $\sin B = \frac{12}{13}$ with B in QII and find

41. $\cos (A - B)$ **42.** $\sin 2B$ **43.** $\cos \dfrac{A}{2}$

44. Find the exact value of $\cos 15°$.

45. Write the following expression as a single trigonometric function:

$$\sin 15° \cos 75° + \cos 15° \sin 75°$$

46. Evaluate the expression below without using a calculator or tables. (Assume any variables represent positive numbers.)

$$\sin (2 \cos^{-1} x)$$

CHAPTER 6 SUMMARY

Examples

1. a. Solve for x:
$$2\cos x - \sqrt{3} = 0.$$

$$2\cos x - \sqrt{3} = 0$$
$$2\cos x = \sqrt{3}$$
$$\cos x = \frac{\sqrt{3}}{2}$$

Solutions between 0° and 360°

In Degrees	In Radians
$x = 30°$ or	$x = \dfrac{\pi}{6}$ or
$x = 330°$	$x = \dfrac{11\pi}{6}$

All Solutions (k is an integer)

In Degrees	In Radians
$x = 30° + 360°k$	$x = \dfrac{\pi}{6} + 2k\pi$
or	or
$x = 330° + 360°k$	$x = \dfrac{11\pi}{6} + 2k\pi$

b. Solve $2\cos^2 t - 9\cos t = 5$, if $0 \leq t \leq 2\pi$.

$$2\cos^2 t - 9\cos t = 5$$
$$2\cos^2 t - 9\cos t - 5 = 0$$
$$(2\cos t + 1)(\cos t - 5) = 0$$
$$2\cos t + 1 = 0 \quad \text{or} \quad \cos t - 5 = 0$$
$$\cos t = -\frac{1}{2} \qquad \cos t = 5$$

The first result, $\cos t = -1/2$, gives us $t = 2\pi/3$ or $t = 4\pi/3$. The second result, $\cos t = 5$, has no solution. For any value of t, $\cos t$ must always be between -1 and 1. It will never be 5.

2. Solve $\cos 2\theta + 3\sin\theta - 2 = 0$, if $0° \leq \theta \leq 360°$.

$$\cos 2\theta + 3\sin\theta - 2 = 0$$
$$1 - 2\sin^2\theta + 3\sin\theta - 2 = 0$$
$$-2\sin^2\theta + 3\sin\theta - 1 = 0$$
$$2\sin^2\theta - 3\sin\theta + 1 = 0$$
$$(2\sin\theta - 1)(\sin\theta - 1) = 0$$
$$2\sin\theta - 1 = 0 \quad \text{or} \quad \sin\theta - 1 = 0$$
$$\sin\theta = \frac{1}{2} \qquad \sin\theta = 1$$
$$\theta = 30°, 150°, 90°$$

Solving Simple Trigonometric Equations [6.1]

We solve linear equations in trigonometry by applying the properties of equality developed in algebra. The two most important properties from algebra are stated as follows:

Addition Property of Equality

For any three algebraic expressions A, B, and C

If $A = B$

then $A + C = B + C$

In Words: Adding the same quantity to both sides of an equation will not change the solution set.

Multiplication Property of Equality

For any three algebraic expressions A, B, and C with $C \neq 0$.

If $A = B$

then $AC = BC$

In Words: Multiplying both sides of an equation by the same nonzero quantity will not change the solution set.

To solve a trigonometric equation that is quadratic in $\sin x$ or $\cos x$, we write it in standard form and then factor it or use the quadratic formula.

Using Identities in Trigonometric Equations [6.2]

Sometimes it is necessary to use identities to make trigonometric substitutions when solving equations. Identities are usually required if the equation contains more than one trigonometric function or if there is more than one angle named in the equation. In the example to the left, we begin by replacing $\cos 2\theta$ with $1 - 2\sin^2\theta$. Doing so gives us a quadratic equation in $\sin\theta$, which we put in standard form and solve by factoring.

3. Find all radian solutions for
$\sin 2x \cos x + \cos 2x \sin x = \frac{1}{\sqrt{2}}$.

Solution

$$\sin 2x \cos x + \cos 2x \sin x = \frac{1}{\sqrt{2}}$$

$$\sin(2x + x) = \frac{1}{\sqrt{2}}$$

$$\sin 3x = \frac{1}{\sqrt{2}}$$

$$3x = \frac{\pi}{4} + 2k\pi \quad \text{or} \quad 3x = \frac{3\pi}{4} + 2k\pi$$

$$x = \frac{\pi}{12} + \frac{2k\pi}{3} \quad \text{or} \quad x = \frac{\pi}{4} + \frac{2k\pi}{3}$$

where k is an integer.

Equations Involving Multiple Angles [6.3]

Sometimes the equations we solve in trigonometry reduce to single equations that contain multiple angles. When this occurs, we have to be careful in the last step that we do not leave out any solutions. For instance, if we are asked to find all solutions between $x = 0$ and $x = 2\pi$, and our final equation contains $2x$, we must find all values of $2x$ between 0 and 4π in order that x remain between 0 and 2π.

4. Eliminate the parameter t from the equations

$$x = 3 + \sin t \quad y = \cos t - 2$$

Solving $\sin t$ and $\cos t$ we have

$$\sin t = x - 3 \quad \text{and} \quad \cos t = y + 2$$

Substituting these expressions into the Pythagorean identity, we have

$$\cos^2 t + \sin^2 t = 1$$
$$(x - 3)^2 + (y + 2)^2 = 1$$

which is the equation of a circle with a radius of 1 and center at $(3, -2)$.

Parametric Equations [6.4]

When the coordinates of point (x, y) are described separately by two equations of the form $x = f(t)$ and $y = g(t)$, then the two equations are called *parametric equations* and t is called the parameter. One way to graph a set of points (x, y) that are given in terms of the parameter t is to eliminate the parameter and obtain an equation in just x and y that gives the same set of points (x, y).

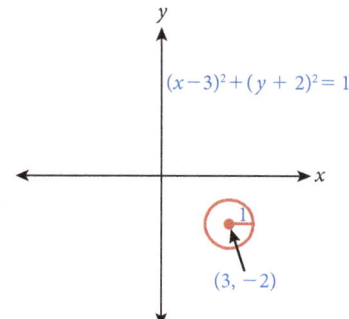

5. Graph one complete cycle of $y = \sin x + \cos x$.

Multiplying and dividing the right side of our original equation by $\sqrt{2}$ gives us

$$y = \sqrt{2}\left(\frac{1}{\sqrt{2}} \sin x + \frac{1}{\sqrt{2}} \cos x\right)$$

Next we substitute $\cos \frac{\pi}{4}$ for the first $\frac{1}{\sqrt{2}}$ and $\sin \frac{\pi}{4}$ for the second $\frac{1}{\sqrt{2}}$.

$$y = \sqrt{2}\left(\sin x \cos \frac{\pi}{4} + \cos x \sin \frac{\pi}{4}\right)$$

Writing the right side in the form $\sin(A + B)$ we have

$$y = \sqrt{2} \sin\left(x + \frac{\pi}{4}\right)$$

the graph of which is a sine curve with amplitude $\sqrt{2}$, period 2π, and phase shift $-\frac{\pi}{4}$.

Graphing Equations of the form $y = a \sin x + b \cos x$ [6.4]

We graph equations of the form $y = a \sin x + b \cos x$ by multiplying and dividing the right side by $\sqrt{a^2 + b^2}$. Two substitutions are made for $a/\sqrt{a^2 + b^2}$ and $b/\sqrt{a^2 + b^2}$ and the equation that results is then written in the form $y = A \sin(Bx + C)$ using a sum formula.

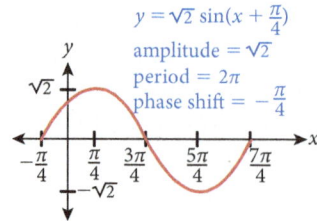

Find all solutions between 0° and 360°, including 0° and 360° if they are solutions.

1. $2 \sin \theta - 1 = 0$

2. $\sqrt{3} \tan \theta + 1 = 0$

3. $\cos \theta - 2 \sin \theta \cos \theta = 0$

4. $\tan \theta - 2 \cos \theta \tan \theta = 0$

5. $4 \cos \theta - 2 \sec \theta = 0$

6. $2 \sin \theta - \csc \theta = 1$

7. $\sin \dfrac{\theta}{2} + \cos \theta = 0$

8. $\cos \dfrac{\theta}{2} - \cos \theta = 0$

9. $2 \cos^2 2\theta - 3 \cos 2\theta = -1$

10. $4 \sin^2 2\theta - 1 = 0$

11. $\sin \theta + \cos \theta = 1$

12. $\sin \theta - \cos \theta = 1$

Find all solutions for the following equations. Write your answers in radians using exact values.

13. $\cos 2x - 3 \cos x = -2$

14. $\sqrt{3} \sin x - \cos x = 0$

15. $\sin 2x \cos x + \cos 2x \sin x = -1$

16. $\sin^3 4x = 1$

Find all solutions between 0° and 360° to the nearest tenth of a degree, including 0° and 360° if they are solutions.

17. $5 \sin^2 \theta - 3 \sin \theta = 2$

18. $4 \cos^2 \theta - 4 \cos \theta = 2$

Eliminate the parameter t from each of the following and then sketch the graph:

19. $x = 3 \cos t \quad y = 3 \sin t$

20. $x = \sec t, \quad y = \tan t$

21. $x = 3 + 2 \sin t \quad y = 1 + 2 \cos t$

22. $x = 3 \cos t - 3 \quad y = 3 \sin t + 1$

Write each equation in terms of a single trigonometric function and then graph using amplitude, period, and phase shift.

23. $y = \sin x - \cos x$

24. $y = \sin x + \sqrt{3} \cos x$

Spotlight on Success

Cynthia, Student Instructor

Each time we face our fear, we gain strength, courage, and confidence in the doing.—Unknown

I must admit, when it comes to math, it takes me longer to learn the material compared to other students. Because of that, I was afraid to ask questions, especially when it seemed like everyone else understood what was going on. Because I wasn't getting my questions answered, my quiz and exam scores were only getting worse. I realized that I was already paying a lot to go to college and that I couldn't afford to keep doing poorly on my exams. I learned how to overcome my fear of asking questions by studying the material before class, and working on extra problem sets until I was confident enough that at least I understood the main concepts. By preparing myself beforehand, I would often end up answering the question myself. Even when that wasn't the case, the professor I tried to answer the question on my own. If you want to be successful, but you are afraid to ask a question, try putting in a little extra time working on problems before you ask your instructor for help. I think you will find, like I did, that it's not as bad as you imagined it, and you will have overcome an obstacle that was in the way of your success.

TRIANGLES

7

WHAT ARE TRIANGLES? A triangle plays a fundamental role as it forms the basis for defining and understanding trigonometric functions and relationships. Specifically, trigonometry deals with right triangles and involves the study of the relationships between the angles and sides of these triangles.

WHY ARE TRIANGLES IMPORTANT? Studying triangles in trigonometry requires students to visualize and manipulate geometric shapes in space. This enhances their spatial reasoning skills, which are valuable not only in mathematics but also in fields such as engineering, design, and architecture.

WHERE MIGHT YOU SEE TRIANGLES? Here are some examples of where you may see triangles in the real world:

ART AND DESIGN Artists and designers use trigonometry to create visually appealing compositions and geometric patterns. Triangles are common elements in art and design, whether they appear as individual shapes or as part of larger compositions. Spatial reasoning allows artists and designers to manipulate shapes, proportions, and perspectives to achieve their desired effects.

ARCHITECTURAL DESIGN Architects use trigonometry to design structures such as buildings, bridges, and towers. They need to consider the angles and dimensions of triangular elements within these structures to ensure stability and aesthetic appeal. Spatial reasoning comes into play when visualizing how different parts of the structure fit together in three-dimensional space.

Scenario: Elevation Angles and Trees

A forester is taking a sample of trees in order to determine the value of a forested property. To do so, he needs to measure the height of a sample tree. He uses a clinometer, which measures the angle to the top of the tree from the horizontal. He takes all measurements from a height of 5 feet on level ground. He stands a distance from the tree and measures an elevation angle of 41°. He then paces 10 feet further from the tree and measures an elevation angle of 37°. Use this information to find the height of the tree, to the nearest foot.

Solution Label his first position A, his second position B, and the top of the tree C. Also label unknown sides d and h as indicated in the figure:

Note that angle BAC = 180° − 41° = 139°, and so angle BCA = 180° − 37° − 139° = 4°.

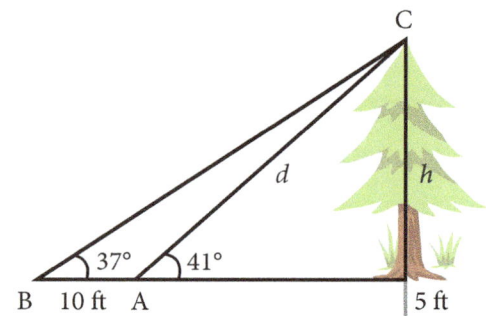

Now, using the law of sines:

$$\frac{\sin 37°}{d} = \frac{\sin 4°}{10}$$

$$d \sin 4° = 10 \sin 37°$$

$$d = \frac{10 \sin 37°}{\sin 4°} = 86.27 \text{ feet}$$

Using the smaller right triangle (from his first position A):

$$\sin 41° = \frac{h}{86.27}$$

$$h = 86.27 \sin 41° \approx 57 \text{ feet}$$

Since this measurement is taken 5 feet off the ground, the tree's height is approximately 57 + 5 = 62 feet.

Learning Objectives

A Solve problems involving triangles using the law of sines.

There are many relationships that exist between the sides and angles in a triangle. One such relationship is called the law of sines, and it states that the ratio of the sine of an angle to the length of the side opposite that angle is constant in any triangle. Here it is stated in symbols:

Law of Sines

$$\frac{\sin A}{a} = \frac{\sin B}{b} = \frac{\sin C}{c}$$

Or, equivalently,

$$\frac{a}{\sin A} = \frac{b}{\sin B} = \frac{c}{\sin C}$$

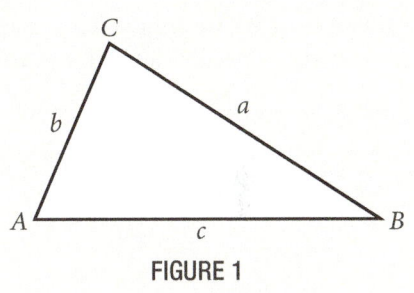

FIGURE 1

Proof 1 The altitude h of the triangle in Figure 2 can be written in terms of $\sin A$ or $\sin B$, depending on which of the two right triangles we are referring to:

$$\sin A = \frac{h}{b} \qquad \sin B = \frac{h}{a}$$

$$h = b \sin A \qquad h = a \sin B$$

Since h is equal to itself, we have

$$h = h$$

$$b \sin A = a \sin B$$

$$\frac{b \sin A}{ab} = \frac{a \sin B}{ab} \qquad \text{Divide both sides by } ab.$$

$$\frac{\sin A}{a} = \frac{\sin B}{b} \qquad \text{Divide out common factors.}$$

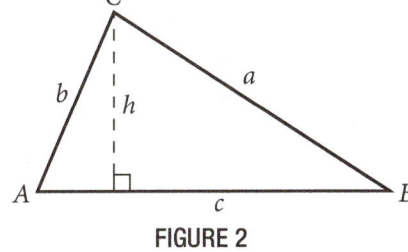

FIGURE 2

If we do the same kind of thing with the altitude that extends from A, we will have the third ratio in the law of sines, $\frac{\sin C}{c}$, equal to the two ratios above.

Note that the derivation of the law of sines will proceed in the same manner if triangle ABC contains an obtuse angle, as in Figure 3.

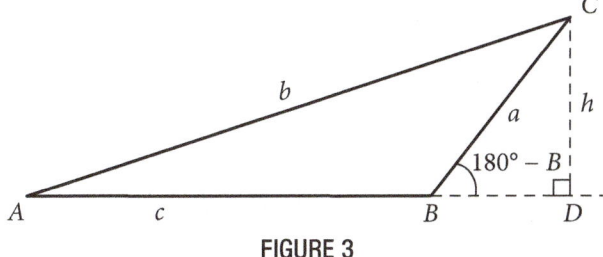

FIGURE 3

In triangle BDC, we have

$$\sin(180° - B) = \frac{h}{a}$$

$$\text{but, } \sin(180° - B) = \sin 180° \cos B - \cos 180° \sin B$$

$$= (0) \cos B - (-1)\sin B$$

$$= \sin B$$

So, $\sin B = \frac{h}{a}$, which is the result we obtained previously. Using triangle ADC we have $\sin A = \frac{h}{b}$. As you can see, these are the same two expressions we began with when deriving the law of sines for acute triangle in Figure 2. From this point on, the derivation would match our previous derivation.

We can use the law of sines to find missing parts of triangles in which we are given two angles and a side.

∏ Two Angles and One Side

In our first example, we are given two angles and the side opposite one of them. (You may recall that, in geometry, these were the parts we needed equal in two triangles in order to prove them congruent using the *AAS* theorem.)

EXAMPLE 1   In triangle ABC, $A = 30°$, $B = 70°$, and $a = 8.0$ centimeters. Use the law of sines to find c.

Solution We begin by drawing a picture of triangle ABC (it does not have to be accurate) and labeling it so that the information we are given is showing. See Figure 4.

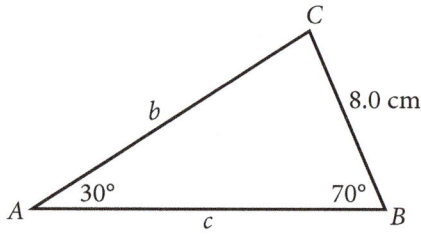

FIGURE 4

When we use the law of sines, we must have one of the ratios given to us. In this case, since we are given a and A, we have the ratio $\frac{a}{\sin A}$. To solve for c we need to first find C. Since the sum of the angles in any triangle is 180°, we have

$$C = 180° - (A + B)$$

$$= 180° - (30° + 70°) = 80°$$

To find c, we use the following two ratios given in the law of sines:

$$\frac{c}{\sin C} = \frac{a}{\sin A}$$

We multiply both sides by sin C and then substitute in the given values:

$$c = \frac{a \sin C}{\sin A} \qquad \textcolor{teal}{\text{Multiply both sides by sin } C.}$$

$$= \frac{8 \sin 80°}{\sin 30°} \qquad \textcolor{teal}{\text{Substitute in given values.}}$$

$$\approx 16 \text{ centimeters} \qquad \textcolor{teal}{\text{Rounded to 2 significant digits}}$$

In our next example we are given two angles and the side included between them (*ASA*) and asked to find all the missing parts.

EXAMPLE 2 Solve triangle ABC if $B = 34°$, $C = 82°$, and $a = 5.6$ centimeters.

Solution We begin drawing a diagram of the situation, Figure 5, and then we find angle A. We find angle A so that we have one of the ratios in the law of sines completed.

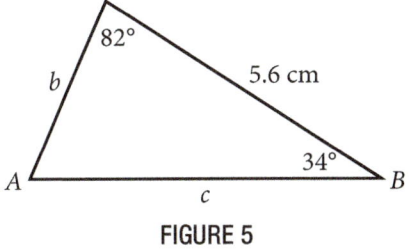

FIGURE 5

Angle A

$$A = 180° - (B + C)$$

$$= 180° - (34° + 82°) = 64°$$

Side b

If
$$\frac{b}{\sin B} = \frac{a}{\sin A}$$

then
$$b = \frac{a \sin B}{\sin A} \qquad \text{Multiply both sides by sin B.}$$

$$= \frac{5.6 \sin 34°}{\sin 64°} \qquad \text{Substitute in given values.}$$

$$\approx 3.5 \text{ centimeters} \qquad \text{Rounded to 2 significant digits}$$

Side c

If
$$\frac{c}{\sin C} = \frac{a}{\sin A}$$

then
$$c = \frac{a \sin C}{\sin A} \qquad \text{Multiply both sides by sin C.}$$

$$= \frac{5.6 \sin 82°}{\sin 64°} \qquad \text{Substitute in given values.}$$

$$\approx 6.2 \text{ centimeters} \qquad \text{Rounded to 2 significant digits}$$

The law of sines, along with some fancy electronic equipment, was used to obtain the results of some of the field events in one of the recent Olympic Games.

Figure 6 is a diagram of a shot-put ring. The shot is tossed (put) from the left and lands at A. A small electronic device is then placed at A (there is usually a dent in the ground where the shot lands, so it is easy to find where to place the device). The device at A sends a signal to a booth in the stands that gives the measures of angles A and B. The distance a is found ahead of time. To find the distance x, the law of sines is used.

$$\frac{x}{\sin B} = \frac{a}{\sin A} \qquad \text{or} \qquad x = \frac{a \sin B}{\sin A}$$

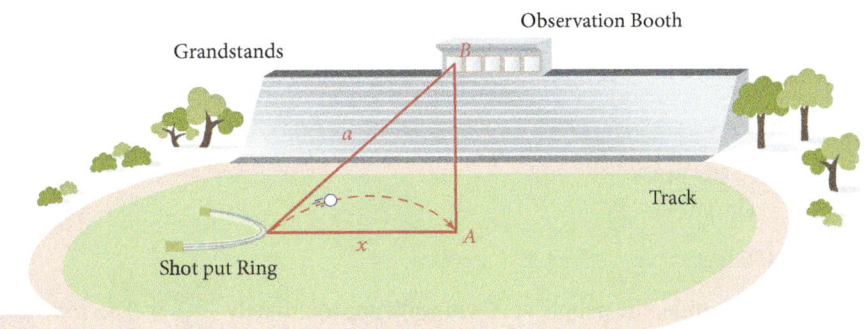

FIGURE 6

EXAMPLE 3   Find x in Figure 6 if $a = 562$ ft, $B = 5.7°$, and $A = 85.3°$.

Solution

$$x = \frac{a \sin B}{\sin A}$$

$$= \frac{562 \sin 5.7°}{\sin 85.3°}$$

$$= 56.0 \text{ ft}$$

EXAMPLE 4   A hot air balloon is flying over a dry lake when the wind stops blowing. The balloon comes to a stop at point D, which is 450 feet above the ground at point C, as shown in Figure 7. A jeep following the balloon runs out of gas at point A. The nearest service station is due north of the jeep at point B. The bearing of the balloon from the jeep at A is N 13° E, while the bearing of the balloon from the service station at B is S 19° E. If the angle of elevation of the balloon from A is 12°, how far will the people in the jeep have to walk to reach the service station at point B?

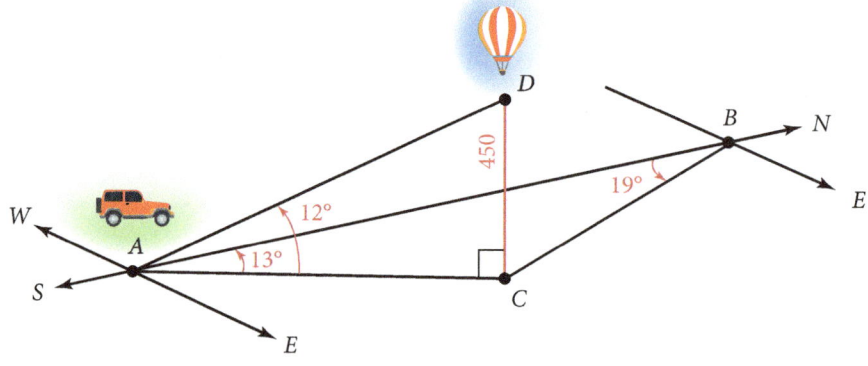

FIGURE 7

Solution   First, we find the distance between C and A, using right triangle trigonometry.

$$\tan 12° = \frac{450}{\overline{AC}}$$

$$\overline{AC} = \frac{450}{\tan 12°}$$

$$\approx 2,117 \text{ feet}$$

Next, we find angle ACB:

$$\angle ACB = 180° - (13° + 19°) = 148°$$

Finally, we find $\overline{AB}$, using the law of sines:

$$\frac{\overline{AB}}{\sin 148°} = \frac{2,117}{\sin 19°}$$

$$\overline{AB} = \frac{2,117 \sin 148°}{\sin 19°}$$

$$\approx 3,450 \text{ feet} \qquad \text{Three-digit accuracy}$$

Since there are 5,280 feet in a mile, the people at A will walk approximately $\frac{3,450}{5,280} \approx 0.65$ mile to get to the service station at B.

Problem Set 7.1

Vocabulary

Use the vocabulary words or text below to fill in the blanks in the sentences. You may not need to use every vocabulary word or text.

cos *B*   ratio   sin *B*   sum   angle   opposite

1. The law of sines states that the _____ of the sine of an angle to the side _____ that angle is constant in any triangle.

2. When we are given two angles in a triangle and one of the sides, the easiest thing to find first is the missing _____ .

3. For any triangle *ABC*, the relationship below is true if *x* is _____ .

$$\frac{a}{\sin A} = \frac{b}{x}$$

4. The _____ of the angles in any triangle is always 180°.

Each problem that follows refers to a triangle *ABC*.

1. If $A = 40°$, $B = 60°$, and $a = 12$ centimeters, find b.
2. If $A = 80°$, $B = 30°$, and $b = 14$ centimeters, find a.
3. If $B = 120°$, $C = 20°$, and $b = 28$ inches, find c.
4. If $B = 110°$, $C = 40°$, and $b = 18$ inches, find c.
5. If $A = 10°$, $C = 100°$, and $a = 12$ yards, find c.
6. If $A = 5°$, $C = 125°$, and $c = 51$ yards, find a.
7. If $A = 50°$, $B = 60°$, and $a = 36$ kilometers, find C and then find c.
8. If $B = 40°$, $C = 70°$, and $c = 42$ kilometers, find A and then find a.
9. If $A = 52°$, $B = 48°$, and $c = 14$ centimeters, find C and then find a.
10. If $A = 33°$, $C = 87°$, and $b = 18$ centimeters, find B and then find c.

The information below refers to a triangle *ABC*. In each case find all the missing parts.

11. $A = 42.5°$, $B = 71.4°$, $a = 210$ inches
12. $A = 110.4°$, $C = 21.8°$, $c = 240$ inches
13. $A = 46°$, $B = 95°$, $c = 6.8$ meters
14. $B = 57°$, $C = 31°$, $a = 7.3$ meters
15. $B = 13.4°$, $C = 24.8°$, $a = 315$ centimeters
16. $A = 105°$, $B = 45°$, $c = 630$ centimeters
17. In triangle *ABC*, $A = 30°$, $b = 20$ feet, and $a = 2$ feet. Show that it is impossible to solve this triangle by using the law of sines to find sin *B*.
18. In triangle *ABC*, $A = 40°$, $b = 20$ feet, and $a = 18$ feet. Use the law of sines to find sin *B*, and then give two possible values for angle *B*.

19. **Angle of Elevation** A man standing near a radio station antenna observes that the angle of elevation to the top of the antenna is 64°. He then walks 100 feet farther away and observes the angle of elevation to the top of the antenna to be 46°. Find the height of the antenna to the nearest foot. (Hint: Find x first. See Figure 8.)

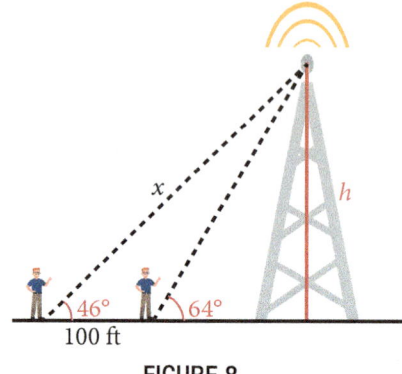

FIGURE 8

20. **Angle of Elevation** A woman standing on the street looks up to the top of a building and finds the angle of elevation is 38°. She then walks one block farther away (440 feet) and finds the angle of elevation to the top of the building is now 28°. How far away from the building is she when she makes her second observation? (See Figure 9.)

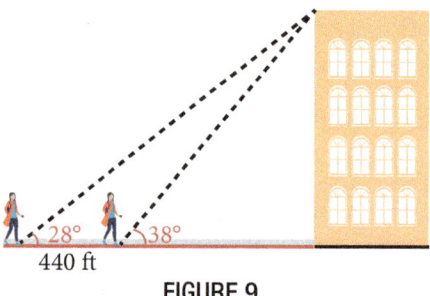

FIGURE 9

21. **Angle of Depression** A man is flying in a hot-air balloon in a straight line at a constant rate of 5 feet per second, while keeping it at a constant altitude. As he approaches the parking lot of a market, he notices the angle of depression from his balloon to a friend's car in the parking lot is 35°. A minute and a half later, after flying directly over his friend's car, he looks back to see his friend getting into the car and observes the angle of depression to be 36°. At that time, what is the distance between him and his friend? (Give your answer to the nearest foot.)

22. **Angle of Elevation** A woman entering an outside glass elevator on the ground floor of a hotel building glances up to the top of the building across the street and notices the angle of elevation to be 48°. She rides the elevator up three floors (60 feet) and finds the angle of elevation to the top of the building across the street is 32°. How tall is the building across the street? (Give you answer to the nearest foot.)

23. **Angle of Elevation** From a point on the ground a person notices that an antenna that is 110 feet tall on the top of a mountain subtends an angle of 0.5°. If the angle of elevation to the bottom of the antenna is 35°, find the height of the mountain. (See Figure 10.)

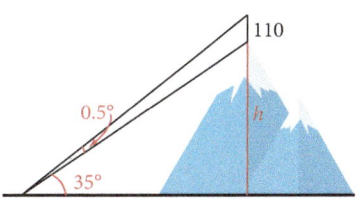

FIGURE 10

24. **Angle of Elevation** An antenna that is 150 feet tall is on top of a tall building. From a point on the ground the angle of elevation to the top of the antenna is 28.5°, while the angle of elevation to the bottom of the antenna from the same point is 27.5°. How tall is the building?

25. **Angle of Elevation** Figure 11 is a diagram that shows how Colleen estimates the height of a tree that is on the other side of a stream. She stands at point *A* facing the tree and finds the angle of elevation from *A* to the top of the tree to be 51°. Then she turns 105° and walks 25 feet to point *B*, where she measures the angle between her path and a line to the base of the tree and finds that angle to be 44°. Use this information to find the height of the tree.

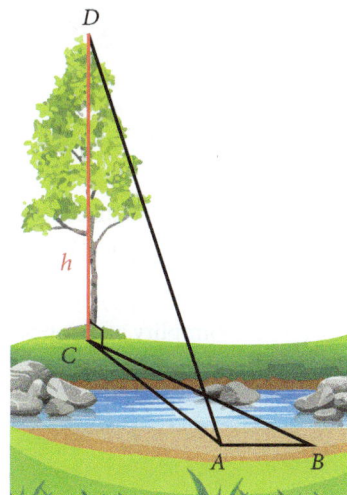

FIGURE 11

26. Distance to an Accident A plane makes a forced landing at sea. The last radio signal received at station C gives the bearing of the plane from C as N 55.4° E at an altitude of 1,050 feet. An observer at C sighted the plane and gives $\angle DCB$ as 22.5°. How far will a rescue boat at A have to travel to reach any survivors at B, if the bearing of B from A is S 56.4° E? (See Figure 12.)

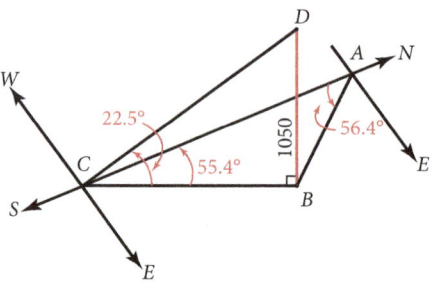

FIGURE 12

27. Bearing A ship is anchored off a long straight shoreline that runs north and south. From two observation points 18 miles apart on shore, the bearings of the ship are N 31° E and S 53° E. What is the distance from the ship to each of the observation points.

28. Bearing Tom and Fred are 3.5 miles apart watching a rocket being launched from Vandenberg Air Force Base. Tom estimates the bearing of the rocket from his position is S 75° W, while Fred estimates the bearing of the rocket from his position is N 65° W. If Fred is due south of Tom, how far is each of them from the rocket?

Getting Ready for the Next Section

Simplify.

29. $180° - (40° + 48°)$ **30.** $180° - (40° + 132°)$

Use a calculator to find each of the following. Round to the nearest ten thousandth.

31. $\sin 92°$ **32.** $\sin 8°$

Simplify and round your answer to the nearest whole number.

33. $\dfrac{54(\sin 92°)}{\sin 40°}$ **34.** $\dfrac{54(\sin 8°)}{\sin 40°}$

learning Objectives

A   Solve problems involving triangles with an ambiguous case.

In this section we will extend the law of sines to solve triangles in which we are given two sides and the angle opposite one of the given sides.

EXAMPLE 1   Find angle B in triangle ABC if $a = 2$, $b = 6$, and $A = 30°$.

Solution   Applying the law of sines, we have

$$\sin B = \frac{b \sin A}{a}$$
$$= \frac{6 \sin 30°}{2}$$
$$= 1.5 \qquad\qquad \text{This is impossible!}$$

For any value of B, $\sin B$ is between -1 and 1. The function $\sin B$ can never be larger than 1. No triangle exists for which $a = 2$, $b = 6$, and $A = 30°$. Figure 1 illustrates what went wrong here.

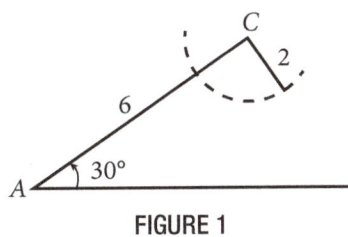

FIGURE 1

When we are given two sides and an angle opposite one of them (SSA), we have several possibilities for the triangle, or triangles, that result. As was the case in Example 1, one of the possibilities is that no triangle will fit the given information. Another possibility is that two different triangles can be obtained from the given information, and a third possibility is that exactly one triangle will fit the given information. Because of these different possibilities, we call the situation where we are solving a triangle in which we are given two sides and the angle opposite one of them the **ambiguous case**.

> *Note* You may recall from geometry that there was no congruence theorem SSA.

EXAMPLE 2   Find the missing parts in triangle ABC if $a = 54$ centimeters, $b = 62$ centimeters, and $A = 40°$.

Soultion   First we solve for $\sin B$, using the law of sines.
Angle B

$$\sin B = \frac{b \sin A}{a}$$
$$= \frac{62 \sin 40°}{54}$$
$$\approx 0.7380$$

> *Note* Entering 0.7380 and pressing [inv] [sin] on a calculator gives us angle B. Then we find angle B' by subtracting B from 180°.

Now, since sin B is positive for any angle in quadrant I or II, we have two possibilities. We will call one of them B and the other B'.

$$B \approx 48° \quad \text{or} \quad B' \approx 180° - 48° = 132°$$

We have two different triangles that can be found with $a = 54$, $b = 62$, and $A = 40°$. Figure 2 shows both shows both of them. One is labeled ABC, while the other is labeled $AB'C$.

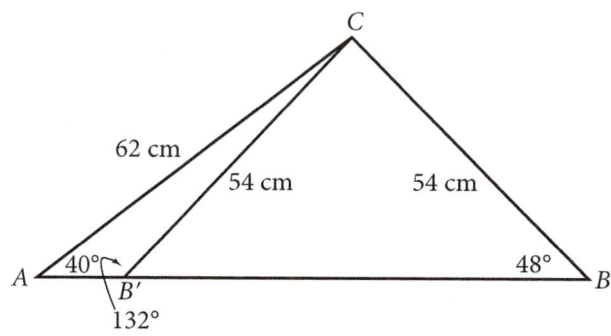

FIGURE 2

Angles C and C'   Since there are two values for B, we have two values for C:

$$C = 180° - (A + B) \qquad \text{and} \qquad C' = 180° - (A + B')$$

$$\approx 180° - (40° + 48°) \qquad\qquad \approx 180° - (40° + 132°)$$

$$\approx 92° \qquad\qquad\qquad\qquad \approx 8°$$

Sides c and c'

$$c = \frac{a \sin C}{\sin A} \qquad \text{and} \qquad c' = \frac{a \sin C'}{\sin A}$$

$$\approx \frac{54 \sin 92°}{\sin 40°} \qquad\qquad \approx \frac{54 \sin 8°}{\sin 40°}$$

$$\approx 84 \text{ centimeters} \qquad\qquad \approx 12 \text{ centimeters}$$

Figure 3 shows both triangles.

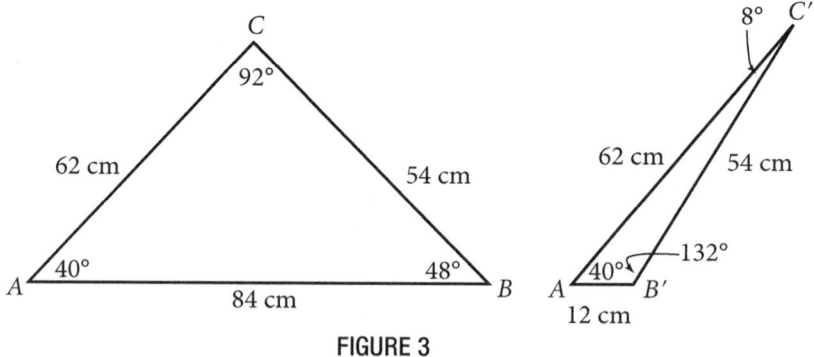

FIGURE 3

EXAMPLE 3   Find the missing parts of triangle ABC if $C = 35.4°$, $a = 205$ feet, and $c = 314$ feet.

Solution Applying the law of sines, we find sin A.

Angle *A*

$$\sin A = \frac{a \sin C}{c}$$

$$= \frac{205 \sin 35.4°}{314}$$

$$\approx 0.3782$$

Since sin A is positive in quadrants I and II, we have two possible values for A:

$$A \approx 22.2° \qquad \text{and} \qquad A' \approx 180° - 22.2°$$
$$\approx 157.8°$$

The second possibility, $A' = 157.8°$, will not work, however, since C is already $35.4°$ and therefore

$$C + A' \approx 35.4° + 157.8°$$
$$\approx 193.2°$$

which is larger than 180°. This result indicates that there is exactly one triangle that fits the given description. In that triangle $A \approx 22.2°$.

Angle *B*

$$B \approx 180° - (35.4° + 22.2°)$$
$$\approx 122.4°$$

Side *b*

$$b = \frac{c \sin B}{\sin C}$$

$$\approx \frac{314 \sin 122.4°}{\sin 35.4°}$$

$$\approx 458 \text{ feet}$$

Figure 4 is a diagram of its triangle.

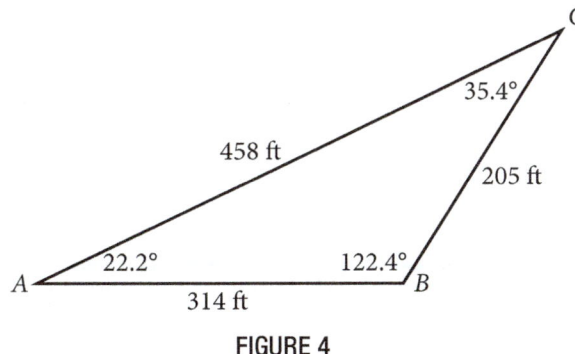

FIGURE 4

The different cases that can occur when we solve the kinds of triangles we have been given in Examples 1, 2, and 3 become apparent in the process of solving for the missing parts. Table 1 summarizes the set of conditions under which we will have 1, 2, or no triangles in the ambiguous case. In Table 1 we are assuming we are given angle A and sides a and b in triangle ABC and that h is the altitude from C.

Table 1

Conditions*	Number of Triangles	Diagram
$A < 90°$ and $a < h$	0	
$A > 90°$ and $a < b$	0	
$A < 90°$ and $a = h$	1	
$A < 90°$ and $a \geq b$	1	
$A > 90°$ and $a > b$	1	
$A < 90°$ and $h < a < b$	2	

*Assuming we are given angle A and sides a and b in triangle ABC and that h is the altitude from C.

Problem Set 7.2

Vocabulary

Use the vocabulary words or text below to fill in the blanks in the sentences.
You may not need to use every vocabulary word or text.

one   two   three   no   ambiguous

1. When we are given two sides in a triangle, and the angle opposite one of those sides, there can be several possible triangles that fit that description. In trigonometry, this situation is called the _____ case.

2. When we are given two sides in a triangle, and the angle opposite one of those sides, there can be one, two, or _____ triangles that fit that description.

3. If we are trying to find triangle ABC, and we find that $\sin B = 0.7380$, then we know that there are _____ possibilities for angle B.

4. The statement $\sin B = 1.5$ is impossible. No angle has a sine larger than _____ .

For each triangle below, solve for B and use the results to explain why the triangle has the given number of solutions.

1. $A = 30°$, $b = 40$ feet, $a = 10$ feet; no solution

2. $A = 150°$, $b = 30$ feet, $a = 10$ feet; no solution

3. $A = 120°$, $b = 20$ centimeters, $a = 30$ centimeters; one solution

4. $A = 30°$, $b = 12$ centimeters, $a = 6$ centimeters; one solution

5. $A = 60°$, $b = 18$ meters, $a = 16$ meters; two solutions

6. $A = 20°$, $b = 40$ meters, $a = 30$ meters; two solutions

Find all solutions to each of the following triangles:

7. $A = 38°$, $a = 41$ feet, $b = 54$ feet

8. $A = 43°$, $a = 31$ feet, $b = 37$ feet

9. $A = 112.2°$, $a = 43$ centimeters, $b = 22$ centimeters

10. $B = 30°$, $b = 42$ centimeters, $a = 84$ centimeters

11. $B = 118°$, $b = 68$ centimeters, $a = 92$ centimeters

12. $A = 124.3°$, $a = 27$ centimeters, $b = 50$ centimeters

13. $A = 142°$, $b = 2.9$ yards, $a = 1.4$ yards

14. $A = 65°$, $b = 7.6$ yards, $a = 7.1$ yards

15. $C = 26.8°$, $c = 36.8$ kilometers, $b = 36.8$ kilometers

16. $C = 73.4°$, $c = 51.1$ kilometers, $b = 92.4$ kilometers

Applying the Concepts

17. Circus Tent A wire 50 feet long running from the top of a tent pole to the ground makes an angle of 58° with the ground. If the length of the tent pole is 44 feet, how far is it from the bottom of the tent pole to the point where the wire is fastened to the ground? Hint: The tent pole is not vertical.

18. Hot-Air Balloon A hot-air balloon is held at a constant altitude by two ropes that are anchored to the ground. One rope is 120 feet long and makes an angle of 65° with the ground. The other rope is 115 feet long. What is the distance between the points on the ground at which the two ropes are anchored?

19. Leaning Windmill After a windstorm a farmer notices that his windmill may be leaning, but he is not sure. The windmill is 32 feet tall. From a point on the ground 30 feet from the base of the windmill the farmer finds the angle of elevation is 48° to the top of the windmill. Is the windmill leaning? If so, what is the acute angle the windmill makes with the ground?

20. Distance to a Road A boy is riding his motorcycle on a road that runs east and west. He leaves the road at a service station and rides for 5.25 miles in the direction N 15.5° E. Then he turns to his right and rides 6.50 miles back to the road, where his motorcycle breaks down. How far will he have to walk to get back to the service station?

Getting Ready for the Next Section

Simplify.

21. $180° - (60° + 42°)$

22. $180° - (126.9° + 25.0°)$

Simplify and round your answer to the nearest tenth.

23. $20^2 + 30^2 - 2(20)(30)(0.5)$

24. $\dfrac{20^2 + 18^2 - 34^2}{2(20)(18)}$

25. $\dfrac{18(\sin 126°)}{34}$

26. $\dfrac{20(\sin 60°)}{26}$

THE LAW OF COSINES

7.3

7.3 Videos

KEY WORDS

law of cosines

Learning Objectives

A Solve problems involving triangles using the law of cosines.

In this section we will derive another relationship that exists between the sides and angles in any triangle. It is called the **law of cosines** and is stated like this:

Law of Cosines

$$a^2 = b^2 + c^2 - 2bc \cos A$$

$$b^2 = a^2 + c^2 - 2ac \cos B$$

$$c^2 = a^2 + b^2 - 2ab \cos C$$

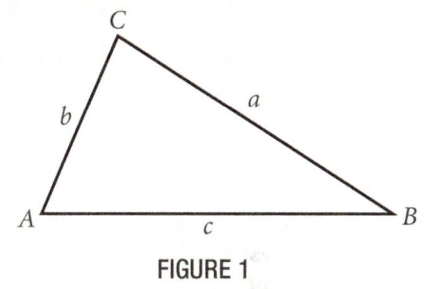

FIGURE 1

Derivation

To derive the formulas stated in the law of cosines, we apply the Pythagorean Theorem and some of the basic trigonometric identities.

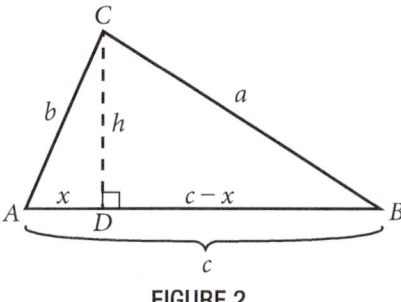

FIGURE 2

Applying the Pythagorean Theorem to right triangle *BCD* in Figure 2, we have

$$a^2 = (c - x)^2 + h^2$$

$$= c^2 - 2cx + x^2 + h^2$$

But, from right triangle *ACD*, we have $x^2 + h^2 = b^2$, so

$$a^2 = c^2 - 2cx + b^2$$

$$= b^2 + c^2 - 2cx$$

Now, since $\cos A = \frac{x}{b}$, we have $x = b \cos A$, so

$$a^2 = b^2 + c^2 - 2bc \cos A$$

Applying the same sequence of substitutions and reasoning to the right triangles formed by the altitudes from vertices *A* and *B* will give us the other two formulas listed in the law of cosines.

We can use the law of cosines to solve triangles in which we are given two sides and the angle included between them (*SAS*) or to solve triangles in which we are given all three sides (*SSS*).

Two Sides and the Included Angle

EXAMPLE 1 Find the missing parts of triangle ABC if $A = 60°$, $b = 20$ inches, and $c = 30$ inches. See Figure 3.

Solution The solution process will include the use of both the law of cosines and the law of sines. We being by using the law of cosines to find a.

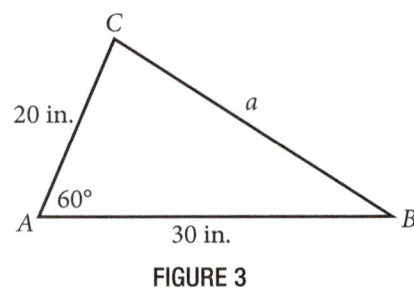

FIGURE 3

Side a

$$a^2 = b^2 + c^2 - 2bc \cos A \qquad \textit{Law of cosines}$$

$$= 20^2 + 30^2 - (2)(20)(30) \cos 60° \qquad \textit{Substitute given values.}$$

$$= 400 + 900 - 1{,}200 \cos 60° \qquad \textit{Calculator}$$

$$a^2 = 700$$

$$a \approx 26 \text{ inches} \qquad \textit{To the nearest integer}$$

Now that we have a, we can use the law of sines to solve for either B or C. When we have a choice of angles to solve for, and we are using the law of sines to do so, it is usually best to solve for the smaller angle. Since b is smaller than c, B will be smaller than C.

Angle B

$$\sin B = \frac{b \sin A}{a}$$

$$\approx \frac{20 \sin 60°}{26}$$

$$\sin B \approx 0.6662$$

So $\qquad B \approx 42°$ *To the nearest degree*

Note that we don't have to check $B' = 180° - 42° = 138°$ because we know B is an acute angle since it is smaller than either A or C.

Angle C

$$C = 180° - (A + B)$$

$$\approx 180° - (60° + 42°) = 78°$$

EXAMPLE 2 The diagonals of a parallelogram are 24.2 centimeters and 35.4 centimeters and intersect at an angle of 65.6°. Find the length of the two shorter sides of the parallelogram.

Solution A diagram of a parallelogram is shown in Figure 4. The variable x represents the length of the shorter sides. Note also that we have labeled the other two sides of the triangle formed by one short side and the diagonals with $\frac{24.2}{2} = 12.1$ centimeters and $\frac{35.4}{2} = 17.7$ centimeters, respectively. (Remember: The diagonals of a parallelogram bisect each other.) Using the law of cosines,

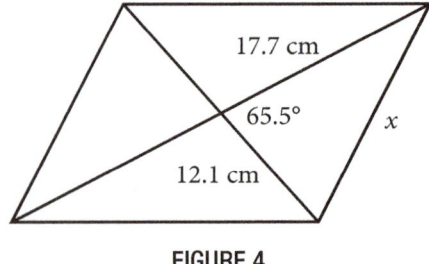

FIGURE 4

$$x^2 = (12.1)^2 + (17.7)^2 - 2(12.1)(17.7) \cos 65.5°$$

$$x^2 \approx 282.07$$

$$x \approx 16.8 \text{ centimeters} \qquad \textit{To the nearest tenth}$$

Next we will see how the law of cosines can be used to find the missing parts of a triangle in which all three sides are given.

Three Sides

To use the law of cosines to solve a triangle in which we are given all three sides, it is convenient to rewrite the equations with the cosines isolated on one side. Here is an equivalent form of the law of cosines:

$$\cos A = \frac{b^2 + c^2 - a^2}{2bc}$$

$$\cos B = \frac{a^2 + c^2 - b^2}{2ac}$$

$$\cos C = \frac{a^2 + b^2 - c^2}{2ab}$$

Here is how we arrived at the first of these formulas.

$$a^2 = b^2 + c^2 - 2bc \cos A$$

$$b^2 + c^2 - 2bc \cos A = a^2 \qquad \text{Exchange sides.}$$

$$-2bc \cos A = -b^2 - c^2 + a^2 \qquad \text{Add } -b^2 \text{ and } -c^2 \text{ to both sides.}$$

$$\cos A = \frac{b^2 + c^2 - a^2}{2bc} \qquad \text{Divide both sides by } -2bc.$$

EXAMPLE 3   Solve triangle ABC if $a = 34$ kilometers, $b = 20$ kilometers, and $c = 18$ kilometers.

Solution We will use the law of cosines to solve for one of the angles and then the law of sines to find one of the remaining angles. We find the largest angle first. If it is obtuse, its cosine will be negative. After that, we can use the law of sines to solve for the other angles, which must be acute angles. Since the longest side is a, we solve for A first. (This will avoid a possible ambiguous case when finding angle C.)

Angle A

$$\cos A = \frac{b^2 + c^2 - a^2}{2bc}$$

$$= \frac{20^2 + 18^2 - 34^2}{(2)(20)(18)}$$

$$\cos A = -0.6000$$

So $A \approx 126.9°$ To the nearest tenth

Now we use the law of sines to find angle C.

Angle C

$$\sin C = \frac{c \sin A}{a}$$

$$\approx \frac{18 \sin 126.9°}{34}$$

$$\sin C \approx 0.4234$$

So $C \approx 25.0°$ To the nearest tenth

Angle B

$$B = 180° - (A + C)$$
$$\approx 180° - (126.9° + 25.0°) = 28.1°$$

Note Example 3 illustrates an important point when using the law of cosines and the law of sines in the same problem. When you are given 3 sides of a triangle, always use the law of cosines to find the largest angle (opposite the largest side), then use the law of sines to find the smallest angle (opposite the smallest side). Doing this will avoid any potential ambiguous case with the law of sines.

EXAMPLE 4 A plane is flying with an airspeed of 185 mph with heading 120°. The wind currents are running at a constant 32 mph at 165° clockwise from due north. Find the true course and ground speed of the plane.

Solution Figure 5 is a diagram of the situation with the vector **V** representing the airspeed and direction of the plane and **W** representing the speed and direction of the wind currents. From Figure 5, $\alpha = 180° - 120° = 60°$ and $\theta = 360° - (60° + 165°) = 135°$.

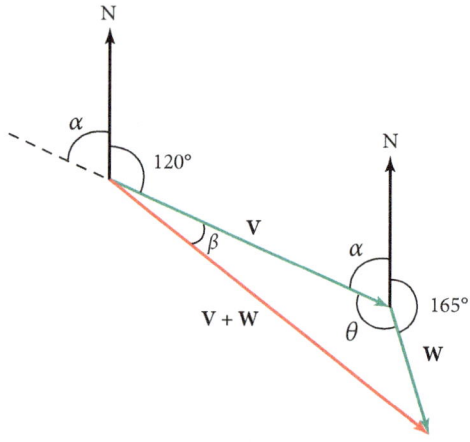

FIGURE 5

The magnitude of **V** + **W** can be found from the law of cosines.

$$|\mathbf{V} + \mathbf{W}|^2 = |\mathbf{V}|^2 + |\mathbf{W}|^2 - 2|\mathbf{V}||\mathbf{W}|\cos\theta$$

$$= 185^2 + 32^2 - 2(185)(32)\cos 135°$$

$$= 43,621$$

so $|\mathbf{V} + \mathbf{W}| = 210$ mph   *To two significant digits*

To find the direction of **V** + **W**, we first find β using the law of sines.

$$\frac{\sin\beta}{32} = \frac{\sin\theta}{210}$$

$$\sin\beta = \frac{32\sin 135°}{210}$$

$$= 0.1077$$

so $\beta = 6°$ *To the nearest degree*

The true course is $120 + \beta = 120° + 6° = 126°$. The speed of the plane with respect to the ground is 210 mph.

Vocabulary

Use the vocabulary words or text below to fill in the blanks in the sentences. You may not need to use every vocabulary word or text.

$\cos A$ $\cos B$ $\cos C$ two   Pythagorean   largest   smallest

1. For any triangle ABC, the law of cosines states that

$$c^2 = a^2 + b^2 - 2ab \cos C$$

When $C = 90°$, the law of cosines becomes the _____ theorem.

2. If we are given three sides in triangle ABC, we can use an alternate form of the law of cosines to find one of the angles first. It looks like this:

$$\underline{\hspace{2cm}} = \frac{a^2 + b^2 - c^2}{2ab}$$

3. If we are given three sides in triangle ABC, we can solve for one of the angles using the formula above. In this situation, it is good practice to solve for the _____ angle first.

Each problem below refers to a triangle ABC. Any answers that involve angles should be rounded to the nearest tenth of a degree.

1. If $a = 100$ inches, $b = 60$ inches, and $C = 60°$, find c.

2. If $a = 100$ inches, $b = 60$ inches, and $C = 120°$, find c.

3. If $a = 5$ yards, $b = 6$ yards, and $c = 8$ yards, find the largest angle.

4. If $a = 10$ yards, $b = 14$ yards, and $c = 8$ yards, find the largest angle.

5. If $b = 4.2$ meters, $c = 6.8$ meters, and $A = 116°$, find a.

6. If $a = 3.7$ meters, $c = 6.4$ meters, and $B = 23°$, find b.

7. If $a = 38$ centimeters, $b = 10$ centimeters, and $c = 31$ centimeters, find the largest angle.

8. If $a = 51$ centimeters, $b = 24$ centimeters, and $c = 31$ centimeters, find the largest angle.

Solve each triangle below

9. $a = 50$ centimeters, $b = 70$ centimeters, $C = 60°$

10. $a = 10$ centimeters, $b = 12$ centimeters, $C = 120°$

11. $a = 4$ inches, $b = 6$ inches, $c = 8$ inches (Remember: Solve for the largest angle first and round to the nearest tenth of a degree).

12. $a = 5$ inches, $b = 10$ inches, $c = 12$ inches

13. $a = 410$ meters, $c = 340$ meters, $B = 151.5°$

14. $a = 76.3$ meters, $c = 42.8$ meters, $B = 16.3°$

15. $a = 0.048$ yards, $b = 0.063$ yards, $c = 0.075$ yards

16. $a = 48$ yards, $b = 75$ yards, $c = 63$ yards

17. $a = 4.38$ feet, $b = 3.79$ feet, $c = 5.22$ feet

18. $a = 832$ feet, $b = 623$ feet, $c = 345$ feet

19. Use the law of cosines to show that, if $A = 90°$, then $a^2 = b^2 + c^2$.

20. Use the law of cosines to show that, if $a^2 = b^2 + c^2$, then $A = 90°$.

21. The diagonals of a parallelogram are 56 inches and 34 inches, and intersect at an angle of 120°. Find the length of the shorter side.

22. The diagonals of a parallelogram are 14 meters and 16 meters, and intersect at an angle of 60°. Find the length of the longer side.

23. Geometry The diagonals of a parallelogram are 56 inches and 34 inches and intersect at an angle of 120°. Find the length of the longer side.

24. Geometry The diagonals of a parallelogram are 14 m and 16 m and intersect at an angle of 60°. Find the length of the shorter side.

25. Distance Between Two Planes Two planes leave an airport at the same time. Their speeds are 130 miles per hour and 150 miles per hour, and the angle between their courses is 36°. How far apart are they after 1.5 hours?

26. Distance Between Two Ships Two ships leave a harbor entrance at the same time. The first ship is traveling at a constant 18 miles per hour, while the second is traveling at a constant 22 miles per hour. If the angle between their courses is 123°, how far apart are they after 2 hours?

27. Distance Between Two Planes Two planes take off at the same time from an airport. The first plane is flying at 246 miles per hour on a course with bearing S 45° E. The second plane is flying in the direction S 5° E at 357 miles per hour. How far apart are they after 2 hours?

28. Distance Between Two Ships Two ships leave the harbor at the same time. The first ship is traveling at 14 miles per hour on a course with bearing S 13° W, while the other is traveling at 12 miles per hour on a course with bearing N 75° E. How far apart are they after 3 hours?

29. True Course and Speed A plane is flying with an airspeed of 160 mph and heading of 150°. The wind currents are running at 35 mph at 165° clockwise from due north. Find the true course and ground speed of the plane.

30. True Course and Speed A plane is flying with an airspeed of 240 mph with heading 273°. The wind currents are running at a constant 46 mph in the direction 262°. Find the ground speed and true course of the plane.

Getting Ready for the Next Section

Find the following to four decimal places, if necessary.

31. $\sin 35.1°$

32. $\sin 35° \, 10'$

33. $24° \, 10' + 120° \, 40'$

34. $180° - (24° \, 10' + 120° \, 40')$

35. $(2.43)(3.57)$

36. $\frac{1}{2}(2.43)(3.57)(0.5750)$

Learning Objectives

A Find the area of a triangle from two sides and a given angle.

B Find the area of a triangle from two angles and a given side

C Find the area of a triangle using Heron's formula.

In this section, we will derive three formulas for the area S of a triangle. We will start by deriving the formula used to find the area of a triangle in which two sides and the included angle are given.

A Two Sides and the Included Angle

To derive our first formula we begin with the general formula for the area of a triangle

$$S = \frac{1}{2}(\text{base})(\text{height})$$

The base of triangle ABC in Figure 1 is c and the height is h. So the formula for S becomes, in this case,

$$S = \frac{1}{2} ch$$

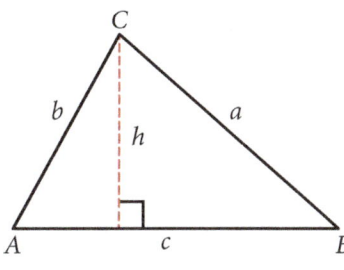

FIGURE 1

Suppose that, for triangle ABC, we are given the lengths of sides b and c and the measure of angle A. Then we can write $\sin A$ as

$$\sin A = \frac{h}{b}$$

or, by solving for h, $\qquad\qquad h = b \sin A$

Substituting this expression for h into the formula

$$S = \frac{1}{2} ch$$

we have

$$S = \frac{1}{2} bc \sin A$$

Applying the same kind of reasoning to the heights drawn from A and C, we also have

$$S = \frac{1}{2} ab \sin C$$

$$S = \frac{1}{2} ac \sin B$$

Each of these three formulas indicates that to find the area of a triangle in which we are given two sides and the angle included between them, we multiply half the product of the two sides times the sine of the angle included between them.

EXAMPLE 1 Find the area of triangle ABC if $A = 35.1°$, $b = 2.43$ centimeters, and $c = 3.57$ centimeters.

Solution Applying the first formula we derived, we have

$$S = \frac{1}{2}\, bc \sin A$$

$$= \frac{1}{2}\, (2.43)(3.57)\sin 35.1°$$

$$= 2.49 \text{ centimeters}^2 \qquad \textit{To three significant digits}$$

The next area formula we will derive is used to find the area of triangles in which we are given two angles and one side.

B Two Angles and One Side

Suppose we were given angles A and B and side a in triangle ABC in Figure 1. We could easily solve for C by subtracting the sum of A and B from $180°$.

To find side b, we use the law of sines

$$\frac{b}{\sin B} = \frac{a}{\sin A}$$

solving this equation for b would give us

$$b = \frac{a \sin B}{\sin A}$$

Substituting this expression for b into the formula

$$S = \frac{1}{2}\, ab \sin C$$

$$S = \frac{1}{2}\, a \left(\frac{a \sin B}{\sin A} \right) \sin C$$

$$= \frac{a^2 \sin B \sin C}{2 \sin A}$$

A similar sequence of steps can be used to derive

$$S = \frac{b^2 \sin A \sin C}{2 \sin B}$$

and

$$S = \frac{c^2 \sin A \sin B}{2 \sin C}$$

The formula we use depends on the side we are given.

EXAMPLE 2   Find the area of triangle ABC if $A = 24°10'$, $B = 120°40'$, and $a = 4.25$ feet.

Solution   We begin by finding C.

$$C = 180° - (24°10' + 120°40')$$

$$= 35°10'$$

Now, applying the formula

$$S = \frac{a^2 \sin B \sin C}{2 \sin A}$$

with $a = 4.25$, $A = 24°10'$, $B = 120°40'$, and $C = 35°10'$, we have

$$S = \frac{(4.25)^2 (\sin 120°40')(\sin 35°10')}{2 \sin 24°10'}$$

$$= 10.93 \text{ feet}^2$$

C Three Sides

The last area formula is called Heron's formula and it is used to find the area of a triangle in which all three sides are known.

Heron's Formula

The area of a triangle with sides of length a, b, and c is given by

$$S = \sqrt{s(s - a)(s - b)(s - c)}$$

where s is half the perimeter of the triangle; that is,

$$s = \frac{1}{2}(a + b + c) \quad \text{or} \quad 2s = a + b + c$$

Proof   We begin our proof by squaring both sides of the formula

$$S = \frac{1}{2} ab \sin C$$

to obtain

$$S^2 = \frac{1}{4} a^2 b^2 \sin^2 C$$

Next, we multiply both sides of the equation by $4/a^2 b^2$ to isolate $\sin^2 C$ on the right side.

$$\frac{4S^2}{a^2 b^2} = \sin^2 C$$

Replacing $\sin^2 C$ with $1 - \cos^2 C$ and then factoring as the difference of two squares, we have

$$\frac{4S^2}{a^2 b^2} = 1 - \cos^2 C$$

$$= (1 + \cos C)(1 - \cos C)$$

From the law of cosines we know that $\cos C = (a^2 + b^2 - c^2)/2ab$.

$$= \left[1 + \frac{a^2 + b^2 - c^2}{2ab} \right]\left[1 - \frac{a^2 + b^2 - c^2}{2ab} \right]$$

$$\frac{4S^2}{a^2b^2} = \left[\frac{2ab + a^2 + b^2 - c^2}{2ab} \right]\left[\frac{2ab - a^2 - b^2 + c^2}{2ab} \right]$$

$$= \left[\frac{(a^2 + 2ab + b^2) - c^2}{2ab} \right]\left[\frac{c^2 - (a^2 - 2ab + b^2)}{2ab} \right]$$

$$= \left[\frac{(a + b)^2 - c^2}{2ab} \right]\left[\frac{c^2 - (a - b)^2}{2ab} \right]$$

Now we factor each numerator as the difference of two squares and multiply the denominators.

$$\frac{4S^2}{a^2b^2} = \frac{[(a + b + c)(a + b - c)][(c + a - b)(c - a + b)]}{4a^2b^2}$$

Now, since $a + b + c = 2s$, it is also true that

$$a + b - c = a + b + c - 2c = 2s - 2c$$

$$c + a - b = a + b + c - 2b = 2s - 2b$$

$$c - a + b = a + b + c - 2a = 2s - 2a$$

Substituting these expressions into our last equation, we have

$$\frac{4S^2}{a^2b^2} = \frac{2s(2s - 2c)(2s - 2b)(2s - 2a)}{4a^2b^2}$$

Factoring out a 2 from each term in the numerator and showing the left side of our equation along with the right side, we have

$$\frac{4S^2}{a^2b^2} = \frac{16s(s - a)(s - b)(s - c)}{4a^2b^2}$$

Multiplying both sides by $a^2b^2/4$ we have

$$S^2 = s(s - a)(s - b)(s - c)$$

Taking the square root of both sides of the equation we have Heron's formula.

$$S = \sqrt{s(s - a)(s - b)(s - c)}$$

EXAMPLE 3   Find the area of triangle ABC if $a = 12$ meters, $b = 14$ meters, and $c = 8$ meters.

Solution   We begin by calculating the formula for s, half the perimeter of ABC.

$$s = \frac{1}{2}(12 + 14 + 8)$$

$$= 17$$

Substituting this value of s into Heron's formula along with the given values of a, b, and c, we have

$$S = \sqrt{17(17 - 12)(17 - 14)(17 - 8)}$$

$$= \sqrt{17(5)(3)(9)}$$

$$= \sqrt{2,295}$$

$$= 47.9 \text{ meters}^2 \qquad \text{To three significant digits}$$

Problem Set 7.4

Vocabulary

Use the vocabulary words or text below to fill in the blanks in the sentences. You may not need to use every vocabulary word or text.

$\sin A$ $\sin B$ $\sin C$ base   perimeter   area

1. The formula for the area of a triangle is $S = \frac{1}{2}$ (_____)(height).

2. Another formula for the area of triangle ABC is $S = \frac{1}{2} ab$ _____ .

3. Heron's formula gives us another formula for the area of triangle ABC. It is stated this way:

$$S = \sqrt{s(s-a)(s-b)(s-c)}$$

where s is half the _____ of the triangle.

Each problem below refers to triangle ABC. In each case find the area of the triangle. Round to three significant digits.

1. $a = 50$ centimeters, $b = 70$ centimeters, $C = 60°$

2. $a = 10$ centimeters, $b = 12$ centimeters, $C = 120°$

3. $a = 41$ meters, $c = 34$ meters, $B = 151.5°$

4. $a = 76.3$ meters, $c = 42.8$ meters, $B = 16.3°$

5. $b = 0.923$ kilometers, $c = 0.387$ kilometers, $A = 43°20'$

6. $b = 63.4$ kilometers, $c = 75.2$ kilometers, $A = 124°40'$

7. $A = 46°$, $B = 95°$, $c = 6.8$ meters

8. $B = 57°$, $C = 31°$, $a = 7.3$ meters

9. $A = 42.5°$, $B = 71.4°$, $a = 210$ inches

10. $A = 110.4°$, $C = 21.8°$, $c = 240$ inches

11. $A = 43°30'$, $C = 120°30'$, $a = 3.48$ feet

12. $B = 14°20'$, $C = 75°40'$, $b = 2.72$ feet

13. $a = 4$ inches, $b = 6$ inches, $c = 8$ inches

14. $a = 5$ inches, $b = 10$ inches, $c = 12$ inches

15. $a = 4.8$ yards, $b = 6.3$ yards, $c = 7.5$ yards

16. $a = 48$ yards, $b = 75$ yards, $c = 63$ yards

17. $a = 4.38$ feet, $b = 3.79$ feet, $c = 5.22$ feet

18. $a = 8.32$ feet, $b = 6.23$ feet, $c = 3.45$ feet

19. Find the area of a parallelogram if the angle between two of the sides is 120° and the two sides are 15 inches and 12 inches.

20. Find the area of a parallelogram if the two sides measure 24.1 inches and 32.4 inches and the longest diagonal is 31.4 inches.

21. The area of a triangle is 40 centimeters². Find the length of the side included between the angles $A = 30°$ and $B = 50°$.

22. The area of a triangle is 80 inches². Find the length of the side included between $A = 25°$ and $C = 110°$.

Maintaining Your Skills

Solve each equation for θ if $0° \leq \theta \leq 360°$. Do not use a calculator.

23. $2 \cos \theta + \sqrt{3} = 0$ **24.** $\sqrt{3} \cot \theta - 1 = 0$

For all equations written in terms of θ, find all degree solutions from $\theta = 0°$ to $\theta = 360°$. If the equation is written in terms of x, write your solutions in radians using exact values for $0 \leq x \leq 2\pi$.

25. $4 \cos x - 3 \sec x = 0$ **26.** $2 \sin \theta + \sin 2\theta = 0$

Find all degree solutions for each of the following:

27. $\sin 2\theta = -\dfrac{\sqrt{3}}{2}$ **28.** $\cos 3\theta = -1$

Eliminate the parameter t in each of the following:

29. $x = 3 \cot t$ **30.** $x = 3 + 5 \tan t$
  $y = 3 \csc t$ $y = 2 + 5 \sec t$

Examples

1. If $A = 30°$, $B = 70°$, and $a = 8$ centimeters in triangle ABC, then by the law of sines,

$$b = \frac{a \sin B}{\sin A}$$

$$= \frac{8 \sin 70°}{\sin 30°}$$

$$\approx 15 \text{ centimeters}$$

2. In triangle ABC, if $a = 54$ centimeters, $b = 62$ centimeters, and $A = 40°$, then

$$\sin B = \frac{b \sin A}{a}$$

$$= \frac{62 \sin 40°}{54}$$

$$\approx 0.7380$$

Since $\sin B$ is positive for any angle in quadrant I or II, we have two possibilities for B:

$$B \approx 48°$$

or

$$B' \approx 180° - 48° = 132°$$

This indicates that two triangles exist, both of which fit the given information.

3. In triangle ABC, if $a = 34$ kilometers, $b = 20$ kilometers, and $c = 18$ kilometers, then we can find A using the law of cosines.

$$\cos A = \frac{b^2 + c^2 - a^2}{2bc}$$

$$= \frac{20^2 + 18^2 - 34^2}{(2)(20)(18)}$$

$$\cos A = -0.6000$$

$$A \approx 126.9°$$

The Law of Sines [7.1]

For any triangle ABC, the following relationships are always true:

$$\frac{\sin A}{a} = \frac{\sin B}{b} = \frac{\sin C}{c}$$

Or, equivalently,

$$\frac{a}{\sin A} = \frac{b}{\sin B} = \frac{c}{\sin C}$$

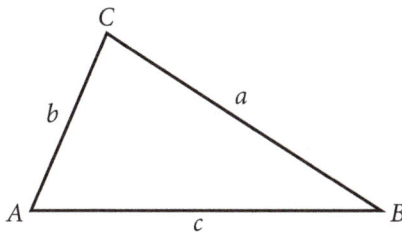

The Ambiguous Case [7.2]

When we are given two sides and an angle opposite one of them (SSA), we have several possibilities for the triangle or triangles that result. One of the possibilities is that no triangle will fit the given information. Another possibility is that two different triangles can be obtained from the given information, and a third possibility is that exactly one triangle will fit the given information. Because of these difference possibilities, we call the situation where we are solving a triangle in which we are given two sides and the angle opposite one of them the **ambiguous case**.

The Law of Cosines [7.3]

In any triangle ABC, the following relationships are always true:

$$a^2 = b^2 + c^2 - 2bc \cos A$$

$$b^2 = a^2 + c^2 - 2ac \cos B$$

$$c^2 = a^2 + b^2 - 2ab \cos C$$

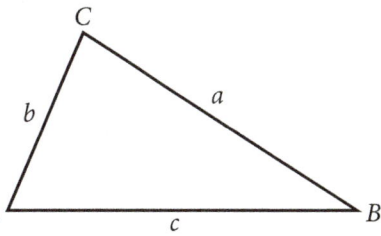

Another form of the law of cosines looks like this:

$$\cos A = \frac{b^2 + c^2 - a^2}{2bc}$$

$$\cos B = \frac{a^2 + c^2 - b^2}{2ac}$$

$$\cos C = \frac{a^2 + b^2 - c^2}{2ab}$$

4. For triangle ABC,

a. If $a = 12$ centimeters, $b = 15$ centimeters, and $C = 20°$, then the area of ABC is

$$S = \frac{1}{2}(12)(15)\sin 20°$$
$$= 30.8 \text{ centimeters}^2$$
$$\text{to the nearest tenth}$$

b. If $a = 24$ inches, $b = 14$ inches, and $C = 18$ inches, then the area of ABC is

$$S = \sqrt{28(28-24)(28-14)(28-18)}$$
$$= \sqrt{28(4)(14)(10)}$$
$$= \sqrt{15{,}680}$$
$$= 125.2 \text{ inches}^2 \text{ to the nearest tenth}$$

c. If $A = 40°$, $B = 72°$, and $c = 45$ meters, then the area of ABC is

$$S = \frac{45^2 \sin 40° \sin 72°}{2 \sin 68°}$$
$$= 667.6 \text{ meters}^2$$
$$\text{to the nearest tenth}$$

The Area of a Triangle [7.4]

The area of a triangle in which we are given two sides and the included angle is given by

$$S = \frac{1}{2} ab \sin C$$

$$S = \frac{1}{2} ac \sin B$$

$$S = \frac{1}{2} bc \sin A$$

The area of a triangle in which we are given all three sides is given by the formula

$$S = \sqrt{s(s-a)(s-b)(s-c)}$$

where $s = \frac{1}{2}(a + b + c)$

The area of a triangle in which we are given two angles and the side included between them is given by

$$S = \frac{a^2 \sin B \sin C}{2 \sin A}$$

$$S = \frac{b^2 \sin C \sin A}{2 \sin B}$$

$$S = \frac{c^2 \sin A \sin B}{2 \sin C}$$

Problems 1 through 14 refer to triangle *ABC* which is not necessarily a right triangle.

1. If $A = 32°$, $B = 70°$, and $a = 3.8$ inches, use the law of sines to find b.

2. If $B = 118°$, $C = 37°$, and $c = 2.9$ inches, use the law of sines to find b.

3. If $A = 38.2°$, $B = 63.4°$, and $c = 42.0$ centimeters, find all the missing parts.

4. If $A = 24.7°$, $C = 106.1°$, and $b = 34.0$ centimeters, find all the missing parts.

5. Use the law of sines to show that no triangle exists for which $A = 60°$, $a = 12$ inches, and $b = 42$ inches.

6. Use the law of sines to show that exactly one triangle exists for which $A = 42°$, $a = 29$ inches, and $b = 21$ inches.

7. Find two triangles for which $A = 51°$, $a = 6.5$ feet, and $b = 7.9$ feet.

8. Find two triangles for which $A = 26°$, $a = 4.8$ feet, and $b = 9.4$ feet.

9. If $C = 60°$, $a = 10$ centimeters, and $b = 12$ centimeters, use the law of cosines to find c.

10. If $C = 120°$, $a = 10$ centimeters, and $b = 12$ centimeters, use the law of cosines to find c.

11. If $a = 5$ kilometers, $b = 7$ kilometers, and $c = 9$ kilometers, use the law of cosines to find C to the nearest tenth of a degree.

12. If $a = 10$ kilometers, $b = 12$ kilometers, and $c = 11$ kilometers, use the law of cosines to find B to the nearest tenth of a degree.

13. Find all the missing parts if $a = 6.4$ meters, $b = 2.8$ meters, and $C = 119°$.

14. Find all the missing parts if $b = 3.7$ meters, $c = 6.2$ meters, and $A = 35°$.

15. **Geometry** The two equal sides of an isosceles triangle are each 38 centimeters. If the base measures 48 centimeters, find the measure of the two equal angles to the nearest tenth of a degree.

16. **Angle of Elevation** A man standing near a building notices the angle of elevation to the top of the building is 64°. He then walks 240 feet farther away from the building and finds the angle of elevation to the top to be 43°. To the nearest foot, how tall is the building?

17. **Geometry** The diagonals of a parallelogram are 26.8 meters and 39.4 meters. If they meet at an angle of 134°, find the length of the shorter side of the parallelogram.

18. Find the area of the triangle in Problem 3.

19. Find the area of the triangle in Problem 4.

20. Find the area of the triangle in Problem 9.

21. Find the area of the triangle in Problem 10.

22. Find the area of the triangle in Problem 11.

23. Find the area of the triangle in Problem 12.

Spotlight on Success

Lauren, Student Instructor

There are a lot of word problems in algebra and many of them involve topics that I don't know much about. I am better off solving these problems if I know something about the subject. So, I try to find something I can relate to. For instance, an example may involve the amount of fuel used by a pilot in a jet airplane engine. In my mind, I'd change the subject to something more familiar, like the mileage I'd be getting in my car and the amount spent on fuel, driving from my hometown to my college. Changing these problems to more familiar topics makes math much more interesting and gives me a better chance of getting the problem right. It also helps me to understand how greatly math affects and influences me in my everyday life. We really do use math more than we would like to admit—budgeting our income, purchasing gasoline, planning a day of shopping with friends—almost everything we do is related to math. So the best advice I can give with word problems is to learn how to associate the problem with something familiar to you.

You should know that I have always enjoyed math. I like working out problems and love the challenges of solving equations like individual puzzles. Although there are more interesting subjects to me, and I don't plan on pursuing a career in math or teaching, I do think it's an important subject that will help you in any profession.

COMPLEX NUMBERS & POLAR COORDINATES 8

WHAT ARE COMPLEX NUMBERS AND POLAR COORDINATES? Complex numbers and polar coordinates are closely related concepts that provide alternative ways to represent and work with points in the plane, particularly when dealing with angles and distances.

WHY ARE COMPLEX NUMBERS AND POLAR COORDINATES IMPORTANT? Understanding complex numbers and polar coordinates are particularly useful in analyzing periodic phenomena, solving differential equations, and understanding the behavior of functions in the complex plane.

WHERE MIGHT YOU SEE COMPLEX NUMBERS AND POLAR COORDINATES? Here are some examples of where you may see complex numbers and polar coordinates in the real world:

FOURIER SERIES AND TRANSFORMATIONS The Fourier series and Fourier transform are mathematical techniques used to decompose and analyze periodic and non-periodic functions, respectively, into their constituent sinusoidal components. Complex numbers and polar coordinates play a crucial role in these techniques, as they allow functions to be represented as sums of complex exponential functions with varying magnitudes and phase angles. This representation simplifies the analysis of signals and facilitates the computation of frequency spectra.

CONTROL SYSTEMS AND SIGNAL PROCESSING In control systems and signal processing, complex numbers and polar coordinates are used to analyze and design filters, controllers, and other systems. For example, the transfer functions of filters and controllers are often represented in the complex frequency domain, where complex numbers are used to characterize their frequency response. Engineers can use polar coordinates to visualize the gain and phase characteristics of these systems, allowing them to optimize performance and stability.

343

In computer graphics, rotation of objects through a fixed angle α (relative to the origin) is necessary for any art software. If a point (x, y) is rotated by angle α, how can we find the location of the rotated point? This series of steps will illustrate how to accomplish this task.

1. Express the point (x, y) in polar coordinates.

2. Find the new point (x', y') after a rotation of angle α, relative to the origin.

3. Use trigonometric identities to find equations in terms of (x, y).

4. Find the new coordinates of the point $(3, 4)$ rotated $45°$.

5. Find the new coordinates of the point $(2, 5)$ rotated $90°$.

Solution

1. Since $x = r \cos \theta$ and $y = r \sin \theta$, we have $(x, y) = (r \cos \theta, r \sin \theta)$.

2. Since the rotation angle is α, the new polar angle is $\theta + \alpha$, so $(x', y') = (r \cos (\theta + \alpha), r \sin (\theta + \alpha))$.

3. Using the addition formulas for sine and cosine, we have

$$x' = r \cos (\theta + \alpha) = r \cos \theta \cos \alpha - r \sin \theta \sin \alpha$$

$$y' = r \sin (\theta + \alpha) = r \cos \theta \sin \alpha + r \sin \theta \cos \alpha$$

Since $x' = r \cos \theta$ and $y' = r \sin \theta$, we have

$$x' = x \cos \alpha - y \sin \alpha \qquad \text{and} \qquad y' = x \sin \alpha + y \cos \alpha$$

These are the desired rotation equations.

4. Using $(x, y) = (3, 4)$ and $\alpha = 45°$:

$$x' = 3 \cos 45° - 4 \sin 45° = 3\left(\frac{1}{\sqrt{2}}\right) - 4\left(\frac{1}{\sqrt{2}}\right) = -\frac{1}{\sqrt{2}}$$

$$y' = 3 \sin 45° + 4 \cos 45° = 3\left(\frac{1}{\sqrt{2}}\right) + 4\left(\frac{1}{\sqrt{2}}\right) = -\frac{7}{\sqrt{2}}$$

The rotated coordinates are $\left(-\frac{1}{\sqrt{2}}, \frac{7}{\sqrt{2}}\right) \approx (-0.71, 4.95)$.

5. Using $(x, y) = (2, 5)$ and $\alpha = 90°$:

$$x' = 2 \cos 90° - 5 \sin 90° = 2(0) - 5(1) = -5$$
$$y' = 2 \sin 90° + 5 \cos 90° = 2(1) + 5(0) = 2$$

The rotated coordinates are $(-5, 2)$.

8.1 Videos

KEY WORDS

number *i*

powers of *i*

square roots of negative numbers

complex number

standard form

real part

imaginary part

complex conjugates

Learning Objectives

A Write square roots in terms of the number *i*.

B Simplify expressions containing the powers of *i*.

C Add, subtract, multiply, and divide complex numbers.

In 2012, two deadly earthquakes struck the Emilia Romagna region of northern Italy. Widespread soil liquefaction damaged much of the region's buildings and infrastructure. Earthquake liquefaction occurs when shaking from a seismic event turns the once-solid ground to a consistency similar to quicksand. Buildings and bridges can literally sink into the ground! Some civil engineers study soil liquefaction using a special group of numbers called complex numbers, which is the focus of this section.

The equation $x^2 = -9$ has no real number solutions because the square of a real number is always positive. We have been unable to work with square roots of negative numbers like $\sqrt{-25}$ and $\sqrt{-16}$ for the same reason. Complex numbers allow us to expand our work with radicals to include square roots of negative numbers and to solve equations like $x^2 = -9$ and $x^2 = -64$.

A Square Roots of Negative Numbers

Our work with complex numbers is based on the following definition.

The Number *i*

The **number *i*** is such that $i = \sqrt{-1}$ (which is the same as saying $i^2 = -1$).

The number *i*, as we have defined it here, is not a real number. Because of the way we have defined *i*, we can use it to simplify square roots of negative numbers.

Square Roots of Negative Numbers

If a is a positive number, then $\sqrt{-a}$ can always be written as $i\sqrt{a}$. That is,

$$\sqrt{-a} = i\sqrt{a} \qquad \text{if } a \text{ is a positive number.}$$

To justify our rule, we simply square the quantity $i\sqrt{a}$ to obtain $-a$. Here is what it looks like when we do so:

$$(i\sqrt{a})^2 = i^2 \cdot (\sqrt{a})^2$$
$$= -1 \cdot a$$
$$= -a$$

Here are some examples that illustrate the use of our new rule.

EXAMPLES Write each square root in terms of the number *i*.

1. $\sqrt{-25} = i\sqrt{25} = i \cdot 5 = 5i$ **2.** $\sqrt{-49} = i\sqrt{49} = i \cdot 7 = 7i$

3. $\sqrt{-12} = i\sqrt{12} = i \cdot 2\sqrt{3} = 2i\sqrt{3}$ **4.** $\sqrt{-17} = i\sqrt{17}$

Note In Examples 3 and 4, we wrote *i* before the radical simply to avoid confusion. If we were to write the answer to 3 as $2\sqrt{3i}$, some people would think the *i* was under the radical sign, but it is not.

B Powers of i

If we assume all the properties of exponents hold when the base is i, we can write any power of i as i, -1, $-i$, or 1. Using the fact that $i^2 = -1$, we have

$$i^1 = i$$

$$i^2 = -1$$

$$i^3 = i^2 \cdot i = -1(i) = -i$$

$$i^4 = i^2 \cdot i^2 = -1(-1) = 1$$

Because $i^4 = 1$, i^5 will simplify to i, and we will begin repeating the sequence i, -1, $-i$, 1 as we simplify higher powers of i: Any power of i simplifies to i, -1, $-i$, or 1. The easiest way to simplify higher powers of i is to write them in terms of i^2. For instance, to simplify i^{21}, we would write it as

$$(i^2)^{10} \cdot i \qquad \text{because} \qquad 2 \cdot 10 + 1 = 21$$

Then, because $i^2 = -1$, we have

$$(-1)^{10} \cdot i = 1 \cdot i = i$$

EXAMPLES Simplify as much as possible.

5. $i^{30} = (i^2)^{15} = (-1)^{15} = -1$

6. $i^{11} = (i^2)^5 \cdot i = (-1)^5 \cdot i = (-1)i = -i$

7. $i^{40} = (i^2)^{20} = (-1)^{20} = 1$

Complex Number

A *complex number* is any number that can be put in the form

$$a + bi$$

where a and b are real numbers and $i = \sqrt{-1}$. The form $a + bi$ is called *standard form* for complex numbers. The number a is called the *real part* of the complex number. The number b is called the *imaginary part* of the complex number.

Every real number is a complex number. For example, 8 can be written as $8 + 0i$. Likewise, $-\frac{1}{2}$, π, $\sqrt{3}$, and 29 are complex numbers because they can all be written in the form $a + bi$:

$$-\frac{1}{2} = -\frac{1}{2} + 0i \qquad \pi = \pi + 0i \qquad \sqrt{3} = \sqrt{3} + 0i \qquad -9 = -9 + 0i$$

Subsets of the Complex Numbers

All numbers of the form $a + bi$ fall into one of the following categories. Each category is a subset of the complex numbers.

Real Numbers	Compound Numbers	Pure Imaginary Numbers
When $a \neq 0$ and $b = 0$ Examples include: $-10, 0, 1, \sqrt{3}, \frac{5}{8}, \pi$	When neither a nor b is 0 Examples include: $5 + 4i, \frac{1}{3} + 4i, \sqrt{5} - i, -6 + i\sqrt{5}$	When $a = 0$ and $b \neq 0$ Examples include: $-4i, i\sqrt{3}, -5i\sqrt{7}, \frac{3}{4}i$

©2009 James Robert Metz

Note: The definition for compound numbers is from Jim Metz of Kapiolani Community College in Hawaii. Some textbooks use the label *imaginary numbers* to represent both the compound numbers and the pure imaginary numbers. In those books, the pure imaginary numbers are a subset of the imaginary numbers. We like the definition from Mr. Metz because it keeps the three subsets from overlapping.

Equality for Complex Numbers

Two complex numbers are equal if and only if their real parts are equal and their imaginary parts are equal. That is, for real numbers a, b, c, and d,

$$a + bi = c + di \quad \text{if and only if} \quad a = c \quad \text{and} \quad b = d$$

EXAMPLE 8 Find x and y if $3x + 4i = 12 - 8yi$.

Solution Because the two complex numbers are equal, their real parts are equal and their imaginary parts are equal:

$$3x = 12 \quad \text{and} \quad 4 = -8y$$
$$x = 4 \qquad\qquad y = -\frac{1}{2}$$

EXAMPLE 9 Find x and y if $(4x - 3) + 7i = 5 + (2y - 1)i$.

Solution The real parts are $4x - 3$ and 5. The imaginary parts are 7 and $2y - 1$:

$$4x - 3 = 5 \quad \text{and} \quad 7 = 2y - 1$$
$$4x = 8 \qquad\qquad 8 = 2y$$
$$x = 2 \qquad\qquad y = 4$$

C Addition and Subtraction with Complex Numbers

To add two complex numbers, add their real parts and their imaginary parts. That is, if a, b, c, and d are real numbers, then

$$(a + bi) + (c + di) = (a + c) + (b + d)i$$

If we assume that the commutative, associative, and distributive properties hold for the number i, then the definition of addition is simply an extension of these properties.

We define subtraction in a similar manner. If a, b, c, and d are real numbers, then

$$(a + bi) - (c + di) = (a - c) + (b - d)i$$

EXAMPLES Add or subtract as indicated.

10. $(3 + 4i) + (7 - 6i) = (3 + 7) + (4 - 6)i = 10 - 2i$

11. $(7 + 3i) - (5 + 6i) = (7 - 5) + (3 - 6)i = 2 - 3i$

12. $(5 - 2i) - (9 - 4i) = (5 - 9) + (-2 + 4)i = -4 + 2i$

Multiplication with Complex Numbers

Because complex numbers have the same form as binomials, we find the product of two complex numbers the same way we find the product of two binomials.

EXAMPLE 13 Multiply $(3 - 4i)(2 + 5i)$.

Solution Multiplying each term in the second complex number by each term in the first, we have

$$
\begin{array}{cccc}
 & F & O & I & L \\
\end{array}
$$
$$(3 - 4i)(2 + 5i) = 3 \cdot 2 + 3 \cdot 5i - 2 \cdot 4i - 4i(5i)$$
$$= 6 + 15i - 8i - 20i^2$$

Combining similar terms and using the fact that $i^2 = -1$, we can simplify as follows:
$$6 + 15i - 8i - 20i^2 = 6 + 7i - 20(-1)$$
$$= 6 + 7i + 20$$
$$= 26 + 7i$$

The product of the complex numbers $3 - 4i$ and $2 + 5i$ is the complex number $26 + 7i$.

EXAMPLE 14 Multiply $2i(4 - 6i)$.

Solution Applying the distributive property gives us

$$2i(4 - 6i) = 2i \cdot 4 - 2i \cdot 6i$$
$$= 8i - 12i^2$$
$$= 12 + 8i$$

EXAMPLE 15 Expand $(3 + 5i)^2$.

Solution We treat this like the square of a binomial. Remember,
$(a + b)^2 = a^2 + 2ab + b^2$:

$$(3 + 5i)^2 = 3^2 + 2(3)(5i) + (5i)^2$$
$$= 9 + 30i + 25i^2$$
$$= 9 + 30i - 25$$
$$= -16 + 30i$$

EXAMPLE 16 Multiply $(2 - 3i)(2 + 3i)$.

Solution This product has the form $(a - b)(a + b)$, which we know results in the difference of two squares, $a^2 - b^2$:

$$(2 - 3i)(2 + 3i) = 2^2 - (3i)^2$$
$$= 4 - 9i^2$$
$$= 4 + 9$$
$$= 13$$

The product of the two complex numbers $2 - 3i$ and $2 + 3i$ is the real number 13. The two complex numbers $2 - 3i$ and $2 + 3i$ are called complex conjugates. The fact that their product is a real number is very useful.

Complex Conjugates

The complex numbers $a + bi$ and $a - bi$ are called *complex conjugates*. One important property they have is that their product is the real number $a^2 + b^2$. Here's why:

$$(a + bi)(a - bi) = a^2 - (bi)^2$$
$$= a^2 - b^2 i^2$$
$$= a^2 - b^2(-1)$$
$$= a^2 + b^2$$

Division with Complex Numbers

The fact that the product of two complex conjugates is a real number is the key to division with complex numbers.

EXAMPLE 17 Divide $\dfrac{2 + i}{3 - 2i}$.

Solution We want a complex number in standard form that is equivalent to the quotient $\frac{2+i}{3-2i}$. We need to eliminate i from the denominator. Multiplying the numerator and denominator by $3 + 2i$ will give us what we want:

$$\frac{2 + i}{3 - 2i} = \frac{2 + i}{3 - 2i} \cdot \frac{3 + 2i}{3 + 2i}$$
$$= \frac{6 + 4i + 3i + 2i^2}{9 - 4i^2}$$
$$= \frac{6 + 7i - 2}{9 + 4}$$
$$= \frac{4 + 7i}{13}$$
$$= \frac{4}{13} + \frac{7}{13}i$$

Dividing the complex number $2 + i$ by $3 - 2i$ gives the complex number $\frac{4}{13} + \frac{7}{13}i$.

EXAMPLE 18 Divide $\dfrac{7 - 4i}{i}$.

Solution The conjugate of the denominator is $-i$. Multiplying the numerator and denominator by this number, we have

$$\frac{7 - 4i}{i} = \frac{7 - 4i}{i} \cdot \frac{-i}{-i}$$
$$= \frac{-7i + 4i^2}{-i^2}$$
$$= \frac{-7i + 4(-1)}{-(-1)}$$
$$= -4 - 7i$$

Problem Set 8.1

Vocabulary

Use the vocabulary words or text below to fill in the blanks in the sentences. You may not need to use every vocabulary word or text.

imaginary   real   standard   complex   binomials   conjugates

1. Complex numbers can be written in the form $a + bi$ where $i = \sqrt{-1}$. This form is called _____ form for a complex number.

2. For the complex number $a + bi$, a is called the _____ part of the complex number, and b is called the _____ part of the complex number.

3. The complex numbers $a + bi$, and $a - bi$ are called complex _____ of each other.

4. Multiplication with complex numbers is very similar to multiplication with _____ in algebra.

Write the following in terms of i, and simplify as much as possible.

1. $\sqrt{-36}$ 2. $\sqrt{-49}$ 3. $-\sqrt{-25}$ 4. $-\sqrt{-81}$

5. $\sqrt{-72}$ 6. $\sqrt{-48}$ 7. $-\sqrt{-12}$ 8. $-\sqrt{-75}$

Write each of the following as i, -1, $-i$, or 1.

9. i^{28} 10. i^{31} 11. i^{26} 12. i^{37}

13. i^{75} 14. i^{42}

Find x and y so that each of the following equations is true.

15. $2x + 3yi = 6 - 3i$ 16. $4x - 2yi = 4 + 8i$

17. $2 - 5i = -x + 10yi$ 18. $4 + 7i = 6x - 14yi$

19. $2x + 10i = -16 - 2yi$ 20. $4x - 5i = -2 + 3yi$

21. $(2x - 4) - 3i = 10 - 6yi$ 22. $(4x - 3) - 2i = 8 + yi$

23. $(7x - 1) + 4i = 2 + (5y + 2)i$ 24. $(5x + 2) - 7i = 4 + (2y + 1)i$

Combine the following complex numbers.

25. $(2 + 3i) + (3 + 6i)$ 26. $(4 + i) + (3 + 2i)$

27. $(3 - 5i) + (2 + 4i)$ 28. $(7 + 2i) + (3 - 4i)$

29. $(5 + 2i) - (3 + 6i)$ 30. $(6 + 7i) - (4 + i)$

31. $(3 - 5i) - (2 + i)$ 32. $(7 - 3i) - (4 + 10i)$

33. $[(3 + 2i) - (6 + i)] + (5 + i)$ **34.** $[(4 - 5i) - (2 + i)] + (2 + 5i)$

35. $[(7 - i) - (2 + 4i)] - (6 + 2i)$ **36.** $[(3 - i) - (4 + 7i)] - (3 - 4i)$

37. $(3 + 2i) - [(3 - 4i) - (6 + 2i)]$ **38.** $(7 - 4i) - [(-2 + i) - (3 + 7i)]$

39. $(4 - 9i) + [(2 - 7i) - (4 + 8i)]$ **40.** $(10 - 2i) - [(2 + i) - (3 - i)]$

Find the following products.

41. $3i(4 + 5i)$ **42.** $2i(3 + 4i)$ **43.** $6i(4 - 3i)$

44. $11i(2 - i)$ **45.** $(3 + 2i)(4 + i)$ **46.** $(2 - 4i)(3 + i)$

47. $(4 + 9i)(3 - i)$ **48.** $(5 - 2i)(1 + i)$ **49.** $(1 + i)^3$

50. $(1 - i)^3$ **51.** $(2 - i)^3$ **52.** $(2 + i)^3$

53. $(2 + 5i)^2$ **54.** $(3 + 2i)^2$ **55.** $(1 - i)^2$

56. $(1 + i)^2$ **57.** $(3 - 4i)^2$ **58.** $(6 - 5i)^2$

59. $(2 + i)(2 - i)$ **60.** $(3 + i)(3 - i)$ **61.** $(6 - 2i)(6 + 2i)$

62. $(5 + 4i)(5 - 4i)$ **63.** $(2 + 3i)(2 - 3i)$ **64.** $(2 - 7i)(2 + 7i)$

65. $(10 + 8i)(10 - 8i)$ **66.** $(11 - 7i)(11 + 7i)$

Find the following quotients. Write all answers in standard form for complex numbers.

67. $\dfrac{2 - 3i}{i}$ **68.** $\dfrac{3 + 4i}{i}$ **69.** $\dfrac{5 + 2i}{-i}$

70. $\dfrac{4 - 3i}{-i}$ **71.** $\dfrac{4}{2 - 3i}$ **72.** $\dfrac{3}{4 - 5i}$

73. $\dfrac{6}{-3 + 2i}$ **74.** $\dfrac{-1}{-2 - 5i}$ **75.** $\dfrac{2 + 3i}{2 - 3i}$

76. $\dfrac{4 - 7i}{4 + 7i}$ **77.** $\dfrac{5 + 4i}{3 + 6i}$ **78.** $\dfrac{2 + i}{5 - 6i}$

79. Electrical Circuits Complex numbers may be applied to electrical circuits. Electrical engineers use the fact that resistance R to electrical flow of the electrical current I and the voltage V are related by the formula $V = RI$. (Voltage is measured in volts, resistance in ohms, and current in amperes.) Find the resistance to electrical flow in a circuit that has a voltage $V = (80 + 20i)$ volts and current $I = (-6 + 2i)$ amps.

80. Electrical Circuits Refer to the information about electrical circuits in Problem 79, and find the current in a circuit that has a resistance of $(4 + 10i)$ ohms and a voltage of $(5 - 7i)$ volts.

Getting Ready for the Next Section

The problems that follow review material we covered in Sections 1.3 and 3.1. Reviewing these problems will help you with some of the material in the next section.

Find $\sin \theta$ and $\cos \theta$ if the given point lies on the terminal side of θ.

81. $(3, -4)$ **82.** $(-5, 12)$

83. (a, b) **84.** $(1, -1)$

Find θ between $0°$ and $360°$ if

85. $\sin \theta = \dfrac{1}{\sqrt{2}}$ and $\cos \theta = -\dfrac{1}{\sqrt{2}}$

86. $\tan \theta = 1$ and θ terminates in QIII

87. $\sin \theta = \dfrac{1}{2}$ and θ terminates in QII

88. $\cos \theta = \dfrac{\sqrt{3}}{2}$ and θ terminates in QIV

Learning Objectives

A   Graph complex numbers.

B   Write equations containing complex numbers in trigonometric form.

We will begin this section with a definition that will give us a way to represent complex numbers graphically. We will then develop a way to relate the work we did in Section 8.1 to some of the concepts in trigonometry.

A Graphing Complex Numbers

KEY WORDS

real axis

imaginary axis

absolute value

modulus

argument

trigonometric form

> **Definition**
> The graph of the complex number $a + bi$ is a vector (arrow) that extends from the origin out to the point (a, b).

EXAMPLE 1   Graph each complex number.

$2 + 4i$, $-2 - 4i$, $2 - 4i$

Solution

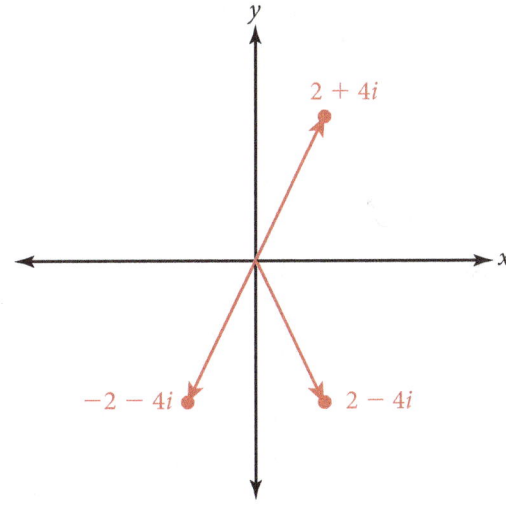

FIGURE 1

Notice how the graph of $2 + 4i$ and $2 - 4i$, which are conjugates, have symmetry about the x-axis. Note also that the graphs of $2 + 4i$ and $-2 - 4i$, which are opposites, have symmetry about the origin.

EXAMPLE 2   Graph the complex numbers 1, i, -1, and $-i$.

Solution Here are the four complex numbers written in standard form and the corresponding graphs of those numbers.

$$1 = 1 + 0i \qquad i = 0 + i$$
$$-1 = -1 + 0i \qquad -i = 0 - i$$

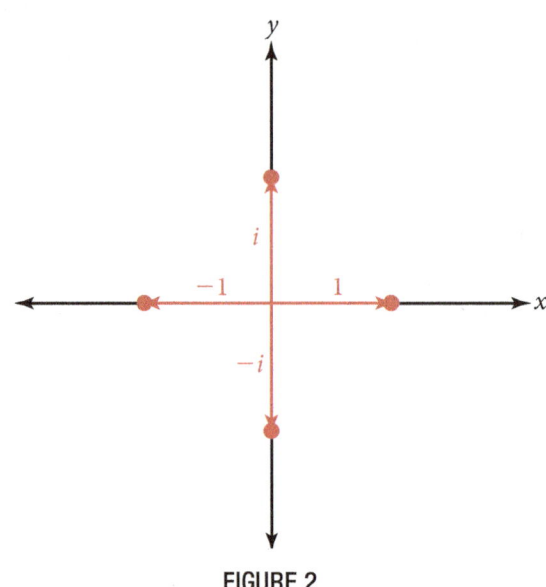

FIGURE 2

If we write a real number as a complex number in standard form, its graph will fall on the x-axis. Therefore, we call the x-axis the *real axis* when we are graphing complex numbers. Likewise, because the imaginary numbers i and $-i$ fall on the y-axis, we call the y-axis the *imaginary axis* when we are graphing complex numbers.

B Trigonometric Form

To study the trigonometric form of complex numbers, we begin with two useful definitions: modulus and argument.

Definition

The *absolute value* or *modulus* of the complex number $z = a + bi$ is the distance from the origin to the point (a, b). If this distance is denoted by r, then

$$r = |z| = |a + bi| = \sqrt{a^2 + b^2}$$

EXAMPLE 3   Find the modulus of each of the complex numbers $5i$, 7, and $3 + 4i$.

Solution   Writing each number in standard form and then applying the definition of modulus, we have

$$\text{For } z = 5i = 0 + 5i, \quad r = |z| = |0 + 5i| = \sqrt{0^2 + 5^2} = 5$$

$$\text{For } z = 7 = 7 + 0i, \quad r = |z| = |7 + 0i| = \sqrt{7^2 + 0^2} = 7$$

$$\text{For } z = 3 + 4i, \quad r = |z| = |3 + 4i| = \sqrt{3^2 + 4^2} = 5$$

Definition

The *argument* of the complex number $z = x + yi$ is the smallest positive angle θ from the positive real axis to the graph of z.

Figure 3 illustrates the relationships between the complex number $z = x + yi$, its graph, and the modulus r and argument θ of z. From Figure 3 we see that

$$\cos \theta = \frac{x}{r} \text{ or } x = r \cos \theta$$

and

$$\sin \theta = \frac{y}{r} \text{ or } y = r \sin \theta$$

We can use this information to write z in terms of r and θ.

$$z = x + yi$$
$$= r \cos \theta + (r \sin \theta)i$$
$$= r \cos \theta + ri \sin \theta$$
$$= r(\cos \theta + i \sin \theta)$$

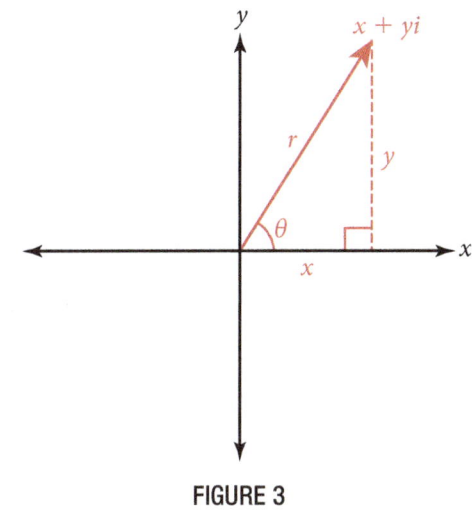

FIGURE 3

This last expression is called the *trigonometric form* for z. The formal definition follows.

Definition

If $z = x + yi$ is a complex number in standard form, then the *trigonometric form* for z is given by

$$z = r(\cos \theta + i \sin \theta)$$

Where r is the modulus of z and θ is the argument of z.

We can convert back and forth between standard form and trigonometric form by using the relationships that follow.

$$\text{For } z = x + yi = r(\cos \theta + i \sin \theta)$$
$$r = \sqrt{x^2 + y^2} \text{ and } \theta \text{ is such that}$$
$$\cos \theta = \frac{x}{r}, \sin \theta = \frac{y}{r}, \text{ and } \tan \theta = \frac{y}{x}$$

EXAMPLE 4 Write $z = -1 + i$ in trigonometric form.

Solution We have $x = -1$ and $y = 1$, therefore

$$r = \sqrt{(-1)^2 + 1^2} = \sqrt{2}$$

Angle θ is the smallest positive angle for which $\cos \theta = \frac{x}{r} = -\frac{1}{\sqrt{2}}$ and $\sin \theta = \frac{y}{r} = \frac{1}{\sqrt{2}}$, (or $\tan \theta = \frac{y}{x} = \frac{1}{-1} = -1$). Therefore, θ must be 135°.

Using these values of r and θ in the formula for trigonometric form we have

$$z = r(\cos \theta + i \sin \theta)$$
$$= \sqrt{2}(\cos 135° + i \sin 135°)$$

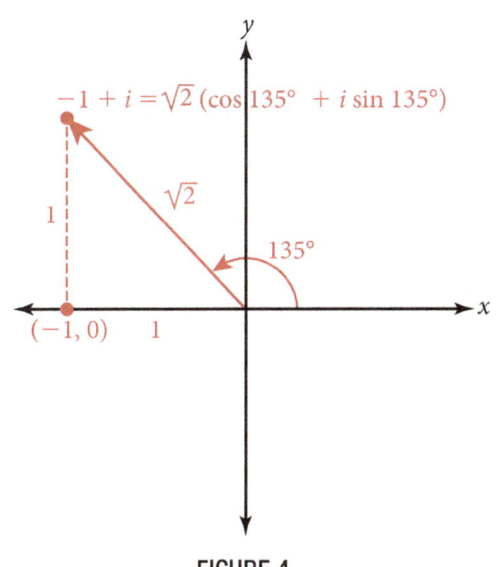

FIGURE 4

EXAMPLE 5 Write $4 - 4i\sqrt{3}$ in *trigonometric* form.

Solution We have $x = 4$ and $y = -4\sqrt{3}$, therefore

$$r = \sqrt{(4)^2 + (-4\sqrt{3})^2} = \sqrt{16 + 48} = 8$$

Angle θ is the smallest positive angle for which $\cos \theta = \frac{x}{r} = \frac{4}{8} = \frac{1}{2}$ and $\sin \theta = \frac{y}{r} = \frac{-4\sqrt{3}}{8} = -\frac{\sqrt{3}}{2}$. Therefore, $\theta = 300°$. Using the values of r and θ in the trigonometric form we have

$$z = r(\cos \theta + i \sin \theta)$$

$$= 8(\cos 300° + i \sin 300°)$$

EXAMPLE 6 Write $z = 2(\cos 60° + i \sin 60°)$ in standard form.

Solution Using exact values for $\cos 60°$ and $\sin 60°$ we have

$$z = 2(\cos 60° + i \sin 60°)$$

$$= 2\left(\frac{1}{2} + i\frac{\sqrt{3}}{2}\right)$$

$$= 1 + i\sqrt{3}$$

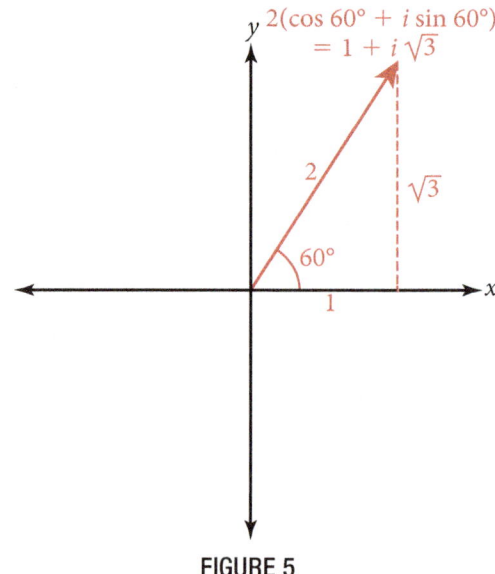

FIGURE 5

As you can see, converting from trigonometric form to standard form is usually more direct than converting from standard form to trigonometric form.

EXAMPLE 7 Use your calculator to write the complex number $20(\cos 105° + i \sin 105°)$ in standard form. Round the numbers in your answer to the nearest hundredth.

Solution

$$20(\cos 105° + i \sin 105°) = 20 \cos 105° + 20(\sin 105°)i$$

$$\approx -5.18 + 19.32i$$

Problem Set 8.2

Vocabulary

Use the vocabulary words or text below to fill in the blanks in the sentences. You may not need to use every vocabulary word or text.

    scientific   argument   arrow   trigonometric   modulus

1. For the complex number $a + bi$, the expression $\sqrt{a^2 + b^2}$ is called the absolute value or _____ of the complex number.

2. To graph the complex number $a + bi$, we draw an _____ from the origin to the point (a, b).

3. When the complex number $x + yi$ is written as $r(\cos \theta + i \sin \theta)$, we say it is written in _____ form.

4. When the complex number $x + yi$ is written as $r(\cos \theta + i \sin \theta)$, we call θ the _____ of the complex number.

Graph each complex number. In each case give the absolute value of the number.

1. $3 + 4i$ 2. $3 - 4i$

3. $1 + i$ 4. $1 - i$

5. $-5i$ 6. $4i$

7. 2 8. -4

9. $-4 - 3i$ 10. $-3 - 4i$

Graph each complex number along with its opposite and conjugate.

11. $2 - i$ 12. $2 + i$

13. $4i$ 14. $-3i$

15. -3 16. 5

17. $-5 - 2i$ 18. $-2 - 5i$

Write each complex number in standard form.

19. $2(\cos 30° + i \sin 30°)$

20. $4(\cos 30° + i \sin 30°)$

21. $4(\cos 120° + i \sin 120°)$

22. $8(\cos 120° + i \sin 120°)$

23. $\cos 210° + i \sin 210°$

24. $\cos 240° + i \sin 240°$

25. $\cos 315° + i \sin 315°$

26. $\sqrt{2}(\cos 315° + i \sin 315°)$

Use your calculator to write each complex number in standard form. Round the numbers in your answers to the nearest hundredth.

27. $10(\cos 12° + i \sin 12°)$

28. $100(\cos 70° + i \sin 70°)$

29. $100(\cos 143° + i \sin 143°)$

30. $100(\cos 171° + i \sin 171°)$

31. $\cos 205° + i \sin 205°$

32. $\cos 261° + i \sin 261°$

33. $10(\cos 342° + i \sin 342°)$

34. $10(\cos 318° + i \sin 318°)$

Write each complex number in trigonometric form. In each case begin by sketching the graph to help with finding the argument θ.

35. $-1 + i$

36. $1 + i$

37. $1 - i$

38. $-1 - i$

39. $3 + 3i$

40. $5 + 5i$

41. $8i$ **42.** $-8i$

43. -9 **44.** 2

45. $-2 + 2i\sqrt{3}$ **46.** $-2\sqrt{3} + 2i$

47. $3 - 3i\sqrt{3}$ **48.** $3\sqrt{3} - 3i$

49. We know that $2i \cdot 3i = 6i^2 = -6$. Change $2i$ and $3i$ to trigonometric form and then show that their product in trigonometric form is still -6.

50. Change $4i$ and 2 to trigonometric form and then multiply. Show that this product is $8i$.

51. Show that $2(\cos 30° + i \sin 30°)$ and $2[\cos(-30°) + i \sin(-30°)]$ are conjugates.

52. Show that $2(\cos 60° + i \sin 60°)$ and $2[\cos(-60°) + i \sin(-60°)]$ are conjugates.

53. Let $z_1 = 2 + 3i$ and $z_2 = z_1 i$ (the product of z_1 and i). Graph both z_1 and z_2. Then find $z_3 = z_2 i$ and graph z_3 on the same coordinate system used to graph z_1 and z_2. Multiplying a complex number by i should produce a complex number with a graph that is rotated 90° from the graph of the original number. Multiply z_3 by i to obtain z_4 and show that the graph of z_4 follows the same pattern.

54. Let $z_1 = 3 - i$ and find $z_2 = z_1 i$, $z_3 = z_2 i$, and $z_4 = z_3 i$. Graph z_1, z_2, z_3, and z_4 on the same coordinate system.

55. Show that if $z = \cos \theta + i \sin \theta$, then $|z| = 1$.

56. Show that if $z = \cos \theta - i \sin \theta$, then $|z| = 1$.

Getting Ready for the Next Section

57. Use the formula for $\cos(A + B)$ to find the exact value of $\cos 75°$.

58. Use the formula for $\sin(A + B)$ to find the exact value of $\sin 75°$.

Let $\sin A = \frac{3}{5}$ with A in QI and $\sin B = \frac{5}{13}$ with B in QI and find

59. $\sin(A + B)$ **60.** $\cos(A + B)$

Simplify each expression to a single trigonometric function.

61. $\sin 30°\cos 90° + \cos 30°\sin 90°$ **62.** $\cos 30°\cos 90° - \sin 30°\sin 90°$

63. $\cos 18°\cos 32° - \sin 18°\sin 32°$ **64.** $\sin 18°\cos 32° + \cos 18°\sin 32°$

Multiply.

65. $4\left(\dfrac{\sqrt{3}}{2} + \dfrac{1}{2}i\right)$ **66.** $(1 + i\sqrt{3})(-\sqrt{3} + i)$

Divide.

67. $\dfrac{1 + i\sqrt{3}}{\sqrt{3} + i}$ **68.** $\dfrac{8 + 4i\sqrt{3}}{4}$

Learning Objectives

A   Multiply complex numbers in trigonometric form.

B   Dividing complex numbers in trigonometric form.

A Multiplying Numbers in Trigonometric Form

Multiplication and division with complex numbers becomes a very simple process when the numbers are written in trigonometric form. Let's state the rule for finding the product of two complex numbers written in trigonometric form as a theorem and then prove the theorem.

Theorem (Multiplication)

If

$$z_1 = r_1(\cos \theta_1 + i \sin \theta_1)$$

and

$$z_2 = r_2(\cos \theta_2 + i \sin \theta_2)$$

are two complex numbers in trigonometric from, then their product, $z_1 z_2$ is

$$z_1 z_2 = [r_1(\cos \theta_1 + i \sin \theta_1)][r_2(\cos \theta_2 + i \sin \theta_2)]$$
$$= r_1 r_2[\cos (\theta_1 + \theta_2) + i \sin (\theta_1 + \theta_2)]$$

In words: To multiply two complex numbers in trigonometric form, multiply each modulus and add each argument.

Proof We begin by multiplying algebraically. Then we simplify our product by using the sum formulas we introduced in Section 5.2.

$$z_1 z_2 = [r_1(\cos \theta_1 + i \sin \theta_1)][r_2(\cos \theta_2 + i \sin \theta_2)]$$
$$= r_1 r_2(\cos \theta_1 + i \sin \theta_1)(\cos \theta_2 + i \sin \theta_2)$$
$$= r_1 r_2(\cos \theta_1 \cos \theta_2 + i \cos \theta_1 \sin \theta_2 + i \sin \theta_1 \cos \theta_2 + i^2 \sin \theta_1 \sin \theta_2)$$
$$= r_1 r_2[\cos \theta_1 \cos \theta_2 + i(\cos \theta_1 \sin \theta_2 + \sin \theta_1 \cos \theta_2) - \sin \theta_1 \sin \theta_2]$$
$$= r_1 r_2[(\cos \theta_1 \cos \theta_2 - \sin \theta_1 \sin \theta_2) + i(\sin \theta_1 \cos \theta_2 + \cos \theta_1 \sin \theta_2)]$$
$$= r_1 r_2[\cos(\theta_1 + \theta_2) + i \sin(\theta_1 + \theta_2)]$$

This completes our proof. As you can see, to multiply two complex numbers in trigonometric form, we multiply absolute values, $r_1 r_2$, and add angles, $\theta_1 + \theta_2$.

EXAMPLE 1 Multiply $3(\cos 40° + i \sin 40°)$ and $5(\cos 10° + i \sin 10°)$.

Solution Applying the formula from our theorem on products, we have

$$[3(\cos 40° + i \sin 40°)][5(\cos 10° + i \sin 10°)]$$
$$= 3 \cdot 5[\cos(40° + 10°) + i \sin(40° + 10°)]$$
$$= 15(\cos 50° + i \sin 50°)$$

EXAMPLE 2   Find the product of $z_1 = 1 + i\sqrt{3}$ and $z_2 = -\sqrt{3} + i$ in standard form, and then write z_1 and z_2 in trigonometric form and find their product again.

Solution Leaving each complex number in standard form and multiplying, we have

$$z_1 z_2 = (1 + i\sqrt{3})(-\sqrt{3} + i)$$
$$= -\sqrt{3} + i - 3i + i^2\sqrt{3}$$
$$= -2\sqrt{3} - 2i$$

Changing z_1 and z_2 to trigonometric form and multiplying looks like this

$$z_1 = 1 + i\sqrt{3} = 2\left(\frac{1}{2} + \frac{\sqrt{3}}{2}i\right) = 2(\cos 60° + i \sin 60°)$$

$$z_2 = -\sqrt{3} + i = 2\left(-\frac{\sqrt{3}}{2} + \frac{1}{2}i\right) = 2(\cos 150° + i \sin 150°)$$

$$z_1 z_2 = [2(\cos 60° + i \sin 60°)][2(\cos 150° + i \sin 150°)]$$
$$= 4(\cos 210° + i \sin 210°)$$

To compare our two products, we convert our product in trigonometric form to standard form.

$$4(\cos 210° + i \sin 210°) = 4\left(-\frac{\sqrt{3}}{2} - \frac{1}{2}i\right)$$
$$= -2\sqrt{3} - 2i$$

As you can see, both methods of multiplying complex numbers produce the same result.

The next theorem is an extension of the work we have done so far with multiplication. We will not give a formal proof of the theorem.

> **DeMoivre's Theorem**
> If $z = r(\cos \theta + i \sin \theta)$ is a complex number in trigonometric form and n is an integer, then
> $$z^n = [r(\cos \theta + i \sin \theta)]^n = r^n(\cos n\theta + i \sin n\theta)$$
> The theorem seems reasonable after the work we have done with multiplication. For example, if n is 2,
> $$[r(\cos \theta + i \sin \theta)]^2 = r(\cos \theta + i \sin \theta)r(\cos \theta + i \sin \theta)$$
> $$= r^2(\cos 2\theta + i \sin 2\theta)$$

EXAMPLE 3   Find $(1 + i)^{10}$.

Solution First we write $1 + i$ in trigonometric form

$$1 + i = \sqrt{2}\left(\frac{1}{\sqrt{2}} + \frac{1}{\sqrt{2}}i\right) = \sqrt{2}(\cos 45° + i \sin 45°)$$

Then we use DeMoivre's theorem to raise this expression to the 10th power.

$$(1 + i)^{10} = [\sqrt{2}(\cos 45° + i \sin 45°)]^{10}$$
$$= (\sqrt{2})^{10}(\cos 10 \cdot 45° + i \sin 10 \cdot 45°)$$
$$= 32(\cos 450° + i \sin 450°)$$

which we can simplify to

$$= 32(\cos 90° + i \sin 90°)$$

Since 90° and 450° are coterminal. In standard form our result is

$$= 32(0 + i)$$

$$= 32i$$

that is,

$$(1 + i)^{10} = 32i$$

B Dividing Numbers in Trigonometric Form

Since multiplication with complex numbers in trigonometric form is accomplished by multiplying absolute values and adding angles, we should expect that division is accomplished by dividing absolute values and subtracting angles.

Theorem (Division)

If

$$z_1 = r_1(\cos \theta_1 + i \sin \theta_1)$$

and

$$z_2 = r_2(\cos \theta_2 + i \sin \theta_2)$$

are two complex numbers in trigonometric form, then the quotient, z_1/z_2, is

$$\frac{z_1}{z_2} = \frac{r_1(\cos \theta_1 + i \sin \theta_1)}{r_2(\cos \theta_2 + i \sin \theta_2)}$$

$$= \frac{r_1}{r_2}\left[\cos(\theta_1 - \theta_2) + i \sin(\theta_1 - \theta_2)\right]$$

Proof As was the case with division of complex numbers in standard form, the major step in this proof is multiplying the numerator and denominator of our quotient by the conjugate of the denominator.

$$\frac{r_1(\cos \theta_1 + i \sin \theta_1)}{r_2(\cos \theta_2 + i \sin \theta_2)}$$

$$= \frac{r_1(\cos \theta_1 + i \sin \theta_1)}{r_2(\cos \theta_2 + i \sin \theta_2)} \cdot \frac{(\cos \theta_2 - i \sin \theta_2)}{(\cos \theta_2 - i \sin \theta_2)}$$

$$= \frac{r_1(\cos \theta_1 + i \sin \theta_1)(\cos \theta_2 - i \sin \theta_2)}{r_2(\cos^2\theta_2 + \sin^2\theta_2)}$$

$$= \frac{r_1}{r_2}(\cos \theta_1\cos \theta_2 - i \cos \theta_1\sin \theta_2 + i \sin \theta_1\cos \theta_2 - i^2 \sin \theta_1\sin \theta_2)$$

$$= \frac{r_1}{r_2}[(\cos \theta_1\cos \theta_2 + \sin \theta_1\sin \theta_2)] + i[(\sin \theta_1\cos \theta_2 - \cos \theta_1\sin \theta_2)]$$

$$= \frac{r_1}{r_2}[\cos (\theta_1 - \theta_2) + i \sin (\theta_1 - \theta_2)]$$

EXAMPLE 4   Divide $20(\cos 75° + i \sin 75°)$ by $4(\cos 40° + i \sin 40°)$.

Solution We divide according to the formula given in our theorem on division.

$$\frac{20(\cos 75° + i \sin 75°)}{4(\cos 40° + i \sin 40°)} = \frac{20}{4}[\cos(75° - 40°) + i \sin(75° - 40°)]$$

$$= 5(\cos 35° + i \sin 35°)$$

EXAMPLE 5 Divide $z_1 = 1 + i\sqrt{3}$ by $z_2 = \sqrt{3} + i$ and leave the answer in standard form. Then change each to trigonometric form and divide again.

Solution Dividing in standard form, we have

$$\frac{z_1}{z_2} = \frac{1 + i\sqrt{3}}{\sqrt{3} + i}$$

$$= \frac{1 + i\sqrt{3}}{\sqrt{3} + i} \cdot \frac{\sqrt{3} - i}{\sqrt{3} - i}$$

$$= \frac{\sqrt{3} - i + 3i - i^2\sqrt{3}}{3 + 1}$$

$$= \frac{2\sqrt{3} + 2i}{4}$$

$$= \frac{\sqrt{3}}{2} + \frac{1}{2}i$$

Changing z_1 and z_2 to trigonometric form and multiplying again, we have

$$z_1 = 1 + i\sqrt{3} = 2\left(\frac{1}{2} + \frac{\sqrt{3}}{2}i\right) = 2(\cos 60° + i \sin 60°)$$

$$z_2 = \sqrt{3} + i = 2\left(\frac{\sqrt{3}}{2} + \frac{1}{2}i\right) = 2(\cos 30° + i \sin 30°)$$

$$\frac{z_1}{z_2} = \frac{2(\cos 60° + i \sin 60°)}{2(\cos 30° + i \sin 30°)}$$

$$= \frac{2}{2}\left[\cos(60° - 30°) + i \sin(60° - 30°)\right]$$

$$= \cos 30° + i \sin 30°$$

which, in standard form, is

$$\frac{\sqrt{3}}{2} + \frac{1}{2}i$$

Problem Set 8.3

Vocabulary

Use the vocabulary words or text below to fill in the blanks in the sentences. You may not need to use every vocabulary word or text.

add   subtract   multiply   divide   modulus   argument

1. To multiply two complex numbers in trigonometric form, we _____ absolute values and _____ angles.

2. Demoivre's theorem tells us that we can raise a complex number to an integer power of n by raising the _____ to the nth power and multiplying the _____ by n.

3. To _____ two complex numbers in trigonometric form, we divide absolute values and _____ angles.

Multiply. Leave all answers in trigonometric form.

1. $3(\cos 20° + i \sin 20°) \cdot 4(\cos 30° + i \sin 30°)$

2. $5(\cos 15° + i \sin 15°) \cdot 2(\cos 25° + i \sin 25°)$

3. $7(\cos 110° + i \sin 110°) \cdot 8(\cos 47° + i \sin 47°)$

4. $9(\cos 115° + i \sin 115°) \cdot 4(\cos 51° + i \sin 51°)$

5. $2(\cos 135° + i \sin 135°) \cdot 2(\cos 45° + i \sin 45°)$

6. $2(\cos 120° + i \sin 120°) \cdot 4(\cos 30° + i \sin 30°)$

Find the product $z_1 z_2$ in standard form. Then write z_1 and z_2 in trigonometric form and find their product again. Finally, convert the answer that is in trigonometric form to standard form to show that the two products are equal.

7. $z_1 = 1 + i, z_2 = -1 + i$ 8. $z_1 = 1 + i, z_2 = 2 + 2i$

9. $z_1 = 1 + i\sqrt{3}, z_2 = -\sqrt{3} + i$ 10. $z_1 = -1 + i\sqrt{3}, z_2 = \sqrt{3} + i$

11. $z_1 = 3i, z_2 = -4i$ 12. $z_1 = 2i, z_2 = -5i$

13. $z_1 = 1 + i, z_2 = 4i$ 14. $z_1 = 1 + i, z_2 = 3i$

15. $z_1 = -5, z_2 = 1 + i\sqrt{3}$ 16. $z_1 = -3, z_2 = \sqrt{3} + i$

Use DeMoivre's theorem to find each of the following. Write your answer in standard form.

17. $[2(\cos 10° + i \sin 10°)]^6$ 18. $[4(\cos 15° + i \sin 15°)]^3$

19. $(\cos 12° + i \sin 12°)^{10}$ 20. $(\cos 18° + i \sin 18°)^{10}$

21. $[3(\cos 60° + i \sin 60°)]^4$ 22. $[3(\cos 30° + i \sin 30°)]^4$

23. $[\sqrt{2}(\cos 45° + i \sin 45°)]^{10}$ 24. $[\sqrt{2}(\cos 70° + i \sin 70°)]^6$

25. $(1 + i)^4$ 26. $(1 + i)^5$

27. $(-\sqrt{3} + i)^4$ 28. $(\sqrt{3} + i)^4$

29. $(1 - i)^6$ 30. $(-1 + i)^8$

31. $(-2 + 2i)^3$ 32. $(-2 - 2i)^3$

Divide. Leave your answers in trigonometric form.

33. $\dfrac{20(\cos 75° + i \sin 75°)}{5(\cos 40° + i \sin 40°)}$ 34. $\dfrac{30(\cos 80° + i \sin 80°)}{10(\cos 30° + i \sin 30°)}$

35. $\dfrac{18(\cos 51° + i \sin 51°)}{12(\cos 32° + i \sin 32°)}$ **36.** $\dfrac{21(\cos 63° + i \sin 63°)}{14(\cos 44° + i \sin 44°)}$

37. $\dfrac{4(\cos 90° + i \sin 90°)}{8(\cos 30° + i \sin 30°)}$ **38.** $\dfrac{6(\cos 120° + i \sin 120°)}{8(\cos 90° + i \sin 90°)}$

Find the quotient z_1/z_2 in standard form. Then write z_1 and z_2 in trigonometric form and find their quotient again. Finally, convert the answer that is in trigonometric form to standard form to show that the two quotients are equal.

39. $z_1 = 2 + 2i, z_2 = 1 + i$ **40.** $z_1 = 2 - 2i, z_2 = 1 - i$

41. $z_1 = \sqrt{3} + i, z_2 = 2i$ **42.** $z_1 = 1 + i\sqrt{3}, z_2 = 2i$

43. $z_1 = 4 + 4i, z_2 = 2 - 2i$ **44.** $z_1 = 6 + 6i, z_2 = -3 - 3i$

45. $z_1 = 8, z_2 = -4$ **46.** $z_1 = -6, z_2 = 3$

Convert all complex numbers to trigonometric form and then simplify each expression. Write all answers in standard form.

47. $\dfrac{(1 + i)^4(2i)^2}{-2 + 2i}$ **48.** $\dfrac{(\sqrt{3} + i)^4(2i)^5}{(1 + i)^{10}}$

49. $\dfrac{(1 + i\sqrt{3})^4(\sqrt{3} - i)^2}{(1 - i\sqrt{3})^3}$ **50.** $\dfrac{(2 + 2i)^5(-3 + 3i)^3}{(\sqrt{3} + i)^{10}}$

51. Show that $x = 2(\cos 60° + i \sin 60°)$ is a solution to the quadratic equation $x^2 - 2x + 4 = 0$ by replacing x with $2(\cos 60° + i \sin 60°)$ and simplifying.

52. Show that $x = 2(\cos 300° + i \sin 300°)$ is a solution to the equation $x^2 - 2x + 4 = 0$.

53. Show that $w = 2(\cos 15° + i \sin 15°)$ is a fourth root of $z = 8 + 8i\sqrt{3}$ by raising w to the fourth power and simplifying to get z. (The number w is a fourth root of z, $w = z^{1/4}$, if the fourth power of w is z, $w^4 = z$.)

54. Show that $x = \frac{1}{2} + \left(\frac{\sqrt{3}}{2}\right)i$ is a cube root of -1.

DeMoivre's theorem can be used to find reciprocals of complex numbers. Recall from algebra that the reciprocal of x is $1/x$, which can be expressed as x^{-1}. Use this fact, along with DeMoivre's theorem, to find the reciprocal of each number below.

55. $1 + i$ **56.** $1 - i$ **57.** $\sqrt{3} - i$

58. $\sqrt{3} + i$ **59.** $1 + i\sqrt{3}$ **60.** $1 - i\sqrt{3}$

Getting Ready for the Next Section

Use a calculator to evaluate to four decimal places.

61. $\sin 15°$ **62.** $\cos 15°$ **63.** $\cos 165°$ **64.** $\sin 165°$

65. $\cos 195°$ **66.** $\sin 195°$ **67.** $\sin 345°$ **68.** $\cos 345°$

69. Evaluate $15° + 90°k$ for $k = 0, 1,$ and 2.

70. Evaluate $60° + 120°k$ for $k = 0, 1,$ and 2.

Write each number in trigonometric form.

71. 8 **72.** $8i$ **73.** $1 + i\sqrt{3}$ **74.** $1 + i$

Find all real and complex solutions.

75. $x^3 + 1 = 0$ **76.** $x^3 - 8 = 0$

8.4 Videos

KEY WORDS

roots theorem

Learning Objectives

A   Solve problems involving complex numbers using the roots theorem.

In this section, we want to develop a formula for finding roots of a complex number. The discussion that follows will lead to the theorem containing the formula for these roots.

Suppose that z and w are complex numbers such that w is the nth root of z. That is,

$$\sqrt[n]{z} = w$$

If we raise both sides of this last expression to the nth power, we have

$$w^n = z$$

Now suppose $z = r(\cos \theta + i \sin \theta)$ and $w = s(\cos \alpha + i \sin \alpha)$. Substituting these expressions into the equation $w^n = z$, we have

$$[s(\cos \alpha + i \sin \alpha)]^n = r(\cos \theta + i \sin \theta)$$

We can rewrite the left side of this last equation using DeMoivre's theorem.

$$s^n(\cos n\alpha + i \sin n\alpha) = r(\cos \theta + i \sin \theta)$$

The only way these two expressions can be equal is if their absolute values are equal and their angles are coterminal.

Equal Absolute Values

$$s^n = r$$

Coterminal Angles

$$n\alpha = \theta + 360°k$$
$$(k = \text{an integer})$$

Solving for s and α, we have

$$s = r^{1/n}$$

$$\alpha = \frac{\theta + 360°k}{n}$$

To summarize, we find the nth roots of a complex number by first finding the real nth root of the absolute value and then adding multiples of 360° to θ and dividing the result by n. The multiples of 360° that we add on will range from $360° \cdot 0$ up to $360° \cdot (n - 1)$. After that we start repeating angles.

Theorem (Roots)

The nth roots of the complex number

$$z = r(\cos \theta + i \sin \theta)$$

are given by

$$w_k = r^{1/n}\left[\cos \frac{\theta + 360°k}{n} + i \sin \frac{\theta + 360°k}{n}\right]$$

where $k = 0, 1, 2, \ldots, n - 1$. That is,

$$w_0 = r^{1/n}\left[\cos \frac{\theta}{n} + i \sin \frac{\theta}{n}\right]$$

$$w_1 = r^{1/n}\left[\cos \frac{\theta + 360°}{n} + i \sin \frac{\theta + 360°}{n}\right]$$

$$w_2 = r^{1/n}\left[\cos \frac{\theta + 720°}{n} + i \sin \frac{\theta + 720°}{n}\right]$$

$$\ldots$$

$$w_{n-1} = r^{1/n}\left[\cos \frac{\theta + 360°(n - 1)}{n} + i \sin \frac{\theta + 360°(n - 1)}{n}\right]$$

EXAMPLE 1 Find the 4 fourth roots of $z = 16(\cos 60° + i \sin 60°)$.

Solution According to the formula given in our theorem on roots, the 4 fourth roots will be

$$w_k = 16^{1/4}\left[\cos\frac{60° + 360°k}{4} + i\sin\frac{60° + 360°k}{4}\right] \quad k = 0, 1, 2, 3$$

$$= 2[\cos(15° + 90°k) + i\sin(15° + 90°k)]$$

Replacing k with 0, 1, 2, and 3, we have

$$w_0 = 2(\cos 15° + i \sin 15°)$$

$$w_1 = 2(\cos 105° + i \sin 105°)$$

$$w_2 = 2(\cos 195° + i \sin 195°)$$

$$w_3 = 2(\cos 285° + i \sin 285°)$$

It is interesting to note the relationships among the graphs of these 4 roots. As Figure 1 indicates, the graphs of the four roots are evenly distributed around the coordinate plane.

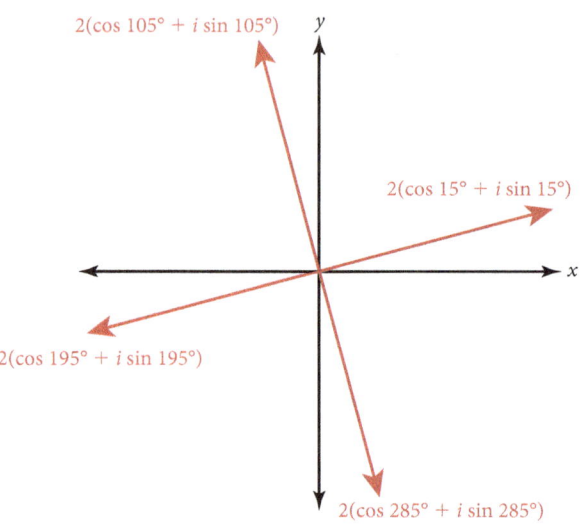

FIGURE 1

EXAMPLE 2 Find the 3 cube roots of -1.

Solution We already know that one of the cube roots of -1 is -1. There are two other cube roots that are imaginary numbers. To find them, we write -1 in trigonometric form and then apply the formula from our theorem on roots. Writing -1 in trigonometric form, we have

$$-1 = 1(\cos 180° + i \sin 180°)$$

The 3 cube roots are given by

$$w_k = 1^{1/3}\left[\cos\frac{180° + 360°k}{3} + i\sin\frac{180° + 360°k}{3}\right]$$

$$= \cos(60° + 120°k) + i\sin(60° + 120°k)$$

where $k = 0, 1,$ and 2. Replacing k with 0, 1, and 2 and then simplifying each result, we have

$$w_0 = \cos 60° + i \sin 60° = \frac{1}{2} + \frac{\sqrt{3}}{2}i$$

$$w_1 = \cos 180° + i \sin 180° = -1$$

$$w_2 = \cos 300° + i \sin 300° = \frac{1}{2} - \frac{\sqrt{3}}{2}i$$

Note that the two complex roots are conjugates. Let's check root w_0 by cubing it.

$$w_0{}^3 = (\cos 60° + i \sin 60°)^3$$

$$= \cos 3 \cdot 60° + i \sin 3 \cdot 60°$$

$$= \cos 180° + i \sin 180°$$

$$= -1$$

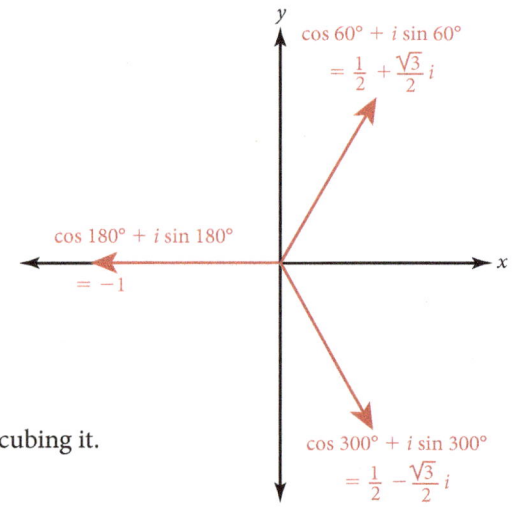

FIGURE 2

EXAMPLE 3   Solve $x^3 - 8 = 0$.

Solution   We begin by factoring the loft side, then we set each factor equal to 0.

$$x^3 - 8 = 0$$
$$(x - 2)(x^2 + 2x + 4) = 0$$
$$x - 2 = 0 \quad \text{or} \quad x^2 + 2x + 4 = 0$$

The first equation gives us $x = 2$ as one solution. We apply the quadratic formula to the second equation to find the other two solutions. Here is the quadratic formula with $a = 1$, $b = 2$, and $c = 4$.

$$x = \frac{-2 \pm \sqrt{4 - 4(1)(4)}}{2(1)}$$

$$= \frac{-2 \pm \sqrt{-12}}{2}$$

$$= \frac{-2 \pm 2i\sqrt{3}}{2}$$

$$= -1 \pm i\sqrt{3}$$

Our equation has three solutions. They are

$$2, -1 + i\sqrt{3}, \text{and} -1 - i\sqrt{3}$$

Now, let's solve the same equation using our knowledge of roots of a complex number. If we add 8 to both sides of our original equation, we have

Writing 8 in trigonometric form

$$x^3 = 8 = 8(1 + 0i) = 8(\cos 0° + i \sin 0°)$$

the solution to which are the cube roots of 8. From our theorem on roots we have the three cube roots of 8 will be

$$w_k = 8^{1/3}\left[\cos \frac{0° + 360°k}{3} + i \sin \frac{0° + 360°k}{3}\right]$$

$$= 2(\cos 120°k + i \sin 120°k)$$

Evaluating this expression for $k = 0$, 1, and 2, we have the three cube roots of 8.

$$w_0 = 2(\cos 0° + i \sin 0°) = 2(1 + 0) = 2$$

$$w_1 = 2(\cos 120° + i \sin 120°) = 2\left(-\frac{1}{2} + i\frac{\sqrt{3}}{2}\right) = -1 + i\sqrt{3}$$

$$w_2 = 2(\cos 240° + i \sin 240°) = 2\left(-\frac{1}{2} - i\frac{\sqrt{3}}{2}\right) = -1 - i\sqrt{3}$$

As you can see, the solutions are the same whether we use algebra to solve the problem or trigonometry.

Here are the graphs of all the roots:

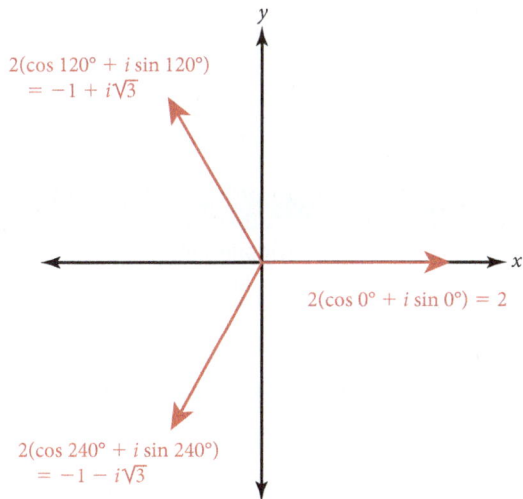

FIGURE 3

EXAMPLE 4   Solve the equation $x^4 - 2\sqrt{3}x^2 + 4 = 0$.

Solution   The equation is quadratic in x^2. We can solve for x^2 by applying the quadratic formula.

$$x^2 = \frac{2\sqrt{3} \pm \sqrt{12 - 4(1)(4)}}{2}$$

$$= \frac{2\sqrt{3} \pm \sqrt{-4}}{2}$$

$$= \frac{2\sqrt{3} \pm 2i}{2}$$

$$= \sqrt{3} \pm i$$

The two solutions for x^2 are $\sqrt{3} + i$ and $\sqrt{3} - i$, which we write in trigonometric form as follows:

$$x^2 = \sqrt{3} + i \qquad\qquad \text{or} \quad x^2 = \sqrt{3} - i$$

$$= 2\left(\frac{\sqrt{3}}{2} + \frac{1}{2}i\right) \qquad\qquad = 2\left(\frac{\sqrt{3}}{2} - \frac{1}{2}i\right)$$

$$= 2(\cos 30° + i \sin 30°) \qquad = 2(\cos 330° + i \sin 330°)$$

Now each of these expressions has two square roots, each of which is a solution to our original equation.

When $x^2 = 2(\cos 30° + i \sin 30°)$

$$x = 2^{1/2}\left[\cos \frac{30° + 360°k}{2} + i \sin \frac{30° + 360°k}{2}\right]$$

for $k = 0$ and 1

When $k = 0$, we have $x = 2^{1/2}(\cos 15° + i \sin 15°)$

When $k = 1$, we have $x = 2^{1/2}(\cos 195° + i \sin 195°)$

When $x^2 = 2(\cos 330° + i \sin 330°)$

$$x = 2^{1/2}\left[\cos \frac{330° + 360°k}{2} + i \sin \frac{330° + 360°k}{2}\right]$$

for $k = 0$ and 1

When $k = 0$, we have $x = 2^{1/2}(\cos 165° + i \sin 165°)$

When $k = 1$, we have $x = 2^{1/2}(\cos 345° + i \sin 345°)$

Using a calculator and rounding to the nearest hundredth, we can write decimal approximations to each of these four solutions.

Solutions		
Trigonometric Form		**Decimal Approximation**
$2^{1/2}(\cos 15° + i \sin 15°)$	$=$	$1.37 + 0.37i$
$2^{1/2}(\cos 165° + i \sin 165°)$	$=$	$-1.37 + 0.37i$
$2^{1/2}(\cos 195° + i \sin 195°)$	$=$	$-1.37 - 0.37i$
$2^{1/2}(\cos 345° + i \sin 345°)$	$=$	$1.37 - 0.37i$

Problem Set 8.4

Find two square roots for each of the following complex numbers. Leave your answers in trigonometric form. In each case, graph the two roots.

1. $4(\cos 30° + i \sin 30°)$ **2.** $16(\cos 30° + i \sin 30°)$

3. $25(\cos 210° + i \sin 210°)$ **4.** $9(\cos 310° + i \sin 310°)$

Find two square roots for each of the following complex numbers. Write your answers in standard form.

5. $2 + 2i\sqrt{3}$ **6.** $-2 + 2i\sqrt{3}$

7. $4i$ **8.** $-4i$

9. -25 **10.** 25

Find three cube roots for each of the following complex numbers. Leave your answers in trigonometric form.

11. $8(\cos 210° + i \sin 210°)$ **12.** $27(\cos 303° + i \sin 303°)$

13. $4\sqrt{3} + 4i$ **14.** $-4\sqrt{3} + 4i$

15. -27 **16.** 8

17. Find 4 fourth roots of $z = 16(\cos 120° + i \sin 120°)$. Write each root in standard form.

18. Find 4 fourth roots of $z = \cos 240° + i \sin 240°$. Leave your answers in trigonometric form.

19. Find 5 fifth roots of $z = 10^5(\cos 15° + i \sin 15°)$. Write each root in trigonometric form and then give a decimal approximation, accurate to the nearest hundredth, for each one.

20. Find 5 fifth roots of $z = 10^{10}(\cos 75° + i \sin 75°)$. Write each root in trigonometric form and then give a decimal approximation, accurate to the nearest hundredth, for each one.

21. Find 6 sixth roots of $z = -1$. Leave your answers in trigonometric form. Graph all six roots on the same coordinate system.

22. Find 6 sixth roots of $z = 1$. Leave your answers in trigonometric form. Graph all six roots on the same coordinate system.

Solve each of the following equations. Leave your solutions in trigonometric form.

23. $x^4 + 2\sqrt{3}x^2 + 4 = 0$ **24.** $x^4 - 4\sqrt{3}x^2 + 16 = 0$

25. $x^4 - 2x^2 + 4 = 0$ **26.** $x^4 + 2x^2 + 4 = 0$

27. $x^4 + 2x^2 + 2 = 0$ **28.** $x^4 - 2x^2 + 2 = 0$

Getting Ready for the Next Section

Sketch each angle in standard position.

29. 45°   **30.** −45°   **31.** 135°   **32.** 225°

Simplify.

33. $4 \cos 30°$ **34.** $4 \sin 30°$ **35.** $-\sqrt{2} \sin 135°$

36. $-\sqrt{2} \cos 135°$ **37.** $\sqrt{9 + 9}$ **38.** $\sqrt{1 + 3}$

Learning Objectives

A   Graph polar coordinates.

B   Convert between polar coordinates and rectangular coordinates.

C   Solve problems involving polar coordinates.

In the rectangular coordinate system, we locate points in the plane by indicating their distance from two perpendicular lines (the x-axis and y-axis).

The *polar coordinate system* also allows us to locate points in the plane but by a slightly different method.

Definition (Polar Coordinates)

The ordered pair (r, θ) names the point that is r units from the origin along the number line (polar axis) that has been rotated through and angle θ from the positive x-axis. The coordinates r and θ are said to be the polar *coordinates* of the point they name. In polar coordinates, the origin is sometimes referred to as the pole.

Graphing in polar coordinates will be a little easier if we revise our coordinate system somewhat. It helps to have angles that are multiples of 15°, along with circles centered at the origin with radii of 1, 2, 3, 4, 5, and 6.

A Graphing Polar Coordinates

EXAMPLE 1   Graph the points $(3, 45°)$, $(2, 120°)$, $(-4, 60°)$, and $(-5, 150°)$ on a polar coordinate system.

Solution   To graph $(3, 45°)$, we locate the point that is 3 units from the origin along the terminal side of 45°.

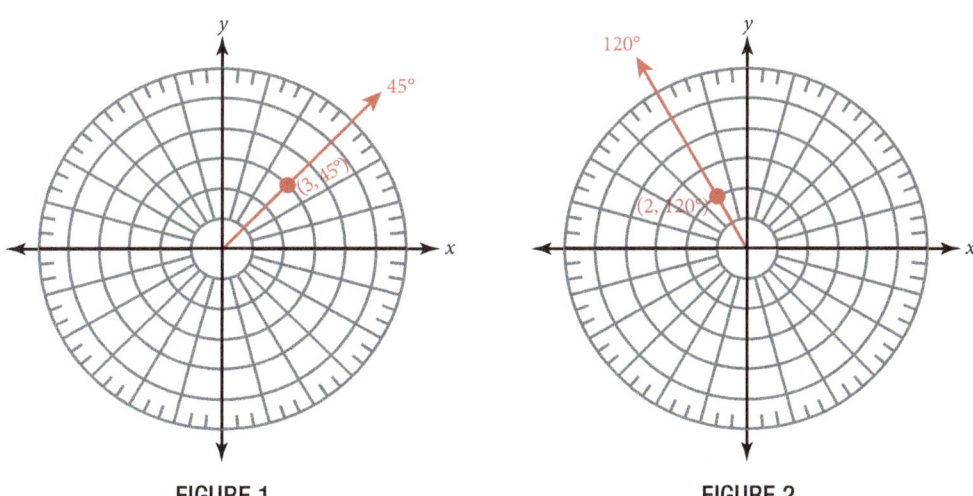

FIGURE 1   FIGURE 2

The point $(2, 120°)$ is 2 units out on the terminal side of 120°. As you can see from Figures 1 and 2, if r is positive, we locate the point (r, θ) along the terminal side of θ. The next two points we will graph have negative values of r. To graph a point (r, θ) in which r is negative, we look for the point on the *projection* of the terminal side of θ through the origin.

To graph $(-4, 60°)$, we locate the point that is 4 units from the origin on the projection of 60° through the origin (Figure 3). To graph $(-5, 150°)$, we look for the point that is 5 units from the origin along the projection of 150° through the origin (Figure 4).

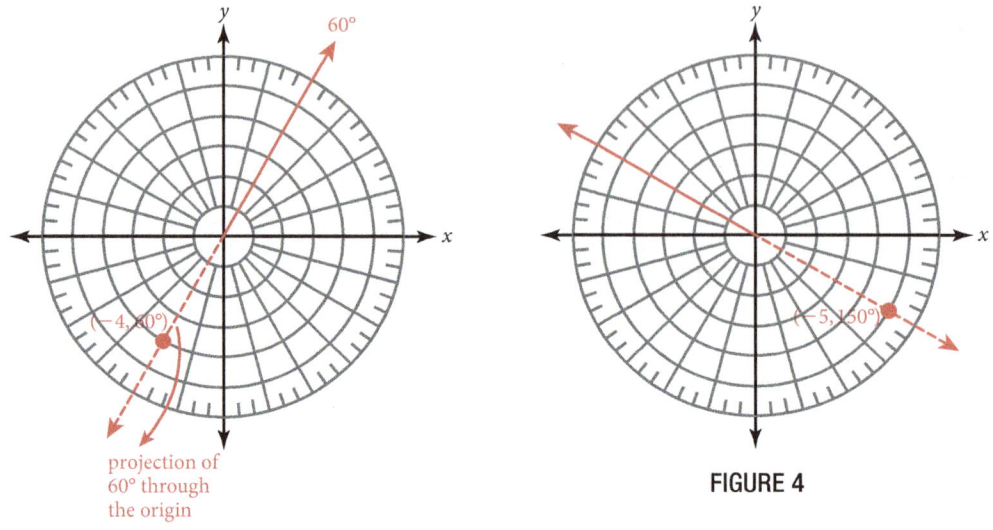

projection of
60° through
the origin

FIGURE 3

FIGURE 4

In rectangular coordinates, each point in the plane is named by a unique ordered pair (x, y). That is, no point can be named by two different ordered pairs. The same is not true of points named by polar coordinates, as Example 2 illustrates.

EXAMPLE 2   Give three other ordered pairs that name the same point as $(3, 60°)$.

Solution As Figure 5 illustrates, the points $(-3, 240°), (-3, -120°)$, and $(3, -300°)$ all name the point $(3, 60°)$.

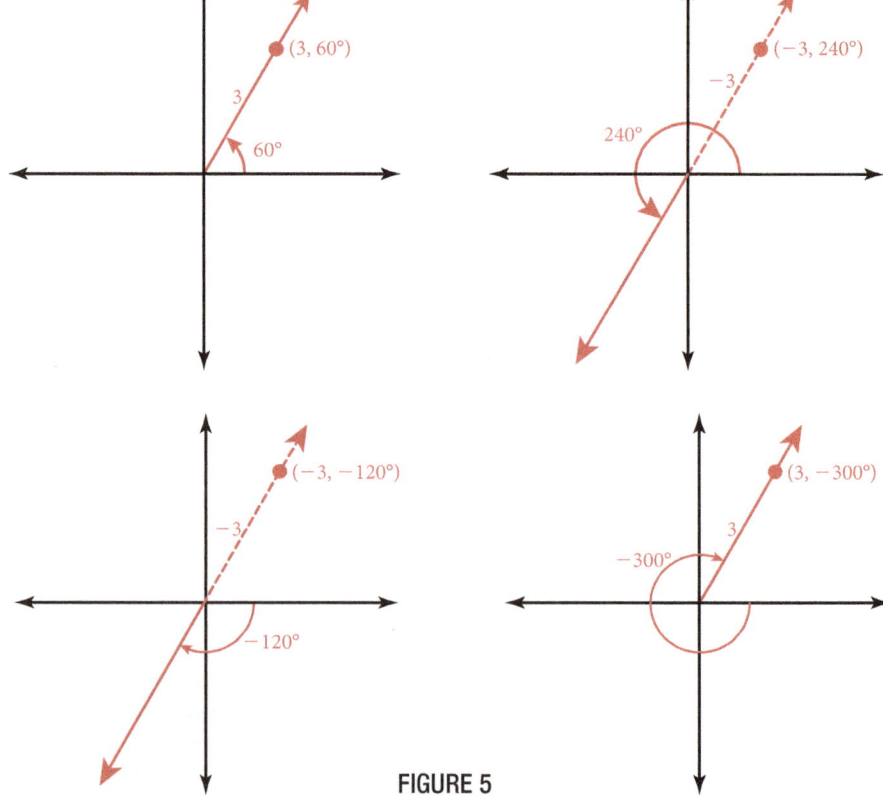

FIGURE 5

There are actually an infinite number of ordered pairs that name the point $(3, 60°)$. Any angle that is coterminal with $60°$ will have its terminal side pass through $(3, 60°)$. Therefore, all points of the form

$$(3, 60° + 360°k) \quad k = \text{an integer}$$

will name the point $(3, 60°)$.

B Polar Coordinates and Rectangular Coordinates

To derive the relationship between polar coordinates and rectangular coordinates, we consider a point P with rectangular coordinates (x, y) and polar coordinates (r, θ).

   To convert back and forth between polar and rectangular coordinates, we simply use the relationships that exist among x, y, r, and θ in Figure 6.

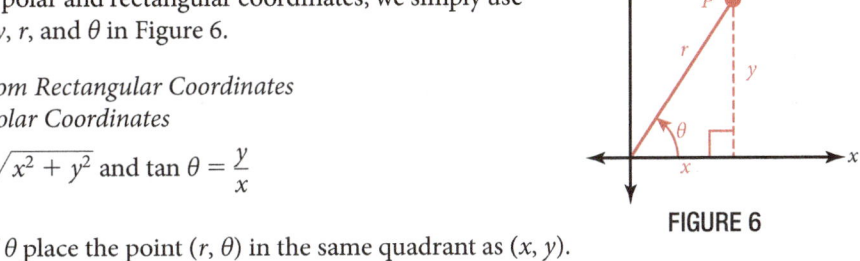

FIGURE 6

To Convert from Rectangular Coordinates to Polar Coordinates

Let $r = \pm\sqrt{x^2 + y^2}$ and $\tan \theta = \dfrac{y}{x}$

Where the sign of r and the choice of θ place the point (r, θ) in the same quadrant as (x, y).

To Convert from Polar Coordinates to Rectangular Coordinates

Let $x = r \cos \theta$ and $y = r \sin \theta$

The process of converting to rectangular coordinates is simply a matter of substituting r and θ into the equations given above. To convert to polar coordinates we have to choose θ and the sign of r so that the point (r, θ) is in the same quadrant as the point (x, y).

EXAMPLE 3   Convert to rectangular coordinates.

a. $(4, 30°)$ **b.** $(-\sqrt{2}, 135°)$ **c.** $(3, 270°)$

Solution To convert from polar coordinates to rectangular coordinates, we substitute the given values of r and θ into the equations

$$x = r \cos \theta \quad \text{and} \quad y = r \sin \theta$$

Here are the conversions for each point along with the graphs in both rectangular and polar coordinates.

a. $x = 4 \cos 30° = 4\left(\dfrac{\sqrt{3}}{2}\right) = 2\sqrt{3}$

$y = 4 \sin 30° = 4\left(\dfrac{1}{2}\right) = 2$

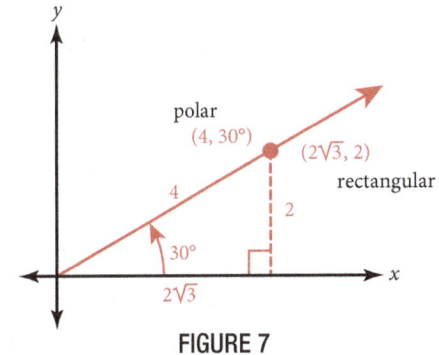

FIGURE 7

The point $(2\sqrt{3}, 2)$ in rectangular coordinates is equivalent to $(4, 30°)$ in polar coordinates.

b. $x = -\sqrt{2}\cos 135° = -\sqrt{2}\left(-\dfrac{1}{\sqrt{2}}\right) = 1$

$y = -\sqrt{2}\sin 135° = -\sqrt{2}\left(\dfrac{1}{\sqrt{2}}\right) = -1$

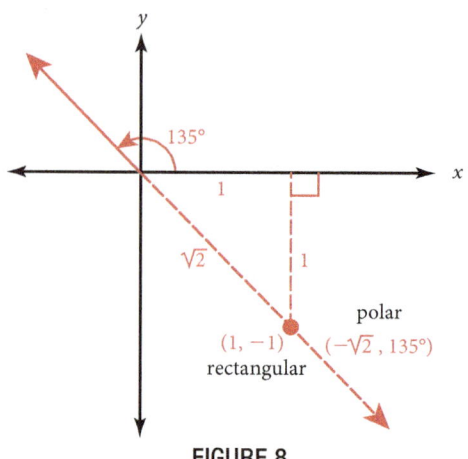

FIGURE 8

The point $(1, -1)$ in rectangular coordinates is equivalent to $(-\sqrt{2}, 135°)$ in polar coordinates.

c. $x = 3\cos 270° = 3(0) = 0$

$y = 3\sin 270° = 3(-1) = -3$

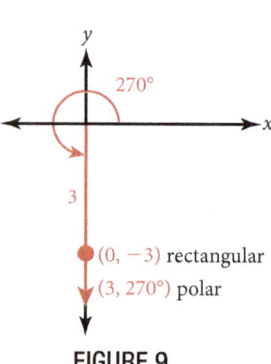

FIGURE 9

The point $(0, -3)$ in rectangular coordinates is equivalent to $(3, 270°)$ in polar coordinates.

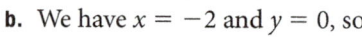

EXAMPLE 4 Convert to polar coordinates.

a. $(3, 3)$ **b.** $(-2, 0)$ **c.** $(-1, \sqrt{3})$

Solution

a. Since x is 3 and y is 3, we have

$$r = \pm\sqrt{9 + 9} = \pm 3\sqrt{2} \text{ and } \tan\theta = \dfrac{3}{3} = 1$$

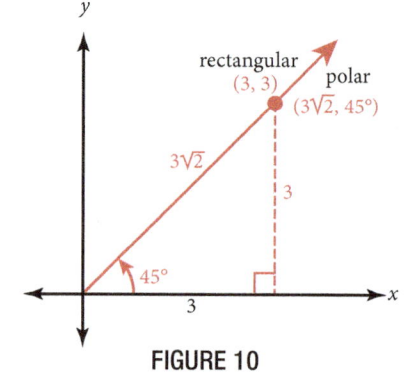

FIGURE 10

Since $(3, 3)$ is in quadrant I, we can choose $r = 3\sqrt{2}$ and $\theta = 45°$. Remember, there are an infinite number of ordered pairs in polar coordinates that name the point $(3, 3)$. The point $(3\sqrt{2}, 45°)$ is just one of them. Generally, we choose r and θ so that θ is between 0° and 360°, and r is positive.

b. We have $x = -2$ and $y = 0$, so

$$r = \pm\sqrt{4 + 0} = \pm 2 \text{ and } \tan\theta = \dfrac{0}{-2} = 0$$

Since $(-2, 0)$ is on the negative x-axis, we can choose $r = 2$ and $\theta = 180°$ to get the point $(2, 180°)$.

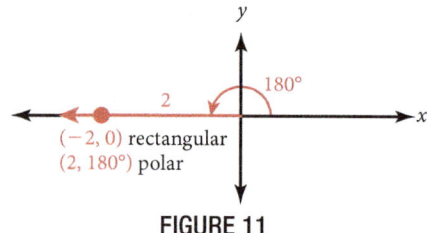

FIGURE 11

c. Since $x = -1$ and $y = \sqrt{3}$, we have

$$r = \pm\sqrt{1 + 3} = \pm2 \text{ and } \tan \theta = \frac{\sqrt{3}}{-1}$$

Since $(-1, \sqrt{3})$ is in quadrant II, we can let $r = 2$ and $\theta = 120°$.
In polar coordinates the point is $(2, 120°)$.

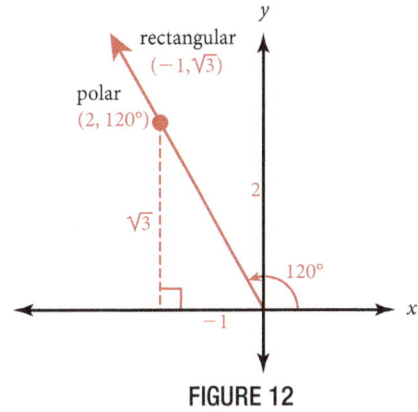

FIGURE 12

C Equations in Polar Coordinates

Equations in polar coordinates have variables r and θ instead of x and y. The conversions we used to change ordered pairs from polar coordinates to rectangular coordinates and from rectangular coordinates to polar coordinates are the same ones we use to convert back and forth between equations given in polar coordinates and those in rectangular coordinates.

EXAMPLE 5 Change $r^2 = 9 \sin 2\theta$ to rectangular coordinates.

Solution Before we substitute to clear the equation of r and θ, we must use a double-angle identity to write $\sin 2\theta$ in terms of $\sin \theta$ and $\cos \theta$.

$$r^2 = 9 \sin 2\theta$$

$$r^2 = 9 \cdot 2 \sin \theta \cos \theta \qquad \textit{Double-angle identity}$$

$$r^2 = 18 \cdot \frac{y}{r} \cdot \frac{x}{r} \qquad \textit{Substitute } \frac{y}{r} \textit{ for } \sin \theta \textit{ and } \frac{x}{r} \textit{ for } \cos \theta$$

$$r^2 = \frac{18xy}{r^2} \qquad \textit{Multiply}$$

$$r^4 = 18xy \qquad \textit{Multiply both sides by } r^2$$

$$(x^2 + y^2)^2 = 18xy \qquad \textit{Substitute } x^2 + y^2 \textit{ for } r^2$$

EXAMPLE 6 Change $x + y = 4$ to polar coordinates.

Solution Since $x = r \cos \theta$ and $y = r \sin \theta$, we have

$$r \cos \theta + r \sin \theta = 4$$

$$r(\cos \theta + \sin \theta) = 4 \qquad \textit{Factor out } r$$

$$r = \frac{4}{\cos \theta + \sin \theta} \qquad \textit{Divide both sides by } \cos \theta + \sin \theta$$

The last equation gives us r in terms of θ.

Problem Set 8.5

Vocabulary

Use the vocabulary words or text below to fill in the blanks in the sentences. You may not need to use every vocabulary word or text.

standard   origin   parametric   polar   rectangular

1. In polar coordinates the graph of the ordered pair (r, θ) is the point that is r units from the _____ along the terminal side of angle θ, in _____ position.

2. Suppose a point is named by (x, y) in the _____ coordinate system and (r, θ) in the _____ coordinate system. The relationship between x, y, r, and θ is given by the _____ equations
$$x = r \cos \theta$$
$$y = r \sin \theta$$

Graph each ordered pair on a polar coordinate system.

1. $(2, 45°)$
2. $(3, 60°)$

3. $(3, 150°)$
4. $(4, 135°)$

5. $(1, -225°)$
6. $(2, -240°)$

7. $(-3, 45°)$
8. $(-4, 60°)$

9. $(-4, -210°)$
10. $(-5, -225°)$

11. $(-2, 0°)$
12. $(-2, 270°)$

For each ordered pair, give three other ordered pairs with θ between $-360°$ and $360°$ that name the same point.

13. $(2, 60°)$
14. $(1, 30°)$

15. $(5, 135°)$
16. $(3, 120°)$

17. $(-3, 30°)$
18. $(-2, 45°)$

Convert to rectangular coordinates. Use exact values.

19. $(2, 60°)$ **20.** $(-2, 60°)$

21. $(3, 270°)$ **22.** $(1, 180°)$

23. $(\sqrt{2}, -135°)$ **24.** $(\sqrt{2}, -225°)$

25. $(-4\sqrt{3}, 30°)$ **26.** $(4\sqrt{3}, -30°)$

Convert to polar coordinates with $r \geq 0$ and θ between 0° and 360°.

27. $(-3, 3)$ **28.** $(-3, -3)$

29. $(-2\sqrt{3}, 2)$ **30.** $(2, -2\sqrt{3})$

31. $(2, 0)$ **32.** $(-2, 0)$

33. $(-\sqrt{3}, -1)$ **34.** $(-1, -\sqrt{3})$

Convert to polar coordinates. Use a calculator or table to find θ to the nearest tenth of a degree. Keep r positive and θ between 0° and 360°.

35. $(3, 4)$ **36.** $(4, 3)$

37. $(-1, 2)$ **38.** $(1, -2)$

39. $(-2, -3)$ **40.** $(-3, -2)$

Write each equation with rectangular coordinates.

41. $r^2 = 9$ **42.** $r^2 = 4$

43. $r = 6 \sin \theta$ **44.** $r = 6 \cos \theta$

45. $r^2 = 4 \sin 2\theta$ **46.** $r^2 = 4 \cos 2\theta$

47. $r(\cos \theta + \sin \theta) = 3$ **48.** $r(\cos \theta - \sin \theta) = 2$

Write each equation in polar coordinates.

49. $x - y = 5$ **50.** $x + y = 5$

51. $x^2 + y^2 = 4$ **52.** $x^2 + y^2 = 9$

53. $x^2 + y^2 = 6x$ **54.** $x^2 + y^2 = 4x$

55. $y = x$ **56.** $y = -x$

Getting Ready for the Next Section

Simplify and write your answer as a decimal accurate to the nearest tenth.

57. $6 \sin 45°$ **58.** $6 \sin 135°$

59. $6 \sin 240°$ **60.** $6 \sin 300°$

Graph one complete cycle of each equation.

61. $y = 6 \sin x$ **62.** $y = 6 \cos x$

63. $y = 4 \sin 2x$ **64.** $y = 2 \sin 4x$

65. $y = 4 + 2 \sin x$ **66.** $y = 4 + 2 \cos x$

EQUATIONS IN POLAR COORDINATES AND THEIR GRAPHS

8.6

8.6 Videos

Learning Objectives

A   Graph polar equations.

In this section we will consider the graphs of polar equations. The solutions to these equations are ordered pairs (r, θ), where r and θ are the polar coordinates we defined in Section 8.5.

EXAMPLE 1   Sketch the graph of $r = 6 \sin \theta$.

Solution   We can find ordered pairs (r, θ) that satisfy the equation by making a table. The table is a little different than the ones we made for rectangular coordinates. With polar coordinates, we substitute in convenient values for θ and then use the equation to find corresponding values of r. Let's use multiples of 30° and 45° for θ.

Table 1			
θ	$r = 6 \sin \theta$	r	(r, θ)
0°	$r = 6 \sin 0° = 0$	0	$(0, 0°)$
30°	$r = 6 \sin 30° = 3$	3	$(3, 30°)$
45°	$r = 6 \sin 45° = 4.2$	4.2	$(4.2, 45°)$
60°	$r = 6 \sin 60° = 5.2$	5.2	$(5.2, 60°)$
90°	$r = 6 \sin 90° = 6$	6	$(6, 90°)$
120°	$r = 6 \sin 120° = 5.2$	5.2	$(5.2, 120°)$
135°	$r = 6 \sin 135° = 4.2$	4.2	$(4.2, 135°)$
150°	$r = 6 \sin 150° = 3$	3	$(3, 150°)$
180°	$r = 6 \sin 180° = 0$	0	$(0, 180°)$
210°	$r = 6 \sin 210° = -3$	−3	$(-3, 210°)$
225°	$r = 6 \sin 225° = -4.2$	−4.2	$(-4.2, 225°)$
240°	$r = 6 \sin 240° = -5.2$	−5.2	$(-5.2, 240°)$
270°	$r = 6 \sin 270° = -6$	−6	$(-6, 270°)$
300°	$r = 6 \sin 300° = -5.2$	−5.2	$(-5.2, 300°)$
315°	$r = 6 \sin 315° = -4.2$	−4.2	$(-4.2, 315°)$
330°	$r = 6 \sin 330° = -3$	−3	$(-3, 330°)$
360°	$r = 6 \sin 360° = 0$	0	$(0, 360°)$

Note If we were to continue past 360° with values of θ, we would simply start to repeat the values of r we have already obtained, since $\sin \theta$ is periodic with period 360°.

Plotting each point on a polar coordinate system and then drawing a smooth curve through them, we have the graph in Figure 1.

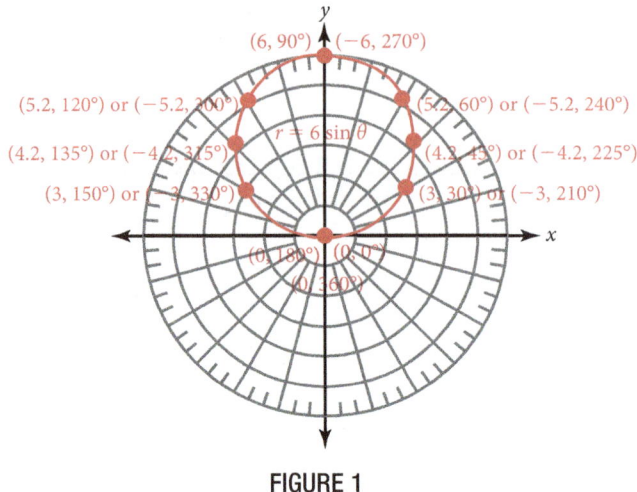

FIGURE 1

Could we have found the graph of $r = 6 \sin \theta$ in Example 1 without making a table? The answer is yes, there are a couple of other ways to do so. One way is to convert to rectangular coordinates and see if we recognize the graph from the rectangular equation. We begin by replacing $\sin \theta$ with y/r.

$$r = 6 \left(\frac{y}{r} \right) \qquad \sin \theta = \frac{y}{r}$$

$$r^2 = 6y \qquad \text{Multiply both sides by } r$$

$$x^2 + y^2 = 6y \qquad r^2 = x^2 + y^2$$

The equation is now written in terms of rectangular coordinates. If we add $-6y$ to both sides and then complete the square on y, we will obtain the rectangular equation of a circle with center at $(0, 3)$ and a radius of 3.

$$x^2 + y^2 - 6y = 0 \qquad \text{Add } -6y \text{ to both sides}$$

$$x^2 + y^2 - 6y + 9 = 9 \qquad \text{Complete the square on } y \text{ by adding 9 to both sides.}$$

$$x^2 + (y - 3)^2 = 3^2 \qquad \text{Standard form for the equation of a circle}$$

This method of graphing, by changing to rectangular coordinates, only works well in some cases. Many of the equations we will encounter in polar coordinates do not have graphs that are recognizable in rectangular form.

In Example 2 we will look at another method of graphing polar equations that does not depend on the use of a table.

EXAMPLE 2   Sketch the graph of $r = 4 \sin 2\theta$.

Solution   One way to visualize the relationship between r and θ as given by the equation $r = 4 \sin 2\theta$ is to sketch the graph of $y = 4 \sin 2x$ on a rectangular coordinate system. (Since we have been using degree measure for our angles in polar coordinates, we will label the x-axis for the graph of $y = 4 \sin 2x$ in degrees rather than radians as we usually do.) The graph of $y = 4 \sin 2x$ will have an amplitude of 4 and a period of $360°/2 = 180°$. Figure 2 shows the graph of $y = 4 \sin 2x$ between 0° and 360°.

As you can see in Figure 2, as x goes from 0° to 45°, y goes from 0 to 4. This means that, for the equation $r = 4 \sin 2\theta$, as θ goes from 0° to 45°, r will go from 0 up to 4. A diagram of this is shown in Figure 3.

As x continues from 45° to 90°, y decreases from 4 down to 0. Likewise, as θ rotates through 45° to 90°, r will decrease from 4 down to 0. A diagram of this is shown in Figure 4. The numbers 1 and 2 in Figure 4 indicate the order in which those sections of the graph are drawn.

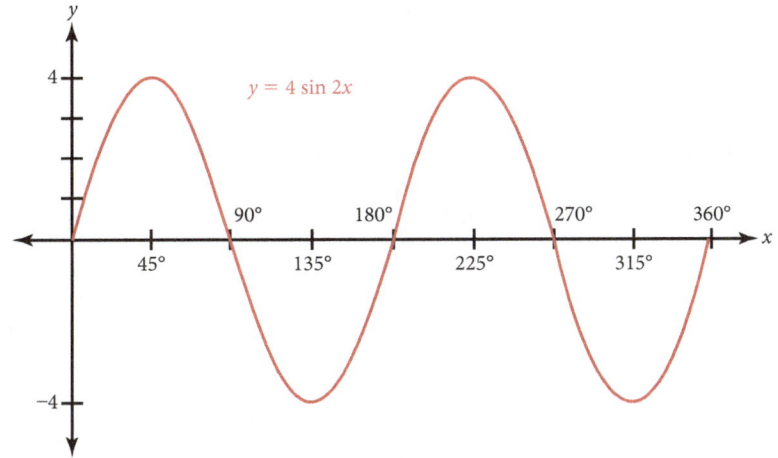

FIGURE 2

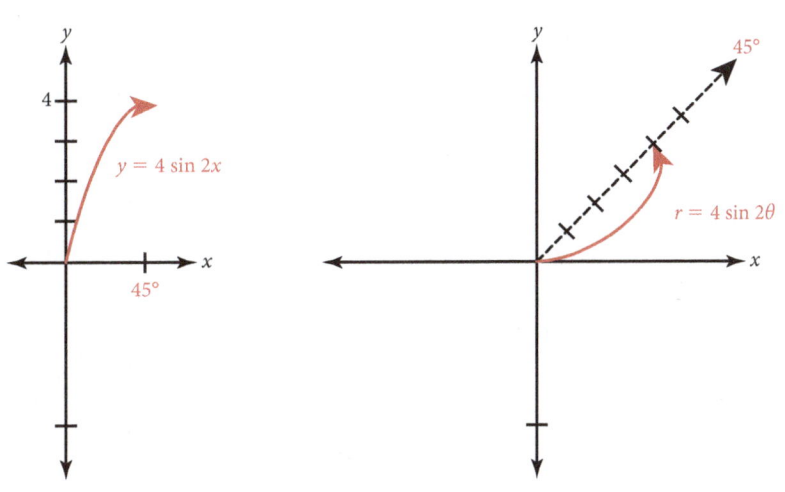

FIGURE 3

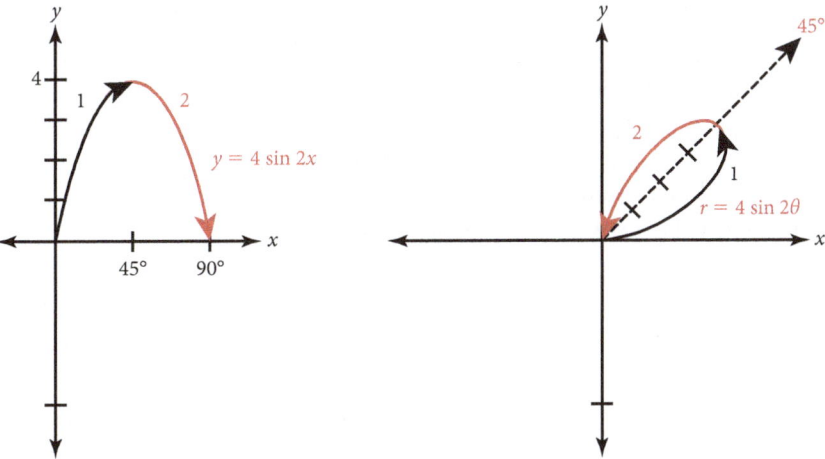

FIGURE 4

If we continue to reason in this manner, we will obtain a good sketch of the graph of $r = 4 \sin 2\theta$ by watching how y is affected by changes in x on the graph of $y = 4 \sin 2x$. Table 2 summarizes this information, and Figure 5 contains both the graph of $y = 4 \sin 2x$ and $r = 4 \sin 2\theta$.

	Table 2	
Reference Number on Graphs	Variations in x (or θ)	Corresponding Variations in y (or r)
1.	0° to 45°	0 to 4
2.	45° to 90°	4 to 0
3.	90° to 135°	0 to −4
4.	135° to 180°	−4 to 0
5.	180° to 225°	0 to 4
6.	225° to 270°	4 to 0
7.	270° to 315°	0 to −4
8.	315° to 360°	−4 to 0

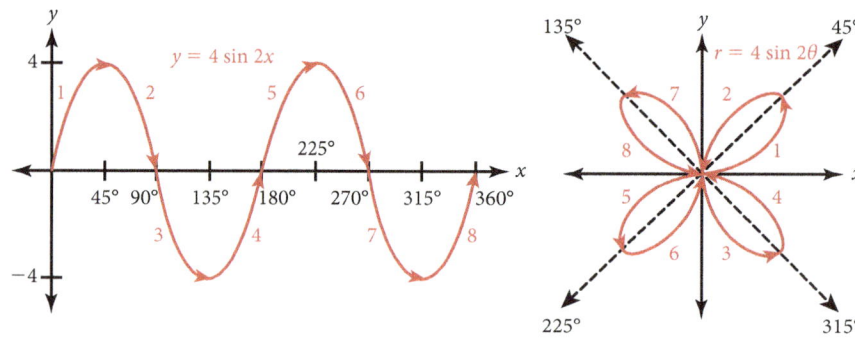

FIGURE 5

EXAMPLE 3 Sketch the graph of $r = 4 + 2 \sin \theta$.

Solution The graph of $r = 4 + 2 \sin \theta$ (Figure 7) is obtained by first graphing $y = 4 + 2 \sin x$ (Figure 6) and then noticing the relationship between variations in x and the corresponding variations in y. These variations are equivalent to those that exist between θ and r.

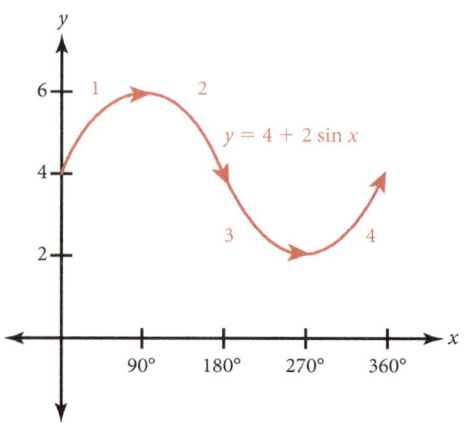

FIGURE 6

Table 3		
Reference Number on Graphs	Variations in x (or θ)	Corresponding Variations in y (or r)
1.	0° to 90°	4 to 6
2.	90° to 180°	6 to 4
3.	180° to 270°	4 to 2
4.	270° to 360°	2 to 4

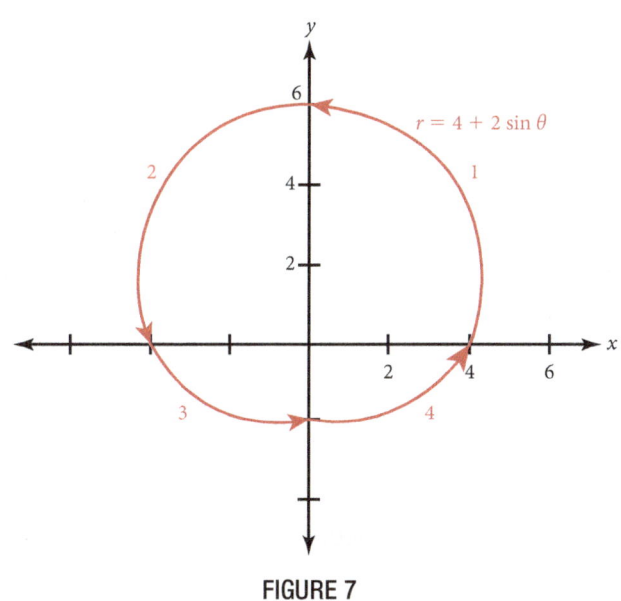

FIGURE 7

Although the method of graphing presented in Examples 2 and 3 is sometimes difficult to comprehend at first, with a little practice it becomes much easier. In any case, the usual alternative is to make a table and plot points until the shape of the curve can be recognized. Probably the best way to graph these equations is to use a combination of both methods.

Here are some other common graphs in polar coordinates along with the equations that produce them. When you start graphing some of the equations in Problem Set 8.6, you may want to keep these graphs handy for reference. It is sometimes easier to get started when you can anticipate the general shape of the curve.

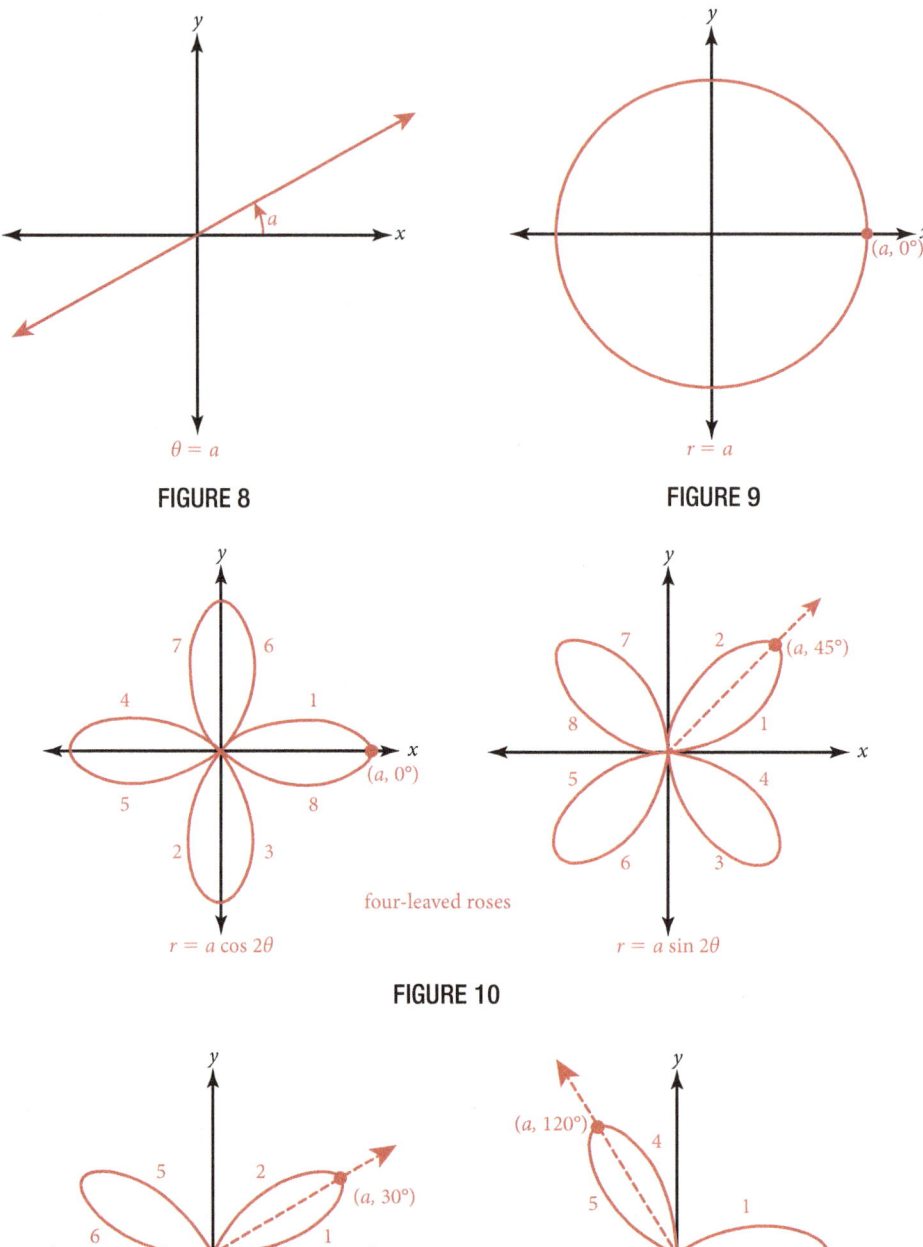

FIGURE 8   FIGURE 9

four-leaved roses

FIGURE 10

three-leaved roses

FIGURE 11

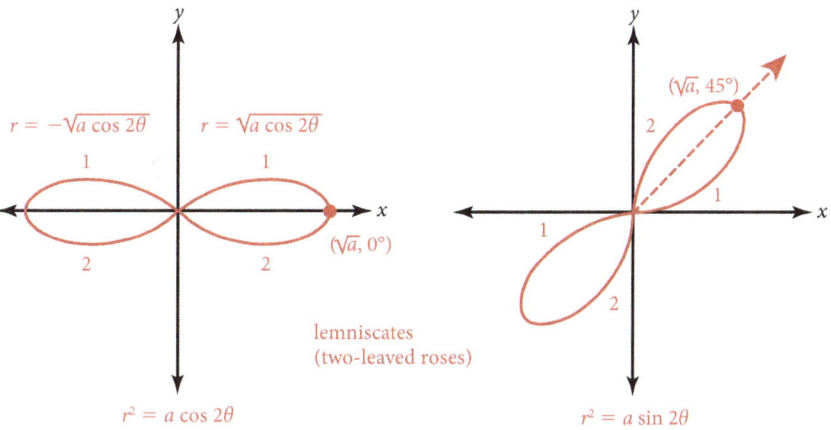

lemniscates
(two-leaved roses)

$r^2 = a \cos 2\theta$

$r^2 = a \sin 2\theta$

FIGURE 12

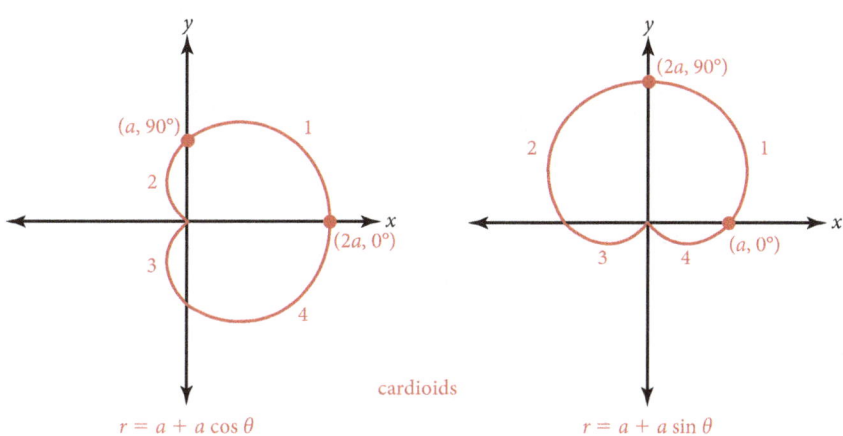

cardioids

$r = a + a \cos \theta$

$r = a + a \sin \theta$

FIGURE 13

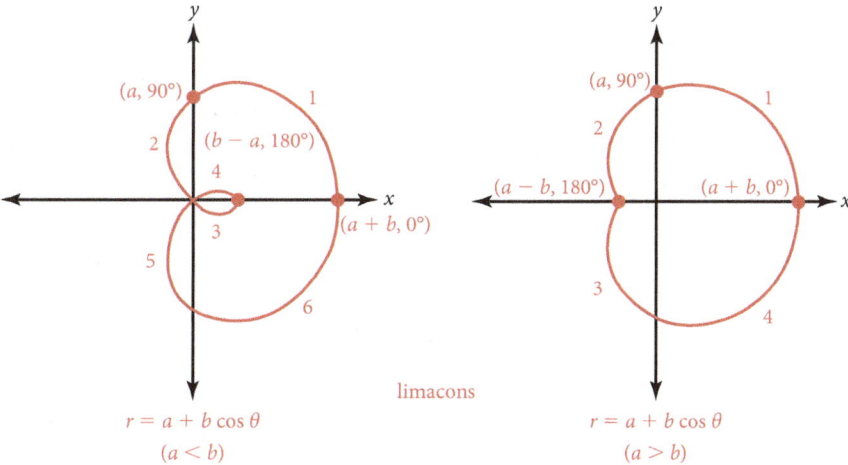

limacons

$r = a + b \cos \theta$
$(a < b)$

$r = a + b \cos \theta$
$(a > b)$

FIGURE 14

Problem Set 8.6

Sketch the graph of each equation by making a table using values of θ that are multiples of 45°.

1. $r = 6 \cos \theta$

2. $r = 4 \sin \theta$

3. $r = \sin 3\theta$

4. $r = \cos 3\theta$

Graph each equation.

5. $r = 3$

6. $r = 2$

7. $\theta = 45°$

8. $\theta = 135°$

9. $r = 3 \sin \theta$

10. $r = 3 \cos \theta$

11. $r = 4 + 2 \sin \theta$

12. $r = 4 + 2 \cos \theta$

13. $r = 2 + 4 \cos \theta$

14. $r = 2 + 4 \sin \theta$

15. $r = 2 + 2 \sin \theta$

16. $r = 2 + 2 \cos \theta$

17. $r^2 = 9 \sin 2\theta$

18. $r^2 = 4 \cos 2\theta$

19. $r = 2 \sin 2\theta$

20. $r = 2 \cos 2\theta$

21. $r = 4 \cos 3\theta$

22. $r = 4 \sin 3\theta$

Convert each equation to polar coordinates and then sketch the graph.

23. $x^2 + y^2 = 16$ **24.** $x^2 + y^2 = 25$

25. $x^2 + y^2 = 6x$ **26.** $x^2 + y^2 = 6y$

27. $(x^2 + y^2)^2 = 2xy$ **28.** $(x^2 + y^2)^2 = x^2 - y^2$

Change each equation to rectangular coordinates and then graph.

29. $r(2 \cos \theta + 3 \sin \theta) = 6$ **30.** $r(3 \cos \theta - 2 \sin \theta) = 6$

31. $r(1 - \cos \theta) = 1$ **32.** $r(1 - \sin \theta) = 1$

33. $r = 4 \sin \theta$ **34.** $r = 6 \cos \theta$

35. Graph $r = 2 \sin \theta$ and $r = 2 \cos \theta$ and then name two rectangular points they have in common.

36. Graph $r = 2 + 2 \cos \theta$ and $r = 2 - 2 \cos \theta$ and name three rectangular points they have in common.

Review Problems

The problems that follow review material we covered in Chapter 7.

37. If $B = 118°$, $C = 37°$, and $c = 2.9$ inches, use the law of sines to find b.

38. Use the law of sines to show that no triangle exists for which $A = 60°$, $a = 12$ inches, and $b = 42$ inches.

39. Find two triangles for which $A = 26°$, $a = 4.8$ feet, and $b = 9.4$ feet.

40. If $C = 120°$, $a = 10$ centimeters, and $b = 12$ centimeters, use the law of cosines to find c.

41. Find all the missing parts if $b = 3.7$ meters, $c = 6.2$ meters, and $A = 35°$.

42. Find the area of the triangle in Problem 40.

Examples

1. Each of the following is a complex number.

$$5 + 4i$$
$$-\sqrt{3} + i$$
$$7i$$
$$-8$$

The number $7i$ is complex because

$$7i = 0 + 7i$$

The number -8 is complex because

$$-8 = -8 + 0i$$

Definitions [8.1]

The number i is such that $i^2 = -1$. If $a > 0$, the expression $\sqrt{-a}$ can be written as $\sqrt{ai^2} = i\sqrt{a}$.

A *complex number* is any number that can be written in the form

$$a + bi$$

where a and b are real numbers and $i^2 = -1$. The number a is called the *real part* of the complex number and b is called the *imaginary part*. The form $a + bi$ is called *standard form*.

All real numbers are also complex numbers since they can be put in the form $a + bi$, where $b = 0$. If $a = 0$ and $b \neq 0$, then $a + bi$ is also called an *imaginary number*.

2. If $3x + 2i = 12 - 4yi$, then

$$3x = 12 \quad \text{and} \quad 2 = -4y$$
$$x = 4 \qquad\quad y = -1/2$$

Equality for Complex Numbers [8.1]

Two complex numbers are equal if and only if their real parts are equal and their imaginary parts are equal. That is,

$$a + bi = c + di \text{ if and only if } a = c \text{ and } b = d$$

3. If $z_1 = 2 - i$ and $z_2 = 4 + 3i$, then

$$z_1 + z_2 = 6 + 2i$$

$$z_1 - z_2 = -2 - 4i$$

$$
\begin{aligned}
z_1 z_2 &= (2 - i)(4 + 3i)\\
&= 8 + 6i - 4i - 3i^2\\
&= 11 + 2i
\end{aligned}
$$

The conjugate of $4 + 3i$ is $4 - 3i$ and $(4 + 3i)(4 - 3i) = 16 + 9 = 25$

$$
\begin{aligned}
\frac{z_1}{z_2} &= \frac{2 - i}{4 + 3i} \cdot \frac{4 - 3i}{4 - 3i}\\
&= \frac{5 - 10i}{25}\\
&= \frac{1}{5} - \frac{2}{5}i
\end{aligned}
$$

Operations on Complex Numbers in Standard Form [8.1]

If $z_1 = a_1 + b_1 i$ and $z_2 = a_2 + b_2 i$ are two complex numbers in standard form, then the following definitions and operations apply.

Addition

$$z_1 + z_2 = (a_1 + a_2) + (b_1 + b_2)i$$

Add real parts, add imaginary parts.

Subtraction

$$z_1 - z_2 = (a_1 - a_2) + (b_1 - b_2)i$$

Subtract real parts, subtract imaginary parts.

Multiplication

$$z_1 z_2 = (a_1 a_2 - b_1 b_2) + (a_1 b_2 + a_2 b_1)i$$

In actual practice, simply multiply as you would multiply two binomials.

Conjugates

The conjugate of $a + bi$ is $a - bi$. Their product is the real number $a^2 + b^2$.

Division

Multiply the numerator and denominator of the quotient by the conjugate of the denominator.

4. $i^{20} = (i^2)^{10} = (-1)^{10} = 1$
$i^{21} = i^{20} \cdot i = 1 \cdot i = i$
$i^{22} = (i^2)^{11} = (-1)^{11} = -1$
$i^{23} = i^{22} \cdot i = -1 \cdot i = -i$

Powers of *i* [8.1]

If n is an integer, then i^n can always be simplified to either i, -1, $-i$, or 1.

5. The graph of $4 + 3i$ is

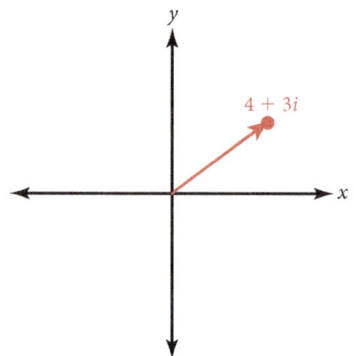

Graphing Complex Numbers [8.2]

The graph of the complex number $z = a + bi$ is the arrow (vector) that extends from the origin to the point (a, b).

6. If $z = \sqrt{3} + i$, then
$|z| = |\sqrt{3} + i| = \sqrt{3 + 1} = 2$

Absolute Value of a Complex Number [8.2]

The *absolute value* (or *modulus*) of the complex number $z = a + bi$ is the distance from the origin to the point (a, b). If this distance is denoted by r then

$$r = |z| = |a + bi| = \sqrt{a^2 + b^2}$$

7. For $z = \sqrt{3} + i$, θ is the smallest positive angle for which

$$\sin \theta = \frac{1}{2} \text{ and } \cos \theta = \frac{\sqrt{3}}{2}$$

which means $\theta = 30°$

Argument of a Complex Number [8.2]

The argument of the complex number $z = a + bi$ is the smallest positive angle from the positive x-axis to the graph of z. If the argument of z is denoted by θ then,

$$\sin \theta = \frac{b}{r}, \cos \theta = \frac{a}{r}, \text{ and } \tan \theta = \frac{b}{a}$$

8. If $z = \sqrt{3} + i$, then in trigonometric form

$$z = 2(\cos 30° + i \sin 30°)$$

Trigonometric Form of a Complex Number [8.2]

The complex number $z = a + bi$ is written in trigonometric form when it is written as

$$z = r(\cos \theta + i \sin \theta)$$

where r is the absolute value of z and θ is the argument of z.

9. If $z_1 = 8(\cos 40° + i \sin 40°)$ and
$z_2 = 4(\cos 10° + i \sin 10°)$,
then

$z_1 z_2 = 32(\cos 50° + i \sin 50°)$

$\dfrac{z_1}{z_2} = 2(\cos 30° + i \sin 30°)$

Products and Quotients in Trigonometric form [8.3]

If $z_1 = r_1(\cos \theta_1 + i \sin \theta_1)$ and $z_2 = r_2(\cos \theta_2 + i \sin \theta_2)$ are two complex numbers in trigonometric form, then

their product is

$$z_1 z_2 = r_1 r_2 [\cos (\theta_1 + \theta_2) + i \sin (\theta_1 + \theta_2)]$$

their quotient is

$$\frac{z_1}{z_2} = \frac{r_1}{r_2}[\cos (\theta_1 - \theta_2) + i \sin (\theta_1 - \theta_2)]$$

DeMoivre's Theorem [8.3]

If $z = r(\cos \theta + i \sin \theta)$ is a complex number in trigonometric form and n is an integer, then

$$z^n = r^n(\cos n\theta + i \sin n\theta)$$

10. If
$z = \sqrt{2}(\cos 30° + i \sin 30°)$,
then
$z^{10} = (\sqrt{2})^{10}(\cos 10 \cdot 30° +$
$i \sin 10 \cdot 30°)$
$= 32(\cos 300° + i \sin 300°)$

Roots of a Complex Number [8.4]

The n nth roots of the complex number $z = r(\cos \theta + i \sin \theta)$ are given by the formula

$$w_k = r^{1/n}\left(\cos \frac{\theta + 360°k}{n} + i \sin \frac{\theta + 360°k}{n}\right)$$

where $k = 0, 1, 2, \ldots, n - 1$. That is, the n nth roots are

$$w_0 = r^{1/n}\left(\cos \frac{\theta}{n} + i \sin \frac{\theta}{n}\right)$$

11. The 3 cube roots of
$z = 8(\cos 60° + i \sin 60°)$
are given by

$w_k = 8^{1/3}\left(\cos \dfrac{60° + 360°k}{3}\right.$

$\left. + i \sin \dfrac{60° + 360°k}{3}\right)$

$= 2[\cos(20° + 120°k)$
$+ i \sin(20° + 120°k)]$

when $k = 0, 1, 2$. That is,

$w_0 = 2(\cos 20° + i \sin 20°)$
$w_1 = 2(\cos 140° + i \sin 140°)$
$w_2 = 2(\cos 260° + i \sin 260°)$

$$w_1 = r^{1/n}\left(\cos \frac{\theta + 360°}{n} + i \sin \frac{\theta + 360°}{n}\right)$$

$$w_2 = r^{1/n}\left(\cos \frac{\theta + 720°}{n} + i \sin \frac{\theta + 720°}{n}\right)$$

$$\cdot$$

$$w_{n-1} = r^{1/n}\left(\cos \frac{\theta + 360°(n - 1)}{n} + i \sin \frac{\theta + 360°(n - 1)}{n}\right)$$

12.

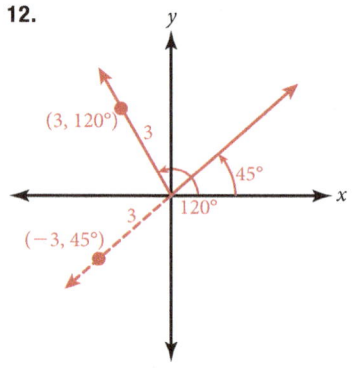

Polar Coordinates [8.5]

The ordered pair (r, θ) names the point that is r units from the origin along the axis rotated through θ degrees from the positive x-axis. The coordinates r and θ are said to be the *polar coordinates* of the point they name.

13. Convert $(-\sqrt{2}, 135°)$ to rectangular coordinates.

To convert from polar coordinates to rectangular coordinates, we substitute the given values of r and θ into the equations

$x = r \cos \theta$ and $y = r \sin \theta$

$x = -\sqrt{2} \cos 135°$

$\quad = -\sqrt{2} \left(-\dfrac{1}{\sqrt{2}} \right) = 1$

$y = -\sqrt{2} \sin 135°$

$\quad = -\sqrt{2} \left(\dfrac{1}{\sqrt{2}} \right) = -1$

Polar Coordinates and Rectangular Coordinates [8.5]

To derive the relationship between polar coordinates and rectangular coordinates, we consider a point P with rectangular coordinates (x, y) and polar coordinates (r, θ).

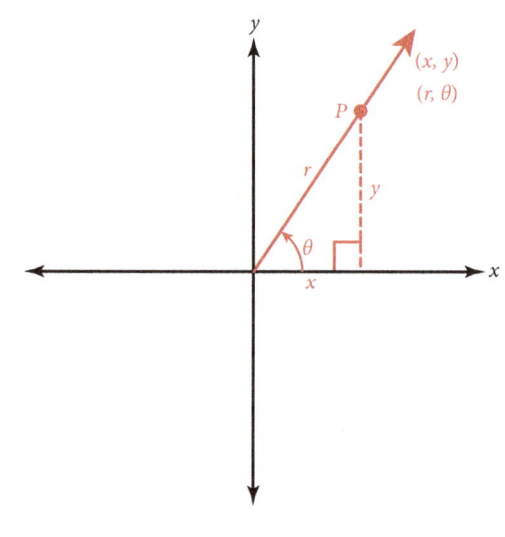

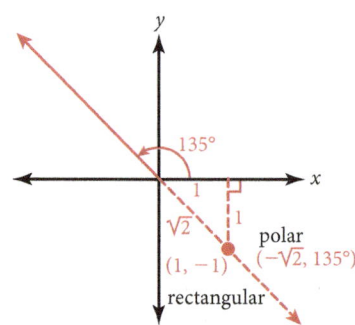

14. Change $x + y = 4$ to polar coordinates.

Since $x = r \cos \theta$ and $y = r \sin \theta$, we have

$r \cos \theta + r \sin \theta = 4$

$r(\cos \theta + \sin \theta) = 4$ *Factor out r*

$r = \dfrac{4}{\cos \theta + \sin \theta}$ *Divide both sides by $\cos \theta + \sin \theta$*

The last equation gives us r in terms of θ.

Equations in Polar Coordinates [8.5]

Equations in polar coordinates have variables r and θ instead of x and y. The conversions we use to change ordered pairs from polar coordinates to rectangular coordinates and from rectangular coordinates to polar coordinates are the same ones we use to convert back and forth between equations given in polar coordinates and those in rectangular coordinates.

Write in terms of i.

1. $\sqrt{-25}$ **2.** $\sqrt{-12}$

Find x and y so that each of the following equations is true.

3. $7x - 6i = 14 - 3yi$ **4.** $(x^2 - 3x) + 16i = 10 + 8yi$

Combine the following complex numbers:

5. $(6 - 3i) + [(4 - 2i) - (3 + i)]$ **6.** $(7 + 3i) - [(2 + i) - (3 - 4i)]$

Simplify each power of i.

7. i^{16} **8.** i^{17}

Multiply. Leave your answer in standard form.

9. $(8 + 5i)(8 - 5i)$ **10.** $(3 + 5i)^2$

Divide. Write all answers in standard form.

11. $\dfrac{5 - 4i}{2i}$ **12.** $\dfrac{6 + 5i}{6 - 5i}$

For each of the following complex numbers give: (a) the absolute value; (b) the opposite; and (c) the conjugate.

13. $3 + 4i$ **14.** $3 - 4i$ **15.** $8i$ **16.** -4

Write each complex number in standard form.

17. $8(\cos 330° + i \sin 330°)$ **18.** $2(\cos 135° + i \sin 135°)$

Write each complex number in trigonometric form.

19. $2 + 2i$ **20.** $-\sqrt{3} + i$ **21.** $5i$ **22.** -3

Multiply or divide as indicated. Leave your answer in trigonometric form.

23. $5(\cos 25° + i \sin 25°) \cdot 3(\cos 40° + i \sin 40°)$

24. $\dfrac{10(\cos 50° + i \sin 50°)}{2(\cos 20° + i \sin 20°)}$ **25.** $[2(\cos 10° + i \sin 10°)]^5$

26. $[3(\cos 20° + i \sin 20°)]^4$

27. Find two square roots of $z = 49(\cos 50° + i \sin 50°)$. Leave your answer in trigonometric form.

28. Find four fourth roots for $z = 2 + 2i\sqrt{3}$. Leave your answer in trigonometric form.

Solve each equation. Write your solutions in trigonometric form.

29. $x^4 - 2\sqrt{3}x^2 + 4 = 0$ **30.** $x^3 = -1$

For each of the following points, name two other ordered pairs in polar coordinates that name the same point and then convert each of the original ordered pairs to rectangular coordinates.

31. $(4, 225°)$ **32.** $(-6, 60°)$

Convert to polar coordinates with r positive and θ between 0° and 360°.

33. $(-3, 3)$ **34.** $(0, 5)$

Write each equation with rectangular coordinates.

35. $r = 6 \sin \theta$ **36.** $r = \sin 2\theta$

Write each equation in polar coordinates.

37. $x + y = 2$ **38.** $x^2 + y^2 = 8y$

Graph each equation.

39. $r = 4$ **40.** $\theta = 45°$

41. $r = 4 + 2 \cos \theta$ **42.** $r = \sin 2\theta$

Spotlight on Success

A Message from Mr. McKeague

Dear Student,

Now that you are close to finishing this course, I want to pass on a couple of things that have helped me a great deal with my career. I'll introduce each one with a quote:

Do something for the person you will be 5 years from now.

I have always made sure that I arranged my life so that I was doing something for the person I would be 5 years later. For example, when I was 20 years old, I was in college. I imagined that the person I would be at 25-year-old, would want to have a college degree, so I made sure I stayed in school. That's all there is to this. It is not a hard, rigid philosophy. It is a soft, behind the scenes, foundation. It does not include ideas such as "Five years from now I'm going to graduate at the top of my class from the best college in the country." Instead, you think, "five years from now I will have a college degree, or I will still be in school working towards it."

This philosophy led to a community college teaching job, writing textbooks, doing videos with the textbooks, then to MathTV and the book you are reading right now. Along the way there were many other options and directions that I didn't take, but all the choices I had were due to keeping the person I would be in 5 years in mind.

It's easier to ride a horse in the direction it is going.

I started my college career thinking that I would become a dentist. I enrolled in all the courses that were required for dental school. When I completed the courses, I applied to a number of dental schools, but wasn't accepted. I kept going to school, and applied again the next year, again, without success. My life was not going in the direction of dental school, even though I had worked hard to put it in that direction. So I did a little inventory of the classes I had taken and the grades I earned, and realized that I was doing well in mathematics. My life was actually going in that direction so I decided to see where that would take me. It was a good decision.

It is a good idea to work hard toward your goals, but it is also a good idea to take inventory every now and then to be sure you are headed in the direction that is best for you.

I wish you good luck with the rest of you college years, and with whatever you decide you want to do as a career.

Pat McKeague

A.1

Learning Objectives

A Know vector notation in component form.

B Add and subtract vectors in component form.

C Add and subtract vectors graphically.

D Use vectors to solve applied problems.

When an airplane flies, its speed and direction characterize the *velocity* of the airplane and cannot be represented by a single real number. In physics and engineering, quantities such as velocity and force are represented by a line segment that has a specified length and points in a specified direction. These directed line segments are called **vectors**. This section will give a brief introduction to vectors in two dimensions and their properties.

A vector is depicted graphically by an arrow, with an **initial point**, the "tail," and a **terminal point**, the "head," as shown in Figure 1. The length of the arrow represents the **magnitude** of the vector. Two vectors can have the same magnitude and the same direction, as shown in Figure 2.

To give a unique representation to a vector, its initial position can be taken to be the origin. This is called the **standard position** of a vector. A vector in standard position is completely and uniquely specified by the coordinates of its terminal point.

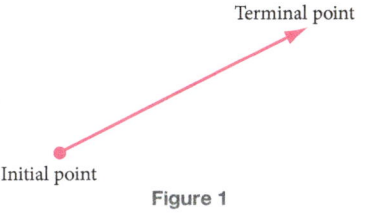

Figure 1

Component Form of a Vector

A vector **v** in two dimensions and in standard position is represented as $\langle v_x, v_y \rangle$, where v_x and v_y are the coordinates of the terminal point. They are referred to as the **x-** and **y-components**, respectively, of **v**. The **zero vector** is given by $\mathbf{0} = \langle 0, 0 \rangle$.

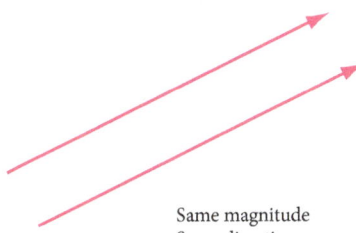

Same magnitude
Same direction

Figure 2

Pictured in Figure 3 are two vectors, **u** and **v**, in component form.

Magnitude and Direction of a Vector

The **magnitude** of a vector $\mathbf{v} = \langle v_x, v_y \rangle$ in standard position is given by

$$\|\mathbf{v}\| = \sqrt{v_x^2 + v_y^2}$$

For **v** nonzero, the **direction angle** of **v** is the angle θ that **v** makes with the positive x-axis. It is given by

$$\cos \theta = \frac{v_x}{\|\mathbf{v}\|} \quad \text{and} \quad \sin \theta = \frac{v_y}{\|\mathbf{v}\|}, \text{ or} \quad \tan \theta = \frac{v_y}{v_x}$$

with θ in the interval $[0°, 360°)$. See Figure 4. The direction of the zero vector is undefined.

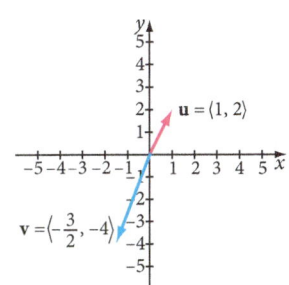

Figure 3

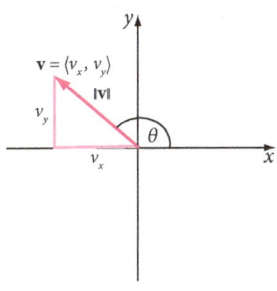

Figure 4

Note Observe that the magnitude of a vector is represented by a single number.

EXAMPLE 1 Finding Magnitude and Direction of a Vector

Find the magnitude and direction of each of the following vectors.

a. $\mathbf{v} = \langle 2, 2 \rangle$ **b.** $\mathbf{u} = \langle 3, -5 \rangle$

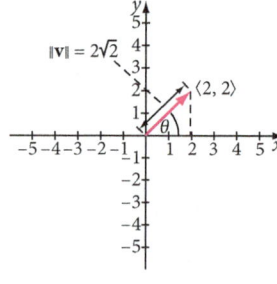

Solution

a. Graph the vector as in Figure 5. By the definition of magnitude, we have

$$\|\mathbf{v}\| = \sqrt{(2)^2 + (2)^2} = \sqrt{4 + 4} = \sqrt{8} = 2\sqrt{2}$$

The direction of $\mathbf{v}$ is defined by the angle θ in standard position such that

$$\tan \theta = \frac{v_y}{v_x} = \frac{2}{2} = 1$$

From our knowledge of trigonometric functions of special angles, the direction angle is $\theta = 45°$, because the vector lies in the first quadrant.

Figure 5

b. Graph the vector as in Figure 6. Using the definition of magnitude, we get

$$\|\mathbf{u}\| = \sqrt{3^2 + (-5)^2} = \sqrt{34}$$

The direction of $\mathbf{u}$ is defined by the angle θ such that

$$\tan \theta = \frac{u_y}{u_x} = \frac{-5}{3}$$

Using the $\tan^{-1}$ key on your calculator, we have

$$\tan^{-1}\left(\frac{-5}{3}\right) \approx -59.04°$$

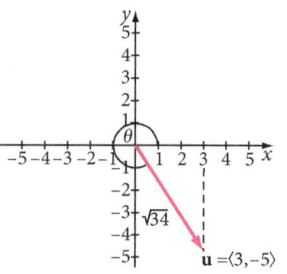

Because the terminal side of θ is in the fourth quadrant, and θ must be in the interval $[0°, 360°)$.

$$\theta = 360° - 59.04° = 300.96°$$

Figure 6

Sometimes, we are given the magnitude of a vector and its direction, and we wish to find its components. By definition of the direction θ of a vector $\mathbf{u} = \langle u_x, u_y \rangle$ with nonzero magnitude,

$$\cos \theta = \frac{u_x}{\|\mathbf{u}\|} \Rightarrow u_x = \|\mathbf{u}\| \cos \theta$$

$$\sin \theta = \frac{u_y}{\|\mathbf{u}\|} \Rightarrow u_y = \|\mathbf{u}\| \sin \theta$$

Components of a Vector

Let $\mathbf{u} = \langle u_x, u_y \rangle$ be a vector with nonzero magnitude. Then $u_x = \|\mathbf{u}\| \cos \theta$ and $u_y = \|\mathbf{u}\| \sin \theta$.

EXAMPLE 2 Finding Components of a Vector

Find the components of the vector $\mathbf{u}$ with magnitude 6 and direction angle 117°.

Solution Using the direction angle θ and the magnitude of $\mathbf{u}$, we have

$$u_x = \|\mathbf{u}\| \cos \theta = 6 \cos 117° \approx -2.72$$

$$u_y = \|\mathbf{u}\| \sin \theta = 6 \sin 117° \approx 5.35$$

Vector Operations

We now discuss addition and subtraction of vectors and scalar multiplication of vectors.

Addition and Subtraction of Vectors
To add or subtract vectors, add or subtract their components.

$$\mathbf{u} + \mathbf{v} = \langle u_x, u_y \rangle + \langle v_x, v_y \rangle = \langle u_x + v_x, u_y + v_y \rangle$$

$$\mathbf{u} - \mathbf{v} = \langle u_x, u_y \rangle - \langle v_x, v_y \rangle = \langle u_x - v_x, u_y - v_y \rangle$$

EXAMPLE 3   Adding and Subtracting Vectors
Let $\mathbf{u} = \langle -1, 3 \rangle$ and $\mathbf{v} = \langle 4, -1 \rangle$. Find the following and illustrate the result graphically.

a. $\mathbf{u} + \mathbf{v}$ **b. $\mathbf{v} - \mathbf{u}$**

Solution

a. Adding the components of $\mathbf{v}$ to the respective components of $\mathbf{u}$ gives

$$\mathbf{u} + \mathbf{v} = \langle -1, 3 \rangle + \langle 4, -1 \rangle = \langle -1 + 4, 3 + (-1) \rangle = \langle 3, 2 \rangle$$

See Figure 7.

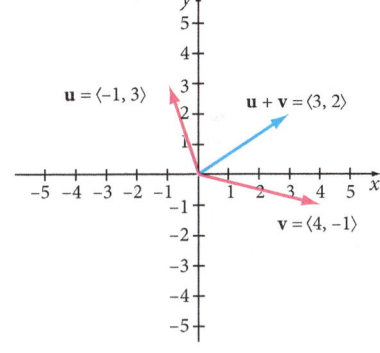

Figure 7

b. Subtracting the components of $\mathbf{u}$ from the respective components of $\mathbf{v}$ gives

$$\mathbf{v} - \mathbf{u} = \langle 4, -1 \rangle - \langle -1, 3 \rangle = \langle 4 - (-1), -1 - 3 \rangle = \langle 5, -4 \rangle$$

See Figure 8.

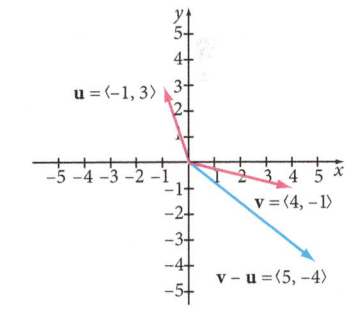

Figure 8

Scalar multiplication
For a real number k and a vector $\mathbf{u}$, the **scalar multiple** $k\mathbf{u}$ is the vector given by

$$k\mathbf{u} = k \langle u_x, u_y \rangle = \langle ku_x, ku_y \rangle$$

Observations
- If $k > 0$, then the vector $k\mathbf{u}$ is in the same direction as $\mathbf{u}$ and has magnitude $k \, \|\mathbf{u}\|$.

- If $k < 0$, then the vector $k\mathbf{u}$ is in the opposite direction as $\mathbf{u}$ and has magnitude $|k| \, \|\mathbf{u}\|$. See Figure 9.

EXAMPLE 4 Scalar Multiplication

Let $\mathbf{u} = \langle 5, -3 \rangle$ and let $\mathbf{v} = \langle -2, 1 \rangle$. Find the following.

a. $3\mathbf{u}$

b. $-2\mathbf{v} + \mathbf{u}$

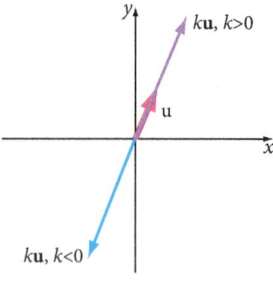

Solution

a. Using the definition of scalar multiplication, we have $3\mathbf{u} = 3\langle 5, -3 \rangle = \langle 15, -9 \rangle$

b. Using the definition of scalar multiplication to get $-2\mathbf{v}$, and then adding $\mathbf{u}$ to the result, gives

$$-2\mathbf{v} + \mathbf{u} = -2\langle -2, 1 \rangle + \langle 5, -3 \rangle = \langle 4, -2 \rangle + \langle 5, -3 \rangle = \langle 4 + 5, -2 - 3 \rangle = \langle 9, -5 \rangle$$

Figure 9

In the special case $k = -1$, we get the vector $-\mathbf{u}$. Note that $\mathbf{u} + (-\mathbf{u}) = \langle 0, 0 \rangle$. The **zero vector**, $\langle 0, 0 \rangle$, has magnitude zero and is denoted by **0**. Thus $-\mathbf{u} = \langle -u_x, -u_y \rangle$, where u_x and u_y are the components of $\mathbf{u}$.

EXAMPLE 5 Finding −u

Let $\mathbf{u} = \langle 3, 4 \rangle$.

a. Find $-\mathbf{u}$ in component form.

b. Graph $\mathbf{u}$ and $-\mathbf{u}$.

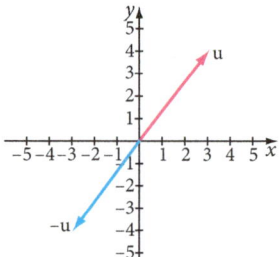

Solution

a. By the definition of $-\mathbf{u}$,

$$-\mathbf{u} = -\langle 3, 4 \rangle = \langle -3, -4 \rangle$$

b. Graph the vectors $\mathbf{u}$ and $-\mathbf{u}$, as shown in Figure 10. Observe that $-\mathbf{u}$ is directed 180° away from $\mathbf{u}$.

Figure 10

Vector operations share many of the same properties as operations on real numbers.

Properties of Vector Addition and Scalar Multiplication

Let $\mathbf{u}$, $\mathbf{v}$, and $\mathbf{w}$ be vectors and let k and m be scalars. Then the following hold:

1. $\mathbf{u} + \mathbf{v} = \mathbf{v} + \mathbf{u}$

2. $\mathbf{u} + (\mathbf{v} + \mathbf{w}) = (\mathbf{u} + \mathbf{v}) + \mathbf{w}$

3. $\mathbf{u} + \mathbf{0} = \mathbf{0} + \mathbf{u}$

4. $\mathbf{u} + (-\mathbf{u}) = \mathbf{0}$

5. $k(\mathbf{u} + \mathbf{v}) = k\mathbf{u} + k\mathbf{v}$

6. $(k + m)\mathbf{u} = k\mathbf{u} + m\mathbf{v}$

7. $k(m\mathbf{v}) = (km)(\mathbf{v})$

Unit Vectors

A **unit vector** is a vector of magnitude 1. For instance, $\left\langle \frac{3}{5}, -\frac{4}{5} \right\rangle$ is a unit vector because

$$\left\| \left\langle \frac{3}{5}, -\frac{4}{5} \right\rangle \right\| = \sqrt{\left(\frac{3}{5}\right)^2 + \left(-\frac{4}{5}\right)^2} = \sqrt{\frac{25}{25}} = 1$$

In many applications of vectors, we are interested in a unit vector in the direction of a nonzero vector **v**.

Unit Vector in the Direction of a Given Vector

Let $\mathbf{v} = \langle v_x, v_y \rangle$ be a nonzero vector. Then the unit vector in the direction of **v** is given by
$$\frac{\mathbf{v}}{\|\mathbf{v}\|}, \text{ where } \|\mathbf{v}\| = \sqrt{v_x^2 + v_y^2}$$

EXAMPLE 6 Calculating a Unit Vector

Find the unit vector **u** in the same direction as $\mathbf{v} = \langle -5, 12 \rangle$. Verify that **u** has magnitude 1.

Solution First calculate $\|\mathbf{v}\|$:

$$\|\mathbf{v}\| = \|\langle -5, 12 \rangle\| = \sqrt{(-5)^2 + 12^2} = \sqrt{169} = 13$$

Then the unit vector **u** in the direction of **v** is given by:

$$\mathbf{u} = \frac{\mathbf{v}}{\|\mathbf{v}\|} = \frac{\langle -5, 12 \rangle}{13} = \left\langle -\frac{5}{13}, \frac{12}{13} \right\rangle$$

We now show that the magnitude of **u** is 1 .

$$\|\mathbf{u}\| = \sqrt{\left(-\frac{5}{13}\right)^2 + \left(\frac{12}{13}\right)^2} = \sqrt{\frac{169}{169}} = 1$$

Other Representations of a Vector

A vector can be represented as the sum of vectors that lie along the axes. By the rules of vector addition,

$$\mathbf{u} = \langle u_x, u_y \rangle = \langle u_x, 0 \rangle + \langle 0, u_y \rangle$$

Observe that the vectors $\langle u_x, 0 \rangle$ and $\langle 0, u_y \rangle$ lie along the x- and y-axes, respectively.

A vector can also be represented in terms of unit vectors that lie along the axes. The **standard unit vectors** $\langle 1, 0 \rangle$ and $\langle 0, 1 \rangle$ lie along the positive x-axis and positive y-axis, respectively. By the rules of vector addition and scalar multiplication,

$$\mathbf{u} = \langle u_x, u_y \rangle = \langle u_x, 0 \rangle + \langle 0, u_y \rangle = u_x\langle 1, 0 \rangle + u_y\langle 0, 1 \rangle$$

The unit vectors $\langle 1, 0 \rangle$ and $\langle 0, 1 \rangle$ are also denoted by **i** and **j**, respectively. Thus, **u** can also be written as

$$\mathbf{u} = u_x \mathbf{i} + u_y \mathbf{j}$$

The unit-vector representation of a vector with a negative x-component and a positive y-component is illustrated in Figure 11.

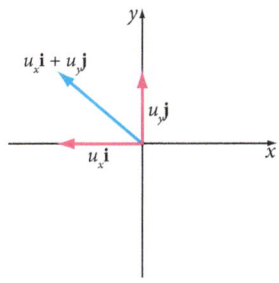

Figure 11

EXAMPLE 7 Representing a Vector in Terms of i and j

Write $\mathbf{u} = \langle -3, 2 \rangle$ in terms of the unit vectors $\mathbf{i}$ and $\mathbf{j}$.

Solution Because $\mathbf{u} = \langle -3, 2 \rangle$, we have $u_x = -3$ and $u_y = 2$. Thus

$$\mathbf{u} = \langle -3, 2 \rangle = \langle -3, 0 \rangle + \langle 0, 2 \rangle = -3\langle 1, 0 \rangle + 2\langle 0, 1 \rangle = -3\mathbf{i} + 2\mathbf{j}$$

Applications of Vectors

Vectors are useful in a number of areas because they can be used for any quantity with direction and magnitude. In navigation, the **velocity** of an airplane or boat gives both speed and direction. Thus, velocity can be represented as a vector, with the speed as its magnitude.

EXAMPLE 8 Wind Velocity

Using a specially designed golf flag, a golfer finds that the wind is blowing through the course at a speed of 12 miles per hour in the direction N40°W. Express the wind velocity in vector form, rounded to two decimal places. (*Source:* directhitgolfflags. com)

Solution Let $\mathbf{w}$ denote the velocity of the wind. Then $\|\mathbf{w}\| = 12$ is the speed of the wind. The direction N40°W, 40° west of north, corresponds to the angle $\theta = 90° + 40° = 130°$. (This notation was introduced in Section 7.1.) See Figure 12.

Compute the x and y velocity components of $\mathbf{w}$ as follows.

$$w_x = \|\mathbf{w}\| \cos \theta = 12 \cos 130° \approx -7.71$$

$$w_y = \|\mathbf{w}\| \sin \theta = 12 \sin 130° \approx 9.19$$

Thus, $\mathbf{w} = \langle -7.71, 9.19 \rangle$.

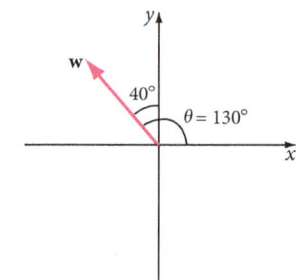

Figure 12

EXAMPLE 9 Combining Boat and Wind Velocities

A boat travels in the direction S18°E at a speed of 15 miles per hour. It encounters a wind blowing north to south at a speed of 10 miles per hour. Find the resulting speed and direction of the boat.

Solution The resulting velocity of the boat is obtained by adding the components of the boat velocity to the wind velocity. The magnitude of the resulting velocity vector is the resulting speed.

Step 1 **Find the components of boat velocity.** Let $\mathbf{v}$ denote the velocity of the boat. Thus, $\|\mathbf{v}\| = 15$, the speed of the boat. The direction S18°E, 18° east of south, corresponds to the angle $\theta = 270° + 18° = 288°$. See Figure 13. We then have

$$v_x = \|\mathbf{v}\| \cos \theta = 15 \cos 288° \approx 4.64$$

$$v_y = \|\mathbf{v}\| \sin \theta = 15 \sin 288° \approx -14.27$$

So $\mathbf{v} = \langle 4.64, -14.27 \rangle$.

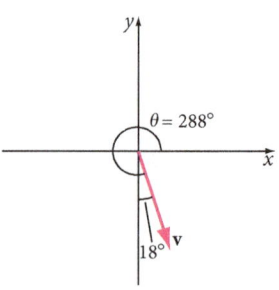

Figure 13

Step 2   **Find the components of wind velocity.**   Let **w** denote the velocity of the wind. Because the wind is blowing north to south at 10 miles per hour, $\|\mathbf{w}\| = 10$, and the direction is $\theta = 270°$. See Figure 14. We then have

$$w_x = \|\mathbf{w}\| \cos \theta = 10 \cos 270° = 0$$

$$w_y = \|\mathbf{w}\| \sin \theta = 10 \sin 270° = -10$$

So $\mathbf{w} = \langle 0, -10 \rangle$.

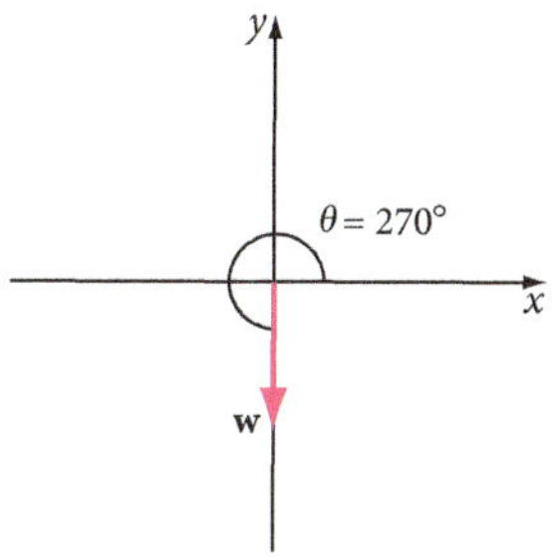

Step 3   **Find the resulting velocity.**   To find the resulting velocity, add **v** and **w**.

$$\mathbf{u} = \mathbf{v} + \mathbf{w} = \langle 4.64, -14.27 \rangle = \langle 0, -10 \rangle = \langle 4.64, -24.27 \rangle$$

Step 4   **Calculate the resulting speed and direction.**

$$\text{Speed: } \|\mathbf{u}\| = \sqrt{(4.64)^2 + (-24.27)^2} \approx 24.71$$

The direction θ is defined by

$$\tan \theta = \frac{u_y}{u_x} = \frac{-24.27}{4.64} \approx -5.23$$

Now, $\tan^{-1}(-5.23) = -79.18°$. Because the terminal side of θ is in the fourth quadrant, and θ must be in the interval $[0°, 360°)$, $\theta = 360° - 79.18° = 280.82°$. So the boat travels at a speed of 24.71 miles per hour with $\theta = 280.82°$, or S10.82°E.

Figure 14

In Exercises 1–6, graph each of the given vectors in the standard position.

1. $\langle 1, 0 \rangle$ **2.** $\langle 4, -1 \rangle$ **3.** $\langle -5, -3 \rangle$ **4.** $\left\langle 0, \frac{1}{2} \right\rangle$ **5.** $\left\langle \frac{4}{3}, -6 \right\rangle$ **6.** $\langle -2, -5.5 \rangle$

In Exercises 7–12, write each of the given vectors in terms of the unit vectors **i** *and* **j**

7. $\mathbf{u} = \langle -4, 6 \rangle$ **8.** $\mathbf{v} = \langle 5, -3 \rangle$ **9.** $\mathbf{w} = \langle -2, -1.5 \rangle$ **10.** $\mathbf{v} = \langle -3.5, 4 \rangle$

11. $\mathbf{u} = \left\langle \frac{1}{3}, \frac{3}{4} \right\rangle$ **12.** $\mathbf{w} = \left\langle -\frac{2}{5}, \frac{1}{6} \right\rangle$

In Exercises 13–24, for each of the following, find $\mathbf{u} - \mathbf{v}$, $\mathbf{u} + 2\mathbf{v}$, *and* $-3\mathbf{u} + \mathbf{v}$.

13. $\mathbf{u} = \langle 3, 0 \rangle, \mathbf{v} = \langle 5, 1 \rangle$ **14.** $\mathbf{u} = \langle 6, -2 \rangle, \mathbf{v} = \langle 3, -1 \rangle$

15. $\mathbf{u} = \langle -4, 5 \rangle, \mathbf{v} = \langle 3, -7 \rangle$ **16.** $\mathbf{u} = \langle -2, 6 \rangle, \mathbf{v} = \langle 7, -3 \rangle$

17. $\mathbf{u} = \langle 1.5, 2.5 \rangle, \mathbf{v} = \langle 0, 1 \rangle$ **18.** $\mathbf{u} = \langle 4, 0 \rangle, \mathbf{v} = \langle -1.5, 2.5 \rangle$

19. $\mathbf{u} = \left\langle \frac{1}{3}, \frac{2}{5} \right\rangle, \mathbf{v} = \langle 1, 2 \rangle$ **20.** $\mathbf{u} = \left\langle \frac{1}{4}, \frac{1}{2} \right\rangle, \mathbf{v} = \left\langle -\frac{1}{2}, \frac{3}{4} \right\rangle$

21. $\mathbf{u} = -2\mathbf{i} + 3\mathbf{j}, \mathbf{v} = 4\mathbf{i} - \mathbf{j}$ **22.** $\mathbf{u} = 6\mathbf{i} - 2\mathbf{j}, \mathbf{v} = -5\mathbf{i} + 3\mathbf{j}$

23. $\mathbf{u} = -1.1\mathbf{i} + 4\mathbf{j}, \mathbf{v} = 4\mathbf{i} + 2.4\mathbf{j}$ **24.** $\mathbf{u} = 2.6\mathbf{i} + 5\mathbf{j}, \mathbf{v} = -2\mathbf{i} + 3.7\mathbf{j}$

In Exercises 25–34, find the magnitude and the direction angle θ *of each of the given vectors, with* $\theta \in [0°, 360°)$.

25. $\mathbf{u} = \langle -1, 2 \rangle$ **26.** $\mathbf{w} = \langle 3, 5 \rangle$ **27.** $\mathbf{v} = \langle 1, 1.5 \rangle$ **28.** $\mathbf{u} = \langle -1.5, 3 \rangle$

29. $\mathbf{v} = \left\langle \frac{4}{3}, \frac{2}{5} \right\rangle$ **30.** $\mathbf{w} = \left\langle -\frac{1}{2}, -\frac{1}{4} \right\rangle$ **31.** $\mathbf{v} = 4\mathbf{i} - 2\mathbf{j}$ **32.** $\mathbf{v} = -3\mathbf{i} + 4\mathbf{j}$

33. $\mathbf{v} = 1\mathbf{i} + 2.5\mathbf{j}$ **34.** $\mathbf{v} = -3.2\mathbf{i} + 2\mathbf{j}$

In Exercises 35–40, find a unit vector in the same direction as the given vector.

35. $\mathbf{u} = \langle 3, 4 \rangle$ **36.** $\mathbf{v} = \langle -12, 5 \rangle$ **37.** $\mathbf{w} = \langle 1, 1 \rangle$ **38.** $\mathbf{u} = \langle 3, 2 \rangle$

39. $\mathbf{v} = -2\mathbf{i} + 1\mathbf{j}$ **40.** $\mathbf{u} = 4\mathbf{i} - 3\mathbf{j}$

In Exercises 41–48, express each vector in component form. Round your answers to four decimal places.

41. Magnitude 19; direction 34°   **42.** Length 7; direction 276°

43. Magnitude 10; direction 190°   **44.** Magnitude 8; direction 145°

45. Magnitude 4.6; points due west   **46.** Length 3.1; direction 16° south of east

47. Length 22; points northwest   **48.** Magnitude 59; direction 108°

In Exercises 49–57, round your answers to two decimal places.

49. Weather The world's largest weathervane is located in Montague, Michigan. On a July day in 2007, it showed that the wind had a speed of 15 miles per hour in the direction S30°E. Express the wind velocity in component form.   (*Source*: www.wunderground.com)

50. Golf A golf ball is hit from a tee with a launch angle of 13.2° and a speed of 140 miles per hour. Express the velocity of the ball in component form.   (*Source*: www.golf.com)

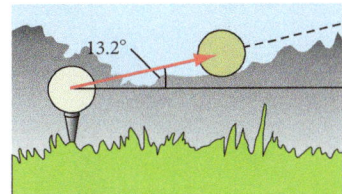

51. Beaufort Scale The Beaufort scale was developed in 1805 by Sir Francis Beaufort of England. It gives a measure for wind intensity based on observed sea and land conditions. For example, a wind speed of 20 knots is classified as a "fresh breeze," and smaller trees sway at this speed. Note that wind speed can also be measured in *knots*, where 1 knot equals 1.15 miles per hour. (*Source*: www.noaa.com)

   a. If the fresh breeze is in the direction S60°W, express the velocity of the breeze in component form. Use knots for the unit of speed.

   b. Express the velocity of the fresh breeze in component form using miles per hour as the unit for speed.

52. Biking Carl rides his bike due east for half an hour at a speed of 12 miles per hour (to be consistent). Then he rides due north for 45 minutes at a speed of 10 mph.

   a. At the end of the trip, how far is Carl from his starting point?

   b. Suppose Carl rides in a single straight-line path, and that his starting point and ending point are the same as before. In what direction does he ride his bike?

53. Distance Wanda goes for a hike. She first walks 2.4 miles in the direction S17°E, and then goes another 1.8 miles in the direction S38°E.

   a. By what east-west distance did Wanda's position change between the time she began the hike and the time she completed it?

   b. By what north-south distance did Wanda's position change?

   c. At the end of the hike, how far is Wanda from her starting point?

   d. Suppose Wanda walks in a single straight-line path and that her starting point and ending point are the same as before. In what direction does she walk?

54. Physics A ball is thrown upward with a velocity of 20 meters per second at an angle of 42° with respect to the horizontal.

a. At the time the ball is thrown, how fast is it moving in the horizontal direction?

b. At the time the ball is thrown, how fast is it moving in the vertical direction?

55. **Sailing** A sailboat travels on White Lake, Michigan, at a speed of 5 miles per hour in the direction N45°E. It encounters a moderate breeze blowing from south to north at a speed of 12 miles per hour. Find the resulting speed and direction of the sailboat.

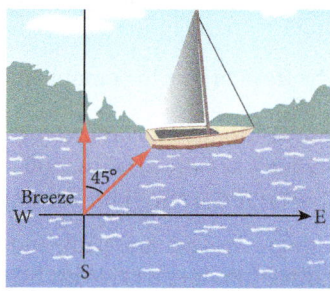

56. **Navigation** The net velocity of a ship is the vector sum of the velocity originating from the ship's engine and the velocity of the wind. The engine propels the ship at a velocity of 20 miles per hour in the direction S35°E.

a. What are the components of the ship's velocity originating from the engine?

b. If the wind is blowing from north to south at 12 miles per hour, find the magnitude and direction of the net velocity of the ship.

c. Rework part (b) for the case where the wind is blowing from north to south at 15 miles per hour.

57. Show that if $\|\mathbf{v}\| = 0$, then $\mathbf{v} = \langle 0, 0 \rangle$.

58. Show that if $\mathbf{u}$ is a nonzero vector, then the vector $\frac{\mathbf{u}}{\|\mathbf{u}\|}$ has magnitude 1.

59. If $\mathbf{u}$ is a nonzero vector, for what values of k does the equation $\|k\mathbf{u}\| = k\|\mathbf{u}\|$ hold? Explain.

A.2 Videos

Learning Objectives

A   Find the dot product of two vectors.

B   Determine the angle between two vectors.

C   Determine whether two vectors are orthogonal.

D   Calculate the projection of a vector onto another vector.

E   Calculate the parallel and perpendicular components of a vector.

F   Solve applied problems using dot products.

We have already defined multiplication of a vector by a scalar, but we have not yet defined multiplication of two vectors. One type of product of two vectors that is quite useful is called the **dot product.**

Dot Product

The **dot product** of vectors $\mathbf{u} = \langle u_x, u_y \rangle$ and $\mathbf{v} = \langle v_x, v_y \rangle$ is defined as

$$\mathbf{u} \cdot \mathbf{v} = u_x v_x + u_y v_y$$

The dot product of two vectors is a scalar, and is sometimes referred to as the *scalar product* or the *inner product*.

EXAMPLE 1   Calculating the Dot Product

Calculate $\langle -4, 3 \rangle \cdot \langle 2, -5 \rangle$.

Solution   Using the definition of the dot product, we have

$$\langle -4, 3 \rangle \cdot \langle 2, -5 \rangle = (-4)(2) + (3)(-5) = -8 - 15 = -23$$

The following are properties of the dot product.

Properties of the Dot Product

Let $\mathbf{u}$, $\mathbf{v}$, and $\mathbf{w}$ be vectors, and let k be a real number. Then the following hold:

1. $\mathbf{u} \cdot \mathbf{v} = \mathbf{v} \cdot \mathbf{u}$

2. $\mathbf{u} \cdot (\mathbf{v} + \mathbf{w}) = \mathbf{u} \cdot \mathbf{v} + \mathbf{u} \cdot \mathbf{w}$

3. $\mathbf{0} \cdot \mathbf{u} = 0$

4. $\mathbf{v} \cdot \mathbf{v} = ||\mathbf{v}||^2$

5. $(k\mathbf{u}) \cdot (\mathbf{v}) = k(\mathbf{u} \cdot \mathbf{v}) = \mathbf{u} \cdot (k\mathbf{v})$

We will prove the first property. The rest are left as exercises at the end of this section. Let $\mathbf{u} = \langle u_x, u_y \rangle$ and $\mathbf{v} = \langle v_x, v_y \rangle$. Then

$$\mathbf{u} \cdot \mathbf{v} = u_x v_x + u_y v_y$$

Now

$$\mathbf{v} \cdot \mathbf{u} = v_x u_x + v_y u_y = u_x v_x + u_y v_y$$

because multiplication is commutative. Thus

$$\mathbf{u} \cdot \mathbf{v} = \mathbf{v} \cdot \mathbf{u}$$

Angle Between Vectors

Another formula for the dot product of nonzero vectors is useful in determining the angle between two vectors and in solving applied problems.

Alternate Formula for Dot Products

$$\mathbf{u} \cdot \mathbf{v} = (\|\mathbf{u}\|)(\|\mathbf{v}\|)\cos\theta,$$

where θ is the smallest nonnegative angle between $\mathbf{u}$ and $\mathbf{v}$, as shown in Figure 1.

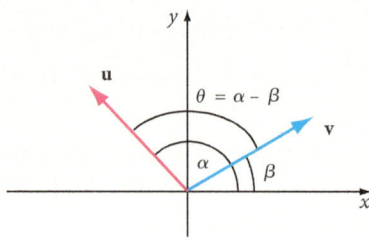

Figure 1

This formula can be derived from the original formula for the dot product:

$$\mathbf{u} \cdot \mathbf{v} = u_x v_x + u_y v_y$$

Let α and β be the directions of $\mathbf{u}$ and $\mathbf{v}$, respectively. Expressing the components of $\mathbf{u}$ and $\mathbf{v}$ in terms of their magnitudes and directions, we have

$$u_x = \|\mathbf{u}\| \cos \alpha \qquad v_x = \|\mathbf{v}\| \cos \beta$$

$$u_y = \|\mathbf{u}\| \sin \alpha \qquad v_y = \|\mathbf{v}\| \sin \beta$$

Substituting these expressions into the formula for the dot product, we obtain

$$\mathbf{u} \cdot \mathbf{v} = (\|\mathbf{u}\| \cos \alpha)(\|\mathbf{v}\| \cos \beta) + (\|\mathbf{u}\| \sin \alpha)(\|\mathbf{v}\| \sin \beta)$$

$$= (\|\mathbf{u}\|)(\|\mathbf{v}\|)(\cos \alpha \cos \beta + \sin \alpha \sin \beta) \qquad \text{Factor out } (\|\mathbf{u}\|)(\|\mathbf{v}\|)$$

$$= (\|\mathbf{u}\|)(\|\mathbf{v}\|)\cos(\alpha - \beta) \qquad \text{Apply difference identity for cosine}$$

$$= (\|\mathbf{u}\|)(\|\mathbf{v}\|) \cos \theta. \qquad \text{Substitute } \theta = \alpha - \beta$$

EXAMPLE 2   Finding the Angle Between Two Vectors

Find the smallest nonnegative angle, in degrees, between the vectors $\langle 3, -7 \rangle$ and $\langle 2, 6 \rangle$. See Figure 2.

Solution   First, use the original formula to compute the dot product.

$$\langle 3, -7 \rangle \cdot \langle 2, 6 \rangle = (3)(2) + (-7)(6) = 6 - 42 = -36$$

Next, use the alternative formula for the dot product.

$$\langle 3, -7 \rangle \cdot \langle 2, 6 \rangle = (\|\langle 3, -7 \rangle\|)(\|\langle 2,6 \rangle\|) \cos\theta$$

where θ is the angle between the vectors. Now, compute the magnitudes of the vectors.

$$\|\langle 3, -7 \rangle\| = \sqrt{3^2 + (-7)^2} = \sqrt{9 + 49} = \sqrt{58}$$

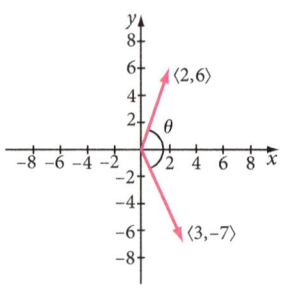

Figure 2

$$\|\langle 2, 6 \rangle\| = \sqrt{2^2 + 6^2} = \sqrt{4 + 36} = \sqrt{40} = \sqrt{4(10)} = 2\sqrt{10}$$

Substituting the magnitudes, we obtain

$$\langle 3, -7 \rangle \cdot \langle 2, 6 \rangle = \left(\sqrt{58} \right) \left(2\sqrt{10} \right) \cos \theta$$

Equating the two expressions for the dot product, we have

$$-36 = \left(\sqrt{58} \right) \left(2\sqrt{10} \right) \cos \theta$$

$$\cos \theta = -\frac{36}{\left(\sqrt{58} \right) \left(2\sqrt{10} \right)} \qquad \text{Isolate } \cos \theta$$

$$\cos \theta = -\frac{18}{\sqrt{580}} \approx -0.7474 \qquad \text{Simplify}$$

$$\theta \approx 138.37° \qquad \text{Use } \boxed{\cos^{-1}} \text{ key on calculator}$$

Parallel and Perpendicular Vectors

Two vectors **v** and **w** are said to be **parallel** if the angle θ between the vectors is 0° or 180°. If $\theta = 0°$, the vectors lie in the same direction. If $\theta = 180°$, the vectors lie in opposite directions.

If $\theta = 90°$, then the vectors are **perpendicular**. The term **orthogonal** is also used to mean perpendicular, especially in reference to vectors. When $\theta = 90°$, note that

$$\mathbf{v} \cdot \mathbf{w} = ||\mathbf{v}|| \, ||\mathbf{w}|| \cos \theta = ||\mathbf{v}|| \, ||\mathbf{w}|| \cos 90° = 0$$

We have the following facts about orthogonal vectors.

The Dot Product and Orthogonal Vectors

Let **v** and **w** be two nonzero vectors.

- If **v** and **w** are perpendicular, or orthogonal, then $\mathbf{v} \cdot \mathbf{w} = 0$.
- If $\mathbf{v} \cdot \mathbf{w} = 0$, then **v** and **w** are perpendicular, or orthogonal.

EXAMPLE 3   Determining Orthogonal Vectors

Determine whether the vectors $\mathbf{v} = \langle -4, 6 \rangle$ and $\mathbf{w} = \langle 3, 2 \rangle$ are orthogonal.

Solution   To determine whether the vectors are perpendicular, check whether their dot product is zero.

$$\mathbf{v} \cdot \mathbf{w} = \langle -4, 6 \rangle \cdot \langle 3, 2 \rangle = (-4)(3) + (6)(2) = -12 + 12 = 0$$

Because the dot product is 0, the vectors are orthogonal, or perpendicular.

Vector Projection

We saw in Section 7.5 that when a vector is written as $\mathbf{v} = a\mathbf{i} + b\mathbf{j}$, the component $a\mathbf{i}$ lies along the horizontal axis, and the component $b\mathbf{j}$ lies along the vertical axis. See Figure 3. Now, we discuss how to find components of a vector **v** along *any* nonzero vector **w** and perpendicular to **w**. We will refer to these components as $\mathbf{v}_1$ and $\mathbf{v}_2$, respectively. See Figure 4.

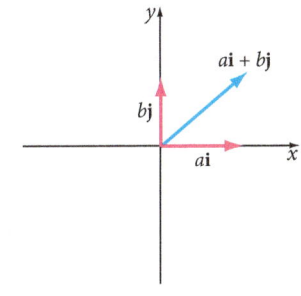

Figure 3

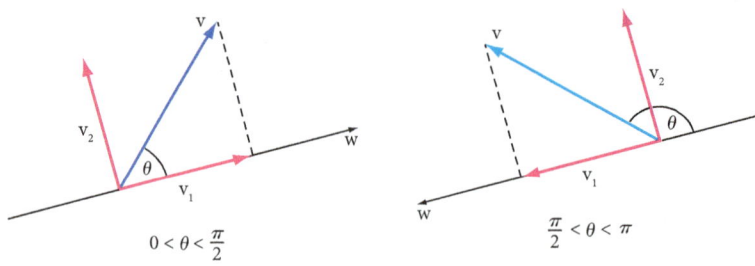

Figure 4

We first calculate $\|\mathbf{v}_1\|$ using trigonometry:

$$\|\mathbf{v}_1\| = \|\mathbf{v}\| \cos \theta$$

Now

$$\mathbf{v} \cdot \mathbf{w} = \|\mathbf{w}\| \, \|\mathbf{v}\| \cos \theta \Rightarrow \cos \theta = \frac{\mathbf{v} \cdot \mathbf{w}}{\|\mathbf{w}\| \, \|\mathbf{v}\|}$$

Substituting the expression for $\cos \theta$ in the expression for $\mathbf{v}_1$, we have

$$\|\mathbf{v}_1\| = \|\mathbf{v}\| \cos \theta = \|\mathbf{v}\| \frac{\mathbf{v} \cdot \mathbf{w}}{\|\mathbf{w}\| \, \|\mathbf{v}\|} = \frac{\mathbf{v} \cdot \mathbf{w}}{\|\mathbf{w}\|}$$

Because $\mathbf{v}_1$ is in the direction of $\mathbf{w}$, $\mathbf{v}_1$ can be obtained by scalar multiplication of $\|\mathbf{v}_1\|$ by a unit vector in the direction of $\mathbf{w}$. Thus

$$\mathbf{v}_1 = \|\mathbf{v}_1\| \left(\frac{\mathbf{w}}{\|\mathbf{w}\|} \right) \qquad \textcolor{teal}{\mathbf{v}_1 \text{ in the same direction as } \mathbf{w}}$$

$$= \left(\frac{\mathbf{v} \cdot \mathbf{w}}{\|\mathbf{w}\|} \right) \left(\frac{\mathbf{w}}{\|\mathbf{w}\|} \right) \qquad \textcolor{teal}{\text{Substitute } \|\mathbf{v}_1\| = \frac{\mathbf{v} \cdot \mathbf{w}}{\|\mathbf{w}\|}}$$

$$= \frac{\mathbf{v} \cdot \mathbf{w}}{\|\mathbf{w}\|^2} \, \mathbf{w}$$

The component $\mathbf{v}_1$ we found above is referred to as the **vector projection** of $\mathbf{v}$ onto $\mathbf{w}$, denoted by $\text{proj}_w \, \mathbf{v}$.

Vector Projection of v on w

Let $\mathbf{v}$ and $\mathbf{w}$ be nonzero vectors. The vector projection of $\mathbf{v}$ on the vector $\mathbf{w}$ is given by

$$\text{proj}_w \, \mathbf{v} = \left(\frac{\mathbf{v} \cdot \mathbf{w}}{\|\mathbf{w}\|^2} \right) \mathbf{w}$$

EXAMPLE 4   Finding the Vector Projection of v onto w

If $\mathbf{v} = \langle 4, 3 \rangle$ and $\mathbf{w} = \langle 3, 1 \rangle$, calculate $\text{proj}_w \mathbf{v}$. See Figure 5.

Solution First compute

$$\mathbf{v} \cdot \mathbf{w} = \langle 4, 3 \rangle \cdot \langle 3, 1 \rangle = 12 + 3 = 15$$

and

$$\|\mathbf{w}\|^2 = ((3)^2 + 1^2) = 10$$

Then

$$\text{proj}_w \mathbf{v} = \left(\frac{\mathbf{v} \cdot \mathbf{w}}{\|\mathbf{w}\|^2} \right) \mathbf{w} = \frac{15}{10} \langle 3, 1 \rangle = \left\langle \frac{9}{2}, \frac{3}{2} \right\rangle$$

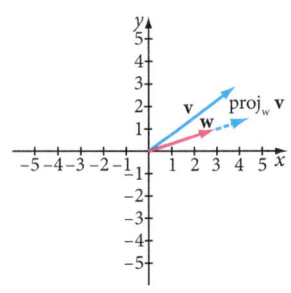

Figure 5

Using the vector projection, we can express **v** as the sum of the two orthogonal vector components.

Orthogonal Decomposition of a Vector

Let **v** and **w** be nonzero vectors. Then **v** can be written as

$$\mathbf{v} = \mathbf{v}_1 + \mathbf{v}_2$$

where $\mathbf{v}_1 = \mathrm{proj}_w\, \mathbf{v}$ and $\mathbf{v}_2 = \mathbf{v} - \mathbf{v}_1$. The vectors $\mathbf{v}_1$ and $\mathbf{v}_2$ are orthogonal.

EXAMPLE 5   **Decomposing a Vector into Components**

Let $\mathbf{v} = \langle 4, 3 \rangle$ and $\mathbf{w} = \langle 3, 1 \rangle$. Decompose **v** into two vectors $\mathbf{v}_1$ and $\mathbf{v}_2$, where $\mathbf{v}_1$ is parallel to **w** and $\mathbf{v}_2$ is orthogonal to **w**. See Figure 6.

Solution   From Example 4,

$$\mathbf{v}_1 = \mathrm{proj}_w\, \mathbf{v} = \left\langle \frac{9}{2}, \frac{3}{2} \right\rangle$$

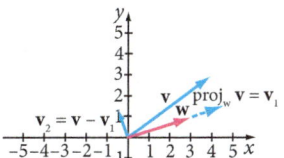

Figure 6

Now compute $\mathbf{v}_2$.

$$\mathbf{v}_2 = \mathbf{v} - \mathbf{v}_1 = \langle 4, 3 \rangle - \left\langle \frac{9}{2}, \frac{3}{2} \right\rangle \qquad \text{Definition of } v_2$$

$$= \left\langle \frac{8}{2}, \frac{6}{2} \right\rangle - \left\langle \frac{9}{2}, \frac{3}{2} \right\rangle \qquad \text{Rewrite with common denominator}$$

$$= \left\langle -\frac{1}{2}, \frac{3}{2} \right\rangle$$

Thus **v** can be decomposed as

$$\mathbf{v} = \mathbf{v}_1 + \mathbf{v}_2 = \left\langle \frac{9}{2}, \frac{3}{2} \right\rangle + \left\langle -\frac{1}{2}, \frac{3}{2} \right\rangle$$

You can quickly check that $\mathbf{v}_1 + \mathbf{v}_2 = \langle 4, 3 \rangle = \mathbf{v}$.

Applications of Dot Products

Dot products have long been used in engineering and physics. Recent applications of them include computer simulations and computer game design.

   In physics, the **work** done by a force in moving an object is given by the product of the distance moved with the magnitude of the force component in the direction of motion. See Figure 7. In terms of vectors and dot products, we have the following definition of work.

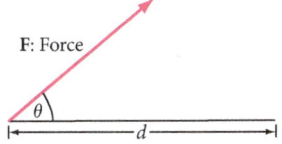

F: Force

Figure 7

Work Done by a Force

The work W done by a constant force **F** as it moves along the vector **v** is given by

$$W = \mathbf{F} \cdot \mathbf{v} = \|\mathbf{F}\|\, d \cos \theta$$

where $d = \|\mathbf{v}\|$.

The quantity $\|\mathbf{F}\| \cos \theta$ is the component of the force vector in the direction of motion. In particular, if the force vector is perpendicular to the direction of motion, no work is done because $\theta = 90°$ and so $W = (\|\mathbf{F}\| \cos \theta)\|\mathbf{v}\| = 0$. The units of work are *foot-pounds* in the English system and *Newton-meters* in the metric system.

> *Note* In most problems involving work, it is usually easier to apply the definition of work using the cosine.

EXAMPLE 6   Work Done by a Force

A man is pushing on a lawn mower handle with a force of 35 pounds. If the angle the handle makes with the horizontal is 40°, how much work is done in moving the lawn mower a distance of 150 feet on level ground?

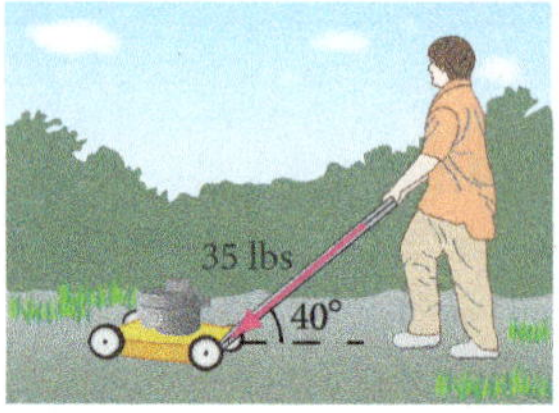

Solution Note that the force is exerted along the handle of the lawn mower, but the direction of motion is horizontal. Thus, only the horizontal component of the force contributes to the work done in moving the lawn mower. See Figure 8. Using the formula for work,

$$W = \|\mathbf{F}\| d \cos \theta$$

$$= (35)(150) \cos(40°) \qquad \|\mathbf{F}\| = 35, d = 150, \theta = 40°$$

$$= 5250(0.7660) \qquad \cos 40° \approx 0.7660$$

$$= 4021.50$$

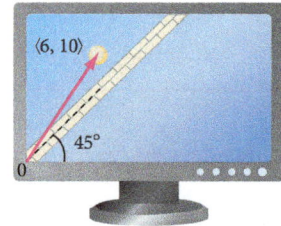

Figure 8

Thus, the work done in moving the lawn mower is 4021.50 foot-pounds.

Dot products are used extensively in computer graphics and in designing computer video games, because they give algebraic formulations of geometric concepts. Thus, dot products can be programmed more easily than geometric constructions. An elementary example from game design is given next.

EXAMPLE 7   Video Game Design

A portion of a computer video game consists of a ball colliding with a wall. The origin is taken to be the left bottom-most corner of the computer screen. The ball's location is given by the vector $\mathbf{v} = \langle 6, 10 \rangle$, and the wall makes an angle of 45° with the horizontal. See Figure 9. What is the perpendicular distance from the ball to the wall?

Solution We decompose $\mathbf{v} = \mathbf{v}_1 + \mathbf{v}_2$, so that $\mathbf{v}_1$ lies along the direction of the wall. Then, the magnitude of the perpendicular vector, $\|\mathbf{v}_2\|$, will give the required distance.

First, compute the projection $\mathbf{v}_1$, the component of $\mathbf{v}$ that lies along the wall. To do so, we need a vector $\mathbf{w}$ that lies along the direction of the wall. Because the wall makes an angle of 45° with the horizontal, the vector

$$\mathbf{w} = \langle \cos 45°, \sin 45° \rangle = \left\langle \frac{\sqrt{2}}{2}, \frac{\sqrt{2}}{2} \right\rangle$$

is a unit vector in the same direction as the wall. See Figure 10.

Figure 9

Thus $\mathbf{v}_1 = \operatorname{proj}_w \mathbf{v} = \left(\dfrac{\mathbf{v} \cdot \mathbf{w}}{\|\mathbf{w}\|^2} \right) \mathbf{w}$ Definition of projection

$\qquad\qquad = 8\sqrt{2} \left\langle \dfrac{\sqrt{2}}{2}, \dfrac{\sqrt{2}}{2} \right\rangle$ $\mathbf{v} \cdot \mathbf{w} = 8\sqrt{2};\ \|\mathbf{w}\| = 1$

$\qquad\qquad = \langle 8, 8 \rangle$ Simplify

Now compute $\mathbf{v}_2$.

$$\mathbf{v}_2 = \mathbf{v} - \mathbf{v}_1 = \langle 6, 10 \rangle - \langle 8, 8 \rangle = \langle -2, 2 \rangle$$

The perpendicular distance is therefore

$$\|\mathbf{v}_2\| = \sqrt{8} = 2\sqrt{2}$$

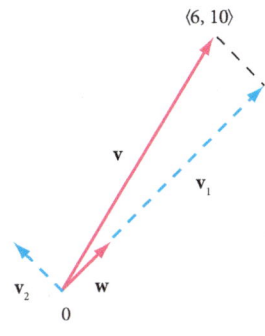

Figure 10

In Exercises 1–8, find $\mathbf{v} \cdot \mathbf{w}$.

1. $\mathbf{v} = \langle -3, 4 \rangle$, $\quad \mathbf{w} = \langle 2, -3 \rangle$ **2.** $\mathbf{v} = \langle 6, -1 \rangle$, $\quad \mathbf{w} = \langle 4, 3 \rangle$ **3.** $\mathbf{v} = \langle 5, -8 \rangle$, $\quad \mathbf{w} = \left\langle -2, \frac{1}{2} \right\rangle$

4. $\mathbf{v} = \left\langle \frac{3}{2}, -1 \right\rangle$, $\quad \mathbf{w} = \langle 4, 0 \rangle$ **5.** $\mathbf{v} = \langle 3, -1 \rangle$, $\quad \mathbf{w} = \langle 1, 3 \rangle$ **6.** $\mathbf{v} = \langle -2, 0 \rangle$, $\quad \mathbf{w} = \langle 0, 4 \rangle$

7. $\mathbf{v} = \left\langle -\frac{5}{3}, \frac{4}{5} \right\rangle$, $\quad \mathbf{w} = \left\langle \frac{2}{5}, \frac{1}{3} \right\rangle$ **8.** $\mathbf{v} = \left\langle \frac{6}{5}, \frac{1}{2} \right\rangle$ $\quad \mathbf{w} = \langle -10, 2 \rangle$

In Exercises 9–16, find the smallest nonnegative angle between the vectors $\mathbf{v}$ and $\mathbf{w}$. Round your answer to the nearest tenth of a degree.

9. $\mathbf{v} = \langle 1, 3 \rangle$, $\quad \mathbf{w} = \langle -2, 0 \rangle$ **10.** $\mathbf{v} = \langle 0, -1 \rangle$, $\quad \mathbf{w} = \langle 2, 3 \rangle$ **11.** $\mathbf{v} = \langle -2, 0 \rangle$, $\quad \mathbf{w} = \langle 0, 3 \rangle$

12. $\mathbf{v} = \langle 2, -4 \rangle$, $\quad \mathbf{w} = \langle 6, 3 \rangle$ **13.** $\mathbf{v} = \langle 4, 3 \rangle$, $\quad \mathbf{w} = \langle 2, -1 \rangle$ **14.** $\mathbf{v} = \langle 2, 4 \rangle$, $\quad \mathbf{w} = \langle -3, 2 \rangle$

15. $\mathbf{v} = \left\langle \frac{1}{3}, 1 \right\rangle$, $\quad \mathbf{w} = \langle 6, -1 \rangle$ **16.** $\mathbf{v} = \left\langle -2, \frac{3}{2} \right\rangle$, $\quad \mathbf{w} = \langle 1, 2 \rangle$

In Exercises 17–24, calculate $\text{proj}_w \mathbf{v}$. Then, decompose $\mathbf{v}$ into $\mathbf{v}_1$ and $\mathbf{v}_2$, where $\mathbf{v}_1$ is parallel to $\mathbf{w}$ and $\mathbf{v}_2$ is orthogonal to $\mathbf{w}$.

17. $\mathbf{v} = \langle 2, -4 \rangle$, $\quad \mathbf{w} = \langle 2, 6 \rangle$ **18.** $\mathbf{v} = \langle 5, -3 \rangle$, $\quad \mathbf{w} = \langle 1, 1 \rangle$ **19.** $\mathbf{v} = \langle 10, 5 \rangle$, $\quad \mathbf{w} = \langle 2, -1 \rangle$

20. $\mathbf{v} = \langle 1, 2 \rangle$, $\quad \mathbf{w} = \langle -3, 3 \rangle$ **21.** $\mathbf{v} = \langle 6, 12 \rangle$, $\quad \mathbf{w} = \langle 3, 1 \rangle$ **22.** $\mathbf{v} = \langle -4, 3 \rangle$, $\quad \mathbf{w} = \langle 1, -3 \rangle$

23. $\mathbf{v} = \langle 4, 5 \rangle$, $\quad \mathbf{w} = \langle -3, 4 \rangle$ **24.** $\mathbf{v} = \langle 6, -3 \rangle$, $\quad \mathbf{w} = \langle 4, 2 \rangle$

In Exercises 25–32, determine whether the given pairs of vectors are orthogonal.

25. $\mathbf{v} = \langle 1, 2 \rangle$, $\quad \mathbf{w} = \langle -4, 1 \rangle$ **26.** $\mathbf{v} = \langle 3, 1 \rangle$, $\quad \mathbf{w} = \langle 0, 1 \rangle$ **27.** $\mathbf{v} = \langle 2, -3 \rangle$, $\quad \mathbf{w} = \langle -6, 4 \rangle$

28. $\mathbf{v} = \langle -5, 2 \rangle$, $\quad \mathbf{w} = \langle 4, -10 \rangle$ **29.** $\mathbf{v} = \left\langle \frac{1}{3}, 2 \right\rangle$, $\quad \mathbf{w} = \left\langle 6, \frac{5}{2} \right\rangle$ **30.** $\mathbf{v} = \langle 3, 5 \rangle$, $\quad \mathbf{w} = \left\langle \frac{5}{6}, \frac{1}{2} \right\rangle$

31. $\mathbf{v} = \langle 1, 0 \rangle$, $\quad \mathbf{w} = \langle 0, 3 \rangle$ **32.** $\mathbf{v} = \langle 2, 0 \rangle$, $\quad \mathbf{w} = \langle 0, 4 \rangle$

In Exercises 33–40, perform each operation, given $\mathbf{u}$, $\mathbf{v}$, and $\mathbf{w}$.

$$\mathbf{u} = \langle 3, 2 \rangle, \mathbf{v} = \langle -1, 4 \rangle, \mathbf{w} = \langle -2, -1 \rangle$$

33. $2\mathbf{u} + 3\mathbf{w}$ **34.** $-4\mathbf{u} + \mathbf{v}$ **35.** $-\mathbf{u} \cdot (\mathbf{v} + \mathbf{w})$ **36.** $-2\mathbf{w} \cdot (\mathbf{v} + \mathbf{w})$

37. $3\mathbf{u} + \mathbf{v} - 2\mathbf{w}$ **38.** $-\mathbf{u} - 2\mathbf{v} + \mathbf{w}$ **39.** $\text{proj}_w(\mathbf{u} - \mathbf{v})$ **40.** $\text{proj}_u(\mathbf{v} + \mathbf{w})$

41. Work A parent pulling a wagon in which her child is riding along level ground exerts a force of 20 pounds on the handle. The handle make an angle of 45° with the horizontal. How much work is done in pulling the wagon 100 feet? Round to the nearest foot-pound.

42. Work A child pulls a wagon along level ground. He exerts a force of 20 pounds on the handle which makes a 30° angle with the horizontal. Find the work done in pulling the wagon 100 feet. Round to the nearest foot-pound.

43. Game Design In a new video game, Mario and Luigi are at positions defined by the vectors $\langle 10, 3 \rangle$ and $\langle x, 15 \rangle$. What must be the value of x so that their position vectors are orthogonal?

44. Design The position vectors of a tower and a small garden from the center of a fountain are given by $\langle 50, 60 \rangle$ and $\langle 40, y \rangle$. Find y so that the two position vectors are orthogonal.

45. Distance Two tourboats, *Swan* and *Dolphin*, depart from a lighthouse, with one directly behind another. However, after some time, *Swan* strays off course and ends up at position ⟨1, 2⟩, relative to the lighthouse. The other boat, *Dolphin*, stays on course with a position vector ⟨5, 3⟩, relative to the lighthouse. What is the shortest distance that *Swan* must travel to get back on the same course as *Dolphin*, assuming that *Dolphin* keeps traveling in its current direction?

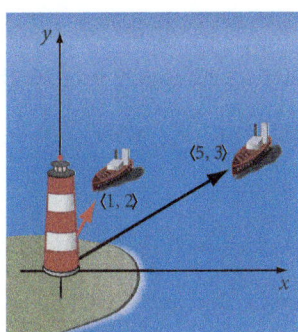

46. Computer Animation An animated figure's location is given by ⟨5, 2⟩. By what angle must the figure be rotated so that its new location is in the direction of ⟨4, 3⟩? Round your answer to the nearest tenth of a degree.

47. Work on Incline A wagon is pulled by a child with a force of 20 pounds, at an angle of 30° with the horizontal. Find the work done in pulling the wagon 100 feet up an incline that makes an angle of 10° with the horizontal. Round your answer to the nearest foot-pound.

48. Work on Incline A box weighing 100 pounds is pushed up a hill. The hill makes an angle of 30° with the horizontal. Find the work done against gravity in pushing the box a distance of 60 feet.

For Exercises 49 and 50, use the following information. The horsepower of an engine pulling a cart is determined by the formula

$$P = \frac{1}{550}(\mathbf{F} \cdot \mathbf{v})$$

where $\mathbf{F}$ is the force, in pounds, exerted on the cart and $\mathbf{v}$ is the velocity, in feet per second, of the cart as it is moved by the engine.

49. **Power** Find the horsepower of an engine that is exerting a force of 1500 pounds at an angle of 30° and is moving the cart horizontally at a speed of 10 feet per sec. Round to the nearest tenth of a horsepower.

50. **Power** Find the horsepower of an engine that is exerting a force of 2000 pounds at an angle of 30° and is moving the cart horizontally at a speed of 15 feet per sec. Round to the nearest tenth of a horsepower.

51. Find a so that $\langle 4, a \rangle$ and $\langle -3, 2 \rangle$ are orthogonal.

52. Show that if $\mathbf{v}$ and $\mathbf{w}$ are nonzero orthogonal vectors, then $\text{proj}_w \mathbf{v} = 0$.

53. Show that for vectors $\mathbf{u}$, $\mathbf{v}$ and $\mathbf{w}$, $\mathbf{u} \cdot (\mathbf{v} + \mathbf{w}) = \mathbf{u} \cdot \mathbf{v} + \mathbf{u} \cdot \mathbf{w}$.

54. Show that for any vector $\mathbf{u}$, $\mathbf{0} \cdot \mathbf{u} = 0$.

55. Show that for any vector $\mathbf{v}$, $\mathbf{v} \cdot \mathbf{v} = \|\mathbf{v}\|^2$. (In advanced mathematics, this relationship is very useful.)

56. Show that for any vector $\mathbf{u}$ and $\mathbf{v}$, and any real number k, $(k\mathbf{u}) \cdot (\mathbf{v}) = k(\mathbf{u} \cdot \mathbf{v}) = \mathbf{u} \cdot (k\mathbf{v})$.

Three-Dimensional Space

In Appendix A.1, vectors in the two-dimensional plane were introduced. We will now turn our attention to vectors in three-dimensional space. First we need a bit of background into three dimensions. Just as the plane identifies points by the coordinates (x, y), space includes a third coordinate z, so that points are identified by the coordinates (x, y, z). Visually, we can think of the xy-plane with a third vertical axis z.

The point $(0, 0, 0)$ is called the **origin**, and represents the point where the three axes intersect.

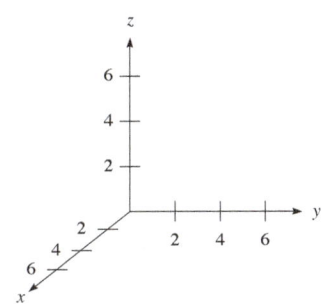

Figure 1

Points are located in space just as they are in the plane, except with a vertical component. For example, the point $(4, 3, 2)$ would be located by moving 4 units along the x-axis, 3 units along the y-axis, and 2 units along the z-axis. We often represent this as a box to give better perspective.

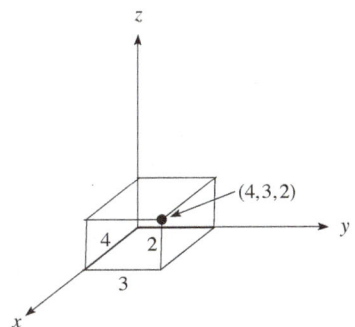

Figure 2

In order to study the magnitude of space vectors, we need an extension of the distance formula.

Distance Formula

If (x_1, y_1, z_1) and (x_2, y_2, z_2) are two points in space, then the distance between these points is given by

$$d = \sqrt{(x_2 - x_1)^2 + (y_2 - y_1)^2 + (z_2 - z_1)^2}$$

Note: If the two points are $(0, 0, 0)$ and (x, y, z), then $d = \sqrt{x^2 + y^2 + z^2}$, a much simpler form.

EXAMPLE 1   Finding the Distance Between Two Points

Find the distance between the following pairs of points.

a. $(-1, 2, 4)$ and $(2, -3, 1)$ **b.** $(0, 0, 0)$ and $(4, 3, 2)$

Solution

a. Using the distance formula:

$$d = \sqrt{(2 - (-1))^2 + (-3 - 2)^2 + (1 - 4)^2} = \sqrt{3^2 + (-5)^2 + (-3)^2}$$

b. Since $(0, 0, 0)$ is one of the points, the distance is:

$$d = \sqrt{4^2 + 3^2 + 2^2} = \sqrt{16 + 9 + 4} = \sqrt{20}$$

Vectors in Space

As in the plane, a vector in space has components (now three).

Component Form of a Space Vector

A vector $\mathbf{v}$ in three dimensions (space) in standard position is represented as $\langle v_x, v_y, v_z \rangle$, where (v_x, v_y, v_z) is the terminal point and $(0, 0, 0)$ is the original point. An alternate notation for a space vector is $\mathbf{v} = vx\mathbf{i} + vy\mathbf{j} + vz\mathbf{k}$.

Note: The unit vector $\langle 0, 0, 1 \rangle$ in the z-direction is designated by $\mathbf{k}$.

Magnitude of a Space Vector

The magnitude of a vector $\mathbf{v} = \langle v_x, v_y, v_z \rangle$ in standard position is given by

$$||\mathbf{v}|| = \sqrt{v_x^2 + v_y^2 + v_z^2}$$

If $||\mathbf{v}|| = 0$, then $\mathbf{v}$ is called a unit vector (same definition as plane vectors).

Note: The magnitude follows from the simplified distance formula since $(0, 0, 0)$ is the original point.

Notice that we have not defined direction of space vectors as was done for plane vectors. The reason for this is that direction of a space vector is a bit more complicated. At the end of this section, you will learn that a space vector has three direction angles corresponding to the angles a vector makes with each of the x-, y-, and z-axes. Thus, for space vectors, we tend to treat them more by their algebraic properties than by their geometric properties.

EXAMPLE 2   Finding the Magnitude of a Space Vector

Find the distance between the following pairs of points.

a. $\mathbf{v} = \langle 1, -2, 4 \rangle$ **b.** $\mathbf{u} = \langle -3, 4, 0 \rangle$

Solution

a. Using the magnitude formula:

$$||\mathbf{v}|| = \sqrt{1^2 + (-2)^2 + 4^2} = \sqrt{1 + 4 + 16} = \sqrt{21}$$

b. Using the magnitude formula:

$$||\mathbf{u}|| = \sqrt{(-3)^2 + 4^2 + 0^2} = \sqrt{9 + 16 + 0} = \sqrt{25} = 5$$

EXAMPLE 3 Finding a Unit Vector

Find a unit vector in the direction of each of the following vectors, and write each vector as a multiple of the unit vector.

a. $\mathbf{v} = \langle 1, -2, 4 \rangle$ **b.** $\mathbf{u} = 2\mathbf{i} + 4\mathbf{j} - 5\mathbf{k}$

Solution

a. In the previous example we found $||\mathbf{v}|| = \sqrt{21}$. If we divide by this magnitude we form the unit vector

$$\frac{\mathbf{v}}{||\mathbf{v}||} = \left\langle \frac{1}{\sqrt{21}}, -\frac{2}{\sqrt{21}}, \frac{4}{\sqrt{21}} \right\rangle$$

This is a unit vector in the direction of $\mathbf{v}$. This allows us to write:

$$\mathbf{v} = ||\mathbf{v}|| \left(\frac{\mathbf{v}}{||\mathbf{v}||} \right) = \sqrt{21} \left\langle \frac{1}{\sqrt{21}}, -\frac{2}{\sqrt{21}}, \frac{4}{\sqrt{21}} \right\rangle$$

b. Using the magnitude formula:

$$||\mathbf{u}|| = \sqrt{2^2 + 4^2 + (-5)^2} = \sqrt{4 + 16 + 25} = \sqrt{45} = 3\sqrt{5}$$

If we divide by this magnitude we form the unit vector $\dfrac{\mathbf{u}}{||\mathbf{u}||} = \dfrac{2}{3\sqrt{5}}\mathbf{i} + \dfrac{4}{3\sqrt{5}}\mathbf{j} - \dfrac{5}{3\sqrt{5}}\mathbf{k}$.

This is a unit vector in the direction of $\mathbf{u}$. This allows us to write:

$$\mathbf{u} = ||\mathbf{u}|| \left(\frac{\mathbf{u}}{||\mathbf{u}||} \right) = 3\sqrt{5} \left(\frac{2}{3\sqrt{5}}\mathbf{i} + \frac{4}{3\sqrt{5}}\mathbf{j} - \frac{5}{3\sqrt{5}}\mathbf{k} \right)$$

The previous example illustrates how we use algebraic properties to describe the geometry of vectors. In fact, this unit vector in the direction of a known vector is called a **direction vector**, which is used to describe the geometry (direction) of a vector through algebraic terms, rather than angles.

Vector Operations

The properties of vector addition and scalar multiplication for space vectors are identical to those for plane vectors. Rather than re-list those properties here, we'll illustrate the properties through the next two examples.

EXAMPLE 4 Addition and Subtraction of Vectors

Let $\mathbf{u} = \langle 3, 6, -4 \rangle$ and $\mathbf{v} = \langle -2, 3, -5 \rangle$. Find the following.

a. $\mathbf{u} + \mathbf{v}$ **b.** $\mathbf{u} - \mathbf{v}$

Solution

a. Adding each component:

$$\mathbf{u} + \mathbf{v} = \langle 3, 6, -4 \rangle + \langle -2, 3, -5 \rangle = \langle 3 + (-2), 6 + 3, -4 + (-5) \rangle = \langle 1, 9, -9 \rangle$$

b. Subtracting each component:

$$\mathbf{u} - \mathbf{v} = \langle 3, 6, -4 \rangle - \langle -2, 3, -5 \rangle = \langle 3 - (-2), 6 - 3, -4 - (-5) \rangle = \langle 5, 3, 1 \rangle$$

EXAMPLE 5   Scalar Multiplication

Let $\mathbf{u} = 3\mathbf{i} - 5\mathbf{j} - \mathbf{k}$ and $\mathbf{v} = 4\mathbf{i} + \mathbf{j} - 6\mathbf{k}$. Find the following.

a. $4\mathbf{u}$

b. $-3\mathbf{u} + 2\mathbf{v}$

Solution

a. Multiplying each component by the scalar:

$$4\mathbf{u} = 4(3\mathbf{i} - 5\mathbf{j} - \mathbf{k}) = 12\mathbf{i} - 20\mathbf{j} - 4\mathbf{k}$$

b. We first do the two scalar multiplications, and then add the components:

$$\begin{aligned} -3\mathbf{u} + 2\mathbf{v} &= -3(3\mathbf{i} - 5\mathbf{j} - \mathbf{k}) + 2(4\mathbf{i} + \mathbf{j} - 6\mathbf{k}) \\ &= (-9\mathbf{i} + 15\mathbf{j} + 3\mathbf{k}) + (8\mathbf{i} + 2\mathbf{j} - 12\mathbf{k}) \\ &= -\mathbf{i} + 17\mathbf{j} - 9\mathbf{k} \end{aligned}$$

The Dot Product

Now let's extend the definition for the dot product of two vectors to the **dot product of two vectors in space**.

> **Dot Product of Two Space Vectors**
>
> Let $\mathbf{u} = \langle u_x, u_y, u_z \rangle$ and $\mathbf{v} = \langle v_x, v_y, v_z \rangle$. The **dot product** of $\mathbf{u}$ and $\mathbf{v}$ is defined as
>
> $$\mathbf{u} \cdot \mathbf{v} = u_x v_x + u_y v_y + u_z v_z$$

Note that, as with plane vectors, the dot product of two space vectors is a scalar (real number), rather than another vector.

EXAMPLE 6   Dot Product

Find each dot product.

a. $\langle 3, 6, -4 \rangle \cdot \langle -2, 3, -5 \rangle$

b. $(3\mathbf{i} - 5\mathbf{j} - \mathbf{k}) \cdot (4\mathbf{i} + \mathbf{j} - 6\mathbf{k})$

Solution

a. Using the dot product definition:

$$\langle 3, 6, -4 \rangle \cdot \langle -2, 3, -5 \rangle = (3)(-2) + (6)(3) + (-4)(-5) = -6 + 18 + 20 = 32$$

b. Using the dot product definition:

$$(3\mathbf{i} - 5\mathbf{j} - \mathbf{k}) \cdot (4\mathbf{i} + \mathbf{j} - 6\mathbf{k}) = (3)(4) + (-5)(1) + (-1)(-6) = 12 - 5 + 6 = 13$$

The properties of the dot product are identical to those given in Section 7.6. We will prove the fourth property for space vectors, since the proof of our next theorem is dependent on it.

> **Property 4**
>
> $$\mathbf{v} \cdot \mathbf{v} = ||\mathbf{v}||^2$$

Proof

Let $\mathbf{v} = \langle v_x, v_y, v_z \rangle$. Then $||\mathbf{v}|| = \sqrt{v_x^2 + v_y^2 + v_z^2}$, and:

$$\mathbf{v} \cdot \mathbf{v} = v_x v_x + v_y v_y + v_z v_z = v_x^2 + v_y^2 + v_z^2 = ||\mathbf{v}||^2$$

This leads us to an important theorem for dot products.

> ### Theorem
> If **u** and **v** are plane or space vectors, and θ is the angle between them, then
> $$\mathbf{u} \cdot \mathbf{v} = ||\mathbf{u}|| \, ||\mathbf{v}|| \cos \theta$$

The proof of this for plane vectors was given in Appendix B.2, but if you look back at that proof you will see it cannot be used for space vectors as it relies on the direction angles of the vectors. This is another example of how we need to look at vectors in space more algebraically.

Proof

Consider the two space vectors **u** and **v**, as well as their connecting vector **v** − **u** as shown in the figure:

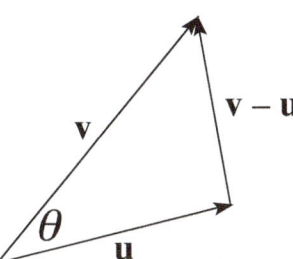

Figure 3

Using the law of cosines (remember the magnitude of a vector is its length):

$$||\mathbf{v} - \mathbf{u}||^2 = ||\mathbf{u}||^2 + ||\mathbf{v}||^2 - 2||\mathbf{u}|| \, ||\mathbf{v}|| \cos \theta$$

Using properties of the dot product:

$$||\mathbf{v} - \mathbf{u}||^2 = (\mathbf{v} - \mathbf{u}) \cdot (\mathbf{v} - \mathbf{u})$$
$$= \mathbf{v} \cdot (\mathbf{v} - \mathbf{u}) - \mathbf{u} \cdot (\mathbf{v} - \mathbf{u})$$
$$= \mathbf{v} \cdot \mathbf{v} - \mathbf{v} \cdot \mathbf{u} - \mathbf{u} \cdot \mathbf{v} + \mathbf{u} \cdot \mathbf{u}$$
$$= ||\mathbf{v}||^2 - 2\mathbf{u} \cdot \mathbf{v} + ||\mathbf{u}||^2$$

Since these expressions must be equal:

$$||\mathbf{v}||^2 - 2\mathbf{u} \cdot \mathbf{v} + ||\mathbf{u}||^2 = ||\mathbf{u}||^2 + ||\mathbf{v}||^2 - 2||\mathbf{u}|| \, ||\mathbf{v}|| \cos \theta$$
$$-2\mathbf{u} \cdot \mathbf{v} = -2||\mathbf{u}|| \, ||\mathbf{v}|| \cos \theta$$
$$2\mathbf{u} \cdot \mathbf{v} = 2||\mathbf{u}|| \, ||\mathbf{v}|| \cos \theta$$

Angle Between Vectors

As long as neither u nor v are the 0 vector, this theorem gives us a formula for finding the angle between two vectors:

$$\cos \theta = \frac{\mathbf{u} \cdot \mathbf{v}}{||\mathbf{u}|| \, ||\mathbf{v}||}$$

Since there are an infinite number of such angles, it is customary to choose the smallest positive angle for this formula.

EXAMPLE 7 Finding the Angle Between Two Vectors

Find the angle between the given pair of vectors.

a. $\langle 2, -3, 1 \rangle$ and $\langle -3, 1, -4 \rangle$ **b.** $-2\mathbf{i} - \mathbf{j} + 3\mathbf{k}$ and $3\mathbf{i} - 2\mathbf{j} + 4\mathbf{k}$

Solution

a. First we compute the required quantities:

$$\langle 2, -3, 1 \rangle \cdot \langle -3, 1, -4 \rangle = (2)(-3) + (-3)(1) + (1)(-4) = -6 - 3 - 4 = -13$$

$$||\langle 2, -3, 1 \rangle|| = \sqrt{2^2 + (-3)^2 + 1^2} = \sqrt{4 + 9 + 1} = \sqrt{14}$$

$$||\langle -3, 1, -4 \rangle|| = \sqrt{(-3)^2 + 1^2 + (-4)^2} = \sqrt{9 + 1 + 16} = \sqrt{26}$$

Now using our formula:

$$\cos\theta = \frac{\langle 2, -3, 1\rangle \cdot \langle -3, 1, -4\rangle}{\left(||\langle 2, -3, 1\rangle||\right)\cdot\left(||\langle -3, 1, -4\rangle||\right)} = -\frac{13}{\sqrt{364}} \approx -0.6814$$

Using a calculator, we find $\theta = \cos^{-1}(-0.6814) \approx 132.95°$.

b. First we compute the required quantities:

$$(-2\mathbf{i} - \mathbf{j} + 3\mathbf{k})\cdot(3\mathbf{i} - 2\mathbf{j} + 4\mathbf{k}) = (-2)(3) + (-1)(-2) + (3)(4) = -6 + 2 + 12 = 8$$

$$||-2\mathbf{i} - \mathbf{j} + 3\mathbf{k}|| = \sqrt{(-2)^2 + (-1)^2 + 3^2} = \sqrt{4 + 1 + 9} = \sqrt{14}$$

$$||3\mathbf{i} - 2\mathbf{j} + 4\mathbf{k}|| = \sqrt{3^2 + (-2)^2 + 4^2} = \sqrt{9 + 4 + 16} = \sqrt{29}$$

Now using our formula:

$$\cos\theta = \frac{(-2\mathbf{i} - \mathbf{j} + 3\mathbf{k})\cdot(3\mathbf{i} - 2\mathbf{j} + 4\mathbf{k})}{\left(||-2\mathbf{i} - \mathbf{j} + 3\mathbf{k}||\right)\left(||3\mathbf{i} - 2\mathbf{j} + 4\mathbf{k}||\right)} = \frac{8}{\left(\sqrt{14}\right)\left(\sqrt{29}\right)} = \frac{8}{\sqrt{406}} \approx 0.3970$$

Using a calculator, we find $\theta = \cos^{-1}(0.3970) \approx 66.61°$.

Also note that if $\mathbf{u}\cdot\mathbf{v} = 0$, then $\cos\theta = 0$ and so the vectors are orthogonal, as with plane vectors. Conversely if $\mathbf{u}\cdot\mathbf{v} \neq 0$, then $\cos\theta \neq 0$ and so the vectors are not orthogonal.

EXAMPLE 8 Determining Orthogonal Vectors

Determine whether the given pair of vectors is orthogonal.

a. $\langle 3, 5, -1\rangle$ and $\langle 7, -5, -4\rangle$ **b.** $4\mathbf{i} - 2\mathbf{j} - 3\mathbf{k}$ and $2\mathbf{i} - 2\mathbf{j} - 4\mathbf{k}$

Solution

a. Finding the dot product:

$$\langle 3, 5, -1\rangle \text{ and } \langle 7, -5, -4\rangle = (3)(7) + (5)(-5) + (-1)(-4) = 21 - 25 + 4 = 0$$

Since the dot product is equal to 0, these vectors are orthogonal.

b. Finding the dot product:

$$(4\mathbf{i} - 2\mathbf{j} - 3\mathbf{k})\cdot(2\mathbf{i} - 2\mathbf{j} - 4\mathbf{k}) = (4)(2) + (-2)(-2) + (-3)(-4) = 8 + 4 + 12 = 24$$

Since the dot product is not equal to 0, these vectors are not orthogonal.

Vector Projection

The formulas and arguments for vector projection extend directly from plane to space vectors, since they don't rely on components or directions of vectors. For convenience we'll restate it here.

Vector Projection of v on w

Let $\mathbf{v}$ and $\mathbf{w}$ be nonzero vectors. The vector projection of $\mathbf{v}$ on the vector $\mathbf{w}$ is given by

$$\text{proj}_{\mathbf{w}}\mathbf{v} = \left(\frac{\mathbf{v}\cdot\mathbf{w}}{||\mathbf{w}||^2}\right)\mathbf{w}$$

EXAMPLE 9 Finding the Vector Projection of v onto w

Let $\mathbf{v} = \langle 1, 6, 4 \rangle$ and $\mathbf{w} = \langle 2, 3, -1 \rangle$. Find $\text{proj}_{\mathbf{w}}\mathbf{v}$ and $\text{proj}_{\mathbf{v}}\mathbf{w}$.

Solution

First we compute the required quantities:

$$\mathbf{v} \cdot \mathbf{w} = \langle 1, 6, 4 \rangle \cdot \langle 2, 3, -1 \rangle = (1)(2) + (6)(3) + (4)(-1) = 2 + 18 - 4 = 16$$

$$||\mathbf{w}||^2 = ||\langle 2, 3, -1 \rangle||^2 = 2^2 + 3^2 + 1^2 = 4 + 9 + 1 = 14$$

$$||\mathbf{v}||^2 = ||\langle 1, 6, 4 \rangle||^2 = 1^2 + 6^2 + 4^2 = 1 + 36 + 16 = 53$$

Now find the projection vectors:

$$\text{proj}_{\mathbf{w}}\mathbf{v} = \left(\frac{\mathbf{v} \cdot \mathbf{w}}{||\mathbf{w}||^2} \right)\mathbf{w} = \frac{16}{14}\langle 2, 3, -1 \rangle = \left\langle \frac{16}{7}, \frac{24}{7}, -\frac{8}{7} \right\rangle$$

$$\text{proj}_{\mathbf{v}}\mathbf{w} = \left(\frac{\mathbf{w} \cdot \mathbf{v}}{||\mathbf{v}||^2} \right)\mathbf{v} = \frac{16}{53}\langle 1, 6, 4 \rangle = \left\langle \frac{16}{53}, \frac{96}{53}, \frac{64}{53} \right\rangle$$

Note that the two projection vectors are not equal.

Vector projection allows us to express a vector as a sum of two orthogonal vectors, as with plane vectors. We'll restate the result here.

> **Orthogonal Decomposition of a Vector**
>
> Let $\mathbf{v}$ and $\mathbf{w}$ be nonzero vectors. Then $\mathbf{v}$ can be written as
>
> $$\mathbf{v} = \mathbf{v}_1 + \mathbf{v}_2$$
>
> where $\mathbf{v}_1 = \text{proj}_{\mathbf{w}}\mathbf{v}$ and $\mathbf{v}_2 = \mathbf{v} - \mathbf{v}_1$. The vectors $\mathbf{v}_1$ and $\mathbf{v}_2$ are orthogonal.

EXAMPLE 10 Decomposing a Vector into Orthogonal Components

Let $\mathbf{v} = \langle 1, 6, 4 \rangle$ and $\mathbf{w} = \langle 2, 3, -1 \rangle$. Decompose $\mathbf{v}$ into two vectors $\mathbf{v}_1$ and $\mathbf{v}_2$, where $\mathbf{v}_1$ is parallel to $\mathbf{w}$ and $\mathbf{v}_2$ is orthogonal to $\mathbf{w}$. Verify that $\mathbf{v}_1$ and $\mathbf{v}_2$ are orthogonal and that $\mathbf{v}_1 + \mathbf{v}_2 = \mathbf{v}$.

Solution

From Example 9 we know that $\mathbf{v}_1 = \text{proj}_{\mathbf{w}}\mathbf{v} = \left\langle \frac{16}{7}, \frac{24}{7}, -\frac{8}{7} \right\rangle$. Therefore:

$$\mathbf{v}_2 = \mathbf{v} - \mathbf{v}_1 = \langle 1, 6, 4 \rangle = \left\langle \frac{16}{7}, \frac{24}{7}, -\frac{8}{7} \right\rangle = \left\langle -\frac{9}{7}, \frac{18}{7}, \frac{36}{7} \right\rangle$$

Let's verify that $\mathbf{v}_1$ and $\mathbf{v}_2$ are orthogonal:

$$\mathbf{v}_1 \cdot \mathbf{v}_2 = \left\langle \frac{16}{7}, \frac{24}{7}, -\frac{8}{7} \right\rangle \cdot \left\langle -\frac{9}{7}, \frac{18}{7}, \frac{36}{7} \right\rangle$$

$$= \left(\frac{16}{7} \right)\left(-\frac{9}{7} \right) + \left(\frac{24}{7} \right)\left(\frac{18}{7} \right) + \left(-\frac{8}{7} \right)\left(\frac{36}{7} \right)$$

$$= -\frac{144}{49} + \frac{432}{49} - \frac{288}{49}$$

$$= 0$$

Since $\mathbf{v}_1 \cdot \mathbf{v}_2 = 0$, the two vectors are orthogonal. Now verifying the sum:

$$\mathbf{v}_1 + \mathbf{v}_2 = \left\langle \frac{16}{7}, \frac{24}{7}, -\frac{8}{7} \right\rangle + \left\langle -\frac{9}{7}, \frac{18}{7}, \frac{36}{7} \right\rangle$$

$$= \frac{16 - 9}{7}, \frac{24 + 18}{7}, \frac{-8 + 36}{7}$$

$$= \langle 1, 6, 4 \rangle$$

$$= \mathbf{v}$$

Direction Angles of a Vector

A space vector has three direction angles corresponding to the angles a vector makes with each of the positive x-, y-, and z-axes. We'll call these angles α, β, and γ. If $\mathbf{v} = \langle v_x, v_y, v_z \rangle$, Figure 4 illustrates these angles in relation to the components of $\mathbf{v}$.

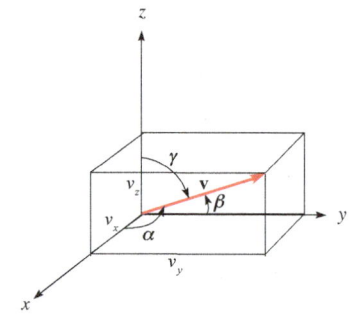

Figure 4

Since $||\mathbf{v}||$ is the length of vector $\mathbf{v}$, using right-triangle trigonometry gives us the following result.

Direction Angles of a Space Vector
If $\mathbf{v} = \langle v_x, v_y, v_z \rangle$, then the **direction angles** α, β, and γ, which $\mathbf{v}$ makes with the positive x-, y-, and z-axes are given by

$$\cos \alpha = \frac{v_x}{||\mathbf{v}||} \qquad \cos \beta = \frac{v_y}{||\mathbf{v}||} \qquad \cos \gamma = \frac{v_z}{||\mathbf{v}||}$$

The quantities $\cos \alpha$, $\cos \beta$, and $\cos \gamma$ are called the **direction cosines** of $\mathbf{v}$.

The direction angles of a vector allow us to have a more geometric interpretation for space vectors. Note that $\langle \cos \alpha, \cos \beta, \cos \gamma \rangle$ is a unit vector in the direction of $\mathbf{v}$, which we can verify:

$$\cos^2 \alpha + \cos^2 \beta + \cos^2 \gamma = \frac{v_1^2}{||\mathbf{v}||^2} + \frac{v_2^2}{||\mathbf{v}||^2} + \frac{v_3^2}{||\mathbf{v}||^2} = \frac{v_1^2 + v_2^2 + v_3^2}{||\mathbf{v}||^2} = \frac{||\mathbf{v}||^2}{||\mathbf{v}||^2} = 1$$

Therefore, any space vector v can be written in terms of its direction angles as

$$\mathbf{v} = ||\mathbf{v}|| \langle \cos \alpha, \cos \beta, \cos \gamma \rangle, \text{ where } \alpha, \beta, \text{ and } \gamma \text{ are the direction angles of } \mathbf{v}$$

EXAMPLE 11   Finding the Direction Angles of a Vector

Let $\mathbf{v} = \langle 1, 6, 4 \rangle$. Find the direction angles of $\mathbf{v}$, then write $\mathbf{v}$ in terms of its direction angles. Round your answers to two decimal places.

Solution

From Example 9 we have $||\mathbf{v}|| = \sqrt{53}$, therefore:

$$\cos \alpha = \frac{v_x}{||\mathbf{v}||} = \frac{1}{\sqrt{53}} \qquad \cos \beta = \frac{v_y}{||\mathbf{v}||} = \frac{6}{\sqrt{53}} \qquad \cos \gamma = \frac{v_z}{||\mathbf{v}||} = \frac{4}{\sqrt{53}}$$

$$\alpha = \cos^{-1}\left(\frac{1}{\sqrt{53}}\right) \approx 82.10° \qquad \beta = \cos^{-1}\left(\frac{6}{\sqrt{53}}\right) \approx 34.50° \qquad \gamma = \cos^{-1}\left(\frac{4}{\sqrt{53}}\right) \approx 56.67°$$

So we can write $\mathbf{v}$ in terms of its direction angles:

$$\mathbf{v} = \langle 1, 6, 4 \rangle = \sqrt{53}\langle \cos 82.10°, \cos 34.50°, \cos 56.67° \rangle$$

In Exercises 1-8, two points P_0 and P_1 are given. Find the distance between the points, and express the vector from P_0 to P_1 in component form.

1. P_0 (1, 2, −1) and P_1 (3, −2, 1)

2. P_0 (−1, −3, 2) and P_1 (2, 3, −4)

3. P_0 (3, 0, −5) and P_1 (2, 2, −6)

4. P_0 (2, −1, −2) and P_1 (1, 2, 3)

5. P_0 (0, 0, 0) and P_1 (2, 3, 4)

6. P_0 (0, 0, 0) and P_1 (−1, −2, −3)

7. P_0 (0, 0, 0) and P_1 (−2, −3, −4)

8. P_0 (0, 0, 0) and P_1 (1, 2, 3)

In Exercises 9-16, a vector is given. Find the magnitude of the vector.

9. $\langle 2, 3, -5 \rangle$

10. $\langle -1, 2, -3 \rangle$

11. $\langle -3, 3, -4 \rangle$

12. $\langle 4, 0, 2 \rangle$

13. $6\mathbf{i} - \mathbf{j} - 2\mathbf{k}$

14. $-2\mathbf{i} + 3\mathbf{j} - \mathbf{k}$

15. $3\mathbf{i} - 4\mathbf{k}$

16. $-5\mathbf{j} + 3\mathbf{k}$

In Exercises 17-24, let $\mathbf{u} = \langle 2, -1, 5 \rangle$, $\mathbf{v} = \langle -1, 4, 2 \rangle$ and $\mathbf{w} = \langle -3, 2, -4 \rangle$. Find the following.

17. $\mathbf{u} + \mathbf{v}$

18. $\mathbf{u} - \mathbf{v}$

19. $4\mathbf{w}$

20. $-3\mathbf{v}$

21. $2\mathbf{u} + 3\mathbf{w}$

22. $3\mathbf{v} - 4\mathbf{w}$

23. $3\mathbf{u} - 2\mathbf{w}$

24. $3\mathbf{u} + 4\mathbf{v}$

In Exercises 25-32, two vectors are given. Find their dot product, then find the smallest nonnegative angle between the two vectors. Round your answer to the nearest tenth of a degree.

25. $\langle 1, 4, -1 \rangle$ and $\langle -1, 1, 2 \rangle$

26. $\langle 3, -2, 5 \rangle$ and $\langle 2, 1, -2 \rangle$

27. $\langle 3, 2, -1 \rangle$ and $\langle 2, 3, 12 \rangle$

28. $\langle 4, 2, 5 \rangle$ and $\langle 3, 4, -4 \rangle$

29. $3\mathbf{i} - \mathbf{j} - \mathbf{k}$ and $2\mathbf{i} + 2\mathbf{j} + 3\mathbf{k}$

30. $\mathbf{i} - 5\mathbf{j} + \mathbf{k}$ and $3\mathbf{i} - 2\mathbf{j} - \mathbf{k}$

31. $2\mathbf{i} - \mathbf{j} + 3\mathbf{k}$ and $-4\mathbf{i} + 2\mathbf{j} - 6\mathbf{k}$

32. $\mathbf{j} - 2\mathbf{j} + \mathbf{k}$ and $3\mathbf{i} - 6\mathbf{j} + 3\mathbf{k}$

In Exercises 33-40, calculate $\text{proj}_{\mathbf{w}}\mathbf{v}$. Then, decompose $\mathbf{v}$ into two vectors $\mathbf{v}_1$ and $\mathbf{v}_2$, where $\mathbf{v}_1$ is parallel to $\mathbf{w}$ and $\mathbf{v}_2$ is orthogonal to $\mathbf{w}$. Then verify that $\mathbf{v}_1$ and $\mathbf{v}_2$ are orthogonal.

33. $\mathbf{v} = \langle 3, 2, 3 \rangle$, $\mathbf{w} = \langle 2, 1, 1 \rangle$

34. $\mathbf{v} = \langle 4, 1, 3 \rangle$, $\mathbf{w} = \langle 5, 2, 2 \rangle$

35. $\mathbf{v} = \langle 1, 2, -4 \rangle$, $\mathbf{w} = \langle -2, 3, -1 \rangle$

36. $\mathbf{v} = \langle 5, -1, -2 \rangle$, $\mathbf{w} = \langle 3, 2, -4 \rangle$

37. $\mathbf{v} = 2\mathbf{i} - \mathbf{j} + 3\mathbf{k}$, $\mathbf{w} = 3\mathbf{i} + \mathbf{j} - 2\mathbf{k}$

38. $\mathbf{v} = 3\mathbf{i} + \mathbf{j} - 2\mathbf{k}$, $\mathbf{w} = 2\mathbf{i} - \mathbf{j} + 3\mathbf{k}$

39. $\mathbf{v} = -2\mathbf{i} + 3\mathbf{j} - \mathbf{k}$, $\mathbf{w} = -\mathbf{i} - \mathbf{j} + 2\mathbf{k}$

40. $\mathbf{v} = -\mathbf{i} - \mathbf{j} + 2\mathbf{k}$, $\mathbf{w} = -2\mathbf{i} + 3\mathbf{j} - \mathbf{k}$

In Exercises 41-48, a vector $\mathbf{v}$ is given. Find the direction angles of $\mathbf{v}$, then write $\mathbf{v}$ in terms of its direction angles. Round your answers to two decimal places.

41. $\mathbf{v} = \langle 3, 2, 3 \rangle$

42. $\mathbf{v} = \langle 4, 1, 3 \rangle$

43. $\mathbf{v} = \langle 1, 2, -4 \rangle$

44. $\mathbf{v} = \langle 5, -1, -2 \rangle$

45. $\mathbf{v} = 2\mathbf{i} - \mathbf{j} + 3\mathbf{k}$

46. $\mathbf{v} = 3\mathbf{i} + \mathbf{j} - 2\mathbf{k}$

47. $\mathbf{v} = -2\mathbf{i} + 3\mathbf{j} - \mathbf{k}$

48. $\mathbf{v} = -\mathbf{i} - \mathbf{j} + 2\mathbf{k}$

A.4 Videos

Determinants

Before we look at our final vector topic, there is a computational tool called the determinant, which will simplify computations for us.

---

2 x 2 and 3 x 3 Determinants

A **2 × 2 determinant**, represented by the notation $\begin{vmatrix} a_1 & a_2 \\ b_1 & b_2 \end{vmatrix}$, can be calculated by the formula

$$\begin{vmatrix} a_1 & a_2 \\ b_1 & b_2 \end{vmatrix} = a_1 b_2 - a_2 b_1$$

A **3 × 3 determinant**, represented by the notation $\begin{vmatrix} x & y & z \\ a_1 & a_2 & a_3 \\ b_1 & b_2 & b_3 \end{vmatrix}$, can be calculated by the formula

$$\begin{vmatrix} x & y & z \\ a_1 & a_2 & a_3 \\ b_1 & b_2 & b_3 \end{vmatrix} = \begin{vmatrix} a_2 & a_3 \\ b_2 & b_3 \end{vmatrix} x - \begin{vmatrix} a_1 & a_3 \\ b_1 & b_3 \end{vmatrix} y + \begin{vmatrix} a_1 & a_2 \\ b_1 & b_2 \end{vmatrix} z$$

As a memory aid, you can think of a 2 × 2 determinant as a forward diagonal product minus a backward diagonal product. For a 3 × 3 determinant, you can use the elements in the top row and "cross out" their corresponding row and column to create the 2 × 2 determinant multiplier. For example, if you "cross out" the row and column containing x, you'll have the 2 × 2 determinant

$$\begin{vmatrix} a_2 & a_3 \\ b_2 & b_3 \end{vmatrix}$$

Likewise, crossing out the row and column containing y results in the 2 × 2 determinant

$$\begin{vmatrix} a_1 & a_3 \\ b_1 & b_3 \end{vmatrix}$$

and similarly for z. This method of calculating 3 × 3 determinants is called a first row expansion.

EXAMPLE 1 Finding 2 × 2 Determinants

Find each of the following determinants.

a. $\begin{vmatrix} -1 & 3 \\ 4 & -5 \end{vmatrix}$

b. $\begin{vmatrix} 6 & 3 \\ 8 & 4 \end{vmatrix}$

Solution

a. Multiplying along the diagonals and subtracting:

$$\begin{vmatrix} -1 & 3 \\ 4 & -5 \end{vmatrix} = (-1)(-5) - (3)(4) = 5 - 12 = -7$$

b. Multiplying along the diagonals and subtracting:

$$\begin{vmatrix} 6 & 3 \\ 8 & 4 \end{vmatrix} = (6)(4) - (3)(8) = 24 - 24 = 0$$

EXAMPLE 2   Finding 3 × 3 Determinants
Find each of the following determinants.

a. $\begin{vmatrix} 1 & -2 & 4 \\ 3 & 1 & -5 \\ -2 & 3 & 2 \end{vmatrix}$ **b.** $\begin{vmatrix} x & y & z \\ 4 & 5 & 6 \\ 7 & 8 & 9 \end{vmatrix}$

Solution
a. Using the first row expansion:

$$\begin{vmatrix} 1 & -2 & 4 \\ 3 & 1 & -5 \\ -2 & 3 & 2 \end{vmatrix} = \begin{vmatrix} 1 & -5 \\ 3 & 2 \end{vmatrix}(1) - \begin{vmatrix} 3 & -5 \\ -2 & 2 \end{vmatrix}(-2) + \begin{vmatrix} 3 & 1 \\ -2 & 3 \end{vmatrix}(4)$$

$$= [(1)(2) - (-5)(3)](1) - [(3)(2) - (-5)(-2)](-2) + [(3)(3) - (1)(-2)](4)$$

$$= (17)(1) - (-4)(-2) + (11)(4)$$

$$= 17 - 8 + 44$$

$$= 53$$

b. Using the first row expansion:

$$\begin{vmatrix} x & y & z \\ 4 & 5 & 6 \\ 7 & 8 & 9 \end{vmatrix} = \begin{vmatrix} 5 & 6 \\ 8 & 9 \end{vmatrix}x - \begin{vmatrix} 4 & 6 \\ 7 & 9 \end{vmatrix}y + \begin{vmatrix} 4 & 5 \\ 7 & 8 \end{vmatrix}z$$

$$= [(5)(9) - (6)(8)]x - [(4)(9) - (6)(7)]y + [(4)(8) - (5)(7)]z$$

$$= -3x + 6y - 3z$$

Cross Products

We are now in a position to define a new vector operation, called the cross product.

Cross Product of Two Space Vectors
Let $\mathbf{u} = \langle u_x, u_y, u_z \rangle$ and $\mathbf{v} = \langle -2, 3, -5 \rangle$. The **cross product** of $\mathbf{u}$ and $\mathbf{v}$ is defined as

$$\mathbf{u} \text{ and } \mathbf{v} = \begin{vmatrix} \mathbf{i} & \mathbf{j} & \mathbf{k} \\ u_x & u_y & u_z \\ v_x & v_y & v_z \end{vmatrix} = \begin{vmatrix} u_y & u_z \\ v_y & v_z \end{vmatrix}\mathbf{i} - \begin{vmatrix} u_x & u_z \\ v_x & v_z \end{vmatrix}\mathbf{j} + \begin{vmatrix} u_x & u_y \\ v_x & v_y \end{vmatrix}\mathbf{k}$$

$$= (u_y v_z - u_z v_y)\mathbf{i} - (u_x v_z - u_z v_x)\mathbf{j} + (u_x v_y - u_y v_x)\mathbf{k}$$

Note that, unlike the dot product, the cross product of two space vectors is a vector, rather than a scalar. Also, note that, although determinant notation is used to find the cross product, the cross product is not a determinant; rather it is a vector.

EXAMPLE 3   Finding Cross Products
Let $\mathbf{u} = \langle 3, 6, -4 \rangle$ and $\mathbf{v} = \langle -2, 3, -5 \rangle$. Find the following cross products.

a. $\mathbf{u} \times \mathbf{v}$ **b.** $\mathbf{v} \times \mathbf{u}$ **c.** $\mathbf{u} \times \mathbf{u}$ **d.** $\mathbf{v} \times \mathbf{v}$

Solution
a. Using the cross-product definition:

$$\mathbf{u} \times \mathbf{v} = \begin{vmatrix} \mathbf{i} & \mathbf{j} & \mathbf{k} \\ 3 & 6 & -4 \\ -2 & 3 & -5 \end{vmatrix} = \begin{vmatrix} 6 & -4 \\ 3 & -5 \end{vmatrix}\mathbf{i} - \begin{vmatrix} 3 & -4 \\ -2 & -5 \end{vmatrix}\mathbf{j} + \begin{vmatrix} 3 & 6 \\ -2 & 3 \end{vmatrix}\mathbf{k}$$

$$= (-30 + 12)\mathbf{i} - (-15 - 8)\mathbf{j} + (9 + 12)\mathbf{k} = 18\mathbf{i} + 23\mathbf{j} + 21\mathbf{k} = \langle -18, 23, 21 \rangle$$

b. Using the cross-product definition:

$$\mathbf{v} \times \mathbf{u} = \begin{vmatrix} \mathbf{i} & \mathbf{j} & \mathbf{k} \\ -2 & 3 & -5 \\ 3 & 6 & -4 \end{vmatrix}$$

$$= \begin{vmatrix} 3 & -5 \\ 6 & -4 \end{vmatrix}\mathbf{i} - \begin{vmatrix} -2 & -5 \\ 3 & -4 \end{vmatrix}\mathbf{j} + \begin{vmatrix} -2 & 3 \\ 3 & 6 \end{vmatrix}\mathbf{k}$$

$$= (-12 + 30)\mathbf{i} - (8 + 15)\mathbf{j} + (-12 - 9)\mathbf{k}$$

$$= 18\mathbf{i} - 23\mathbf{j} - 21\mathbf{k}$$

$$= \langle 18, -23, -21 \rangle$$

c. Using the cross-product definition:

$$\mathbf{u} \times \mathbf{u} = \begin{vmatrix} \mathbf{i} & \mathbf{j} & \mathbf{k} \\ 3 & 6 & -4 \\ 3 & 6 & -4 \end{vmatrix}$$

$$= \begin{vmatrix} 6 & -4 \\ 6 & -4 \end{vmatrix}\mathbf{i} - \begin{vmatrix} 3 & -4 \\ 3 & -4 \end{vmatrix}\mathbf{j} + \begin{vmatrix} 3 & 6 \\ 3 & 6 \end{vmatrix}\mathbf{k}$$

$$= (-24 + 24)\mathbf{i} - (-12 + 12)\mathbf{j} + (18 - 18)\mathbf{k}$$

$$= 0\mathbf{i} + 0\mathbf{j} + 0\mathbf{k}$$

$$= \langle 0, 0, 0 \rangle$$

d. Using the cross-product definition:

$$\mathbf{v} \times \mathbf{v} = \begin{vmatrix} \mathbf{i} & \mathbf{j} & \mathbf{k} \\ -2 & 3 & -5 \\ -2 & 3 & -5 \end{vmatrix}$$

$$= \begin{vmatrix} 3 & -5 \\ 3 & -5 \end{vmatrix}\mathbf{i} - \begin{vmatrix} -2 & -5 \\ -2 & -5 \end{vmatrix}\mathbf{j} + \begin{vmatrix} -2 & 3 \\ -2 & 3 \end{vmatrix}\mathbf{k}$$

$$= (-15 + 15)\mathbf{i} - (-10 - 10)\mathbf{j} + (-6 + 6)\mathbf{k}$$

$$= 0\mathbf{i} + 0\mathbf{j} + 0\mathbf{k}$$

$$= \langle 0, 0, 0 \rangle$$

Properties of the Cross Product

The previous example illustrates a couple of properties of the cross product, namely that $\mathbf{u} \times \mathbf{v} = -(\mathbf{v} \times \mathbf{u})$ and $\mathbf{u} \times \mathbf{u} = 0$. We will illustrate another property with a follow-up example.

EXAMPLE 4 Orthogonal Cross Products

Let $\mathbf{u} = \langle 3, 6, -4 \rangle$ and $\mathbf{v} = \langle -2, 3, -5 \rangle$. Find the following dot products.

a. $\mathbf{u} \cdot (\mathbf{u} \times \mathbf{v})$ **b.** $\mathbf{v} \cdot (\mathbf{u} \times \mathbf{v})$

Solution

a. From Example 3a, we have $\mathbf{u} \times \mathbf{v} = \langle -18, 23, 21 \rangle$. Now finding the dot product:

$$\mathbf{u} \cdot (\mathbf{u} \times \mathbf{v}) = \langle 3, 6, -4 \rangle \cdot \langle -18, 23, 21 \rangle$$

$$= (3)(-18) + (6)(23) + (4)(21)$$

$$= -54 + 138 - 84$$

$$= 0$$

b. From Example 3a, we have $\mathbf{u} \times \mathbf{v} = \langle -18, 23, 21 \rangle$. Now finding the dot product:

$$\mathbf{v} \cdot (\mathbf{u} \times \mathbf{v}) = \langle -2, 3, -5 \rangle \cdot \langle -18, 23, 21 \rangle$$
$$= (-2)(-18) + (3)(23) + (-5)(21)$$
$$= 36 + 69 - 105$$
$$= 0$$

This example indicates an important property of the cross product, namely that the cross product of two space vectors is a vector orthogonal to the two vectors.

> **Property: Orthogonality of the Cross Product**
> Let $\mathbf{u}$ and $\mathbf{v}$ be two space vectors. Then $\mathbf{u} \times \mathbf{v}$ is orthogonal to both $\mathbf{u}$ and $\mathbf{v}$.

Proof

Let $\mathbf{u} = \langle u_x, u_y, u_z \rangle$ and $\mathbf{v} = \langle v_x, v_y, v_z \rangle$. Using the cross product definition:

$$\mathbf{u} \times \mathbf{v} = \begin{vmatrix} \mathbf{i} & \mathbf{j} & \mathbf{k} \\ u_x & u_y & u_z \\ v_x & v_y & v_z \end{vmatrix} = \begin{vmatrix} u_y & u_z \\ v_y & v_z \end{vmatrix} \mathbf{i} - \begin{vmatrix} u_x & u_z \\ v_x & v_z \end{vmatrix} \mathbf{j} + \begin{vmatrix} u_x & u_y \\ v_x & v_y \end{vmatrix} \mathbf{k}$$

$$= (u_y v_z - u_z v_y)\mathbf{i} - (u_x v_z - u_z v_x)\mathbf{j} + (u_x v_y - u_y v_x)\mathbf{k}$$

Now find the two dot products:

$$\mathbf{u} \cdot (\mathbf{u} \times \mathbf{v}) = u_x(u_y v_z - u_z v_y) - u_y(u_x v_z - u_z v_x) + u_z(u_x v_y - u_y v_x)$$
$$= u_x u_y v_z - u_x u_z v_y - u_x u_y v_z + u_y u_z v_x + u_x u_z v_y - u_y u_z v_x$$
$$= 0$$

$$\mathbf{v} \cdot (\mathbf{u} \times \mathbf{v}) = v_x(u_y v_z - u_z v_y) - v_y(u_x v_z - u_z v_x) + v_z(u_x v_y - u_y v_x)$$
$$= u_y v_x v_z - u_z v_x v_y - u_x v_y v_z + u_z v_x v_y + u_x v_y v_z - u_y v_x v_z$$
$$= 0$$

Note: The vector $(\mathbf{v} \times \mathbf{u}) = -(\mathbf{u} \times \mathbf{v})$ is another orthogonal vector to both $\mathbf{u}$ and $\mathbf{v}$.

EXAMPLE 5 Finding a Vector Orthogonal to Two Space Vectors

Let $\mathbf{u} = 3\mathbf{i} - 5\mathbf{j} - \mathbf{k}$ and $\mathbf{v} = 4\mathbf{i} + \mathbf{j} - 6\mathbf{k}$. Find two vectors orthogonal to both $\mathbf{u}$ and $\mathbf{v}$.

Solution

We have to find the cross product:

$$\mathbf{u} \times \mathbf{v} = \begin{vmatrix} \mathbf{i} & \mathbf{j} & \mathbf{k} \\ 3 & -5 & -1 \\ 4 & 1 & -6 \end{vmatrix}$$

$$= \begin{vmatrix} -5 & -1 \\ 1 & -6 \end{vmatrix} \mathbf{i} - \begin{vmatrix} 3 & -1 \\ 4 & -6 \end{vmatrix} \mathbf{j} + \begin{vmatrix} 3 & -5 \\ 4 & 1 \end{vmatrix} \mathbf{k}$$

$$= (30 + 1)\mathbf{i} - (-18 + 4)\mathbf{j} + (3 + 20)\mathbf{k}$$

$$= 31\mathbf{i} + 14\mathbf{j} + 23\mathbf{k}$$

This vector is orthogonal to both **u** and **v**. A second orthogonal vector is

$$\mathbf{v} \times \mathbf{u} = -31\mathbf{i} - 14\mathbf{j} - 23\mathbf{k}$$

We can visualize the vectors $\mathbf{u} \times \mathbf{v}$ and $\mathbf{v} \times \mathbf{u}$ through the following figure. Note that the cross product follows the right-hand rule. Imagine placing your right hand along u, then curling your fingers towards v. Your thumb will point in the direction of **v**. Note that $\mathbf{v} \times \mathbf{u} = -(\mathbf{u} \times \mathbf{v})$ is in exactly the opposite direction.

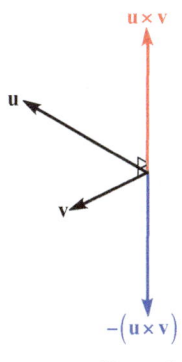

Figure 1

The Cross Product and Area

We now introduce a theorem that connects the magnitude of the cross product with the area of the parallelogram formed by the two vectors.

> **Theorem**
> If **u** and **v** are two nonparallel space vectors, then $||\mathbf{u} \times \mathbf{v}|| = ||\mathbf{u}|| \, ||\mathbf{v}|| \sin \theta$, where θ is the smallest positive angle between them. This quantity, $||\mathbf{u} \times \mathbf{v}||$, is the area of the parallelogram formed by vectors **u** and **v**.

Proof
Let $\mathbf{u} = \langle u_x, u_y, u_z \rangle$ and $\mathbf{u} = \langle v_x, v_y, v_z \rangle$. Then:

$$||\mathbf{u} \times \mathbf{v}|| = \sqrt{(u_y v_z - u_z v_y)^2 + (u_x v_z - u_z v_x)^2 + (u_x v_y - u_y v_x)^2}$$

$$= \sqrt{(u_x^2 + u_y^2 + u_z^2)(v_x^2 + v_y^2 + v_z^2) - (u_x v_x - u_y v_y - u_z v_z)^2}$$

$$= \sqrt{||\mathbf{u}||^2 \, ||\mathbf{v}||^2 - (\mathbf{u} \cdot \mathbf{v})^2}$$

$$= \sqrt{||\mathbf{u}||^2 \, ||\mathbf{v}||^2 - ||\mathbf{u}||^2 \, ||\mathbf{v}||^2 \cos^2 \theta}$$

$$= \sqrt{||\mathbf{u}||^2 \, ||\mathbf{v}||^2 (1 - \cos^2 \theta)}$$

$$= \sqrt{||\mathbf{u}||^2 \, ||\mathbf{v}||^2 \sin^2 \theta}$$

$$= ||\mathbf{u}||^2 \, ||\mathbf{v}||^2 \sin \theta$$

Note that the second step requires a great deal of algebra, which you are welcome to verify (if you are brave).

For the second statement in the theorem, note the following figure:

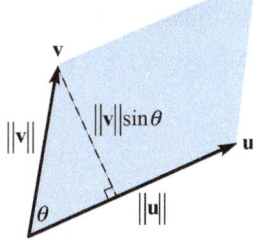

Figure 2

Since the base of the parallelogram is $||\mathbf{u}||$ and the height of the parallelogram is $||\mathbf{v}||\sin\theta$, then its area (base × height) is $||\mathbf{u}||\,||\mathbf{v}||\sin\theta = ||\mathbf{u}\times\mathbf{v}||$. We can apply this most recent result in a couple of examples.

EXAMPLE 6   Finding the Area of a Parallelogram

Find the area of the parallelogram with vertices $A(0, 0, 0)$, $B(2, 3, 1)$, $C(1, 1, 2)$, and $D(3, 4, 3)$.

Solution

The vectors $\vec{AB} = \langle 2, 3, 1\rangle$ and $\vec{AC} = \langle 1, 1, 2\rangle$ form two adjacent (not parallel) sides. Now find their cross product:

$$\vec{AB}\times\vec{AC} = \begin{vmatrix} \mathbf{i} & \mathbf{j} & \mathbf{k} \\ 2 & 3 & 1 \\ 1 & 1 & 2 \end{vmatrix} = \begin{vmatrix} 3 & 1 \\ 1 & 2 \end{vmatrix}\mathbf{i} - \begin{vmatrix} 2 & 1 \\ 1 & 2 \end{vmatrix}\mathbf{j} + \begin{vmatrix} 2 & 3 \\ 1 & 1 \end{vmatrix}\mathbf{k}$$

$$= (6-1)\mathbf{i} - (4-1)\mathbf{j} + (2-3)\mathbf{k}$$

$$= \langle 5, -3, -1\rangle$$

Therefore the area of the parallelogram is:

$$||\langle 5, -3, -1\rangle|| = \sqrt{5^2 + (-3)^2 + (-1)^2}$$

$$= \sqrt{25 + 9 + 1}$$

$$= \sqrt{35}$$

EXAMPLE 7   Finding the Area of a Triangle

Find the area of the triangle with vertices $A(1, 3, 1)$, $B(2, -4, -2)$, and $C(0, 5, 6)$.

Solution

The vectors $\vec{AB} = \langle 1, -7, -3\rangle$ and $\vec{AC} = \langle -1, 2, 5\rangle$ form two adjacent (not parallel) sides of a parallelogram. Now find their cross product:

$$\vec{AB}\times\vec{AC} = \begin{vmatrix} \mathbf{i} & \mathbf{j} & \mathbf{k} \\ 1 & -7 & -3 \\ -1 & 2 & 5 \end{vmatrix} = \begin{vmatrix} -7 & -3 \\ 2 & 5 \end{vmatrix}\mathbf{i} - \begin{vmatrix} 1 & -3 \\ -1 & 5 \end{vmatrix}\mathbf{j} + \begin{vmatrix} 1 & -7 \\ -1 & 2 \end{vmatrix}\mathbf{k}$$

$$= (-35+6)\mathbf{i} - (5-3)\mathbf{j} + (2-7)\mathbf{k}$$

$$= \langle -29, -2, -5\rangle$$

Since the triangle area is half of the parallelogram area, the area of the triangle is

$$\frac{1}{2}||\langle -29, -2, -5\rangle|| = \frac{1}{2}\sqrt{(-29)^2 + (-2)^2 + (-5)^2} = \frac{1}{2}\sqrt{870} \approx 14.75$$

Scalar Triple Products and Volume

The last topic of this section connects the dot and cross products with the volume of the parallelepiped formed by three vectors $\mathbf{u}$, $\mathbf{v}$, and $\mathbf{w}$. First we'll define the **scalar triple product**.

Scalar Triple Product of Three Space Vectors

Let $\mathbf{u} = \langle u_x, u_y, u_z\rangle$, $\mathbf{v} = \langle v_x, v_y, v_z\rangle$, and $\mathbf{w} = \langle w_x, w_y, w_z\rangle$. The **scalar triple product** of $\mathbf{u}$, $\mathbf{v}$, and $\mathbf{w}$ is defined as

$$\mathbf{u}\cdot(\mathbf{v}\times\mathbf{w}) = \begin{vmatrix} u_x & u_y & u_z \\ v_x & v_y & v_z \\ w_x & w_y & w_z \end{vmatrix}$$

Note that the scalar triple product is simply the 3×3 determinant with the vector components forming the rows of the determinant. This is fairly easy to verify, since

$$\mathbf{v} \times \mathbf{w} = \begin{vmatrix} \mathbf{i} & \mathbf{j} & \mathbf{k} \\ v_x & v_y & v_z \\ w_x & w_y & w_z \end{vmatrix} = \begin{vmatrix} v_y & v_z \\ w_y & w_z \end{vmatrix} \mathbf{i} - \begin{vmatrix} v_x & v_z \\ w_x & w_z \end{vmatrix} \mathbf{j} + \begin{vmatrix} v_x & v_y \\ w_x & w_y \end{vmatrix} \mathbf{k}$$

Therefore

$$\mathbf{u} \cdot (\mathbf{v} \times \mathbf{w}) = \begin{vmatrix} v_y & v_z \\ w_y & w_z \end{vmatrix} u_x - \begin{vmatrix} v_x & v_z \\ w_x & w_z \end{vmatrix} u_y + \begin{vmatrix} v_x & v_y \\ w_x & w_y \end{vmatrix} u_z = \begin{vmatrix} u_x & u_y & u_z \\ v_x & v_y & v_z \\ w_x & w_y & w_z \end{vmatrix}$$

Also note that the scalar triple product can be either positive or negative, depending on whether the angle between $\mathbf{u}$ and $\mathbf{v} \times \mathbf{w}$ is less than 90° or greater than 90°.

EXAMPLE 8   Finding the Scalar Triple Product

Let $\mathbf{u} = 3\mathbf{i} + \mathbf{j} + 2\mathbf{k}$, $\mathbf{v} = 2\mathbf{i} - \mathbf{j} - 4\mathbf{k}$, and $\mathbf{w} = \mathbf{i} + 3\mathbf{j} - \mathbf{k}$. Find $\mathbf{u} \cdot (\mathbf{v} \times \mathbf{w})$.

Solution

Following the definition of the scalar triple product:

$$\mathbf{u} \cdot (\mathbf{v} \times \mathbf{w}) = \begin{vmatrix} 3 & 1 & 2 \\ 2 & -1 & -4 \\ 1 & 3 & -1 \end{vmatrix}$$

$$= \begin{vmatrix} -1 & -4 \\ 3 & -1 \end{vmatrix}(3) - \begin{vmatrix} 2 & -4 \\ 1 & -1 \end{vmatrix}(1) + \begin{vmatrix} 2 & -1 \\ 1 & 3 \end{vmatrix}(2)$$

$$= (1 + 12)(3) - (-2 + 4)(1) + (6 + 1)(2)$$

$$= 39 - 2 + 14$$

$$= 51$$

The connection between the scalar triple product and volume is found in the following theorem.

Theorem

Let $\mathbf{u}$, $\mathbf{v}$ and $\mathbf{w}$ be three nonzero space vectors. The parallelepiped formed by $\mathbf{u}$, $\mathbf{v}$ and $\mathbf{w}$ as adjacent edges has a volume V given by $V = |\mathbf{u} \cdot (\mathbf{v} \times \mathbf{w})|$, where $V = 0$ indicates that $\mathbf{u}$, $\mathbf{v}$, and $\mathbf{w}$ lie in the same plane.

Proof

We'll draw a parallelepiped with the base formed by $\mathbf{v}$ and $\mathbf{w}$ and slanted height formed by u:

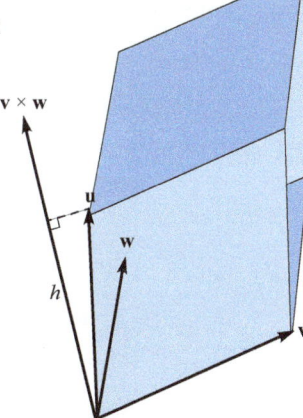

Figure 3

Note that the base area is a parallelogram, which we know has an area of $A = ||\mathbf{v} \times \mathbf{w}||$. Now note that the height of the parallelepiped is the length of the projection vector of $\mathbf{u}$ onto $\mathbf{v} \times \mathbf{w}$, which we know from the previous section is

$$\text{proj}_{\mathbf{v} \times \mathbf{w}} \mathbf{u} = \left(\frac{\mathbf{u} \cdot (\mathbf{v} \times \mathbf{w})}{||\mathbf{v} \times \mathbf{w}||^2} \right) \mathbf{v} \times \mathbf{w}$$

Therefore the height h of the parallelepiped is given by

$$h = \left|\left| \text{proj}_{\mathbf{v} \times \mathbf{w}} \mathbf{u} \right|\right| = \left(\frac{|\mathbf{u} \cdot (\mathbf{v} \times \mathbf{w})|}{||\mathbf{v} \times \mathbf{w}||^2} \right) ||\mathbf{v} \times \mathbf{w}|| = \frac{|\mathbf{u} \cdot (\mathbf{v} \times \mathbf{w})|}{||\mathbf{v} \times \mathbf{w}||}$$

The absolute value is necessary since $\mathbf{u} \cdot (\mathbf{v} \times \mathbf{w})$ could be negative. Since the volume is the base area multiplied by the height:

$$V = Ah = ||\mathbf{v} \times \mathbf{w}|| \frac{|\mathbf{u} \cdot (\mathbf{v} \times \mathbf{w})|}{||\mathbf{v} \times \mathbf{w}||} = |\mathbf{u} \cdot (\mathbf{v} \times \mathbf{w})|$$

EXAMPLE 9   Finding the Volume of a Parallelepiped

Find the volume of the parallelepiped formed by the vectors $\mathbf{u} = \langle 1, 2, 4 \rangle$, $\mathbf{v} = \langle -2, 2, 1 \rangle$, and $\mathbf{w} = \langle 3, 1, -3 \rangle$.

Solution

We find the scalar triple product:

$$\mathbf{u} \cdot (\mathbf{v} \times \mathbf{w}) = \begin{vmatrix} 1 & 2 & 4 \\ -2 & 2 & 1 \\ 3 & 1 & -3 \end{vmatrix}$$

$$= \begin{vmatrix} 2 & 1 \\ 1 & -3 \end{vmatrix}(1) - \begin{vmatrix} -2 & 1 \\ 3 & -3 \end{vmatrix}(2) + \begin{vmatrix} -2 & 2 \\ 3 & 1 \end{vmatrix}(4)$$

$$= (-6 - 1)(1) - (6 - 3)(2) + (-2 - 6)(4)$$

$$= -7 - 6 - 32$$

$$= -45$$

Therefore the volume is: $V = |\mathbf{u} \cdot (\mathbf{v} \times \mathbf{w})| = |-45| = 45$

Problem Set A.4

In Exercises 1-10, find the determinant.

1. $\begin{vmatrix} -3 & 2 \\ 2 & 4 \end{vmatrix}$

2. $\begin{vmatrix} 4 & -2 \\ 2 & -3 \end{vmatrix}$

3. $\begin{vmatrix} 4 & -5 \\ -1 & -3 \end{vmatrix}$

4. $\begin{vmatrix} -5 & 1 \\ 3 & -2 \end{vmatrix}$

5. $\begin{vmatrix} 4 & 2 & 1 \\ 3 & -1 & -2 \\ 1 & -3 & 5 \end{vmatrix}$

6. $\begin{vmatrix} 3 & 5 & -4 \\ -2 & 3 & 4 \\ -3 & 1 & 3 \end{vmatrix}$

7. $\begin{vmatrix} 1 & 2 & 3 \\ 4 & 5 & 6 \\ 7 & 8 & 9 \end{vmatrix}$

8. $\begin{vmatrix} 2 & 1 & 3 \\ 3 & -2 & 5 \\ 2 & 1 & 3 \end{vmatrix}$

9. $\begin{vmatrix} x & y & z \\ 2 & 3 & -1 \\ 3 & 4 & 2 \end{vmatrix}$

10. $\begin{vmatrix} x & y & z \\ 4 & -2 & 3 \\ 1 & 3 & -2 \end{vmatrix}$

In Exercises 11-30, use vectors **u**, **v**, *and* **w** *to find the given expression.*

$$\mathbf{u} = \langle 1, -2, 3 \rangle \qquad \mathbf{v} = \langle -2, 2, 1 \rangle \qquad \mathbf{w} = \langle 3, 2, -1 \rangle$$

11. $\mathbf{v} \times \mathbf{w}$

12. $\mathbf{u} \times \mathbf{w}$

13. $\mathbf{w} \times \mathbf{u}$

14. $\mathbf{w} \times \mathbf{v}$

15. $\mathbf{u} \times \mathbf{v}$

16. $\mathbf{v} \times \mathbf{u}$

17. $\mathbf{u} \times (\mathbf{v} \times \mathbf{w})$

18. $\mathbf{w} \times (\mathbf{u} \times \mathbf{v})$

19. $\mathbf{v} \times (\mathbf{u} \times \mathbf{w})$

20. $\mathbf{u} \times (\mathbf{u} \times \mathbf{w})$

21. $\mathbf{v} \times (\mathbf{v} \times \mathbf{w})$

22. $\mathbf{w} \times (\mathbf{u} \times \mathbf{w})$

23. $\mathbf{u} \cdot (\mathbf{v} \times \mathbf{w})$

24. $\mathbf{w} \cdot (\mathbf{u} \times \mathbf{v})$

25. $\mathbf{v} \cdot (\mathbf{w} \times \mathbf{u})$

26. $\mathbf{u} \cdot (\mathbf{w} \times \mathbf{v})$

27. $\mathbf{v} \cdot (\mathbf{v} \times \mathbf{w})$

28. $\mathbf{w} \cdot (\mathbf{u} \times \mathbf{w})$

29. $\mathbf{w} \cdot (\mathbf{w} \times \mathbf{v})$

30. $\mathbf{u} \cdot (\mathbf{v} \times \mathbf{u})$

In Exercises 31-36, use vectors **u**, **v**, *and* **w** *to find the required vectors.*

$$\mathbf{u} = 3\mathbf{i} + 2\mathbf{j} - \mathbf{k} \qquad \mathbf{v} = -\mathbf{i} + 3\mathbf{j} + 2\mathbf{k} \qquad \mathbf{w} = 2\mathbf{i} - 2\mathbf{j} + 3\mathbf{k}$$

31. Two vectors orthogonal to both **u** and **v**.

32. Two vectors orthogonal to both **u** and **w**.

33. Two vectors orthogonal to both **v** and **w**.

34. Two unit vectors orthogonal to both **w** and **u**.

35. Two vectors orthogonal to both **v** and **u**.

36. Two vectors orthogonal to both **w** and **v**.

In Exercises 37-40, find the area of the parallelogram with vertices A, B, C, and D.

37. $A(1, 2, 1)$, $B(3, 4, 0)$, $C(2, 5, 2)$, $D(4, 7,1)$

38. $A(-1, 1, 2)$, $B(2, 4, 5)$, $C(-2, -3, 1)$, $D(1, 0, 4)$

39. $A(-2, 3 ,2)$, $B(-4, 1, 5)$, $C(2, 0, 4)$, $D(0, -2, 7)$

40. $A(0, -3, -1)$, $B(2, -5, -3)$, $C(3, 1, -4)$, $D(5, -1, -6)$

In Exercises 41-44, find the area of the triangle with vertices A, B, and C. Give the exact answer, then round your answer to two decimal places.

41. $A(2, 4, 1)$, $B(3, 6, 2)$, $C(1, 1, 3)$

42. $A(5, 2, -1)$, $B(8, 4, -3)$, $C(1, 1, -3)$

43. $A(-4, -2, -5)$, $B(-1, 1, 2)$, $C(1, 1, 0)$

44. $A(-5, -6, -2)$, $B(-2, -1, 1)$, $C(2, -1, 3)$

In Exercises 45-50, find the volume of the parallelepiped formed by vectors **u**, **v**, *and* **w**.

45. $\mathbf{u} = \langle 3, 1, 2 \rangle, \mathbf{v} = \langle 4, 5, -1 \rangle, \mathbf{w} = \langle 1, -2, 3 \rangle$

46. $\mathbf{u} = \langle 2, 4, 5 \rangle, \mathbf{v} = \langle -1, 1, -1 \rangle, \mathbf{w} = \langle -3, 2, 2 \rangle$

47. $\mathbf{u} = \langle -2, 6, -3 \rangle, \mathbf{v} = \langle -5, 1, 1 \rangle, \mathbf{w} = \langle 2, -3, -4 \rangle$

48. $\mathbf{u} = \langle -3, 7, 4 \rangle, \mathbf{v} = \langle -5, 2, -2 \rangle, \mathbf{w} = \langle -1, -2, -3 \rangle$

49. $\mathbf{u} = \langle 0, 7, -5 \rangle, \mathbf{v} = \langle 3, -4, -6 \rangle, \mathbf{w} = \langle -2, 3, 6 \rangle$

50. $\mathbf{u} = \langle 6, -6, 0 \rangle, \mathbf{v} = \langle -5, 7, -4 \rangle, \mathbf{w} = \langle -4, 3, 9 \rangle$

Chapter 1

PROBLEM SET 1.1

Vocabulary

1. acute **2.** supplement **3.** hypotenuse
4. twice **5.** equal

Problems

1. Acute; complement = 80°, supplement = 170°

3. Acute; complement = 45°, supplement = 135°

5. Obtuse; complement = −30°, supplement = 60°

7. Neither; complement = 90° − x, supplement = 180° − x

9. 60° **11.** 45° **13.** 50° **15.** $\sqrt{41}$ **17.** 45° **19.** 120°

21. 180° **23.** 5 **25.** 24 **27.** 5

29. 2 (Note that this must be a 30°-60°-90° triangle.)

31. $3\sqrt{2}$ (Note that this must be a 45°-45°-90° triangle.)

33. 4 (This is another 30°-60°-90° triangle.) **35.** x = 1

37. 25 feet **39.** $\sqrt{3}, 2$ **41.** $4, 4\sqrt{3}$ **43.** $\frac{1}{3}, \frac{\sqrt{3}}{3}$

45. 3,6 **47.** $2\sqrt{3}, 4\sqrt{3}$ **49.** $\sqrt{2}$ **51.** $\frac{4\sqrt{2}}{5}$

53. 8, 8 **55.** $2\sqrt{2}, 2\sqrt{2}$

57. a. $35\sqrt{3} = 61$ feet **b.** $\sqrt{9,300} = 96$ feet

59. a. 65.8 feet **b.** 31.6 feet **61.** $1,000\sqrt{2}$ feet

63.

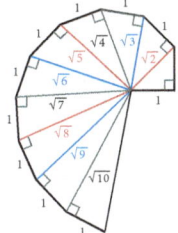

65. a. $\frac{5}{2}$ **b.** 5 **c.** 2 **d.**

e. $y = -\frac{2}{5}x + 2$

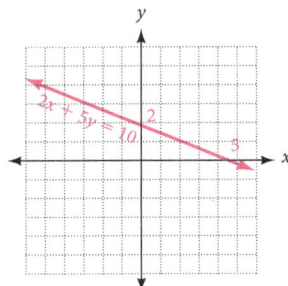

PROBLEM SET 1.2

Vocabulary

1. x-axis **2.** 2 **3.** vertex **4.** radius **5.** standard

Problems

1.

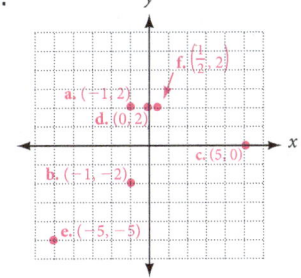

3. A. (4, 1) **B.** (−4, 3) **C.** (−2, −5) **D.** (2, −2) **E.** (0, 5)
F. (−4, 0) **G.** (1, 0) **5. b** **7. b**

9. **11.**

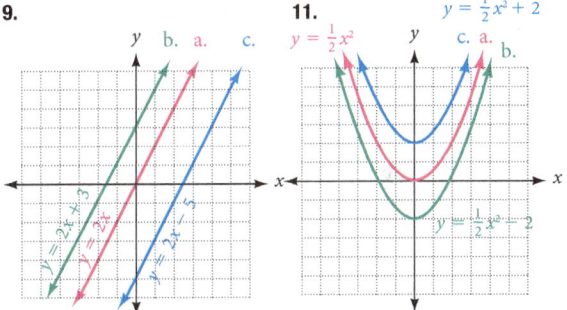

13. **15.** center = (0, 0); radius = 2

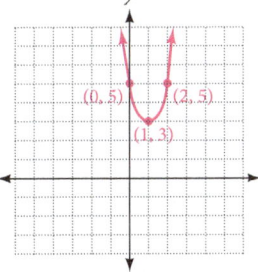

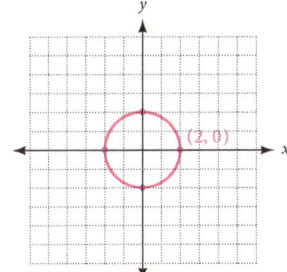

17. center = (1, 3); radius = 5 **19.** center = (−2, 4); radius = $2\sqrt{2}$

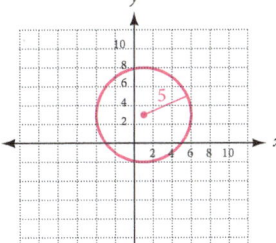

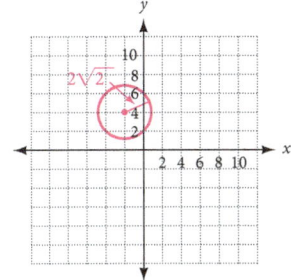

21. A. $\left(x - \frac{1}{2}\right)^2 + (y - 1)^2 = \frac{1}{4}$

B. $(x - 1)^2 + (y - 1)^2 = 1$

C. $(x - 2)^2 + (y - 1)^2 = 4$

23. 5 **25.** $\sqrt{106}$ **27.** $\sqrt{61}$ **29.** $\sqrt{130}$ **31.** 3 or −1

33. 0 or 6 **35.** 6,864 ft

37. home: $(0, 0)$, first: $(60, 0)$, second: $(60, 60)$, third: $(0, 60)$

39.

Point on the Wheel	Time in Minutes	Height in Feet
A	0	12
B	4	132
C	8	252
D	12	132
A	16	12

41. $(-1, 1)$, $-225°$

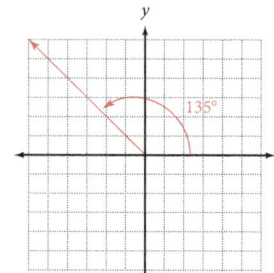

43. $(-1, -1)$, $-135°$

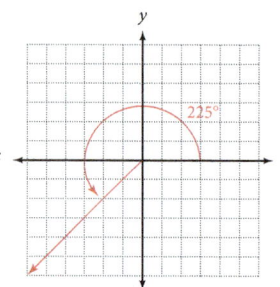

45. $(0, 1)$, $-270°$

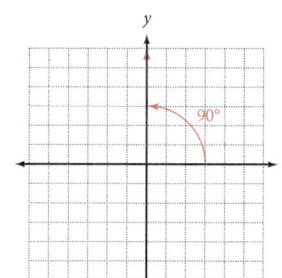

47. $(1, -1)$, $315°$

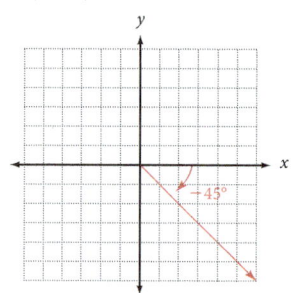

49. **51.**

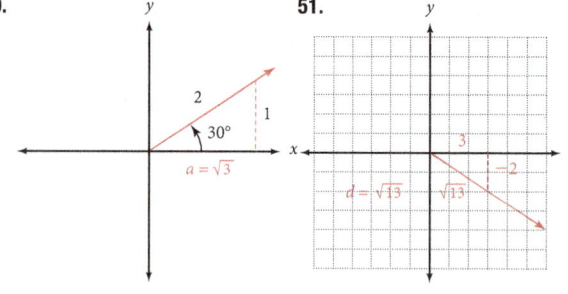

53.

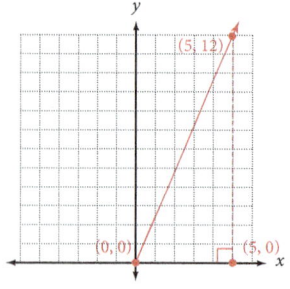

55. -1 **57.** undefined   **59.** -1 **61.** $\frac{3\sqrt{13}}{13}$ **63.** $\sqrt{2}$

65. ± 12 **67.** $\frac{1}{2}$ **69.** $\frac{3}{4}$ **71.** $\frac{b}{a}$

PROBLEM SET 1.3

Vocabulary

1. standard position   **2.** reciprocals   **3.** Quadrant I
4. rationalized

Problems

1. $\sin\theta = \frac{4}{5}$, $\cos\theta = \frac{3}{5}$, $\tan\theta = \frac{4}{3}$, $\cot\theta = \frac{3}{4}$, $\sec\theta = \frac{5}{3}$, $\csc\theta = \frac{5}{4}$

3. $\sin\theta = \frac{12}{13}$, $\cos\theta = -\frac{5}{13}$, $\tan\theta = -\frac{12}{5}$, $\cot\theta = -\frac{5}{12}$, $\sec\theta = -\frac{13}{5}$, $\csc\theta = \frac{13}{12}$

5. $\sin\theta = -\frac{3}{\sqrt{10}}$, $\cos\theta = -\frac{1}{\sqrt{10}}$, $\tan\theta = 3$, $\cot\theta = \frac{1}{3}$, $\sec\theta = -\sqrt{10}$, $\csc\theta = -\frac{\sqrt{10}}{3}$

7. $\sin\theta = \frac{b}{\sqrt{a^2 + b^2}}$, $\cos\theta = \frac{a}{\sqrt{a^2 + b^2}}$, $\tan\theta = \frac{b}{a}$, $\cot\theta = \frac{a}{b}$, $\sec\theta = \frac{\sqrt{a^2 + b^2}}{a}$, $\csc\theta = \frac{\sqrt{a^2 + b^2}}{b}$

9. $\sin 180° = 0$, $\cos 180° = -1$, $\tan 180° = 0$

11. $\sin 135° = \frac{1}{\sqrt{2}}$, $\cos 135° = -\frac{1}{\sqrt{2}}$, $\tan 135° = -1$

13. $\sin(-45°) = -\frac{1}{\sqrt{2}}$, $\cos(-45°) = \frac{1}{\sqrt{2}}$, $\tan(-45°) = -1$

15. $\sin 0° = 0$, $\cos 0° = 1$, $\tan 0° = 0$

17. I, IV   **19.** III, IV   **21.** III   **23.** I, IV

25. $\cos\theta = \frac{5}{13}$, $\tan\theta = \frac{12}{5}$, $\cot\theta = \frac{5}{12}$, $\sec\theta = \frac{13}{5}$, $\csc\theta = \frac{13}{12}$

27. $\sin\theta = \frac{\sqrt{3}}{2}$, $\tan\theta = -\sqrt{3}$, $\cot\theta = -\frac{1}{\sqrt{3}}$, $\sec\theta = -2$, $\csc\theta = \frac{2}{\sqrt{3}}$

29. $\sin\theta = -\frac{4}{5}$, $\tan\theta = \frac{4}{3}$, $\cot\theta = \frac{3}{4}$, $\sec\theta = -\frac{5}{3}$, $\csc\theta = -\frac{5}{4}$

31. $\sin\theta = -\frac{4}{5}$, $\cos\theta = \frac{3}{5}$, $\sec\theta = \frac{5}{3}$, $\csc\theta = -\frac{5}{4}$, $\cot\theta = -\frac{3}{4}$

33. $\sin\theta = -\frac{3}{5}$, $\cos\theta = \frac{4}{5}$, $\tan\theta = -\frac{3}{4}$, $\csc\theta = -\frac{5}{3}$, $\cot\theta = -\frac{4}{3}$

35. $\sin\theta = \frac{2}{\sqrt{5}}$, $\cos\theta = -\frac{1}{\sqrt{5}}$, $\tan\theta = -2$, $\sec\theta = -\sqrt{5}$, $\csc\theta = \frac{\sqrt{5}}{2}$

37. $\sin\theta = \frac{1}{3}$, $\cos\theta = -\frac{2\sqrt{2}}{3}$, $\tan\theta = -\frac{\sqrt{2}}{4}$, $\sec\theta = -\frac{3\sqrt{2}}{4}$, $\cot\theta = -2\sqrt{2}$

39. $\sin\theta = -\frac{5}{13}$, $\cos\theta = -\frac{12}{13}$, $\cot\theta = \frac{12}{5}$, $\sec\theta = -\frac{13}{12}$, $\csc\theta = -\frac{13}{5}$

41. $\sin\theta = \frac{2}{\sqrt{5}}$, $\cos\theta = \frac{1}{\sqrt{5}}$

43. $\cos(-45°) = \frac{1}{\sqrt{2}} = \cos 45°$

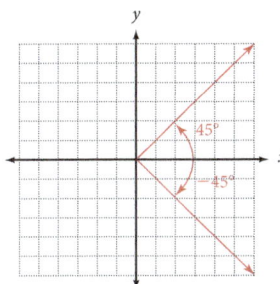

45. $\tan\theta = 2\sqrt{2}$ **47.** $\sec\theta = \frac{5}{3}$

49.

Point on the Wheel	Time in Minutes	Height in Feet
A	0	10
B	0.25	24.6
C	0.5	60
D	0.75	95.4
E	1	110
F	1.25	95.4
G	1.5	60
H	1.75	24.6
A	2	10

51. 2   **53.** $\frac{5}{3}$ **55.** $\frac{r}{x}$ **57.** $\frac{x}{y}$ **59.** $-\frac{3}{4}$ **61.** $-\sqrt{3}$

63. $\frac{9}{25}$ **65.** $\frac{4}{5}$ **67.** $\frac{4}{5}$ **69.** $\frac{1}{2}$ **71.** $-\frac{4}{5}$

PROBLEM SET 1.4

Vocabulary

1. identity   **2.** Pythagorean   **3.** reciprocal   **4.** ratio

Problems

1. $\frac{5}{4}$ **3.** $-\frac{1}{2}$ **5.** $\frac{1}{a}$ **7.** $-\frac{3}{4}$ **9.** $\frac{12}{5}$ **11.** $\frac{4}{5}$

13. $-\frac{2\sqrt{2}}{3}$ **14.**   **15.** $-\frac{3}{5}$ **17.** $\frac{1}{2}$ **19.** $-\frac{2}{\sqrt{5}}$

21. $\sin\theta = \frac{5}{13}$, $\tan\theta = \frac{5}{12}$, $\sec\theta = \frac{13}{12}$, $\csc\theta = \frac{13}{5}$, $\cot\theta = \frac{12}{5}$

23. $\sin\theta = -\frac{2\sqrt{2}}{3}$, $\tan\theta = 2\sqrt{2}$, $\sec\theta = -3$, $\csc\theta = -\frac{3\sqrt{2}}{4}$, $\cot\theta = \frac{\sqrt{2}}{4}$

25. $\cos\theta = \frac{12}{13}$, $\tan\theta = \frac{5}{12}$, $\sec\theta = \frac{13}{12}$, $\csc\theta = \frac{13}{5}$, $\cot\theta = \frac{12}{5}$

27. $\cos\theta = \frac{\sqrt{3}}{2}$, $\tan\theta = -\frac{1}{\sqrt{3}}$, $\sec\theta = \frac{2}{\sqrt{3}}$, $\csc\theta = -2$, $\cot\theta = -\sqrt{3}$

29. $\sin\theta = \frac{3}{5}$, $\cos\theta = \frac{4}{5}$, $\sec\theta = \frac{5}{4}$, $\csc\theta = \frac{5}{3}$, $\cot\theta = \frac{4}{3}$

31. $\sin\theta = \frac{3}{\sqrt{10}}$, $\cos\theta = -\frac{1}{\sqrt{10}}$, $\sec\theta = -\sqrt{10}$, $\csc\theta = \frac{\sqrt{10}}{3}$, $\cot\theta = -\frac{1}{3}$

33. $\sin\theta = \frac{\sqrt{3}}{2}$, $\cos\theta = \frac{1}{2}$, $\tan\theta = \sqrt{3}$, $\csc\theta = \frac{2}{\sqrt{3}}$, $\cot\theta = \frac{1}{\sqrt{3}}$

35. $\sin\theta = -\frac{2\sqrt{2}}{3}$, $\cos\theta = -\frac{1}{3}$, $\tan\theta = 2\sqrt{2}$, $\csc\theta = -\frac{3}{2\sqrt{2}}$, $\cot\theta = \frac{1}{2\sqrt{2}}$

37. $\sin\theta = \frac{4}{5}$, $\cos\theta = -\frac{3}{5}$, $\tan\theta = -\frac{4}{3}$, $\sec\theta = -\frac{5}{3}$, $\cot\theta = -\frac{3}{4}$

39. $\sin\theta = -\frac{5}{13}$, $\cos\theta = -\frac{12}{13}$, $\tan\theta = \frac{5}{12}$, $\sec\theta = -\frac{13}{12}$, $\cot\theta = \frac{12}{5}$

41. $\sin\theta = -\frac{2}{\sqrt{5}}$, $\cos\theta = -\frac{1}{\sqrt{5}}$, $\tan\theta = 2$, $\sec\theta = -\sqrt{5}$, $\csc\theta = -\frac{\sqrt{5}}{2}$

43. $\sin\theta = \frac{1}{\sqrt{5}}$, $\cos\theta = -\frac{2}{\sqrt{5}}$, $\tan\theta = -\frac{1}{2}$, $\sec\theta = -\frac{\sqrt{5}}{2}$, $\csc\theta = \sqrt{5}$

45. $\frac{2}{\sqrt{5}}$ **47.** $\frac{3}{4}$ **49.** $\frac{\sqrt{29}}{2}$ **51.** $-\frac{5}{4}$ **53.** $-\frac{3}{5}$

55. $-\frac{2\sqrt{2}}{3}$ **57.** 3   **59.** m **61.** $\sin\theta = \pm\sqrt{1 - \cos^2\theta}$

63. $\frac{5}{6}$ **65.** $x^2 - 3x - 10$ **67.** $(x+2)(x-2)$

PROBLEM SET 1.5

1. $\frac{\cos\theta}{\sin^2\theta}$ **3.** $\frac{1}{\cos\theta}$ **5.** $\frac{\sin\theta}{\cos\theta}$ **7.** $\sin^2\theta$ **9.** $\frac{\sin\theta + 1}{\cos\theta}$

11. $2\cos\theta$ **13.** $\frac{\sin^2\theta + \cos\theta}{\sin\theta\cos\theta}$ **15.** $\frac{\cos\theta - \sin\theta}{\sin\theta\cos\theta}$

17. $\frac{\sin\theta\cos\theta + 1}{\cos\theta}$ **19.** $\frac{\cos^2\theta}{\sin\theta}$ **21.** $\sin^2\theta + 7\sin\theta + 12$

23. $8\cos^2\theta + 2\cos\theta - 15$ **25.** $\cos^2\theta$ **27.** $1 - \tan^2\theta$

29. $1 - 2\sin\theta\cos\theta$ **31.** $\sin^2\theta - 8\sin\theta + 16$

33. $(x+5)(x-5)$ **35.** $(1+x)(1-x)$

37. $(\sin\theta + 5)(\sin\theta - 5)$ **39.** $(1 + \cos\theta)(1 - \cos\theta)$

41. $(\sin\theta + \cos\theta)(\sin\theta - \cos\theta)$

43. $\cos\theta\tan\theta = \cos\theta \cdot \dfrac{\sin\theta}{\cos\theta} = \sin\theta$

45. $\sin\theta\sec\theta\cot\theta = \sin\theta \cdot \dfrac{1}{\cos\theta} \cdot \dfrac{\cos\theta}{\sin\theta} = 1$

47. $\dfrac{\csc\theta}{\cot\theta} = \dfrac{\frac{1}{\sin\theta}}{\frac{\cos\theta}{\sin\theta}} = \dfrac{1}{\sin\theta} \cdot \dfrac{\sin\theta}{\cos\theta} = \sec\theta$

49. $\dfrac{\sec\theta}{\csc\theta} = \dfrac{\frac{1}{\cos\theta}}{\frac{1}{\sin\theta}} = \dfrac{1}{\cos\theta} \cdot \dfrac{\sin\theta}{\cos\theta}$

49. $\dfrac{\sec\theta}{\csc\theta} = \dfrac{\frac{1}{\cos\theta}}{\frac{1}{\sin\theta}} = \dfrac{1}{\cos\theta} \cdot \dfrac{\sin\theta}{\cos\theta} = \tan\theta$

51. $\sin\theta\tan\theta + \cos\theta = \sin\theta \cdot \dfrac{\sin\theta}{\cos\theta} + \cos\theta$

$= \dfrac{\sin^2\theta}{\cos\theta} + (\cos\theta)\dfrac{\cos\theta}{\cos\theta} = \dfrac{\sin^2\theta}{\cos\theta} + \dfrac{\cos^2\theta}{\cos\theta}$

$= \dfrac{\sin^2\theta + \cos^2\theta}{\cos\theta} = \sec\theta$

53. $\tan\theta + \cot\theta = \dfrac{\sin\theta}{\cos\theta} + \dfrac{\cos\theta}{\sin\theta} = \dfrac{\sin^2\theta}{\sin\theta\cos\theta} + \dfrac{\cos^2\theta}{\sin\theta\cos\theta}$

$= \dfrac{1}{\sin\theta\cos\theta} = \sec\theta\csc\theta$

55. $\csc\theta - \sin\theta = \dfrac{1}{\sin\theta} - \sin\theta = \dfrac{1}{\sin\theta} - \dfrac{\sin^2\theta}{\sin\theta}$

$= \dfrac{1 - \sin^2\theta}{\sin\theta} = \dfrac{\cos^2\theta}{\sin\theta}$

57. $1 - \sin^2\theta = \cos^2\theta$

59. $(1 - \cos\theta)(1 + \cos\theta) = 1 + \cos\theta - \cos\theta - \cos^2\theta$

$= 1 - \cos^2\theta = \sin^2\theta$

61. $(\sin\theta - \cos\theta)^2 - 1 = (\sin\theta - \cos\theta)(\sin\theta - \cos\theta) - 1$

$= \sin^2\theta - \sin\theta\cos\theta - \sin\theta\cos\theta + \cos^2\theta - 1$

$= \sin^2\theta + \cos^2\theta - 1 - 2\sin\theta\cos\theta = 1 - 1 - 2\sin\theta\cos\theta$

$= -2\sin\theta\cos\theta$

63. $\sin\theta(\sec\theta + \csc\theta) = \sin\theta\sec\theta + \sin\theta\csc\theta$

$= \sin\theta\left(\dfrac{1}{\cos\theta}\right) + \sin\theta\left(\dfrac{1}{\sin\theta}\right) = \dfrac{\sin\theta}{\cos\theta} + 1 = \tan\theta + 1$

65. $3\sec\theta$ **67.** $6\cos\theta$ **69.** $5\tan\theta$ **71.** $5\sec\theta$

73. $4\cos\theta$ **75.** $\sqrt{5}\tan\theta$

CHAPTER 1 TEST

1. $5\sqrt{3}, 10$ **2.** $6\sqrt{2}, 6\sqrt{2}$

3. $\sin\theta = -\dfrac{3}{5}$, $\cos\theta = \dfrac{4}{5}$, $\tan\theta = -\dfrac{3}{4}$, $\cot\theta = -\dfrac{4}{3}$,

$\sec\theta = \dfrac{5}{4}$, $\csc\theta = -\dfrac{5}{3}$

4. QIII

5. $\sin\theta = \dfrac{1}{2}$, $\cos\theta = -\dfrac{\sqrt{3}}{2}$, $\tan\theta = -\dfrac{1}{\sqrt{3}}$ $\cot\theta = -\sqrt{3}$,

$\sec\theta = -\dfrac{2}{\sqrt{3}}$ $\csc\theta = 2$

6. $-\dfrac{1}{2}$ **7.** $-\dfrac{1}{2}$ **8.** $3\sqrt{3}$ **9.** 3   **10.** 7.5   **11.** 40

12.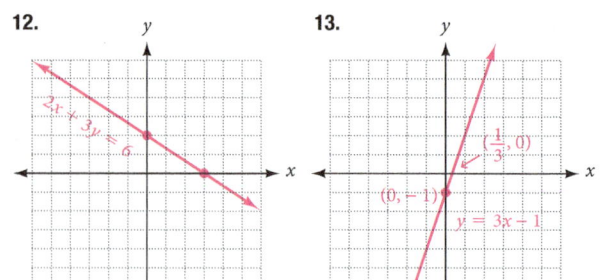

13.

14. 13   **15.** $\sqrt{a^2 + b^2}$ **16.** $-5, 1$

17. $\sin 90° = 1$; $\cos 90° = 0$; $\tan 90° =$ undefined

18. $\sin(-45°) = -\dfrac{1}{\sqrt{2}}$; $\cos(-45°) = \dfrac{1}{\sqrt{2}}$; $\tan(-45°) = -1$

19. QIV   **20.** QII

21. $\sin\theta = \dfrac{4}{5}$, $\cos\theta = -\dfrac{3}{5}$, $\tan\theta = -\dfrac{4}{3}$, $\cot\theta = -\dfrac{3}{4}$,

$\sec\theta = -\dfrac{5}{3}$, $\csc\theta = \dfrac{5}{4}$

22. $\sin\theta = -\dfrac{1}{\sqrt{10}}$, $\cos\theta = -\dfrac{3}{\sqrt{10}}$, $\tan\theta = \dfrac{1}{3}$, $\cot\theta = 3$,

$\sec\theta = -\dfrac{\sqrt{10}}{3}$, $\csc\theta = -\sqrt{10}$

23. $\sin\theta = \dfrac{1}{2}$, $\cos\theta = -\dfrac{\sqrt{3}}{2}$, $\tan\theta = -\dfrac{1}{\sqrt{3}}$, $\cot\theta = -\sqrt{3}$,

$\sec\theta = -\dfrac{2}{\sqrt{3}}$, $\csc\theta = 2$

24. $\sin\theta = -\dfrac{12}{13}$, $\cos\theta = -\dfrac{5}{13}$, $\tan\theta = \dfrac{12}{5}$, $\cot\theta = \dfrac{5}{12}$,

$\sec\theta = -\dfrac{13}{5}$, $\csc\theta = -\dfrac{13}{12}$

25. $\sin\theta = -\dfrac{2}{\sqrt{5}}$; $\cos\theta = \dfrac{1}{\sqrt{5}}$ **26.** $-\dfrac{4}{3}$

27. $\dfrac{2\sqrt{2}}{3}$ and $\dfrac{3}{2\sqrt{2}}$ **28.** $\dfrac{1}{2\sqrt{2}}$ and $2\sqrt{2}$ **29.** 9   **30.** $\dfrac{9}{8}$

31. 2   **32.** 53   **33.** $\pm\dfrac{\sqrt{1 - \cos^2\theta}}{\cos^2\theta}$

34. $\sin^2\theta - 4\sin\theta - 21$ **35.** $1 - 2\sin\theta\cos\theta$ **36.** $\dfrac{\cos^2\theta}{\sin\theta}$

37-42. See Solutions Manual

43. $7\sec\theta$ **44.** $5\cos\theta$ **45.** $\tan\theta$ **46.** $3\sec\theta$

47. a. First ring: $x^2 + y^2 = 25$, second ring: $(x - 5)^2 + y^2 = 25$

b. First ring: $(x + 5)^2 + y^2 = 25$, second ring: $x^2 + y^2 = 25$

Chapter 2

PROBLEM SET 2.1

Vocabulary

1. hypotenuse   **2.** cofunctions   **3.** complement

4. opposite   **5.** adjacent

Problems

1. $\sin A = \frac{4}{5}$, $\cos A = \frac{3}{5}$, $\tan A = \frac{4}{3}$, $\cot A = \frac{3}{4}$,
   $\sec A = \frac{5}{3}$, $\csc A = \frac{5}{4}$

3. $\sin A = \frac{2}{\sqrt{5}}$, $\cos A = \frac{1}{\sqrt{5}}$, $\tan A = 2$, $\cot A = \frac{1}{2}$,
   $\sec A = \sqrt{5}$, $\csc A = \frac{\sqrt{5}}{2}$

5. $\sin A = \frac{1}{2}$, $\cos A = \frac{\sqrt{3}}{2}$, $\tan A = \frac{1}{\sqrt{3}}$, $\cot A = \sqrt{3}$,
   $\sec A = \frac{2}{\sqrt{3}}$, $\csc A = 2$ 7. $\sin A = \frac{2}{3} = \cos B$,
   $\cos A = \frac{\sqrt{5}}{3} = \sin B$

9. $\sin A = \frac{1}{\sqrt{2}} = \cos B$, $\cos A = \frac{1}{\sqrt{2}} = \sin B$

11. $\sin A = \frac{\sqrt{3}}{2} = \cos B$, $\cos A = \frac{1}{2} = \sin B$

13. $80°$ 15. $82°$ 17. $17°$ 19. $90° - x$ 21. x

23. 2 25. 3 27. $\frac{1}{2\sqrt{2}}$ or $\frac{\sqrt{2}}{4}$ 29. $\frac{2+\sqrt{3}}{2}$

31. 0 33. $4 + 2\sqrt{3}$ 35. 1 37. $2\sqrt{3}$

39. $\frac{-3\sqrt{3}}{2}$ 41. $\sqrt{2}$ 43. $\frac{2}{\sqrt{3}}$ 45. $\frac{2}{\sqrt{3}}$ 47. 1

49. $\sqrt{2}$ 51. 0.8660 53. 0.8660 55. 0.7071

57. 0.7071 59. 1 61. 120 63. 65 65. 15

67. $\frac{3}{4} = 0.75$ 69. 0.93 71. 0.8 73. 0.67

PROBLEM SET 2.2

Vocabulary

1. rotation   2. minute   3. degrees   4. decimal   5. multiply

Problems

1. $64° \, 9'$ 3. $89° \, 40'$ 5. $106° \, 49'$ 7. $55° \, 48'$

9. $59° \, 43'$ 11. $53° \, 50'$ 13. $39° \, 50'$ 15. $35° \, 24'$

17. $16° \, 15'$ 19. $92° \, 33'$ 21. $19° \, 54'$ 23. $28° \, 21'$

25. $45.2°$ 27. $62.6°$

29. $17.\overline{3}°$ (the $\overline{3}$ means the 3's repeat indefinitely)   31. $48.45°$

33. 0.4571 35. 0.9511 37. 21.3634 39. 1.6643

41. 1.5003 43. 4.0906 45. 0.9100 47. 0.9057

49. 0.8355 51. 0.3365 53. 0.0204 55. 1.5959

57. 1.8460 59. 0.6950 61. 5.5308 63. $12.3°$

65. $34.5°$ 67. $78.9°$ 69. $44.7°$ 71. $65.7°$ 73. $50.1°$

75. $14.7°$ 77. $10.2°$ 79. $9.2°$ 81. 0.75 83. 9

85. 1 87. 1 89. 9.19 91. 0.80 93. 24.16

95. 11.55 97. 4.37 99. $39°$

PROBLEM SET 2.3

Vocabulary

1. two   2. minutes, tenth   3. significant   4. two, significant

Problems

1. 3 3. 2 5. 5 7. 4 9. 2 11. 2 13. 3

15. 4 17. 4 19. 10 cm   21. 33 ft   23. $26.6°$

25. $53.1°$ 27. $B = 70°$, $a = 8.2$, $b = 23$ in.

29. $A = 14°$, $a = 1.4$ mi., $b = 5.6$ mi.

31. $A = 19.8°$, $B = 70.2°$, $c = 103$ m

33. $A = 41.4°$, $B = 48.6°$, $a = 132$ cm

35. $A = 63° \, 30'$, $a = 650$ mm, $c = 726$ mm

37. $A = 66.55°$, $b = 2.356$ mi., $c = 5.921$ mi.

39. $A = 42.8°$, $B = 47.2°$, $b = 2.97$ cm

41. $A = 61.42°$, $B = 28.58°$, $a = 22.41$ in.

43. $45°$ 45. $43°$ 47. $x = 12.87$, $y = 7.10$

49. $x = 18.12$, $h = 16.89$ 51. $59°$ 53. 7.5

55. $y = 17.1$, $x = 8.9$ 57. $r = 36$ 59. 18.9

61. 169.2 63. $60.2°$ 65. $73.7°$

PROBLEM SET 2.4

Vocabulary

1. elevation   2. depression   3. bearing

Problems

1. Height $= 39.2$ cm; base angles $= 69.1°$ 3. 38.6 ft

5. $36.6°$ 7. $55.1°$ 9. 6.4 ft   11. 78.9 ft   13. 61 cm

15. 8.5 miles   17. 17. a. 800 ft b. 200 ft c. $14.0°$

19. 39 miles   21. 31 mi, N $78°$ E   23. 63.5 mi N, 47.8 mi W

25. 83 feet   27. $79°$ 29. 6.2 31. $1,300$ 33. 29

35. 63 37. 10

PROBLEM SET 2.5

Vocabulary

1. magnitude, scalars   2. horizontal, components
3. velocity

Problems

1.

3.

5.

7.

9. 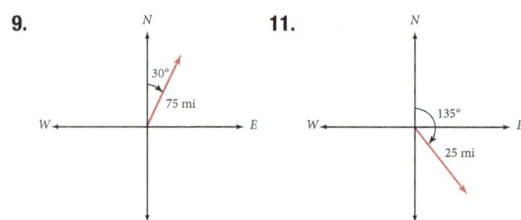 **11.**

13. 12.4 miles per hour at N 76.0° E

15. 202 miles per hour at N 21.5° E   **17.** 2.6 miles per hour

19. $|V_x| = 35, |V_y| = 20$ **21.** $|V_x| = 12.6, |V_y| = 5.66$

23. $|V_x| = 339, |V_y| = 248$ **25.** $|V_x| = 55, |V_y| = 32$

27. $|V| = 50$ **29.** $|V| = 43.6$ **31.** $|V| = 5.9$

33. $|V_x| = 850$ feet per second, $|V_y| = 850$ feet per second

35. 2,550 feet   **37.** 38.1 feet per second at an inclination of 23.2°

39. 100 km south, 90.3 km east

41. 245 miles north, 145 miles east

43. $\sin 135° = \frac{1}{\sqrt{2}}, \cos 135° = -\frac{1}{\sqrt{2}}, \tan 135° = -1$

45. $\sin \theta = \frac{2}{\sqrt{5}}, \cos \theta = \frac{1}{\sqrt{5}}$ **47.** $x = \pm 6$

CHAPTER 2 TEST

1. $\sin A = \frac{1}{\sqrt{5}}, \cos A = \frac{2}{\sqrt{5}}, \tan A = \frac{1}{2}, \sin B = \frac{2}{\sqrt{5}},$
$\cos B = \frac{1}{\sqrt{5}}, \tan B = 2$

2. $\sin A = \frac{\sqrt{3}}{2}, \cos A = \frac{1}{2}, \tan A = \sqrt{3}$

3. $\sin A = \frac{3}{5}, \cos A = \frac{4}{5}, \tan A = \frac{3}{4}, \sin B = \frac{4}{5},$
$\cos B = \frac{3}{5}, \tan B = \frac{4}{3}$

4. $\sin A = \frac{5}{13}, \cos A = \frac{12}{13}, \tan A = \frac{5}{12}, \sin B = \frac{12}{13},$
$\cos B = \frac{5}{13}, \tan B = \frac{12}{5}$

5. 76°   **6.** 17°   **7.** $\frac{5}{4}$ **8.** 2   **9.** 0   **10.** $\frac{\sqrt{3}}{2}$

11. 73° 10′   **12.** 9° 43′   **13.** 73° 12′   **14.** 16° 27′

15. 2.8°   **16.** 79.5°   **17.** 0.4120   **18.** 0.7902

19. 2.0353   **20.** 0.3378   **21.** 4.7°   **22.** 58.7°   **23.** 71.2°

24. 13.3°   **25.** 4   **26.** 2   **27.** $A = 33.2°, B = 56.8°, c = 124$

28. $A = 30.3°, B = 59.7°, b = 41.5$

29. $A = 65.1°, a = 657, c = 724$

30. $A = 54° 30′, a = 0.268, b = 0.376$ **31.** 86 centimeters

32. 5.8 feet   **33.** 112 feet   **34.** 96.4 miles west, 84.3 miles south

35. $|V_x| = 4.3, |V_y| = 2.5$ **36.** 72°

37. $|V_x| = 400$ feet per second, $|V_y| = 690$ feet per second

38. 60 miles south, 104 miles east

Chapter 3

PROBLEM SET 3.1

Vocabulary

 1. reference, acute   **2.** reference   **3.** reference, sign
 4. coterminal

Problems

 1. 30° 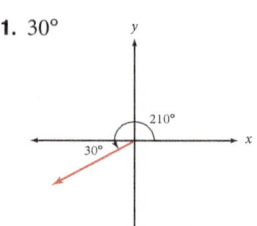   **3.** 36.6°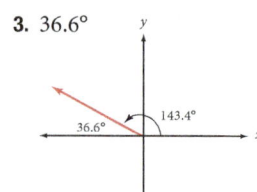

 5. 48.3° 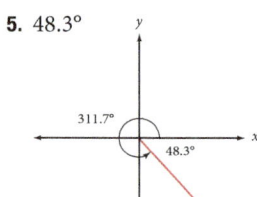   **7.** 15° 10′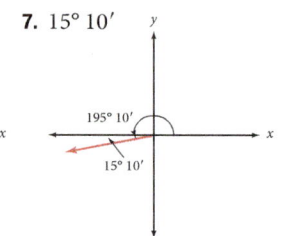

 9. 28° 40′ 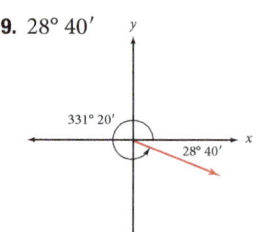   **11.** 60°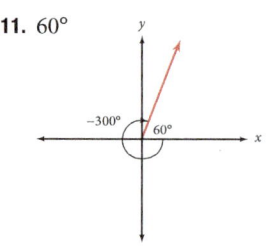

 13. 60°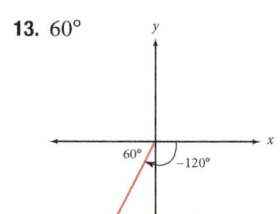

15. 0.9744   **17.** −0.2045
19. −0.7427   **21.** 1.5032
23. 1.7321   **25.** −0.8151
27. −0.5000   **29.** 0.7071
31. 0.2711   **33.** 1.4225
35. 1.3022   **37.** −0.8660

39. 0.5851   **41.** 198.0°   **43.** 140.0°   **45.** 210.5°

47. 74.7°   **49.** 105.2°   **51.** 310.9°   **53.** 119.4°

55. 202.1°   **57.** 144.2°   **59.** 302.8°   **61.** $\frac{\sqrt{3}}{2}$

63. −1   **65.** $-\frac{1}{2}$ **67.** −2   **69.** 2   **71.** $\frac{1}{2}$

73. $-\frac{1}{\sqrt{3}}$ **75.** 240°   **77.** 135°   **79.** 300°

81. 240°   **83.** 65.3°

85.

θ	85°	87°	89°	89.9°	89.99°
$\tan \theta$	11.4	19.1	57.3	573.0	5,730

87. 57.3   **89.** 0.785   **91.** 60   **93.** $\frac{\pi}{4}$ **95.** $\frac{3\pi}{2}$ **97.** $\frac{\pi}{2}$

PROBLEM SET 3.2

Vocabulary

1. radians, degrees   2. center   3. arc   4. degrees
5. radians

Problems

1. 3   3. $\frac{1}{2}$ 5. 3π 7. 2   9. $\frac{9}{80} = 0.1125$ radians

11. $\frac{\pi}{6} \cong 0.52$ 13. $\frac{\pi}{2} \cong 1.57$ 15. $\frac{13\pi}{9} \cong 4.54$

17. $-\frac{5\pi}{6} \cong -2.62$ 19. $\frac{7\pi}{3} \cong 7.33$ 21. $-\frac{3\pi}{4} \cong -2.36$

23. 2.11   25. 0.000291   27. 1.16 miles   29. 60°

31. 120°   33. −210°   35. 300°   37. 720°   39. 15°

41. −70°   43. 74.5°   45. 43.0°   47. 286.5°

49. $-\frac{\sqrt{3}}{2}$ 51. $\frac{1}{\sqrt{3}}$ 53. −2   55. 2   57. $-2\sqrt{2}$

59. $-\frac{1}{\sqrt{2}}$ 61. $\sqrt{3}$ 63. $\frac{\sqrt{3}}{2}$ 65. 0   67. $\frac{\sqrt{3}}{2}$

69. −2   71. $(0,0) \left(\frac{\pi}{4}, \frac{1}{\sqrt{2}}\right) \left(\frac{\pi}{2}, 1\right) \left(\frac{3\pi}{4}, \frac{1}{\sqrt{2}}\right) (\pi, 0)$

73. $(0,0) \left(\frac{\pi}{2}, 2\right) (\pi, 0) \left(\frac{3\pi}{2}, -2\right) (2\pi, 0)$

75. $(0,0) \left(\frac{\pi}{4}, 1\right) \left(\frac{\pi}{2}, 0\right) \left(\frac{3\pi}{4}, -1\right) (\pi, 0)$

77. $\left(\frac{\pi}{2}, 0\right) (\pi, 1) \left(\frac{3\pi}{2}, 0\right) (2\pi, -1) \left(\frac{5\pi}{2}, 0\right)$

79. $\left(-\frac{\pi}{4}, 0\right) (0, 3) \left(\frac{\pi}{4}, 0\right) \left(\frac{\pi}{2}, -3\right) \left(\frac{3\pi}{4}, 0\right)$

81. $\sin \theta = 0$, $\cos \theta = 1$, $\tan \theta = 0$, $\cot \theta =$ undefined, $\sec \theta = 1$, $\csc \theta =$ undefined

83. $\sin \theta = \frac{\sqrt{3}}{2}$, $\cos \theta = \frac{1}{2}$, $\tan \theta = \sqrt{3}$, $\cot \theta = \frac{1}{\sqrt{3}}$, $\sec \theta = 2$, $\csc \theta = \frac{2}{\sqrt{3}}$

85. 150°   87.

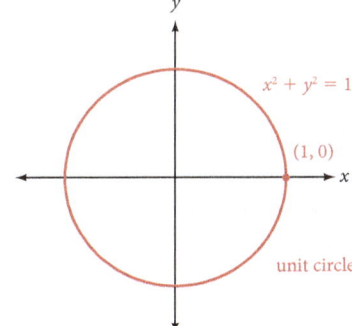

$x^2 + y^2 = 1$
(1, 0)
unit circle

PROBLEM SET 3.3

Vocabulary

1. cosine   2. odd   3. even   4. radians

Problems

1. $\sin 150° = \frac{1}{2}$, $\cos 150° = -\frac{\sqrt{3}}{2}$, $\tan 150° = -\frac{1}{\sqrt{3}}$, $\cot 150° = -\sqrt{3}$, $\sec 150° = -\frac{2}{\sqrt{3}}$, $\csc 150° = 2$

3. $\sin \frac{11\pi}{6} = -\frac{1}{2}$, $\cos \frac{11\pi}{6} = \frac{\sqrt{3}}{2}$, $\tan \frac{11\pi}{6} = -\frac{1}{\sqrt{3}}$, $\cot \frac{11\pi}{6} = -\sqrt{3}$, $\sec \frac{11\pi}{6} = \frac{2}{\sqrt{3}}$, $\csc \frac{11\pi}{6} = -2$

5. $\sin 180° = 0$, $\cos 180° = -1$, $\tan 180° = 0$, $\cot 180° =$ undefined, $\sec 180° = -1$, $\csc 180° =$ undefined

7. $\sin \frac{3\pi}{4} = \frac{1}{\sqrt{2}}$, $\cos \frac{3\pi}{4} = -\frac{1}{\sqrt{2}}$, $\tan \frac{3\pi}{4} = -1$, $\cot \frac{3\pi}{4} = -1$, $\sec \frac{3\pi}{4} = -\sqrt{2}$, $\csc \frac{3\pi}{4} = \sqrt{2}$

9. $\frac{1}{2}$ 11. $-\frac{\sqrt{3}}{2}$ 13. $-\frac{1}{2}$ 15. $-\frac{1}{\sqrt{2}}$

17. $\frac{\pi}{6}, \frac{5\pi}{6}$ 19. $\frac{5\pi}{6}, \frac{7\pi}{6}$ 21. $\frac{2\pi}{3}, \frac{5\pi}{3}$

23. $\sin \theta = -\frac{2}{\sqrt{5}}$, $\cos \theta = \frac{1}{\sqrt{5}}$, $\tan \theta = -2$ 25. $\frac{1}{3}$

27. 1   29. $\frac{1}{2}$ 31. 1   33. $\frac{1}{2}$

35.

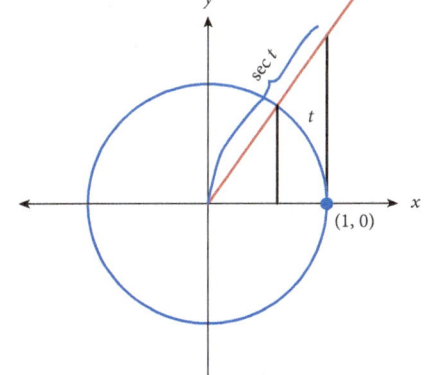

$P(-x, y)$ $180° - \theta$ $P(x, y)$
θ θ
O
$\sin(180° - \theta) = y = \sin \theta$

37. $\tan(-\theta) = \frac{\sin(-\theta)}{\cos(-\theta)} = \frac{-\sin \theta}{\cos \theta} = -\frac{\sin \theta}{\cos \theta} = -\tan \theta$

45.

y
$\sec t$
t
$(1, 0)$
x

47. 2.51   49. 29,300   51. 707

PROBLEM SET 3.4

Vocabulary

1. $s = r\theta$ 2. sector, $A = \frac{1}{2}r^2\theta$ 3. $s = 2\left(\frac{\pi}{2}\right)$

4. $A = \frac{1}{2} \cdot 4 \cdot \frac{\pi}{2}$

Problems

1. 6 inches   3. 2.25 feet   5. 2π centimeters $\cong 6.28$ centimeters

7. $\frac{4\pi}{3}$ millimeters $\cong 4.19$ millimeters

9. $\frac{40\pi}{3}$ inches $\cong 41.9$ inches

11. 1.6π centimeters $= 5.03$ centimeters

13. 1400π miles $\cong 4,400$ miles   15. $\frac{4\pi}{9}$ feet $\cong 1.40$ feet

17. 2,100 miles   **19.** 0.5 feet   **21.** 3 inches

23. 4 centimeters   **25.** 1 meter   **27.** 31,000 feet

29. 9 centimeters^2 **31.** 19.2 inches^2

33. $\frac{9\pi}{10}$ meters$^2 \cong 2.83$ meters^2

35. $\frac{25\pi}{24}$ meters$^2 \cong 3.27$ meters^2 **37.** 4 inches^2

39. 2 centimeters   **41.** $r = \frac{4}{\sqrt{3}}$ inches $\cong 2.31$ inches

43. 900π feet$^2 \cong 2,830$ feet^2 **45.** $\frac{33\pi}{2}$ inches$^2 = 51.8$ inches^2

47. $\frac{55\pi}{3}$ in.$^2 \approx 57.6$ in.^{2} **49.** $\frac{\pi}{15}$ **51.** $\frac{\pi}{3}$ **53.** 195

55. 848   **57.** 0.107

PROBLEM SET 3.5
Vocabulary

1. angular, $\omega = \frac{\theta}{t}$ **2.** linear, $v = \frac{s}{t}$ **3.** revolutions

Problems

1. 1.5 feet per minute   **3.** 3 centimeters per second

5. 15 miles per hour   **7.** 80 feet   **9.** 22.5 miles

11. 7 miles

13. $\frac{2\pi}{15}$ radians per second $\cong 0.419$ radians per second

15. 4 radians per minute

17. $\frac{8}{3}$ radians per second $\cong 2.67$ radians per second

19. 37.5π radians per hour $= 118$ radians per hour

21. $d = 100 \tan\left(\frac{\pi}{2}t\right)$: when $t = \frac{1}{2}$, $d = 100$ feet; when $t = \frac{3}{2}$, $d = -100$ feet; when $t = 1$, d is undefined because the light rays are parallel to the wall.

23. 40 inches   **25.** 180π meters $\cong 565$ meters   **27.** 4,500 feet

29. 20π radians per minute $\cong 62.8$ radians per minute

31. $\frac{200\pi}{3}$ radians per minute $\cong 209$ radians per minute

33. 11.6π radians per minute $\cong 36.4$ radians per minute

35. 10 inches per second   **37.** 0.5 radians per second

39. 80π feet per minute $\cong 251$ feet per minute

41. $\frac{\pi}{12}$ radians per hour $\cong 0.262$ radians per hour

43. 10π feet $\cong 31.4$ feet   **45.** $\frac{33,000}{\pi}$ rpm $\cong 10,500$ rpm

47. 576π centimeters $\cong 1,810$ centimeters

49. 889 radians per minute (53,300 radians per hour)

51. $\omega = \frac{\pi}{8}$ radians/minute; $v = 38.7$ feet/minute ≈ 0.44 miles/hour

53. $\sin A = \frac{\sqrt{3}}{2}$, $\cos A = \frac{1}{2}$, $\tan A = \sqrt{3}$ **55.** $\frac{\sqrt{3}}{2}$

57. 79.5°   **59.** $A = 33.2°$, $B = 56.8°$, $c = 124$

61. 60°   **63.** 2.2 feet

CHAPTER 3 TEST

1. 55°   **2.** 62.2°

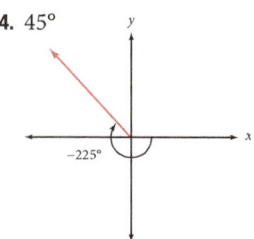

3. 50° 20′   **4.** 45°

5. -0.8391 **6.** -0.4663 **7.** -0.5490 **8.** -0.6115

9. -1.2991 **10.** -6.5121 **11.** 174.0°   **12.** 241.5°   **13.** 226.0°

14. 310.3°   **15.** $-\frac{1}{\sqrt{2}}$ **16.** $-\frac{1}{\sqrt{2}}$ **17.** $-\frac{1}{\sqrt{3}}$

18. $\frac{2}{\sqrt{3}}$ **19.** $\frac{25\pi}{18}$ **20.** $-\frac{13\pi}{6}$ **21.** 240°

22. 105°   **23.** $\frac{\sqrt{3}}{2}$ **24.** $-\frac{1}{2}$ **25.** $-2\sqrt{2}$

26. 1   **27.** $-\frac{2}{\sqrt{3}}$ **28.** 2   **29.** 0   **30.** $2\sqrt{2}$

31. $\cot(-\theta) = \dfrac{\cos(-\theta)}{\sin(-\theta)} = \dfrac{\cos\theta}{-\sin\theta} = -\cot\theta$

32. First use odd and even functions to write everything in terms of θ instead of $-\theta$.

33. 2π meters $\cong 6.28$ meters   **34.** 2π feet $\cong 6.28$ feet

35. 4 centimeters   **36.** 3/8 centimeters $\cong 0.375$ centimeters

37. 4π inches$^2 \cong 12.6$ inches^2 **38.** 10.8 centimeters^2

39. 2π cm ≈ 6.28 cm   **40.** 8 inches^2 **41.** 90 feet

42. 3,960 feet   **43.** 72 inches   **44.** 120π feet $\cong 377$ feet

45. 12π radians per minute $\cong 37.7$ radians per minute

46. 4π radians per minute $\cong 12.6$ radians per minute

47. 0.5 radians per second   **48.** $\frac{5}{3}$ radians per second

49. 80π feet per minute $\cong 251$ feet per minute

50. 20π feet per minute $\cong 62.8$ feet per minute

51. 4 radians per second for the 6 cm pulley and 3 radians per second for the 8 cm pulley

52. 2700π feet per minute $\cong 8,480$ feet per minute

Chapter 4

PROBLEM SET 4.1
Vocabulary

1. period   **2.** amplitude   **3.** asymptotes   **4.** range   **5.** domain

Problems

1.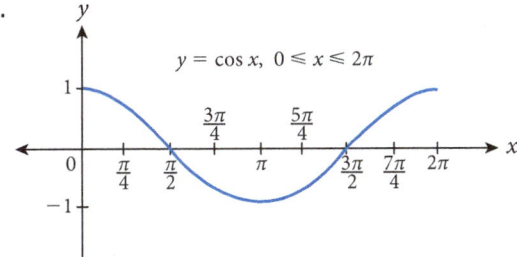

3.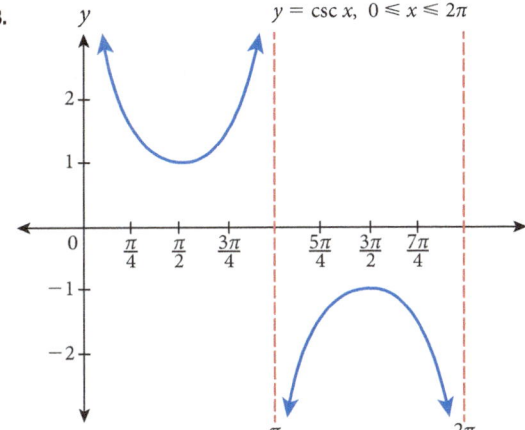

5.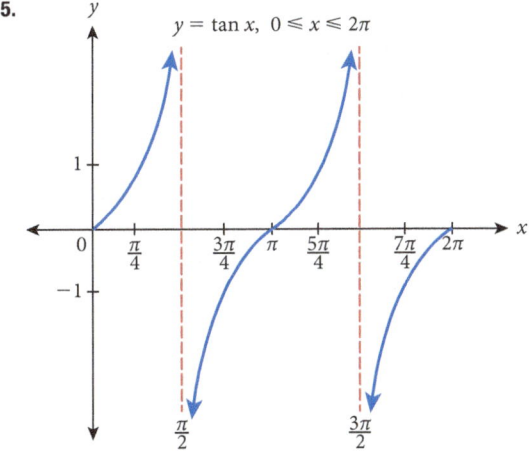

7. $\frac{\pi}{3}, \frac{5\pi}{3}$ **9.** $\frac{3\pi}{2}$ **11.** $\frac{3\pi}{4}, \frac{5\pi}{4}$ **13.** $\frac{\pi}{4}, \frac{5\pi}{4}$ **15.** $\frac{\pi}{6}, \frac{11\pi}{6}$

17.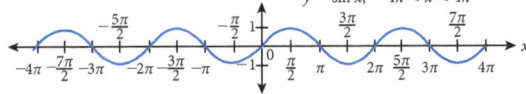

19.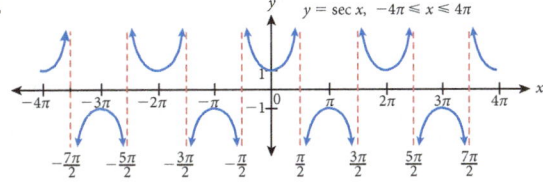

21.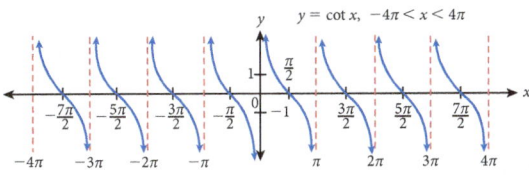

23. $\frac{\pi}{3}, \frac{2\pi}{3}, \frac{7\pi}{3}, \frac{8\pi}{3}$ **25.** $\pi, 3\pi$ **27.** $\frac{7\pi}{6}, \frac{11\pi}{6}, \frac{19\pi}{6}, \frac{23\pi}{6}$

29. $\frac{\pi}{4}, \frac{5\pi}{4}, \frac{9\pi}{4}, \frac{13\pi}{4}$ **31.** $\frac{5\pi}{4}, \frac{7\pi}{4}, \frac{13\pi}{4}, \frac{15\pi}{4}$

33. $\frac{\pi}{2} + k\pi, k =$ any integer **35.** $\frac{\pi}{2} + 2k\pi, k =$ any integer

37. $k\pi, k =$ any integer

39.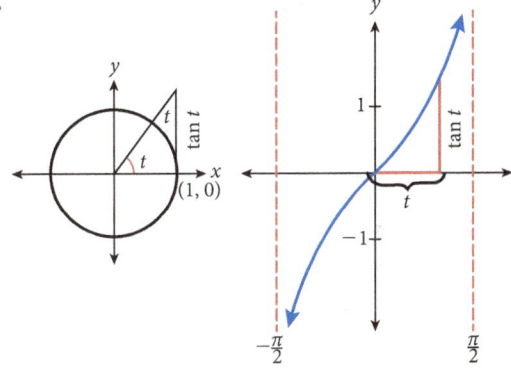

41. Amplitude = 3, period = π **43.** Amplitude = 2, period = 2
45. Amplitude = 3, period = π
47. After you have tried the problem yourself, look at
Example 1 in Section 4.2.
49. After you have tried the problem yourself, look at
Example 4 in Section 4.2.
51. 2 **53.** 4π **55.** 2π **57.** 0 **59.** 2 **61.** -1

PROBLEM SET 4.2
Vocabulary
 1. amplitude **2.** period **3.** negative
 4. period, amplitude **5.** period, domain, range

Problems

1.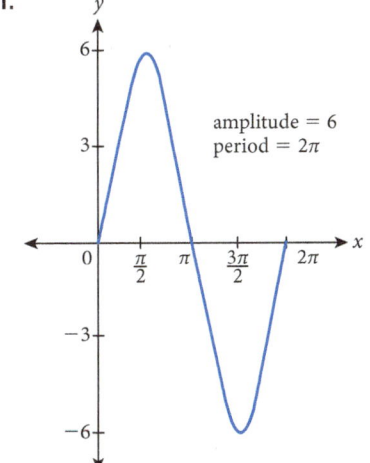

3.

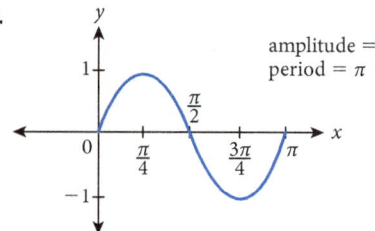

amplitude = 1
period = π

5.

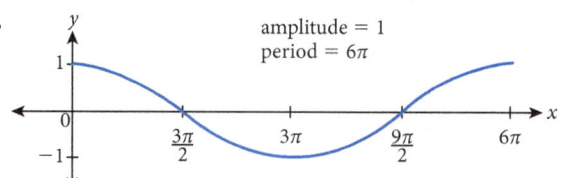

amplitude = 1
period = 6π

7.

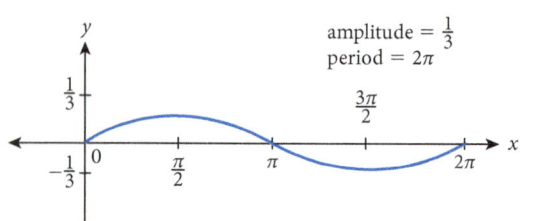

amplitude = $\frac{1}{3}$
period = 2π

9.

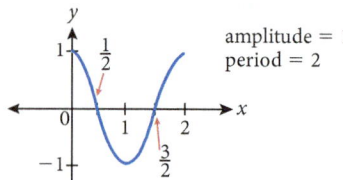

amplitude = 1
period = 2

11.

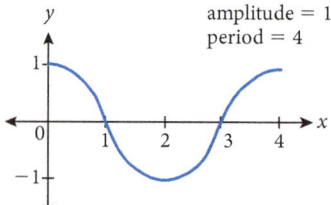

amplitude = 1
period = 4

13.

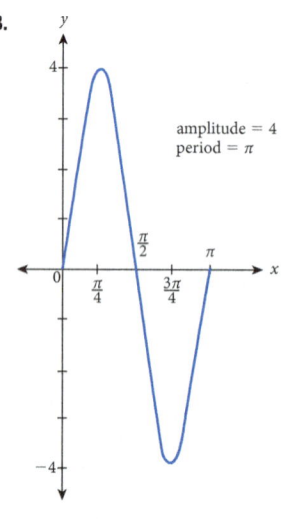

amplitude = 4
period = π

15.

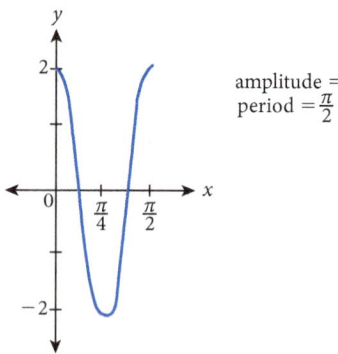

amplitude = 2
period = $\frac{\pi}{2}$

17.

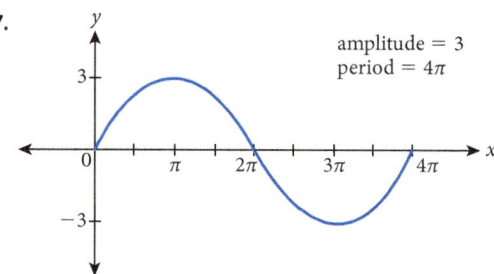

amplitude = 3
period = 4π

19.

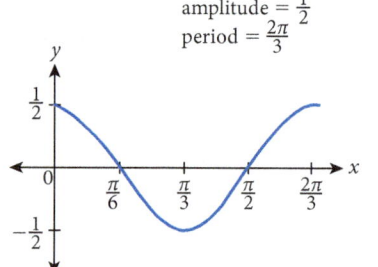

amplitude = $\frac{1}{2}$
period = $\frac{2\pi}{3}$

21.

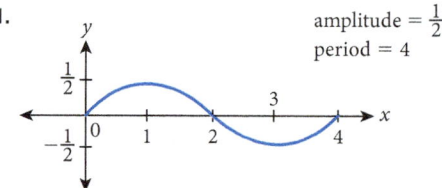

amplitude = $\frac{1}{2}$
period = 4

23.

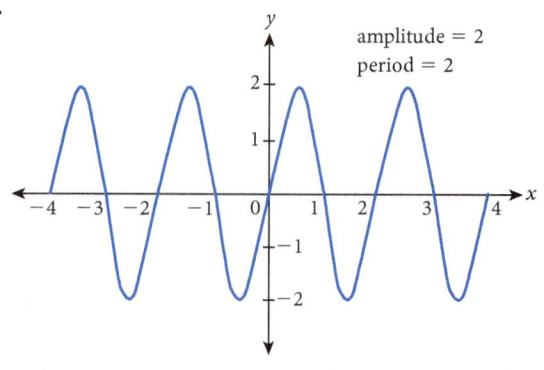

amplitude = 2
period = 2

25.

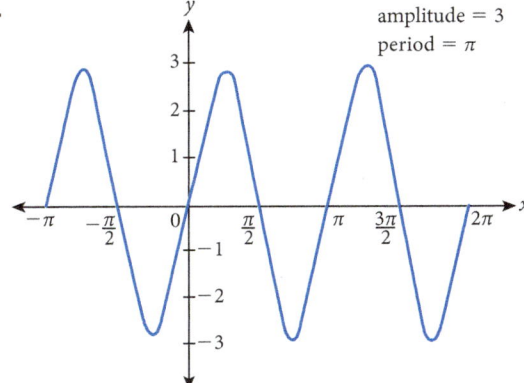

amplitude = 3
period = π

27.

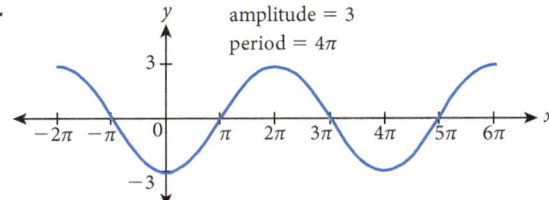

amplitude = 3
period = 4π

29.

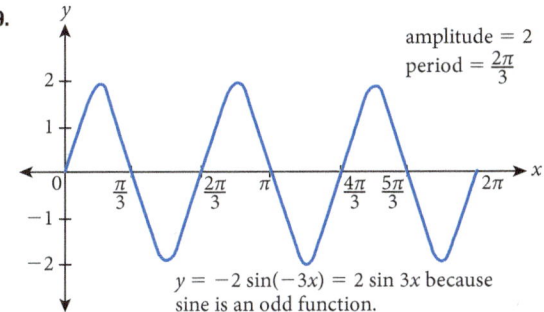

amplitude = 2
period = $\frac{2\pi}{3}$

$y = -2\sin(-3x) = 2\sin 3x$ because
sine is an odd function.

31. Maximum value of I is 20 amperes, one complete cycle takes $\frac{2\pi}{120\pi} = \frac{1}{60}$ seconds.

33. 2 cycles

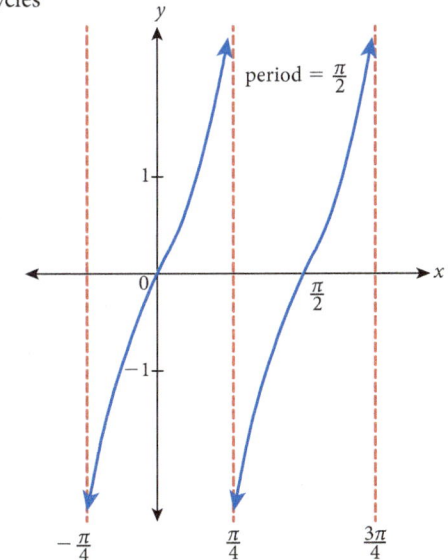

period = $\frac{\pi}{2}$

35.

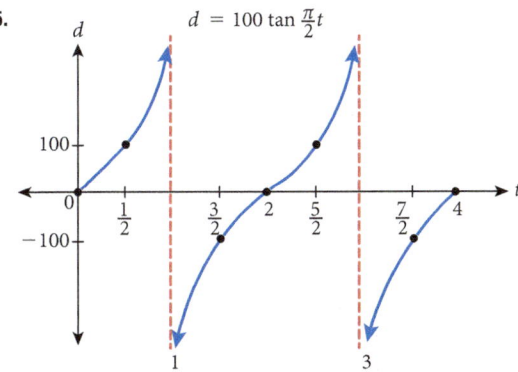

$d = 100 \tan \frac{\pi}{2}t$

37.

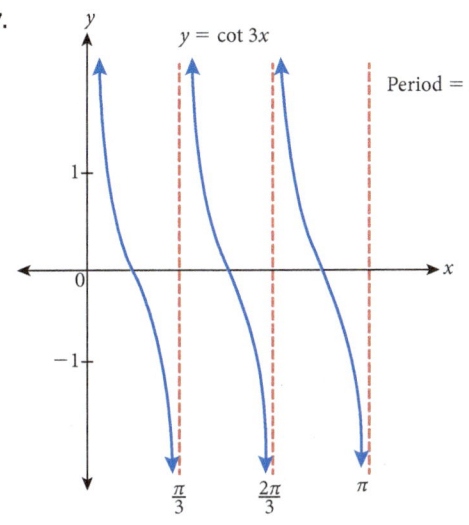

$y = \cot 3x$

Period = $\frac{\pi}{3}$

39.

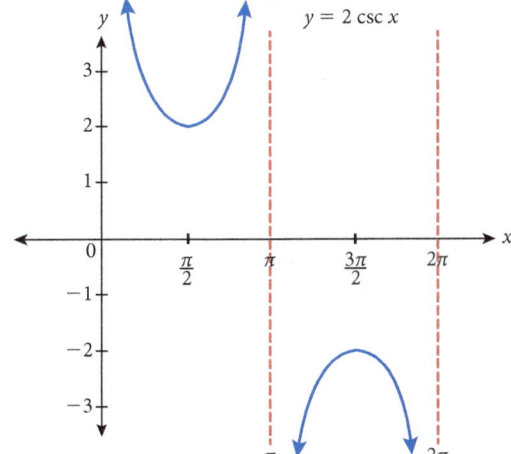

$y = 2\csc x$

41.

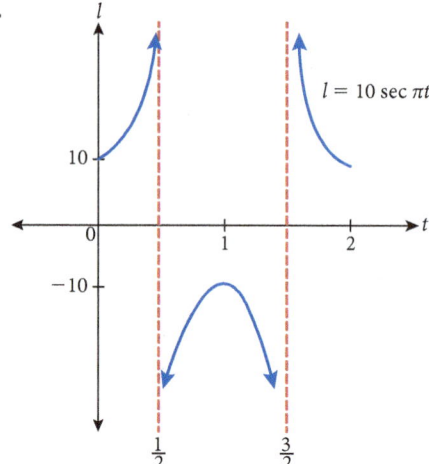

$l = 10 \sec \pi t$

43.

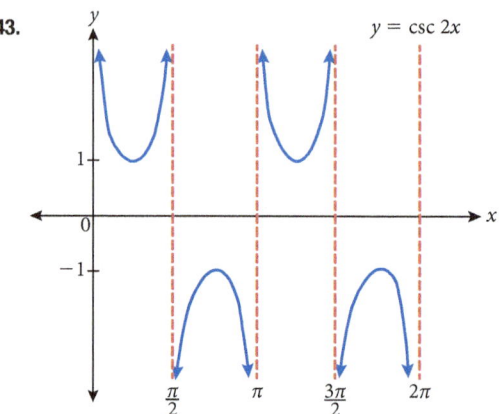

$y = \csc 2x$

45.

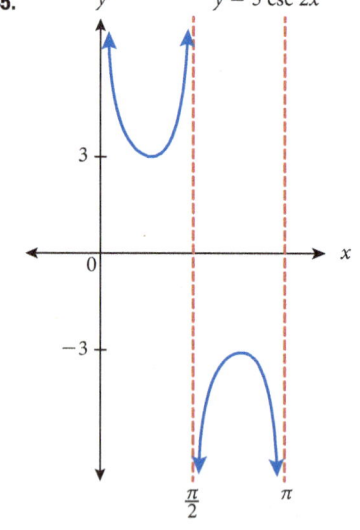

$y = 3 \csc 2x$

47.

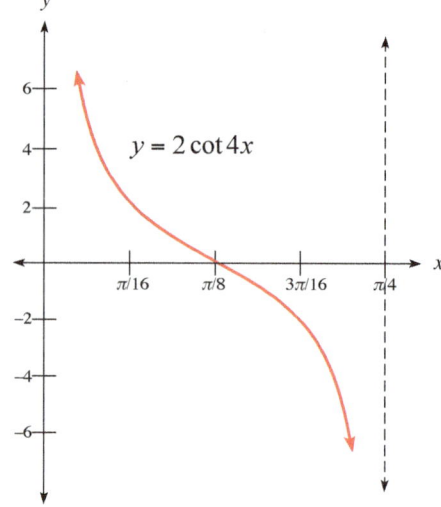

$y = 2 \cot 4x$

49.

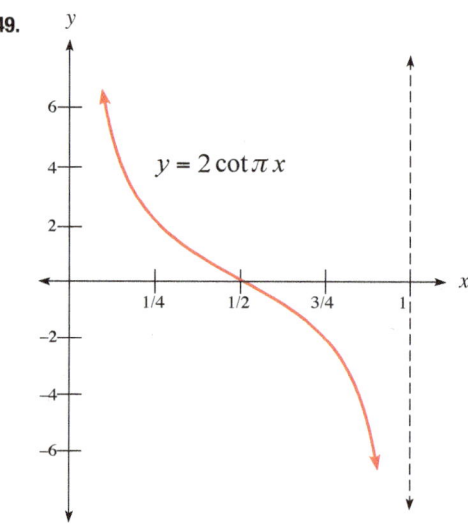

$y = 2 \cot \pi x$

51.

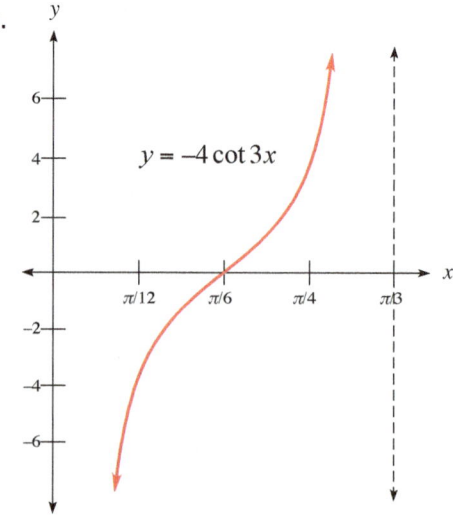

$y = -4 \cot 3x$

53.

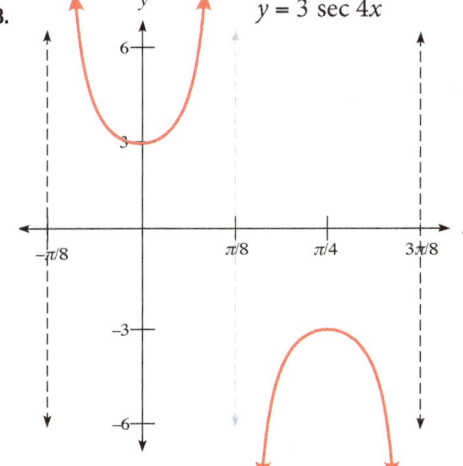

$y = 3 \sec 4x$

55.

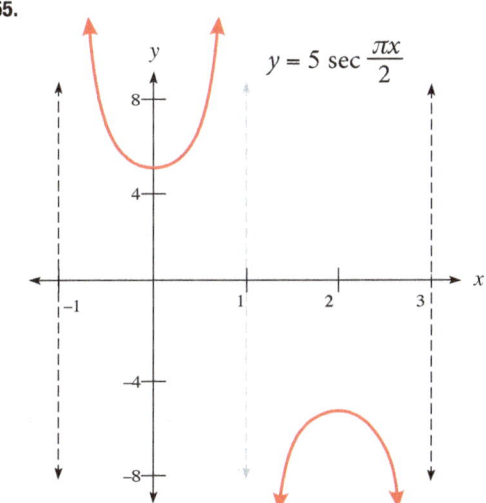

$y = 5 \sec \dfrac{\pi x}{2}$

57.

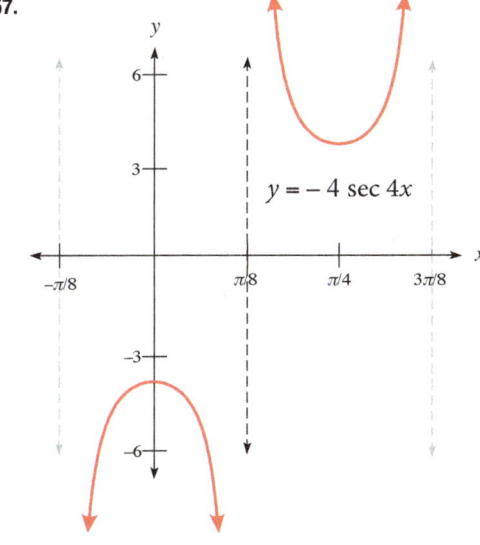

$y = -4 \sec 4x$

59.

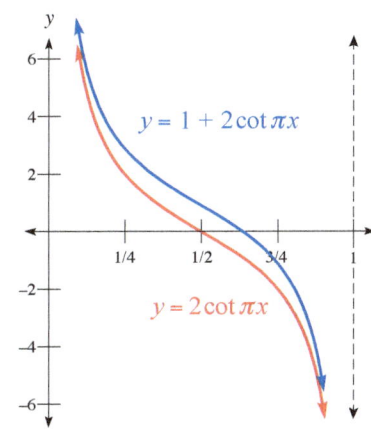

$y = 1 + 2\cot \pi x$

$y = 2\cot \pi x$

61.

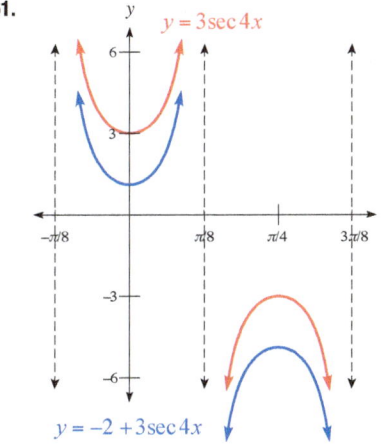

$y = 3\sec 4x$

$y = -2 + 3\sec 4x$

63.

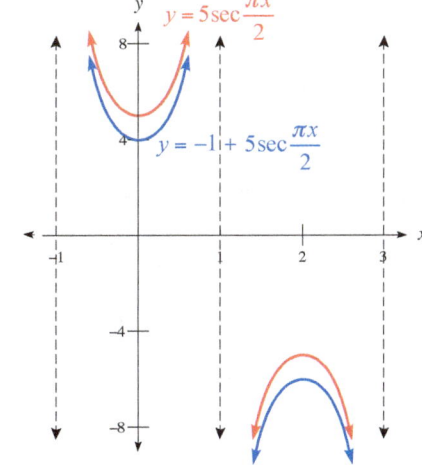

$y = 5\sec \dfrac{\pi x}{2}$

$y = -1 + 5\sec \dfrac{\pi x}{2}$

65. 0 **67.** 1 **69.** $\dfrac{\sqrt{3}}{2}$ **71.** $\dfrac{3}{2}$ **73.** $\dfrac{3\pi}{2}$

75. $-\dfrac{\pi}{4}$ **77.** $\dfrac{5\pi}{12}$

PROBLEM SET 4.3

Vocabulary

1. horizontal   **2.** phase   **3.** vertical

4. cycle, amplitude, phase

Problems

1.

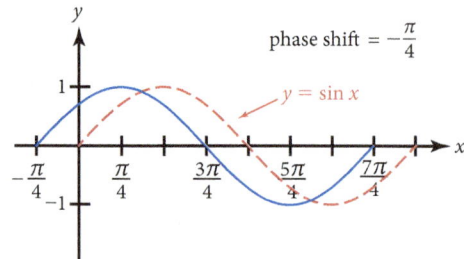

phase shift $= -\dfrac{\pi}{4}$

$y = \sin x$

3.

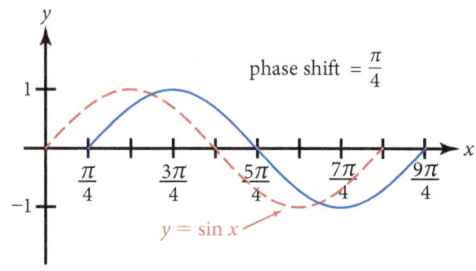

phase shift $= \dfrac{\pi}{4}$

$y = \sin x$

5.

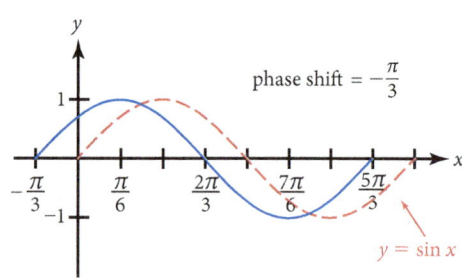

phase shift $= -\dfrac{\pi}{3}$

$y = \sin x$

7.

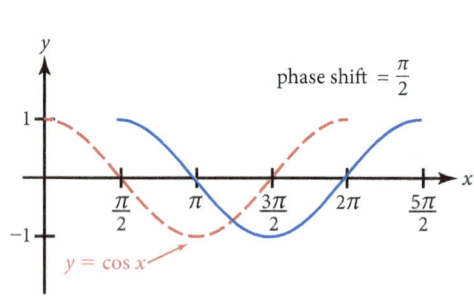

phase shift $= \dfrac{\pi}{2}$

$y = \cos x$

9.

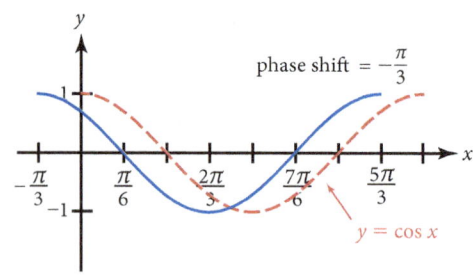

phase shift $= -\dfrac{\pi}{3}$

$y = \cos x$

11.

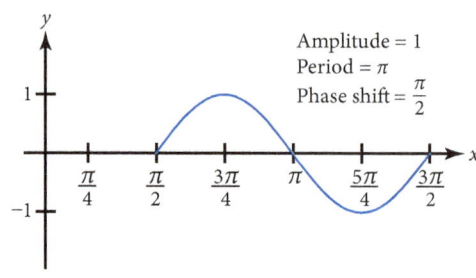

Amplitude $= 1$

Period $= \pi$

Phase shift $= \dfrac{\pi}{2}$

13.

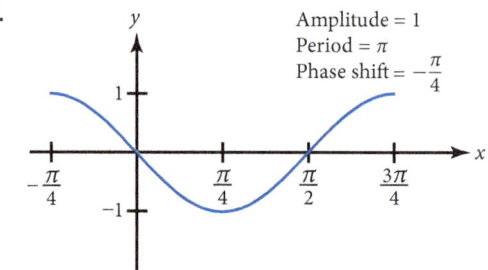

Amplitude $= 1$

Period $= \pi$

Phase shift $= -\dfrac{\pi}{4}$

15.

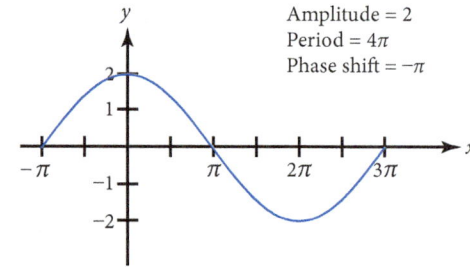

Amplitude $= 2$

Period $= 4\pi$

Phase shift $= -\pi$

17.

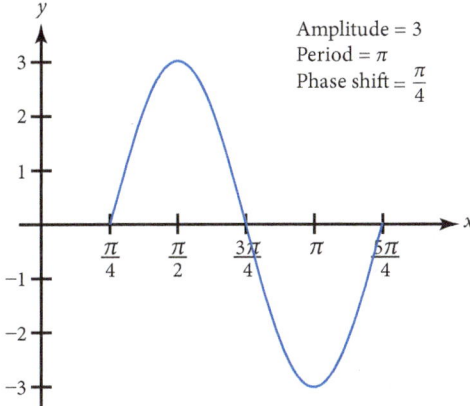

Amplitude = 3
Period = π
Phase shift = $\dfrac{\pi}{4}$

19.

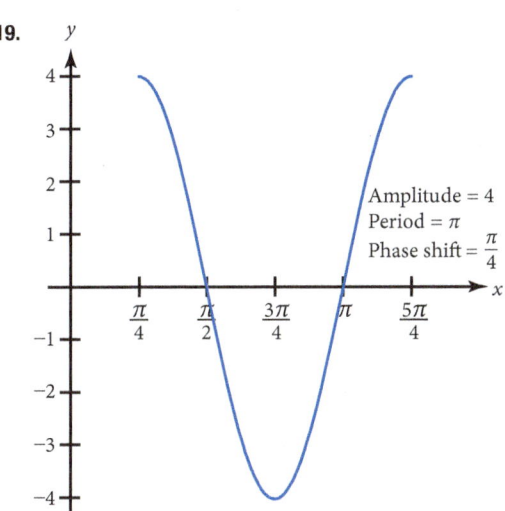

Amplitude = 4
Period = π
Phase shift = $\dfrac{\pi}{4}$

21.

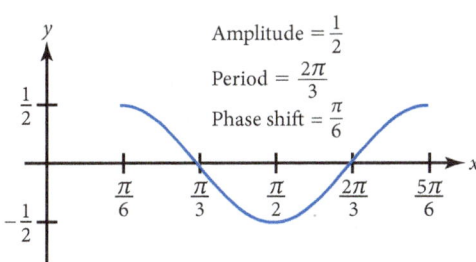

Amplitude = $\dfrac{1}{2}$
Period = $\dfrac{2\pi}{3}$
Phase shift = $\dfrac{\pi}{6}$

23.

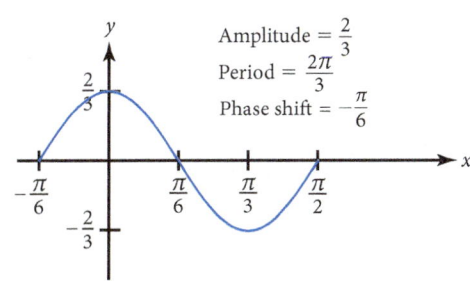

Amplitude = $\dfrac{2}{3}$
Period = $\dfrac{2\pi}{3}$
Phase shift = $-\dfrac{\pi}{6}$

25.

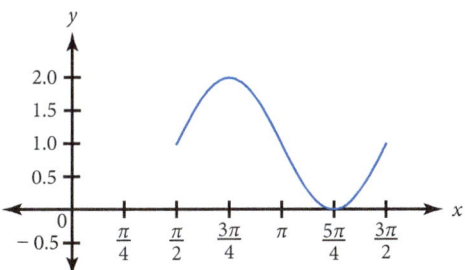

27.

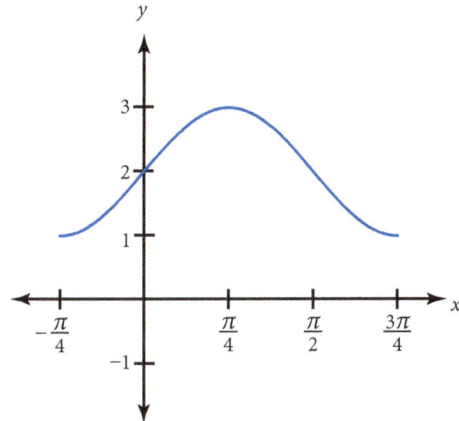

29.

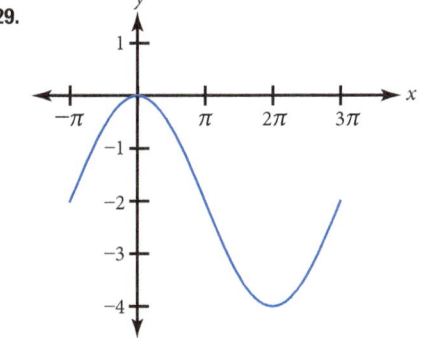

31.

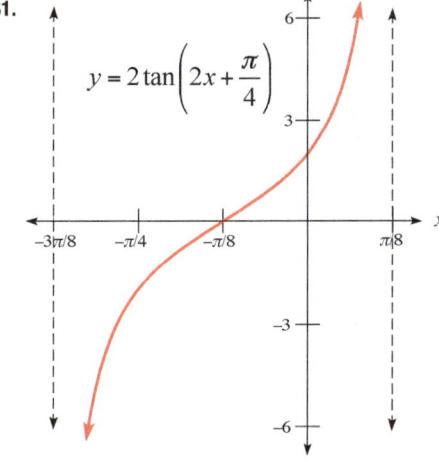

$y = 2\tan\left(2x + \dfrac{\pi}{4}\right)$

33.

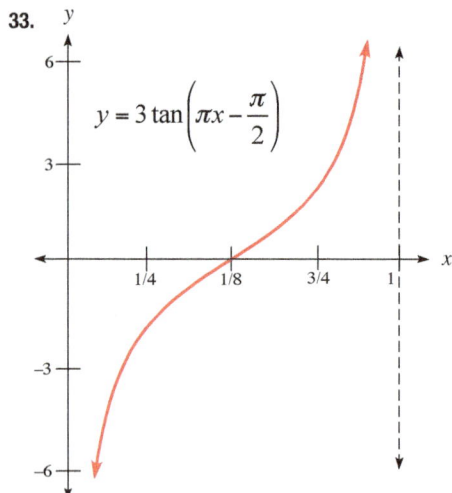

$$y = 3\tan\left(\pi x - \frac{\pi}{2}\right)$$

35.

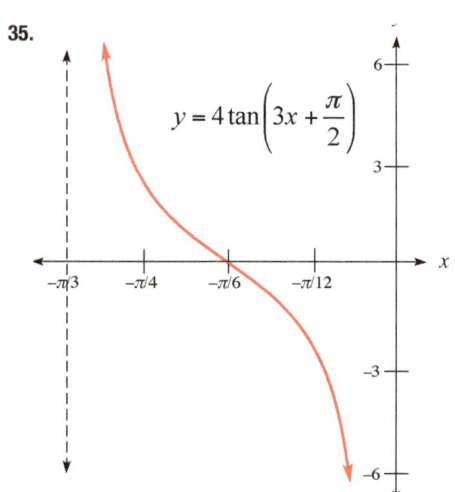

$$y = 4\tan\left(3x + \frac{\pi}{2}\right)$$

37.

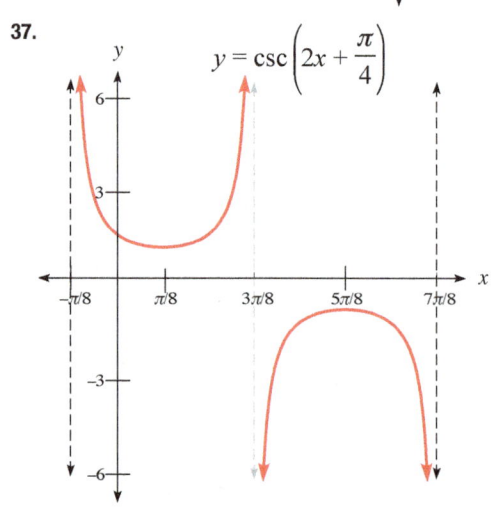

$$y = \csc\left(2x + \frac{\pi}{4}\right)$$

39.

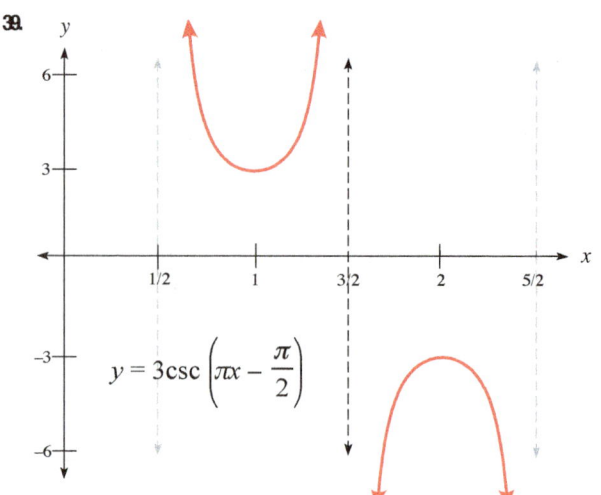

$$y = 3\csc\left(\pi x - \frac{\pi}{2}\right)$$

41.

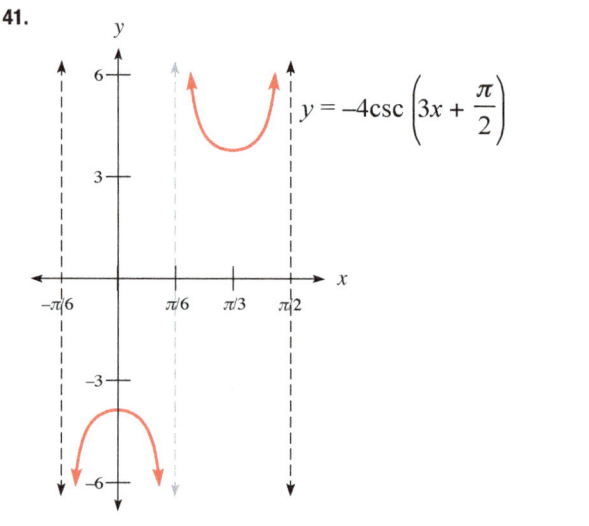

$$y = -4\csc\left(3x + \frac{\pi}{2}\right)$$

43.

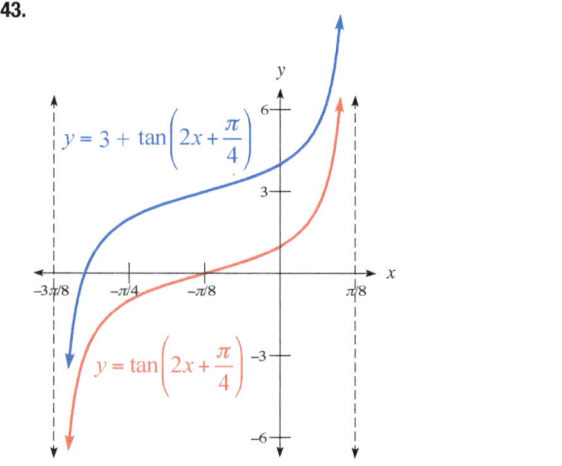

$$y = 3 + \tan\left(2x + \frac{\pi}{4}\right)$$

$$y = \tan\left(2x + \frac{\pi}{4}\right)$$

45.

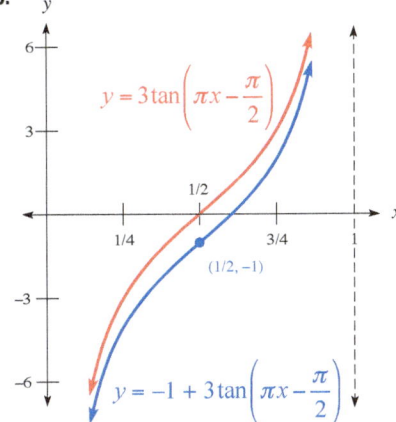

$y = 3\tan\left(\pi x - \dfrac{\pi}{2}\right)$

$y = -1 + 3\tan\left(\pi x - \dfrac{\pi}{2}\right)$

$(1/2, -1)$

47.

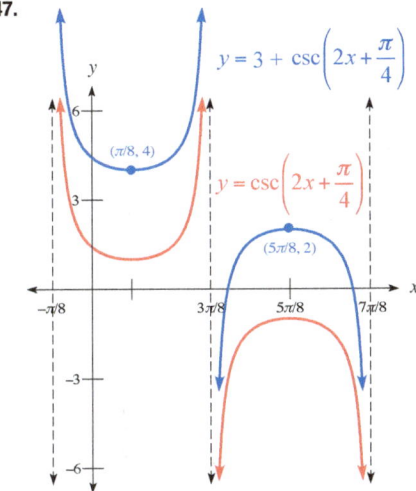

$y = 3 + \csc\left(2x + \dfrac{\pi}{4}\right)$

$y = \csc\left(2x + \dfrac{\pi}{4}\right)$

$(\pi/8, 4)$

$(5\pi/8, 2)$

49. Slope $= \dfrac{1}{2}$, y-intercept $= -4$, $y = \dfrac{1}{2}x - 4$

51. Slope $= -\dfrac{2}{3}$, y-intercept $= 3$, $y = -\dfrac{2}{3}x + 3$

PROBLEM SET 4.4

1. $y = \dfrac{1}{2}x + 1$ **3.** $y = 2x - 3$ **5.** $y = \sin x$

7. $y = 3\cos x$ **9.** $y = -3\cos x$ **11.** $y = \sin 3x$

13. $y = \sin \dfrac{1}{3}x$ **15.** $y = 2\cos 3x$ **17.** $y = 4\sin \pi x$

19. $y = -4\sin \pi x$ **21.** $y = 2 - 4\sin \pi x$

23. $y = 3\cos\left(2x + \dfrac{\pi}{2}\right)$ **25.** $y = -2\cos\left(3x + \dfrac{\pi}{2}\right)$

27. $h = 110.5 - 98.5\cos\left(\dfrac{2\pi t}{15}\right)$

t	h
0 min	12 ft
1.875 min	40.8 ft
3.75 min	110.5 ft
5.625 min	180.2 ft
7.5 min	209 ft
9.375 min	180.2 ft
11.25 min	110.5 ft
13.125 min	40.8 ft
15 min	12 ft

29. 2.57   **31.** 0   **33.** 6.28

PROBLEM SET 4.5

1.

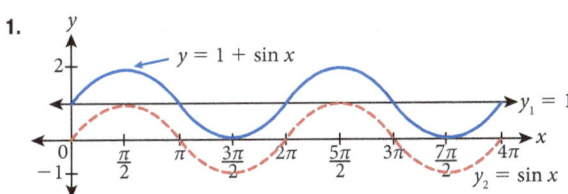

$y = 1 + \sin x$

$y_1 = 1$

$y_2 = \sin x$

3.

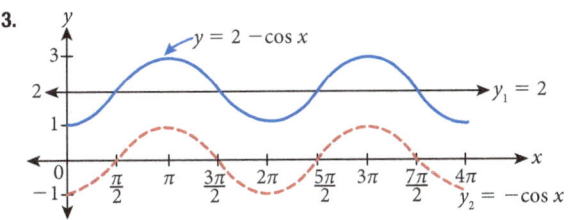

$y = 2 - \cos x$

$y_1 = 2$

$y_2 = -\cos x$

5.

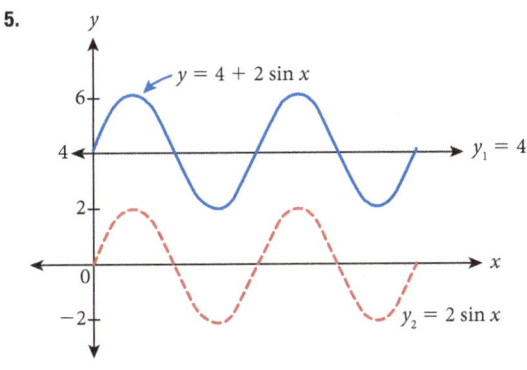

$y = 4 + 2\sin x$

$y_1 = 4$

$y_2 = 2\sin x$

7.

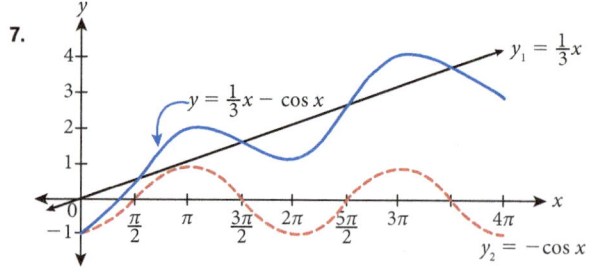

9.

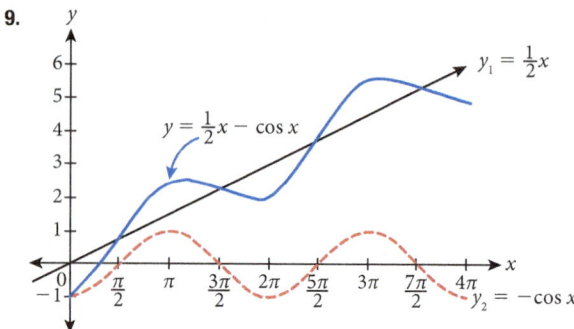

11.

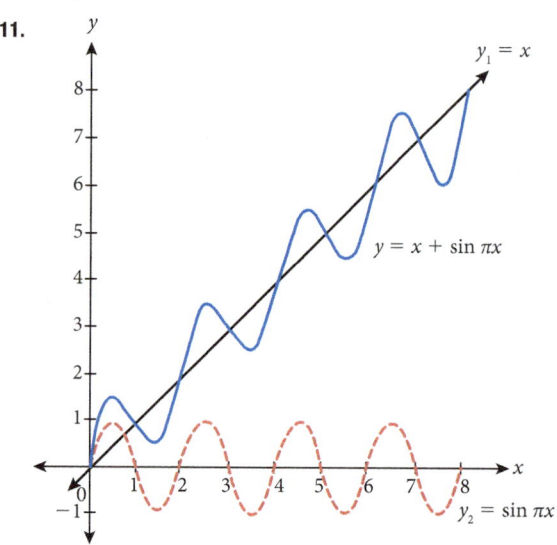

13.

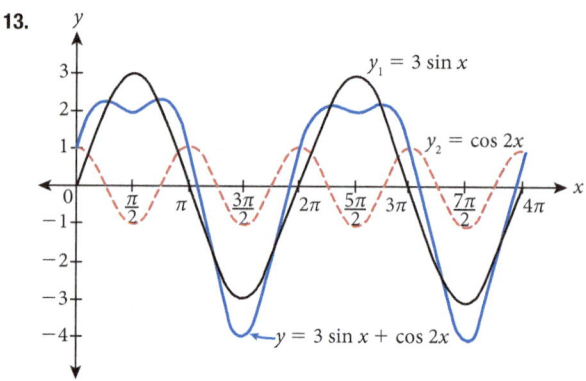

15.

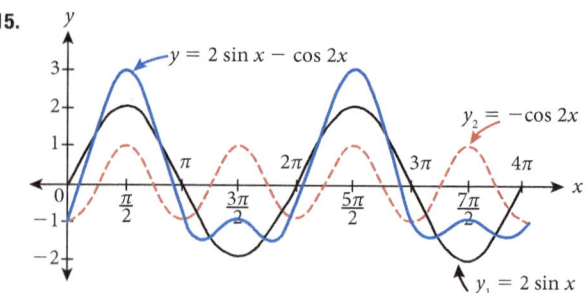

17.

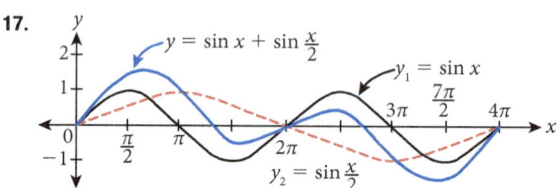

19.

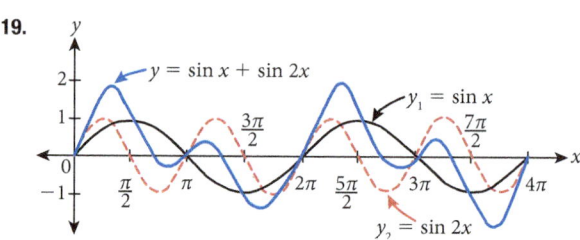

21.

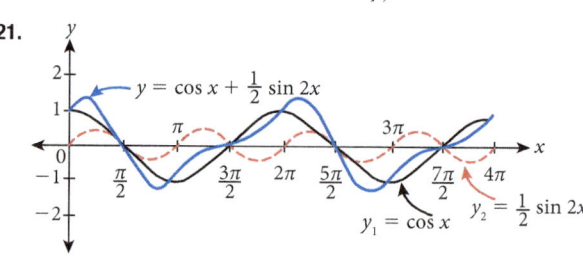

23.

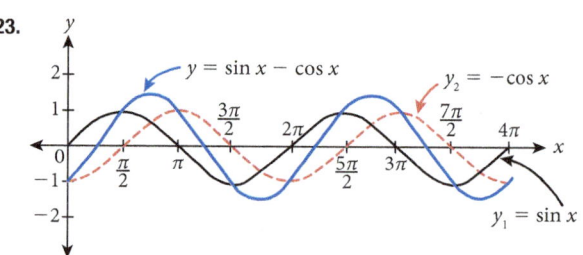

25.

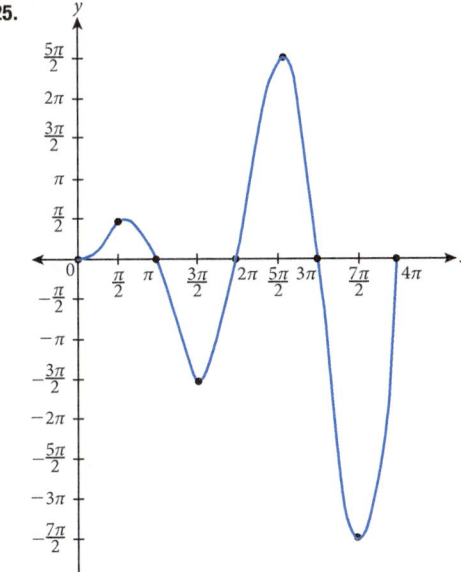

27. $y = \frac{x+3}{2}$ **29.** $y = \pm\sqrt{x-4}$ **31.** $45°$ **33.** $30.5°$

35. $130.0°$

PROBLEM SET 4.6

Vocabulary

1. horizontal   **2.** function   **3.** inverse   **4.** restrict

Problems

1.
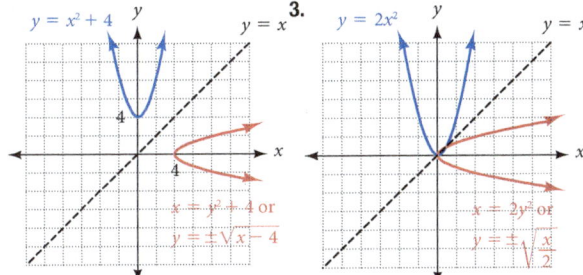

5.
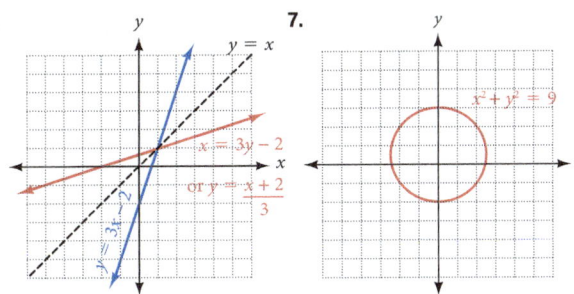

9.
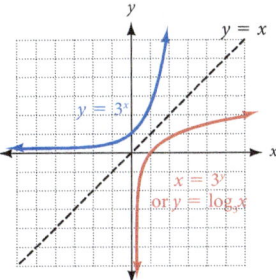

11.
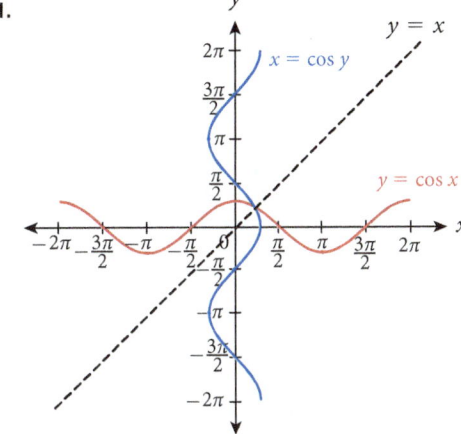

13. We have not included the graph of $y = \tan x$ with the graph of $x = \tan y$ because placing them both on the same coordinate system makes the diagram too complicated. It is best to graph $y = \tan x$ lightly in pencil and then reflect that graph about the line $y = x$ to get the graph of $x = \tan y$.

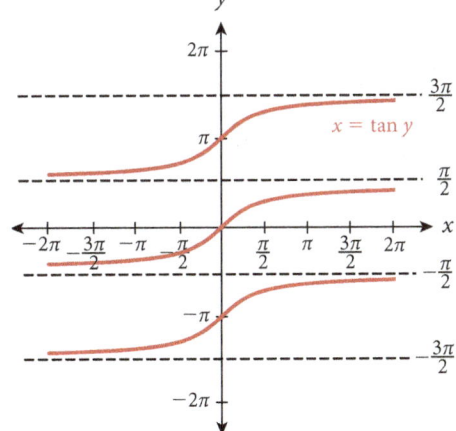

15. $-1 \le x \le 1$ **17.** $0 \le y \le \pi$ **19.** $\frac{\pi}{3}$ **21.** π **23.** $\frac{\pi}{4}$

25. $\frac{3\pi}{4}$ **27.** $-\frac{\pi}{6}$ **29.** $\frac{\pi}{3}$ **31.** 0 **33.** $-\frac{\pi}{6}$ **35.** $\frac{2\pi}{3}$ **37.** $\frac{\pi}{6}$

39. $9.8°$ **41.** $147.4°$ **43.** $20.8°$ **45.** $74.3°$ **47.** $117.8°$

49. $-70.0°$ **51.** $-50.0°$ **53.** $\frac{4}{5}$ **55.** $\frac{3}{4}$ **57.** $\sqrt{5}$ **59.** $\frac{\sqrt{3}}{2}$

61. 2 **63.** $\frac{3}{5}$ **65.** $\frac{1}{2}$ **67.** $\frac{1}{2}$ **69.** x **71.** $\sqrt{1-x^2}$ **73.** $\frac{x}{\sqrt{x^2+1}}$

75. 55°   **77.** 240°   **79.** $-\frac{1}{2}$ **81.** 2π meters $\cong$ 6.28 meters

83. 8 inches^2

CHAPTER 4 TEST

1.
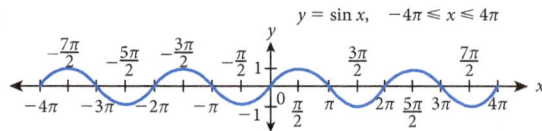
$y = \sin x, \quad -4\pi \leqslant x \leqslant 4\pi$

2.
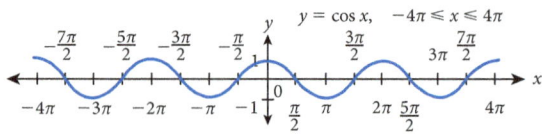
$y = \cos x, \quad -4\pi \leqslant x \leqslant 4\pi$

3.
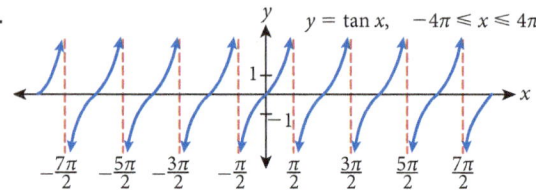
$y = \tan x, \quad -4\pi \leqslant x \leqslant 4\pi$

4.
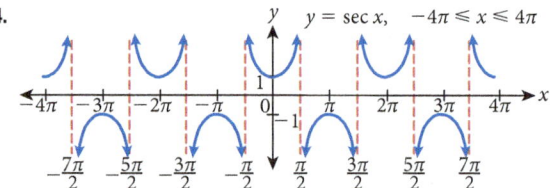
$y = \sec x, \quad -4\pi \leqslant x \leqslant 4\pi$

5. 4 cycles   **6.** 8 cycles   **7.** $-3\pi, -\pi, \pi, 3\pi$

8. $-\frac{11\pi}{3}, -\frac{7\pi}{3}, -\frac{5\pi}{3}, -\frac{\pi}{3}, \frac{\pi}{3}, \frac{5\pi}{3}, \frac{7\pi}{3}, \frac{11\pi}{3}$

9.
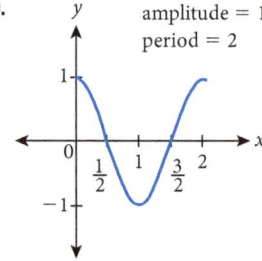
amplitude = 1
period = 2

10.
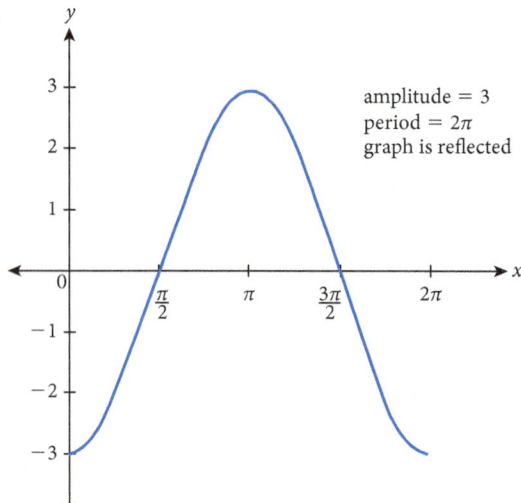
amplitude = 3
period = 2π
graph is reflected

11.
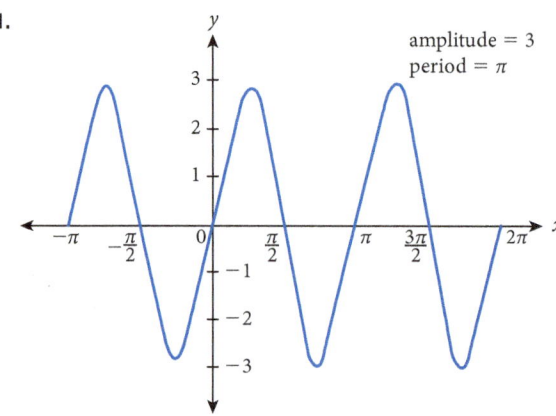
amplitude = 3
period = π

12.
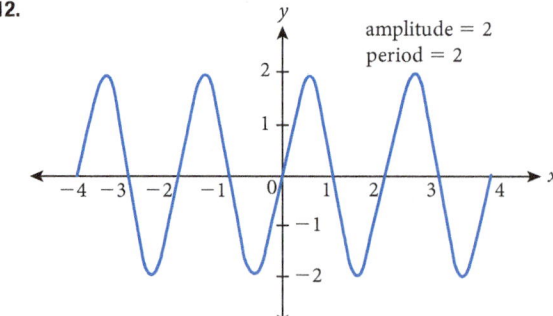
amplitude = 2
period = 2

13.
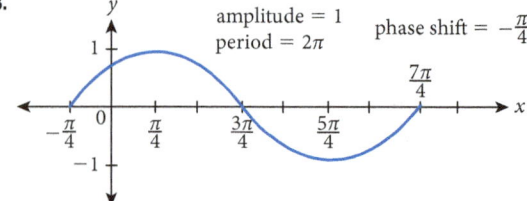
amplitude = 1
period = 2π
phase shift = $-\frac{\pi}{4}$

14.

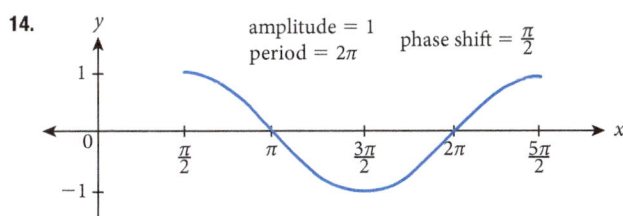

amplitude = 1
period = 2π phase shift = $\frac{\pi}{2}$

15.

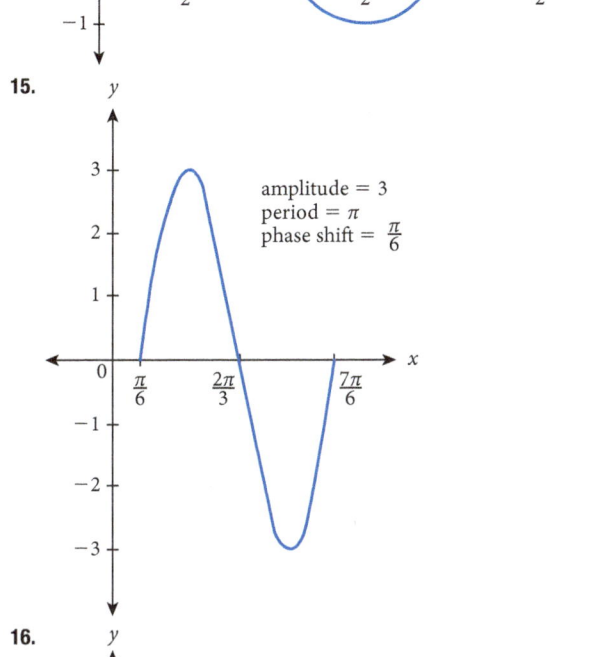

amplitude = 3
period = π
phase shift = $\frac{\pi}{6}$

16.

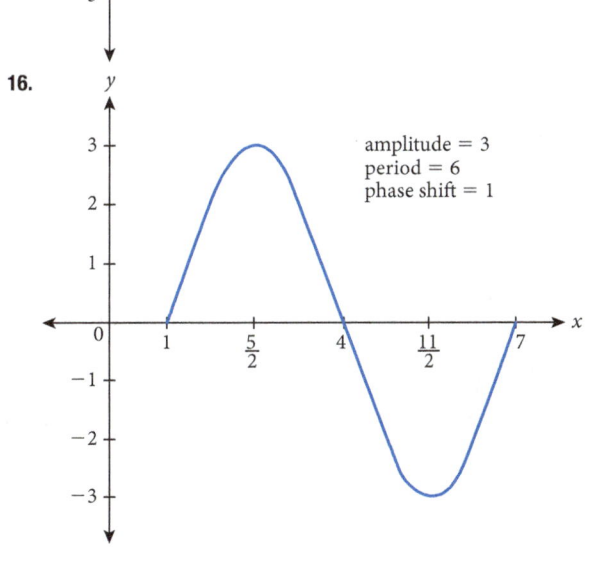

amplitude = 3
period = 6
phase shift = 1

17.

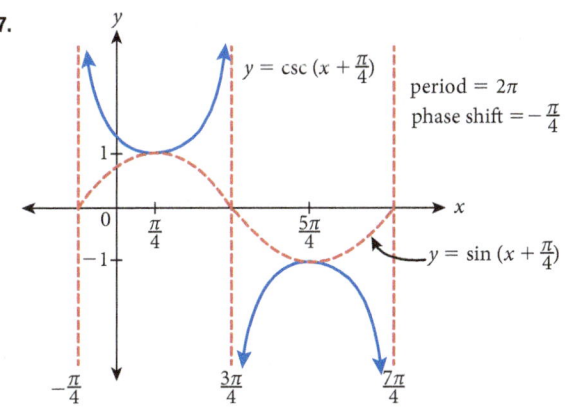

$y = \csc\left(x + \frac{\pi}{4}\right)$

period = 2π
phase shift = $-\frac{\pi}{4}$

$y = \sin\left(x + \frac{\pi}{4}\right)$

18.

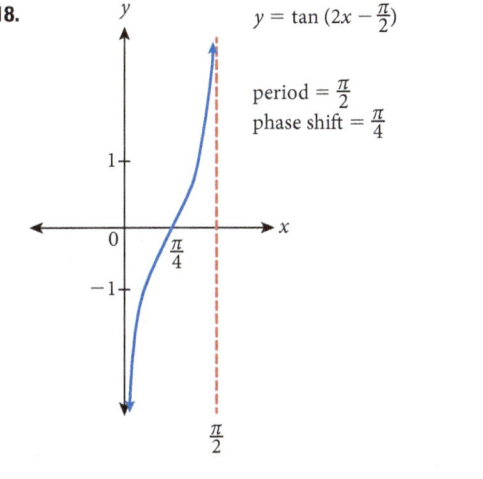

$y = \tan\left(2x - \frac{\pi}{2}\right)$

period = $\frac{\pi}{2}$
phase shift = $\frac{\pi}{4}$

19.

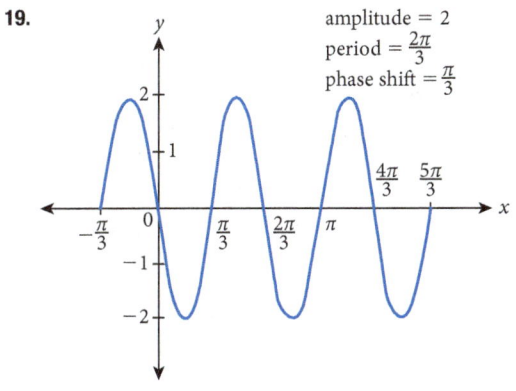

amplitude = 2
period = $\frac{2\pi}{3}$
phase shift = $\frac{\pi}{3}$

20.

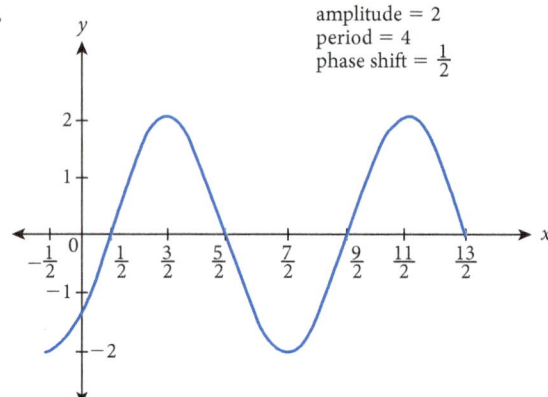

amplitude = 2
period = 4
phase shift = $\frac{1}{2}$

21.

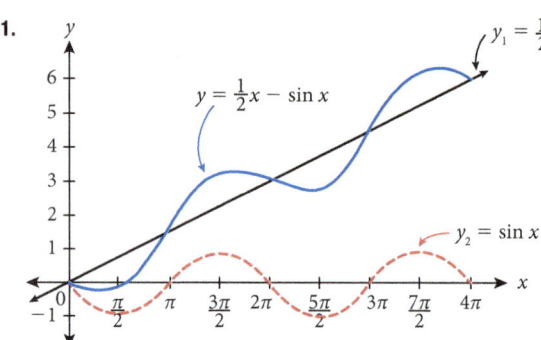

22.

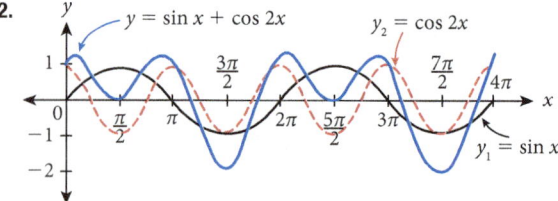

23. 0   **24.** π

25.

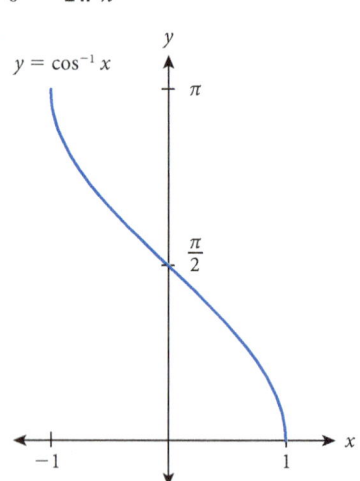

26.

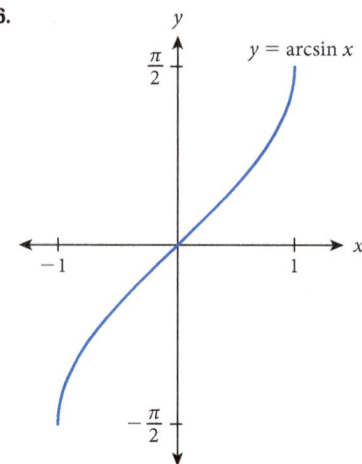

27. $\frac{\pi}{6}$ **28.** $\frac{5\pi}{6}$ **29.** $-\frac{\pi}{4}$ **30.** $\frac{\pi}{2}$ **31.** 36.4°

32. $-39.7°$ **33.** 134.3°   **34.** $-12.5°$ **35.** $\frac{\sqrt{5}}{2}$

36. $\frac{3}{\sqrt{13}}$ **37.** $\sqrt{1-x^2}$ **38.** $\frac{x}{\sqrt{1-x^2}}$

Chapter 5

PROBLEM SET 5.1

Vocabulary

1. identity   **2.** reciprocal   **3.** ratio   **4.** Pythagorean
5. complicated

Problems

45. $\sin(30° + 60°) = \sin 90° = 1$ but
$\sin 30° + \sin 60° = \frac{1}{2} + \frac{\sqrt{3}}{2} \neq 1$

47. $\cos A = \frac{4}{5}$, $\tan A = \frac{3}{4}$ **49.** $\frac{\sqrt{3}}{2}$ **51.** $\frac{\sqrt{3}}{2}$ **53.** 15°

55.

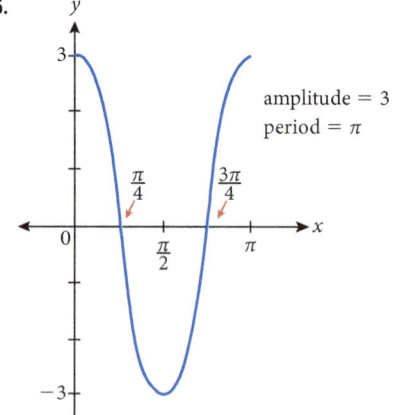

amplitude = 3
period = π

PROBLEM SET 5.2

Vocabulary

1. not   **2.** $\cos A$ **3.** $\sin A, \cos A$ **4.** $\sin A$ **5.** $\cos A, \sin B$

Problems

1. $\frac{\sqrt{6} - \sqrt{2}}{4}$ **3.** $\frac{\sqrt{6} + \sqrt{2}}{4}$ **5.** $\frac{\sqrt{3} - 1}{\sqrt{3} + 1} = 2 - \sqrt{3}$

7. $\dfrac{\sqrt{6}+\sqrt{2}}{4}$ **9.** $-\dfrac{\sqrt{6}+\sqrt{2}}{4}$

11.-20. Expand the left side of each using the appropriate sum or difference formula and then simplify.

21. $\sin 5x$ **23.** $\cos 6x$ **25.** $\sin(45° + \theta)$

27. $\sin(30° + \theta)$ **29.** $\cos 90° = 0$

31. $\sin(A + B) = -\dfrac{16}{65}$, $\cos(A + B) = \dfrac{63}{65}$,

$\tan(A + B) = -\dfrac{16}{63}$

33. $\tan(A + B) = 2$, $\cot(A + B) = \dfrac{1}{2}$ **35.** 1

37. $\sin 2x = 2\sin x \cos x$

39. $y = \sin\left(x + \dfrac{\pi}{4}\right)$

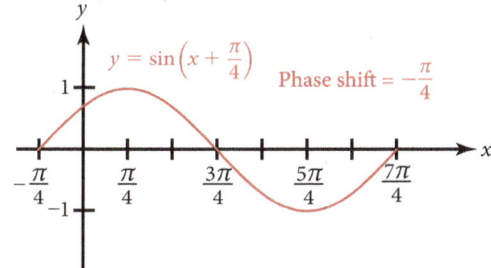

41. $y = 2\sin\left(x + \dfrac{\pi}{3}\right)$

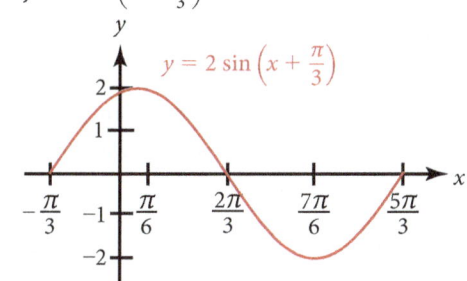

47. $-\dfrac{4}{5}$ **49.** $\dfrac{4}{5}$ **51.** $\dfrac{13}{5}$

PROBLEM SET 5.3

Vocabulary

1. is not   **2.** is   **3.** false   **4.** true

Problems

1. $\dfrac{24}{25}$ **3.** $\dfrac{24}{7}$ **5.** $-\dfrac{4}{5}$ **7.** $\dfrac{4}{3}$ **9.** $\dfrac{120}{169}$ **11.** $\dfrac{169}{120}$

13. $\dfrac{3}{5}$ **15.** $\dfrac{5}{3}$ **17.** $\cos 100° = 1 - 2\sin^2 50°$

19. $\sin \dfrac{\pi}{5} = 2\sin \dfrac{\pi}{10}\cos \dfrac{\pi}{10}$ **29.** $\dfrac{\sqrt{2-\sqrt{3}}}{2}$

45.

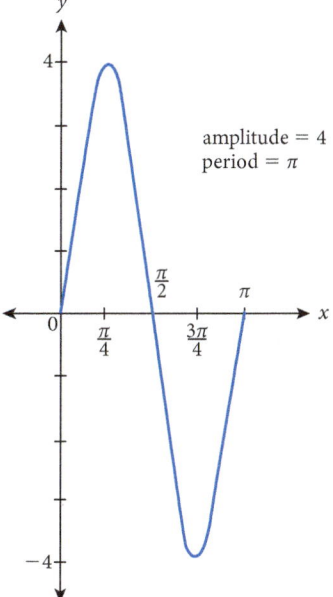

amplitude = 4
period = π

47.

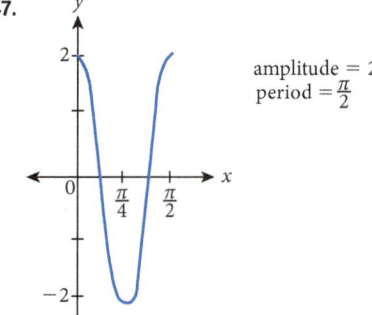

amplitude = 2
period = $\dfrac{\pi}{2}$

49.

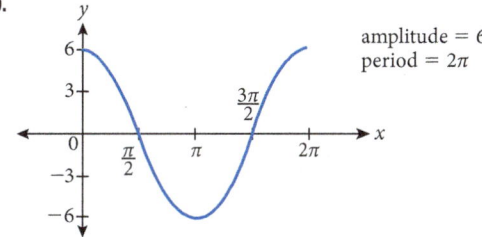

amplitude = 6
period = 2π

51. $\dfrac{5}{13}$ **53.** $\dfrac{1}{\sqrt{5}}$ **55.** $\dfrac{1}{2}$

PROBLEM SET 5.4

Vocabulary

1. is   **2.** is not   **3.** false   **4.** true

Problems

1. $\dfrac{1}{2}$ **3.** 2   **5.** $-\dfrac{1}{\sqrt{10}}$ **7.** $-\dfrac{1}{3}$ **9.** $-\sqrt{10}$

11. $\dfrac{2}{\sqrt{5}}$ **13.** $-\dfrac{7}{25}$ **15.** $-\dfrac{25}{7}$ **17.** $\dfrac{3}{\sqrt{10}}$ **19.** $\dfrac{7}{25}$

21. 0   **23.** $\dfrac{\sqrt{2+\sqrt{3}}}{2}$ **25.** $\dfrac{\sqrt{2+\sqrt{3}}}{2}$ **27.** $-\dfrac{\sqrt{2-\sqrt{3}}}{2}$

39. $\dfrac{3}{5}$ **41.** $\dfrac{1}{\sqrt{5}}$ **43.** $\dfrac{x}{\sqrt{x^2+1}}$ **45.** $\dfrac{x}{\sqrt{1-x^2}}$

PROBLEM SET 5.5

Problems

1. $-\frac{1}{\sqrt{5}}$ **3.** $\frac{2\sqrt{3}-1}{2\sqrt{5}}$ **5.** $\frac{4}{5}$ **7.** $\frac{x}{\sqrt{1-x^2}}$ **9.** $2x\sqrt{1-x^2}$

11. $2x^2-1$ **15.** $5(\sin 8x + \sin 2x)$ **17.** $\frac{1}{2}(\cos 10x + \cos 6x)$

19. $\frac{3}{4}$ **21.** 0 **25.** $2\sin 5x \cos 2x$

27. $2\cos 30° \cos 15° = \sqrt{3}\cos 15°$

29. $2\cos\frac{\pi}{3}\sin\frac{\pi}{4} = \frac{1}{\sqrt{2}}$

37.

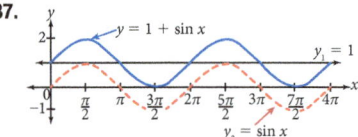

39.

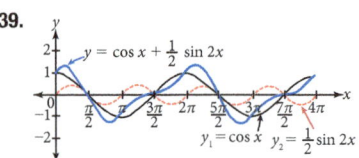

41.

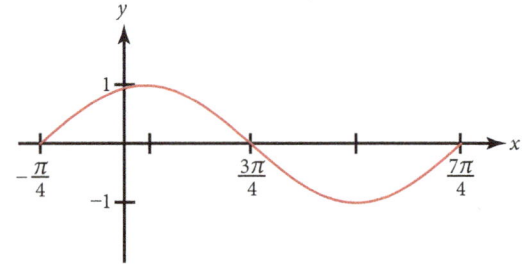

43.

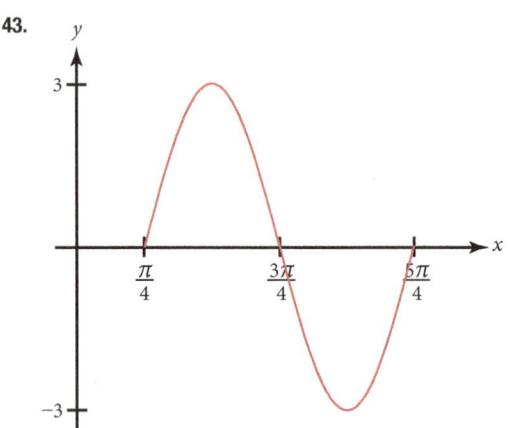

CHAPTER 5 TEST

13. $\frac{63}{65}$ **14.** $-\frac{56}{65}$ **15.** $-\frac{119}{169}$ **16.** $-\frac{120}{169}$ **17.** $\frac{1}{\sqrt{10}}$

18. $-\frac{3}{\sqrt{10}}$ **19.** $\frac{\sqrt{6}+\sqrt{2}}{4} = \frac{\sqrt{2+\sqrt{3}}}{2}$

20. $\frac{\sqrt{6}+\sqrt{2}}{4} = \frac{\sqrt{2+\sqrt{3}}}{2}$ **21.** $\frac{\sqrt{3}-1}{\sqrt{3}+1}$

22. $\frac{\sqrt{3}+1}{\sqrt{3}-1}$ **23.** $\cos 9x$ **24.** $\sin 90° = 1$

25. $\frac{3}{5}$, $-\sqrt{\frac{5-2\sqrt{5}}{10}}$ **26.** $\frac{3}{5}$, $\sqrt{\frac{10-\sqrt{10}}{20}}$ **27.** 1

28. $\pm\frac{\sqrt{3}}{2}$ **29.** $\frac{11}{5\sqrt{5}}$ **30.** $\frac{11}{5\sqrt{5}}$ **31.** $1-2x^2$

32. $2x\sqrt{1-x^2}$ **33.** $\frac{1}{2}(\cos 2x - \cos 10x)$

34. $2\cos 45° \cos(-30°) = \frac{\sqrt{6}}{2}$

Chapter 6

PROBLEM SET 6.1

Vocabulary
1. nonzero   **2.** multiples   **3.** two, one   **4.** no

Problems
1. $30°, 150°$ **3.** $30°, 330°$ **5.** $135°, 315°$ **7.** $\frac{\pi}{3}, \frac{2\pi}{3}$

9. $\frac{\pi}{6}, \frac{11\pi}{6}$ **11.** $\frac{\pi}{2}$ **13.** $48.6°, 131.4°$ **15.** no solution

17. $228.6°, 311.4°$ **19.** $\frac{\pi}{6}, \frac{\pi}{2}, \frac{5\pi}{6}$ **21.** $0, \frac{\pi}{4}, \pi, \frac{5\pi}{4}$

23. $0, \frac{2\pi}{3}, \pi, \frac{4\pi}{3}$ **25.** $\frac{\pi}{2}, \frac{7\pi}{6}, \frac{11\pi}{6}$ **27.** $120°, 150°, 210°, 240°$

29. $0°, 60°, 120°, 180°$ **31.** $120°, 240°$ **33.** $201.5°, 338.5°$

35. $51.8°, 308.2°$ **37.** $30°, 150°$

39. $30° + 360°k, 150° + 360°k$, where k is any integer

41. $\frac{\pi}{3} + 2\pi k, \frac{2\pi}{3} + 2\pi k$, where k is any integer

43. $\frac{\pi}{2} + 2\pi k$, where k is any integer

45. $48.6° + 360°k, 131.4° + 360°k$, where k is any integer

47. $h = -16t^2 + 750t$ **49.** $1,436$ feet   **51.** $15.7°$

53. $\sin 2A = 2\sin A \cos A$ **55.** $\cos 2A = 2\cos^2 A - 1$

57. $\frac{1}{\sqrt{2}}\sin\theta + \frac{1}{\sqrt{2}}\cos\theta$ **59.** $\frac{\sqrt{6}-\sqrt{2}}{4}$

PROBLEM SET 6.2

Vocabulary
1. reciprocal   **2.** radians, degrees   **3.** double angle   **4.** extraneous

Problems
1. $30°, 330°$ **3.** $\frac{5\pi}{4}, \frac{7\pi}{4}$ **5.** $\frac{\pi}{4}, \frac{3\pi}{4}, \frac{5\pi}{4}, \frac{7\pi}{4}$ **7.** $30°, 150°$

9. $30°, 90°, 150°, 270°$ **11.** $60°, 180°, 300°$ **13.** $\frac{7\pi}{6}, \frac{3\pi}{2}, \frac{11\pi}{6}$

15. $0°, 120°, 240°, 360°$ **17.** $\frac{\pi}{2}, \frac{7\pi}{6}, \frac{11\pi}{6}$ **19.** $\frac{\pi}{3}, \frac{5\pi}{3}$

21. $\frac{2\pi}{3}, \frac{4\pi}{3}$ **23.** $45°$ **25.** $30°, 90°$ **27.** $60°, 180°$

29. $\frac{\pi}{3}, \frac{5\pi}{3}$ **31.** $\frac{2\pi}{3}, \pi$

33. $\frac{5\pi}{4} + 2k\pi, \frac{7\pi}{4} + 2k\pi$, where k is any integer

35. $45° + 360°k$, where k is any integer

37. $\frac{2\pi}{3} + 2k\pi, \pi + 2k\pi$, where k is any integer

40. $86.4°$ **41.** $30°, 330°, 390°, 690°$ **43.** $\frac{\pi}{4}, \frac{5\pi}{4}, \frac{9\pi}{4}, \frac{13\pi}{4}$

PROBLEM SET 6.3

1. $30°, 60°, 210°, 240°$ **3.** $67.5°, 157.5°, 247.5°, 337.5°$

5. $60°, 180°, 300°$ **7.** $\frac{\pi}{8}, \frac{3\pi}{8}, \frac{9\pi}{8}, \frac{11\pi}{8}$ **9.** $\frac{\pi}{3}, \pi, \frac{5\pi}{3}$

11. $\frac{\pi}{6}, \frac{2\pi}{3}, \frac{7\pi}{6}, \frac{5\pi}{3}$

13. $15° + 180°k$, $75° + 180°k$, where k is any integer

15. $30° + 60°k$, where k is any integer

17. $6° + 36°k$, $12° + 36°k$, where k is any integer

19. $\frac{\pi}{18}, \frac{5\pi}{18}, \frac{13\pi}{18}, \frac{17\pi}{18}, \frac{25\pi}{18}, \frac{29\pi}{18}$ **21.** $\frac{5\pi}{18}, \frac{7\pi}{18}, \frac{17\pi}{18}, \frac{19\pi}{18}, \frac{29\pi}{18}, \frac{31\pi}{18}$

23. $\frac{\pi}{10} + \frac{2k\pi}{5}$, where k is any integer

25. $\frac{\pi}{8} + \frac{k\pi}{4}$, where k is any integer

27. $\frac{\pi}{5} + \frac{2k\pi}{5}$, where k is any integer

29. $10° + 120°k$, $50° + 120°k$, $90° + 120°k$, where k is any integer

31. $60° + 180°k$, $90° + 180°k$, $120° + 180°k$, where k is any integer

33. $20° + 60°k$, $40° + 60°k$, where k is any integer

35. $0°, 270°, 360°$ **37.** $180°, 270°$ **39.** $n = 6$

41. $\frac{1}{4}$ second (and each second after)   **43.** $t = \frac{1}{12}$

45.

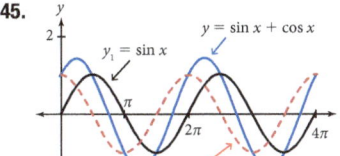

47. $\sin (x + 45°)$

49.

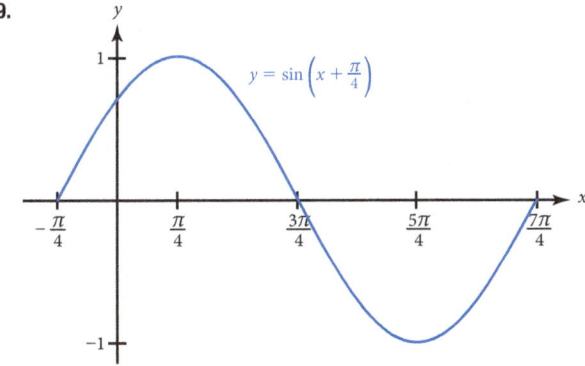

PROBLEM SET 6.4

Vocabulary

1. parametric, parameter   **2.** eliminating
3. identities

Problems

1.

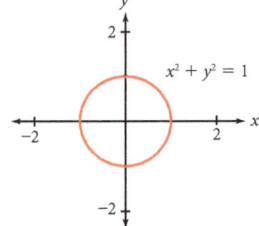

3.

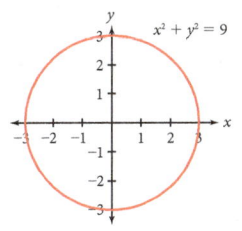

5.

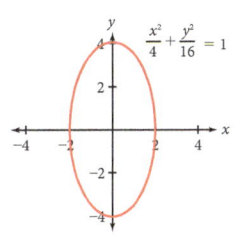

7.

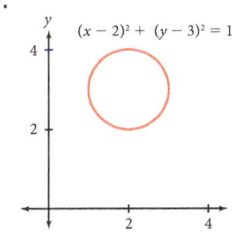

9.

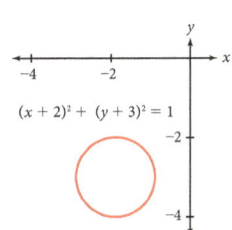

11.

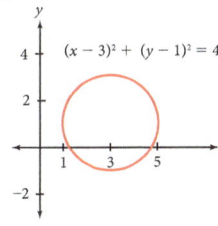

13.

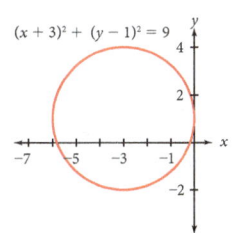

15. $x^2 - y^2 = 1$ **17.** $\frac{x^2}{9} - \frac{y^2}{9} = 1$

19. $\frac{(y-4)^2}{9} - \frac{(x-2)^2}{9} = 1$ **21.** $x = 1 - 2y^2$ **23.** $y = x$

25. $2x = 3y$

27.

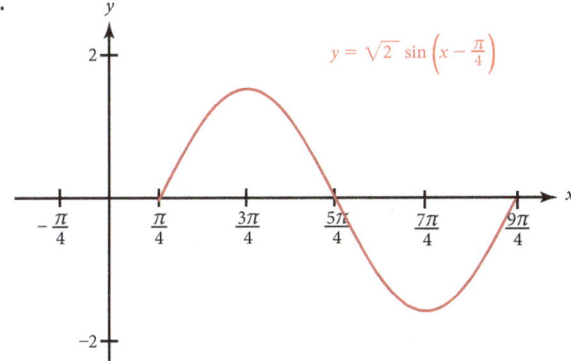

29.

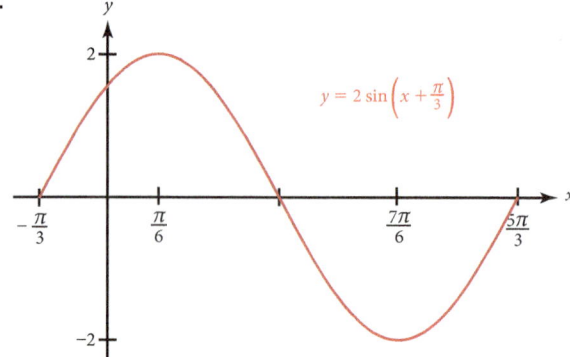

$$y = 2\sin\left(x + \frac{\pi}{3}\right)$$

31.

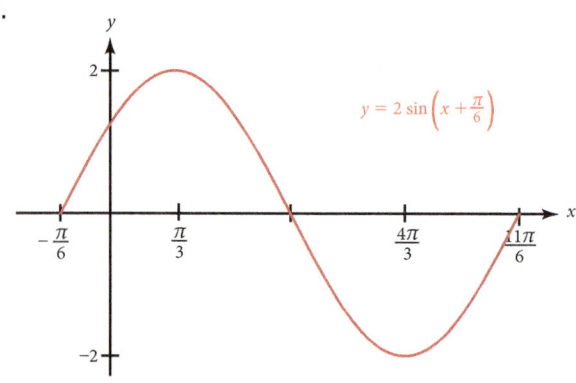

$$y = 2\sin\left(x + \frac{\pi}{6}\right)$$

33. $x = 275\sin\left(\frac{\pi}{15}t\right)$, $y = 305 - 275\cos\left(\frac{\pi}{15}t\right)$

35 a. 3.3 seconds   **b.**

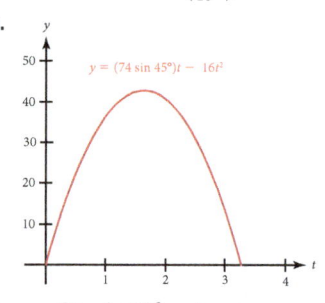

$$y = (74\sin 45°)t - 16t^2$$

41. $-\frac{56}{65}$ **43.** $-\frac{3}{\sqrt{10}}$ **45.** $\sin 90° = 1$

CHAPTER 6 TEST

1. 30°, 150°   **2.** 150°, 330°   **3.** 30°, 90°, 150°, 270°

4. 0°, 60°, 180°, 300°, 360°   **5.** 45°, 135°, 225°, 315°

6. 90°, 210°, 330°   **7.** 180°   **8.** 0°, 240°

9. 0°, 30°, 150°, 180°, 210°, 330°, 360°

10. 15°, 75°, 105°, 165°, 195°, 255°, 285°, 345°

11. 0°, 90°, 360°   **12.** 90°, 180°

13. $2k\pi$, $\frac{\pi}{3} + 2k\pi$, $\frac{5\pi}{3} + 2k\pi$, where k is any integer

14. $\frac{\pi}{6} + k\pi$, where k is any integer

15. $\frac{\pi}{2} + \frac{2k\pi}{3}$, where k is any integer

16. $\frac{\pi}{8} + \frac{k\pi}{2}$, where k is any integer

17. 90°, 203.6°, 336.4°   **18.** 111.5°, 248.5°

19.

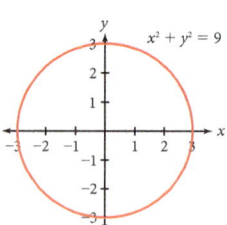

$x^2 + y^2 = 9$

20.

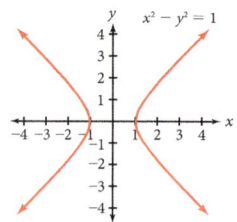

$x^2 - y^2 = 1$

21.

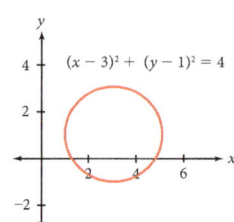

$(x - 3)^2 + (y - 1)^2 = 4$

22.

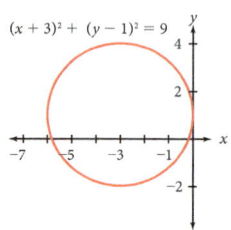

$(x + 3)^2 + (y - 1)^2 = 9$

23.

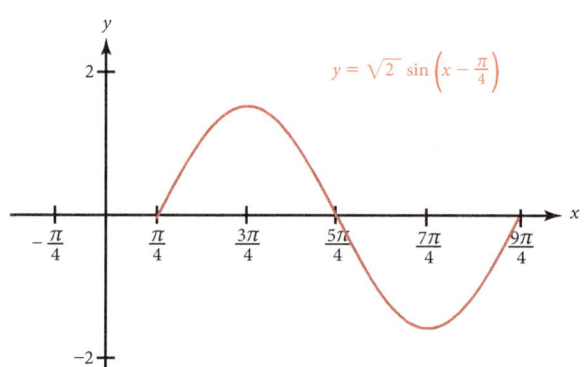

$$y = \sqrt{2}\,\sin\left(x - \frac{\pi}{4}\right)$$

24.

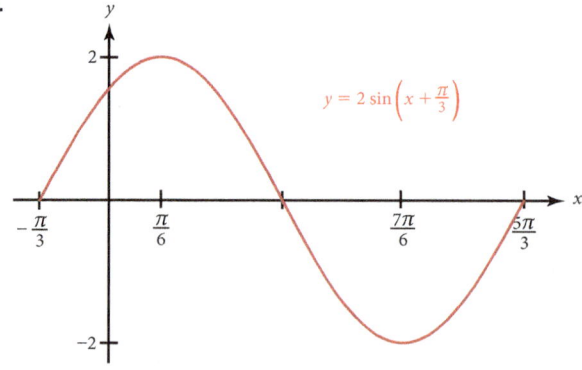

$$y = 2\sin\left(x + \frac{\pi}{3}\right)$$

Chapter 7

PROBLEM SET 7.1

Vocabulary
1. ratio, opposite   **2.** angle   **3.** $\sin B$ **4.** sum

Problems
1. 16 cm   **3.** 11 in.   **5.** 68 yd

7. $C = 70°, c = 44$ km   **9.** $C = 80°, a = 11$ cm

11. $C = 66.1°, b = 295$ in., $c = 284$ in.

13. $C = 39°, a = 7.8$ m, $b = 10.8$ m

15. $A = 141.8°, b = 118$ cm, $c = 214$ cm

17. $\sin B = 5$, which is impossible   **19.** 209 ft   **21.** 273 ft

23. 5,890 ft   **25.** 42 ft   **27.** 14 mi and 9.3 mi   **29.** 92°

31. 0.9994   **33.** 84

PROBLEM SET 7.2

Vocabulary
1. ambiguous   **2.** no   **3.** two   **4.** one

Problems
1. $\sin B = 2$ is impossible

3. $B = 35.3°$ is the only possibility

5. $B = 77°$ or $B' = 103°$

7. $B = 54°, C = 88°, c = 67$ ft or $B' = 126°, C' = 16°, c' = 18$ ft

9. $B = 28.3°, C = 39.5°, c = 30$ cm

11. No solution   **13.** No solution

15. $B = 26.8°, A = 126.4°, a = 65.7$ km   **17.** 15 ft or 38 ft

19. Yes, it makes an angle of 88° with the ground.

21. 78°   **23.** 700   **25.** 0.4

PROBLEM SET 7.3

Vocabulary
1. Pythagorean   **2.** $\cos C$ **3.** largest

Problems
1. 87 in.   **3.** $C = 92.9°$ **5.** 9.4 m   **7.** $a = 128.2°$

9. $A = 44°, B = 76°, c = 62$ cm

11. $A = 29.0°, B = 46.5°, C = 104.5°$

13. $A = 15.6°, C = 12.9°, b = 727$ m

15. $A = 39.5°, B = 56.7°, C = 83.8°$

17. $A = 55.4°, B = 45.5°, C = 79.1°$

19. $a^2 = b^2 + c^2 - 2bc \cos 90° = b^2 + c^2 - 2bc(0) = b^2 + c^2$

21. 24 in.   **23.** 39 in.   **25.** 133 mi.   **27.** 462 mi.

29. true course = 152.7°, ground speed = 194 mph

31. 0.5750   **33.** 144° 50′   **35.** 8.6751

PROBLEM SET 7.4

Vocabulary
1. base   **2.** $\sin C$ **3.** perimeter

Problems
1. 1,520 centimeters^2 **3.** 333 meters^2

5. 0.123 kilometers^2 **7.** 26.3 meters^2

9. 28,300 inches^2 **11.** 2.09 feet^2 **13.** 11.6 inches^2

15. 15.0 yards^2 **17.** 8.15 feet^2 **19.** 156 inches^2

21. 14.3 centimeters   **23.** 150°, 210°   **25.** $\frac{\pi}{6}, \frac{5\pi}{6}, \frac{7\pi}{6}, \frac{11\pi}{6}$

27. 120°, 150°, 300°, 330°   **29.** $\frac{y^2}{9} - \frac{x^2}{9} = 1$

CHAPTER 7 TEST

1. 6.7 inches

2. 4.3 inches

3. $C = 78.4°, a = 26.5$ centimeters, $b = 38.3$ centimeters

4. $B = 49.2°, a = 18.8$ centimeters, $c = 43.2$ centimeters

5. $\sin B = 3.0311$, which is impossible

6. $B = 29°$ is the only possibility for B

7. $B = 71°, C = 58°, c = 7.1$ feet or $B' = 109°, C' = 20°, c' = 2.9$ feet

8. $B = 59°, C = 95°, c = 11$ feet or $B' = 121°, C' = 33°, c' = 6.0$ feet

9. 11 centimeters   **10.** 19 centimeters   **11.** 95.7°

12. 69.5°   **13.** $A = 44°, B = 17°, c = 8.1$ centimeters

14. $B = 34°, C = 111°, a = 3.8$ meters   **15.** 50.8°

16. 411 feet   **17.** 14.2 meters   **18.** 498 centimeters^2

19. 307 centimeters^2 **20.** 52 centimeters^2

21. 52 centimeters^2 **22.** 17 kilometers^2

23. 52 kilometers^2

Chapter 8

PROBLEM SET 8.1

Vocabulary
1. standard   **2.** real, imaginary   **3.** conjugates
4. binomials

Problems
1. $6i$ **3.** $-5i$ **5.** $6i\sqrt{2}$ **7.** $-2i\sqrt{3}$ **9.** 1

11. -1 **13.** $-i$ **15.** $x = 3, y = -1$

17. $x = -2, y = -\frac{1}{2}$ **19.** $x = -8, y = -5$

21. $x = 7, y = \frac{1}{2}$ **23.** $x = \frac{3}{7}, y = \frac{2}{5}$ **25.** $5 + 9i$

27. $5 - i$ **29.** $2 - 4i$ **31.** $1 - 6i$ **33.** $2 + 2i$

35. $-1 - 7i$ **37.** $6 + 8i$ **39.** $2 - 24i$ **41.** $-15 + 12i$

43. $18 + 24i$ **45.** $10 + 11i$ **47.** $21 + 23i$

49. $-2 + 2i$ **51.** $2 - 11i$ **53.** $-21 + 20i$ **55.** $-2i$

57. $-7 - 24i$ **59.** 5 **61.** 40 **63.** 13 **65.** 164

67. $-3 - 2i$ **69.** $-2 + 5i$ **71.** $\dfrac{8}{13} + \dfrac{12}{13}i$

73. $-\dfrac{18}{13} - \dfrac{12}{13}i$ **75.** $-\dfrac{5}{13} + \dfrac{12}{13}i$ **77.** $\dfrac{13}{15} - \dfrac{2}{5}i$

79. $-11 - 7i$ ohms   **81.** $\sin\theta = -\dfrac{4}{5}$, $\cos\theta = \dfrac{3}{5}$

83. $\sin\theta = \dfrac{b}{\sqrt{a^2 + b^2}}$, $\cos\theta = \dfrac{a}{\sqrt{a^2 + b^2}}$ **85.** $135°$

87. $150°$

PROBLEM SET 8.2

Vocabulary

1. modulus   **2.** arrow   **3.** trigonometric   **4.** argument

Problems

1. $|3 + 4i| = 5$ **3.** $|1 + i| = \sqrt{2}$

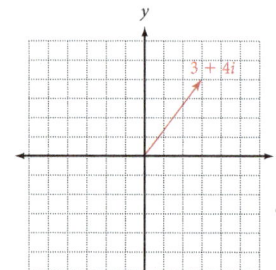   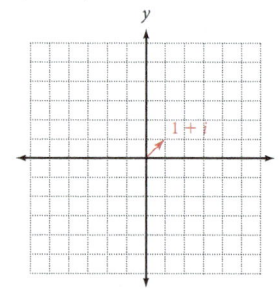

5. $|-5i| = 5$ **7.** $|2| = 2$

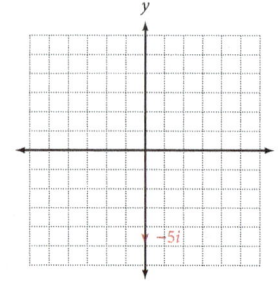   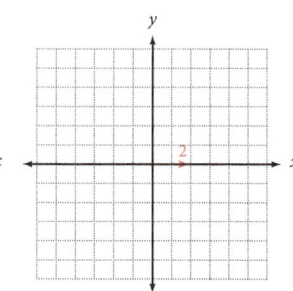

9. $|-4 - 3i| = 5$ **11.**

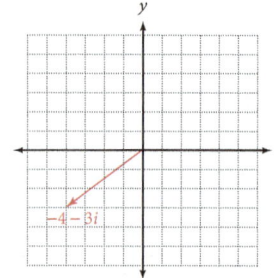   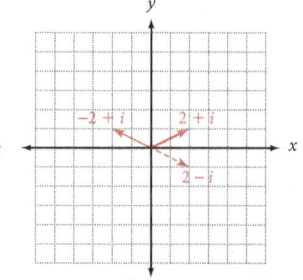

13.   **15.**

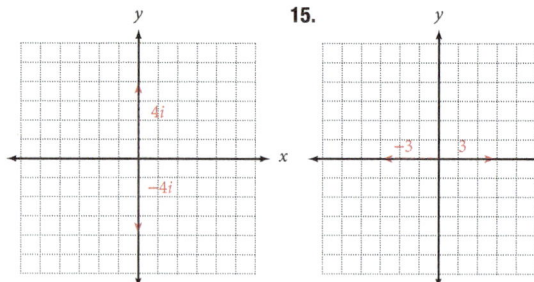

17.

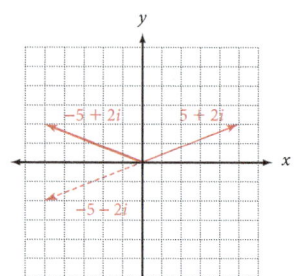

19. $\sqrt{3} + i$ **21.** $-2 + 2i\sqrt{3}$ **23.** $-\dfrac{\sqrt{3}}{2} - \dfrac{1}{2}i$

25. $\dfrac{\sqrt{2}}{2} - \dfrac{\sqrt{2}}{2}i$ **27.** $9.78 + 2.08i$ **29.** $-79.86 + 60.18i$

31. $-0.91 - 0.42i$ **33.** $9.51 - 3.09i$

35. $\sqrt{2}(\cos 135° + i\sin 135°)$

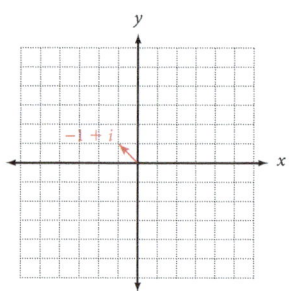

37. $\sqrt{2}(\cos 315° + i\sin 315°)$

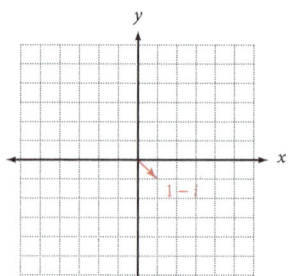

39. $3\sqrt{2}(\cos 45° + i \sin 45°)$

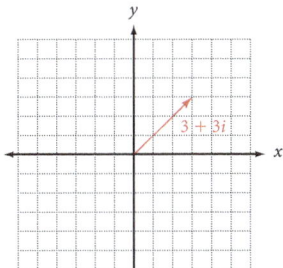

41. $8(\cos 90° + i \sin 90°)$

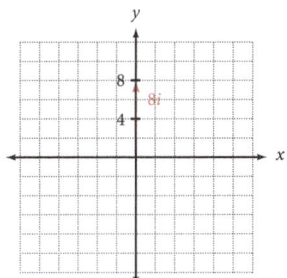

43. $9(\cos 180° + i \sin 180°)$

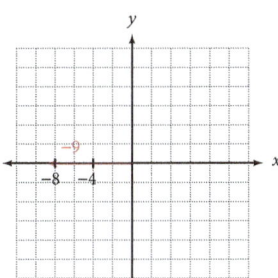

45. $4(\cos 120° + i \sin 120°)$

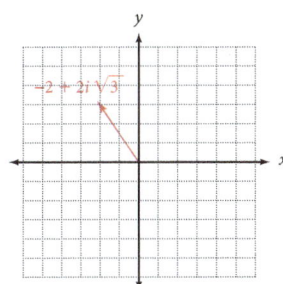

47. $6(\cos 300° + i \sin 300°)$

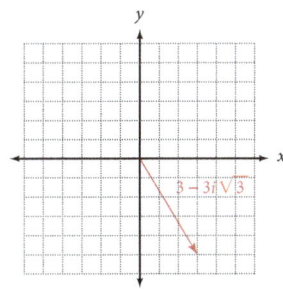

53. $z_1 = 2 + 3i, \ z_2 = -3 + 2i, \ z_3 = -2 - 3i, \ z_4 = 3 - 2i$

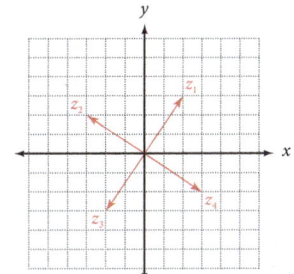

57. $\dfrac{\sqrt{6} - \sqrt{2}}{4}$ **59.** $\dfrac{56}{65}$ **61.** $\sin 120° = \dfrac{\sqrt{3}}{2}$

63. $\cos 50°$ **65.** $2\sqrt{3} + 2i$ **67.** $\dfrac{\sqrt{3}}{2} + \dfrac{1}{2}i$

PROBLEM SET 8.3

Vocabulary

  1. multiply, add   **2.** modulus, argument
  3. divide, subtract

Problems

  1. $12(\cos 50° + i \sin 50°)$ **3.** $56(\cos 157° + i \sin 157°)$

  5. $4(\cos 180° + i \sin 180°)$ **7.** $2(\cos 180° + i \sin 180°) = -2$

  9. $4(\cos 210° + i \sin 210°) = -2\sqrt{3} - 2i$

  11. $12(\cos 360° + i \sin 360°) = 12$

  13. $4\sqrt{2}(\cos 135° + i \sin 135°) = -4 + 4i$

  15. $10(\cos 240° + i \sin 240°) = -5 - 5i\sqrt{3}$

  17. $32 + 32i\sqrt{3}$ **19.** $-\dfrac{1}{2} + \dfrac{\sqrt{3}}{2}i$ **21.** $-\dfrac{81}{2} - \dfrac{81\sqrt{3}}{2}i$

  23. $32i$ **25.** -4 **27.** $-8 - 8i\sqrt{3}$ **29.** $8i$ **31.** $16 + 16i$

  33. $4(\cos 35° + i \sin 35°)$ **35.** $1.5(\cos 19° + i \sin 19°)$

  37. $0.5(\cos 60° + i \sin 60°)$ **39.** $2(\cos 0° + i \sin 0°) = 2$

  41. $\cos(-60°) + i \sin(-60°) = \dfrac{1}{2} - \dfrac{\sqrt{3}}{2}i$

  43. $2[\cos(-270°) + i \sin(-270°)] = 2i$

  45. $2[\cos(-180°) + i \sin(-180°)] = -2$ **47.** $-4 - 4i$

  49. 8 **55.** $\dfrac{1}{2} - \dfrac{1}{2}i$ **57.** $\dfrac{\sqrt{3}}{4} + \dfrac{1}{4}i$ **59.** $\dfrac{1}{4} - \dfrac{\sqrt{3}}{4}i$

  61. 0.2588 **63.** -0.9659 **65.** -0.9659 **67.** -0.2588

69. $15°, 105°, 195°$ **71.** $8(\cos 0° + i \sin 0°)$

73. $2(\cos 60° + i \sin 60°)$ **75.** $-1, \frac{1}{2} + \frac{\sqrt{3}}{2}i, \frac{1}{2} - \frac{\sqrt{3}}{2}i$

PROBLEM SET 8.4

Problems

1. $2(\cos 15° + i \sin 15°), 2(\cos 195° + i \sin 195°)$

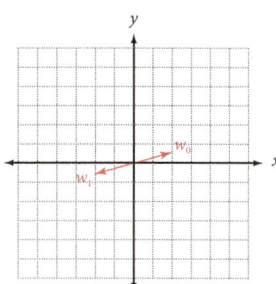

3. $5(\cos 105° + i \sin 105°), 5(\cos 285° + i \sin 285°)$

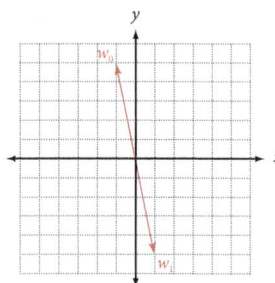

5. $\sqrt{3} + i, -\sqrt{3} - i$ **7.** $\sqrt{2} + i\sqrt{2}, -\sqrt{2} - i\sqrt{2}$

9. $5i, -5i$

11. $2(\cos 70° + i \sin 70°), 2(\cos 190° + i \sin 190°),$
$2(\cos 310° + i \sin 310°)$

13. $2(\cos 10° + i \sin 10°), 2(\cos 130° + i \sin 130°),$
$2(\cos 250° + i \sin 250°)$

15. $3(\cos 60° + i \sin 60°), 3(\cos 180° + i \sin 180°),$
$3(\cos 300° + i \sin 300°)$

17. $\sqrt{3} + i, -1 + i\sqrt{3}, -\sqrt{3} - i, 1 - i\sqrt{3}$

19. $10(\cos 3° + i \sin 3°) \cong 9.99 + 0.52i$

$10(\cos 75° + i \sin 75°) \cong 2.59 + 9.66i$

$10(\cos 147° + i \sin 147°) \cong -8.39 + 5.45i$

$10(\cos 219° + i \sin 219°) \cong -7.77 - 6.29i$

$10(\cos 291° + i \sin 291°) \cong 3.58 - 9.34i$

21. $\cos 30° + i \sin 30°, \cos 90° + i \sin 90°,$
$\cos 150° + i \sin 150°, \cos 210° + i \sin 210°,$
$\cos 270° + i \sin 270°, \cos 330° + i \sin 330°$

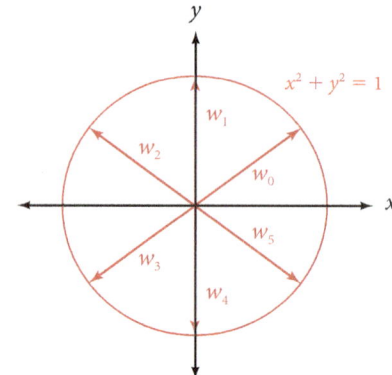

23. $\sqrt{2}(\cos \theta + i \sin \theta)$ where $\theta = 75°, 105°, 255°, 285°$

25. $\sqrt{2}(\cos \theta + i \sin \theta)$ where $\theta = 30°, 150°, 210°, 330°$

27. $\sqrt[4]{2}(\cos \theta + i \sin \theta)$ where $\theta = 67.5°, 112.5°, 247.5°, 292.5°$

29. $(1, 1)$ **31.** $(-1, 1)$

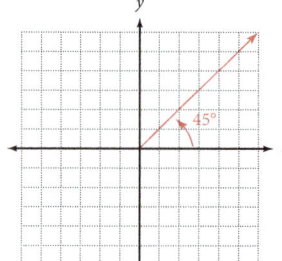
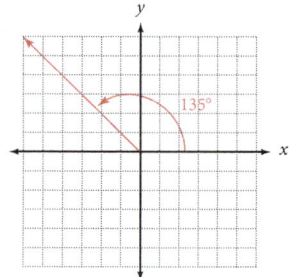

33. $2\sqrt{3}$ **35.** -1 **37.** $3\sqrt{2}$

PROBLEM SET 8.5

Vocabulary

1. origin, standard   **2.** rectangular, polar, parametric

Problems

1.

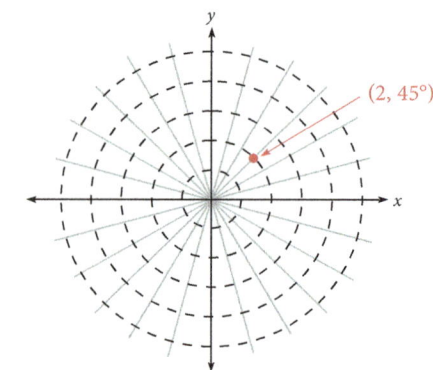

$(2, 45°)$

3.

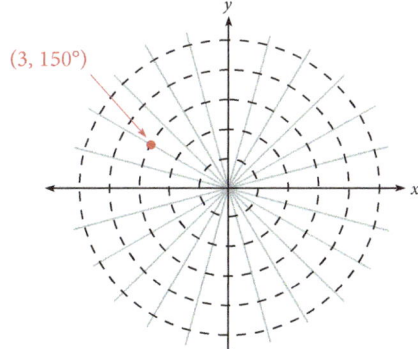

(3, 150°)

5.

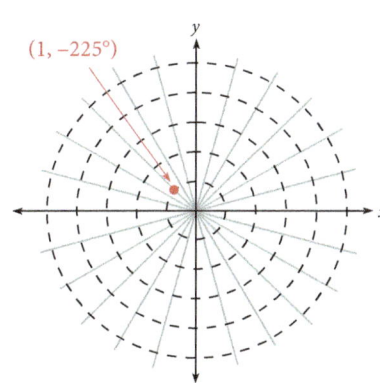

(1, −225°)

7.

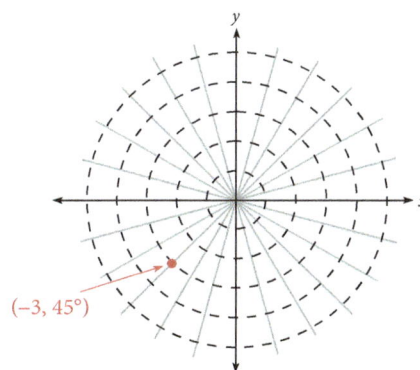

(−3, 45°)

9.

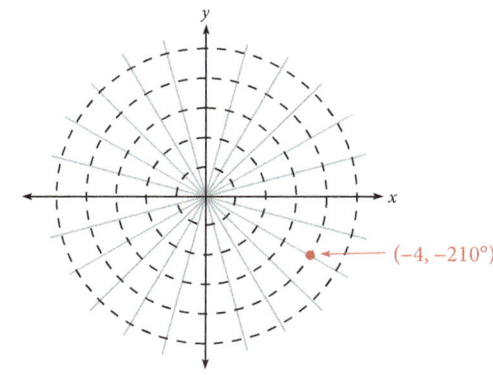

(−4, −210°)

11.

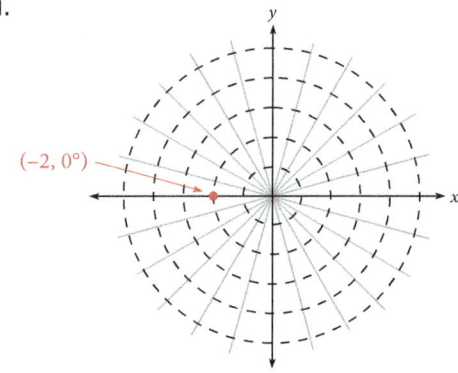

(−2, 0°)

13. $(2, -300°), (-2, 240°), (-2, -120°)$

15. $(5, -225°), (-5, 315°), (-5, -45°)$

17. $(-3, -330°), (3, 210°), (3, -150°)$

19. $(1, \sqrt{3})$ **21.** $(0, -3)$ **23.** $(-1, -1)$

25. $(-6, -2\sqrt{3})$ **27.** $(3\sqrt{2}, 135°)$ **29.** $(4, 150°)$

31. $(2, 0°)$ **33.** $(2, 210°)$ **35.** $(5, 53.1°)$

37. $(\sqrt{5}, 116.6°)$ **39.** $(\sqrt{13}, 236.3°)$ **41.** $x^2 + y^2 = 9$

43. $x^2 + y^2 = 6y$ **45.** $(x^2 + y^2)^2 = 8xy$ **47.** $x + y = 3$

49. $r = \dfrac{5}{\cos \theta - \sin \theta}$ **51.** $r = 2$ **53.** $r = 6 \cos \theta$

55. $\theta = 45°$ **57.** 4.2 **59.** -5.2

61.

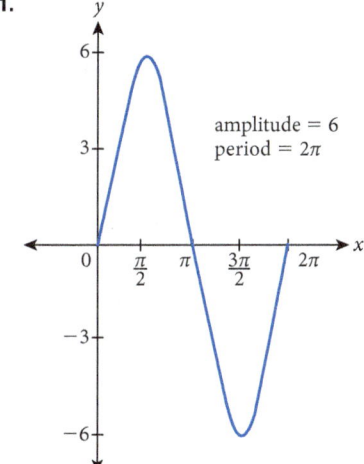

amplitude = 6
period = 2π

63.

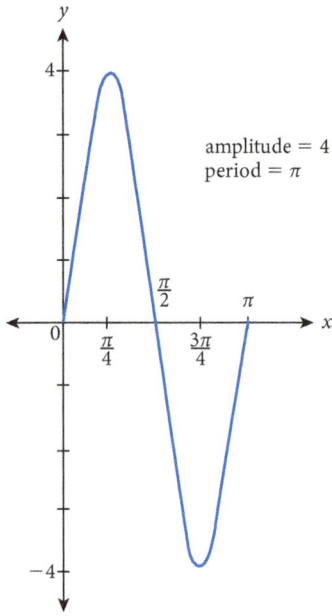

amplitude = 4
period = π

65.

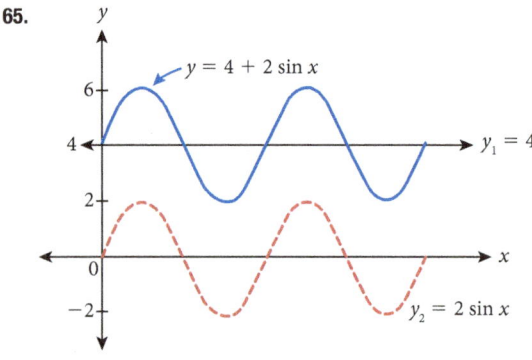

$y = 4 + 2 \sin x$

$y_1 = 4$

$y_2 = 2 \sin x$

PROBLEM SET 8.6

Problems

1.

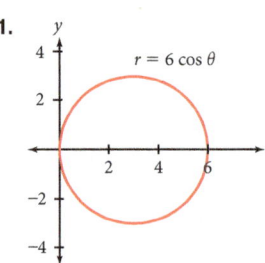

$r = 6 \cos \theta$

3.

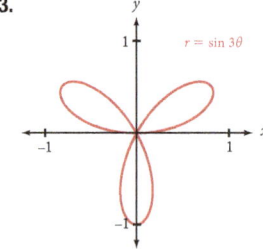

$r = \sin 3\theta$

5.

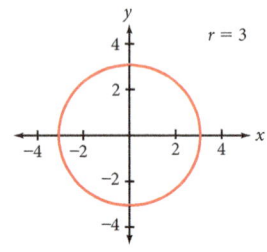

$r = 3$

7.

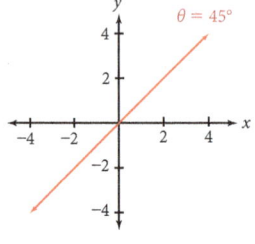

$\theta = 45°$

9.

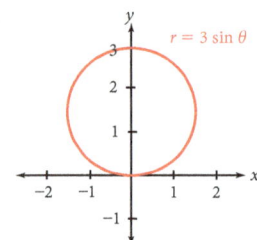

$r = 3 \sin \theta$

11.

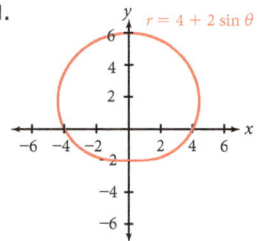

$r = 4 + 2 \sin \theta$

13.

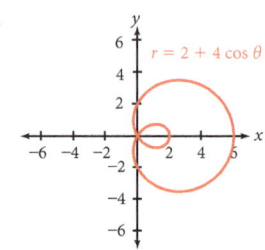

$r = 2 + 4 \cos \theta$

15.

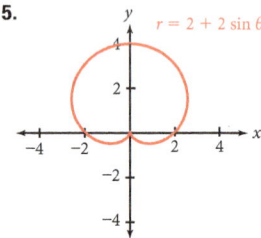

$r = 2 + 2 \sin \theta$

17.

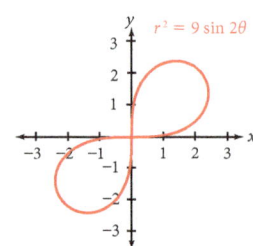

$r^2 = 9 \sin 2\theta$

19.

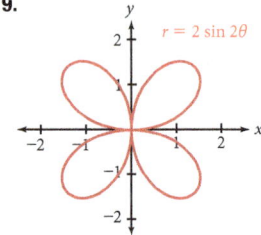

$r = 2 \sin 2\theta$

21.

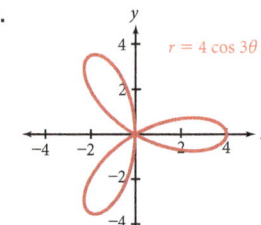

$r = 4 \cos 3\theta$

23.

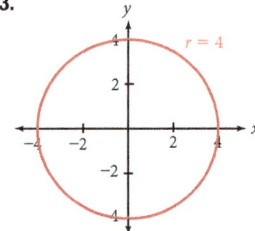

$r = 4$

25.

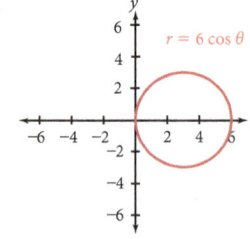

$r = 6 \cos \theta$

27.

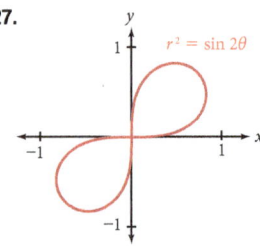

$r^2 = \sin 2\theta$

29.

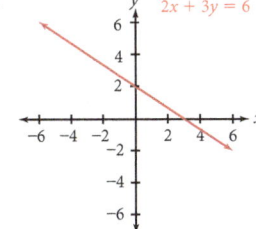

31.

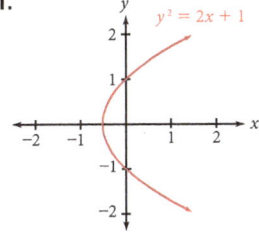

33.

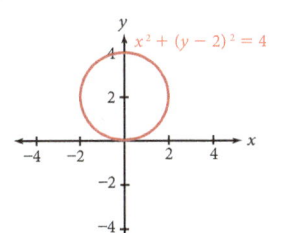

35. $(1, 1), (0, 0)$

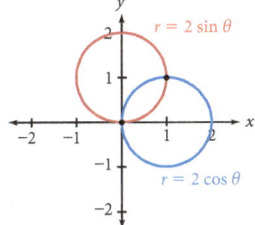

37. 4.3 inches

39. $B = 59°, C = 95°, c = 11$ feet or $B' = 121°, C' = 33°$, $c' = 6.0$ feet

41. $B = 34°, C = 111°, a = 3.8$ meters

CHAPTER 8 TEST

1. $5i$ **2.** $2i\sqrt{3}$ **3.** $x = 2, y = 2$

4. $x = -2$ or $5, y = 2$ **5.** $7 - 6i$ **6.** $8 - 2i$ **7.** 1

8. i **9.** 89   **10.** $-16 + 30i$ **11.** $-2 - \dfrac{5}{2}i$

12. $\dfrac{11}{61} + \dfrac{60}{61}i$ **13.** a. 5   b. $-3 - 4i$ c. $3 - 4i$

14. a. 5   b. $-3 + 4i$ c. $3 + 4i$ **15.** a. 8   b. $-8i$ c. $-8i$

16. a. 4   b. 4   c. -4 **17.** $4\sqrt{3} - 4i$ **18.** $-\sqrt{2} + i\sqrt{2}$

19. $2\sqrt{2}(\cos 45° + i \sin 45°)$ **20.** $2(\cos 150° + i \sin 150°)$

21. $5(\cos 90° + i \sin 90°)$ **22.** $3(\cos 180° + i \sin 180°)$

23. $15(\cos 65° + i \sin 65°)$ **24.** $5(\cos 30° + i \sin 30°)$

25. $32(\cos 50° + i \sin 50°)$ **26.** $81(\cos 80° + i \sin 80°)$

27. $7(\cos 25° + i \sin 25°), 7(\cos 205° + i \sin 205°)$

28. $\sqrt{2}(\cos \theta + i \sin \theta)$ where $\theta = 15°, 105°, 195°, 285°$

29. $\sqrt{2}(\cos \theta + i \sin \theta)$ where $\theta = 15°, 165°, 195°, 345°$

30. $\cos \theta + i \sin \theta$ where $\theta = 60°, 180°, 300°$

31. $(-4, 45°), (4, -135°); (-2\sqrt{2}, -2\sqrt{2})$

32. $(6, 240°), (6, -120°); (-3, -3\sqrt{3})$ **33.** $(3\sqrt{2}, 135°)$

34. $(5, 90°)$ **35.** $x^2 + y^2 = 6y$ **36.** $(x^2 + y^2)^{3/2} = 2xy$

37. $r = \dfrac{2}{\cos \theta + \sin \theta}$ **38.** $r = 8 \sin \theta$

39.

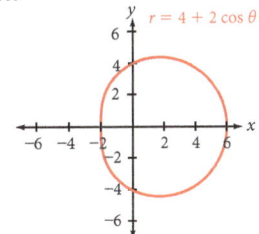

40.

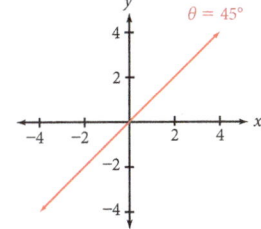

41.

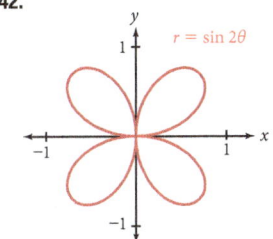

42.